AF323783

FROM ORDERED TO CHAOTIC MOTION IN CELESTIAL MECHANICS

FROM ORDERED TO CHAOTIC MOTION IN CELESTIAL MECHANICS

Yi-Sui Sun
Li-Yong Zhou
Nanjing University, China

World Scientific

NEW JERSEY · LONDON · SINGAPORE · BEIJING · SHANGHAI · HONG KONG · TAIPEI · CHENNAI · TOKYO

Published by

World Scientific Publishing Co. Pte. Ltd.

5 Toh Tuck Link, Singapore 596224

USA office: 27 Warren Street, Suite 401-402, Hackensack, NJ 07601

UK office: 57 Shelton Street, Covent Garden, London WC2H 9HE

Library of Congress Cataloging-in-Publication Data

Sun, Yi-Sui, author.

    From ordered to chaotic motion in celestial mechanics / Yi-Sui Sun, Li-Yong Zhou, Nanjing University, China.

     pages cm

    Includes bibliographical references and index.

    ISBN 978-9814630542 (hardcover : alk. paper)

    1. Few-body problem. 2. Celestial mechanics. I. Zhou, Li-Yong, author. II. Title.

    QB362.F47S87 2015

    521--dc23

                            2015001993

British Library Cataloguing-in-Publication Data

A catalogue record for this book is available from the British Library.

Cover image credit: ESO/M. Kornmesser.

Typeset by Stallion Press

Email: enquiries@stallionpress.com

Printed in Singapore by B & Jo Enterprise Pte Ltd

Contents

Preface

This book provides a brief introduction to some basic but important problems in celestial mechanics. It is based on the main results in some of the authors' research works, which are related to the qualitative method of celestial mechanics and nonlinear dynamics. Some of these works are the interdisciplinary courses, involving the celestial mechanics and nonlinear dynamics, sometimes called nonlinear celestial mechanics.

Since the era of Sir Issac Newton (1643–1727) when the universal gravitation was discovered and the calculus was invented, the celestial mechanics has become a real science. In 300 years afterwards, numerous scientists devoted their great efforts to the problem of motion of celestial bodies affected (only) by their mutual gravitational attraction. The first great success in celestial mechanics, owing to Newton, is the solution of the two-body problem, where the trajectory of a planet traveling around the Sun was fully depicted and understood.

When the number of celestial objects increases by only one to three, the problem becomes much more complicated, as many great scientists like Joseph-Louis Lagrange (1736–1813), Pierre-Simon Laplace (1749–1827), Henri Poincaré (1854–1912), etc. had illustrated and proven.

The new version of the gravitation theory by Albert Einstein (1879–1955), which will introduce some very tiny but important correction to Newton's theory, has not been taken into account in this book.

Nevertheless, this book covers a wide variety of topics. The objects of research presented in this book passes an extensive variety including the comets, asteroids, planetary rings, Trojan asteroids, etc. Many applicable methods in celestial mechanics, such as the nonlinear dynamical method, mapping method, symplectic integrator, spectral analysis etc have been shown, mainly in a practical way as in our research works. The diverse content in this book is organized as follows. Chapter 1 presents some qualitative analyses on the behaviors of motion in three(N)-body system. Then we devote Chapter 2 to the motion of small bodies in the planetary system. Chapter 3 concerns the chaotic motion of orbits in celestial mechanics. And finally Chapter 4 focuses on the orbit diffusion in phase space.

This book is not a textbook. We will not introduce a whole course comprehensively. When a subject will be discussed, a brief overall introduction and the necessary background about the related subject will be given, and more often our readers may be directed to related references to obtain a thorough understanding of a subject. This book is neither a simple collection of the authors' work. It is the reorganization of related results according to subjects.

Most of the results presented in this book were the outcomes of long term pleasurable cooperations with our colleagues. The pleasant recollections invoked by this book made the writing process a happy journey to the past. We would like to thank all the co-authors of the papers on which this book is based. Particularly, Prof. Sun sincerely cherishes the memory of working with Prof. Chen Xiang-Yan.

The scientific work of the authors would be impossible without the financial supports from the Natural Sciences Foundation of China (NSFC), Ministry of Science and Technology of China and Ministry of Education of China. The authors are grateful for their continuous supports. Great thanks also go to Nanjing University where the authors get education, work, and live.

Finally, we would like to thank the Editor of this book, Mr. Ng Kah Fee. His efficient efforts and kind encouragement to us made this book a reality.

Chapter 1

Qualitative analysis on motion in 3-body system

For quite a long time in history, the qualitative theory was the main stream of celestial mechanics. Using analytical and qualitative methods, people try to attack the fundamental and important problems, such as the solution of general N-body problem and the stability of the Solar System. In this chapter, we would like to present our efforts in some related topics, including the permissible and forbidden regions of motion, the variation range of orbits of three bodies, the evolution of moment of inertia, the elliptic restricted 3-body problem, the Sitnikov motion, the central configuration of 4-body problem, etc. Some fundamental concepts of 3-body problem will also be introduced as the necessary background and we assume that the readers have known very well about the solution of the 2-body problem.

1.1 Equations of motion and invariants

The 3-body problem studies the motions of three bodies under their mutual gravitation. At the beginning, the usual formulations of the equations of motion and the invariants of the motion are presented.

1.1.1 *Classical formulation*

Let m_1, m_2, m_3 be the masses of three bodies P_1, P_2, P_3 respectively, $\boldsymbol{r}_1, \boldsymbol{r}_2, \boldsymbol{r}_3$ their radius-vectors, $\boldsymbol{r}_{ij}$ is the vector from P_i to P_j, and $r_{ij} = \|\boldsymbol{r}_{ij}\|$, where $\|\cdot\|$ denotes the norm of vector. Obviously,

$$\boldsymbol{r}_{ij} = \boldsymbol{r}_j - \boldsymbol{r}_i.$$

With these notations, the equations of the 3-body motion are as follows.

$$\frac{\mathrm{d}^2 \boldsymbol{r}_i}{\mathrm{d}t^2} = G\left(m_j \frac{\boldsymbol{r}_{ij}}{r_{ij}^3} + m_k \frac{\boldsymbol{r}_{ik}}{r_{ik}^3} \right), \quad (i, j, k = 1, 2, 3; i \neq j \neq k), \qquad (1.1)$$

where G is the gravitational constant. Equation (1.1) can be written in an alternative way, as

$$m_i \frac{\mathrm{d}^2 \boldsymbol{r}_i}{\mathrm{d}t^2} = \frac{\partial U}{\partial \boldsymbol{r}_i}, \quad (i = 1, 2, 3), \qquad (1.2)$$

where

$$U = U(r_{12}, r_{23}, r_{31}) = G \sum_{i<j} \frac{m_i m_j}{r_{ij}} \quad (i, j = 1, 2, 3), \qquad (1.3)$$

is the potential function.

1.1.2   *Jacobi formulation*

In the Jacobi coordinate system, two vectors $\boldsymbol{r}$ and $\boldsymbol{R}$ are introduced and used as the variables of equations of motion, where $\boldsymbol{r}$ is the vector from P_1 to P_2 and $\boldsymbol{R}$ is the vector from the center of mass O' of P_1 and P_2 to P_3 (see Fig. 1.1). Thus $\boldsymbol{R}$ passes through the barycenter O of three bodies. Formally, $\boldsymbol{r}$ and $\boldsymbol{R}$ are

$$\boldsymbol{r} = \boldsymbol{r}_2 - \boldsymbol{r}_1,$$

$$\boldsymbol{R} = \boldsymbol{r}_3 - \frac{m_1 \boldsymbol{r}_1 + m_2 \boldsymbol{r}_2}{m_1 + m_2} = \frac{m_1 + m_2 + m_3}{m_1 + m_2} \boldsymbol{r}_3. \qquad (1.4)$$

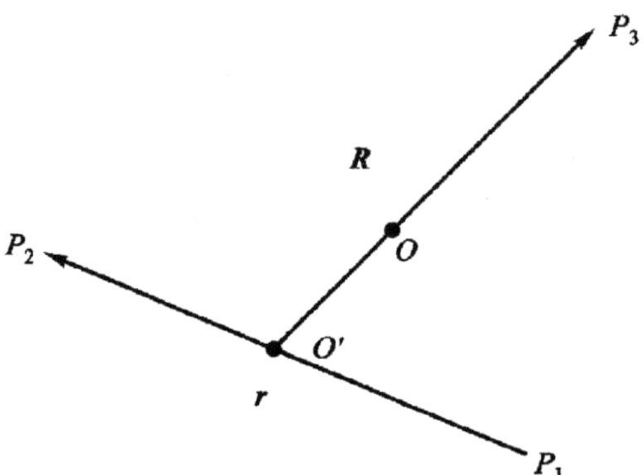

Fig. 1.1   Schematic of Jacobi coordinate.

From Eq. (1.2), one obtains:

$$\frac{\mathrm{d}^2 \boldsymbol{r}}{\mathrm{d}t^2} = \frac{1}{m_2}\frac{\partial U}{\partial \boldsymbol{r}_2} - \frac{1}{m_1}\frac{\partial U}{\partial \boldsymbol{r}_1} = \frac{m_1 + m_2}{m_1 m_2}\frac{\partial U}{\partial \boldsymbol{r}},$$
$$\frac{\mathrm{d}^2 \boldsymbol{R}}{\mathrm{d}t^2} = \frac{m_1 + m_2 + m_3}{m_3\left(m_1 + m_2\right)}\frac{\partial U}{\partial \boldsymbol{R}}. \tag{1.5}$$

By introducing the "reduced masses"

$$\mu_1 = \frac{m_1 m_2}{m_1 + m_2}, \quad \mu_2 = \frac{m_3\left(m_1 + m_2\right)}{m_1 + m_2 + m_3},$$

the equations of motion in the Jacobi coordinate system are simplified as:

$$\mu_1 \frac{\mathrm{d}^2 \boldsymbol{r}}{\mathrm{d}t^2} = \frac{\partial U}{\partial \boldsymbol{r}},$$
$$\mu_2 \frac{\mathrm{d}^2 \boldsymbol{R}}{\mathrm{d}t^2} = \frac{\partial U}{\partial \boldsymbol{R}}. \tag{1.6}$$

1.1.3 *Hamiltonian formulation*

As usual, and hereafter, in this whole book, a dot over a variable represents a derivative with respect to time, while two dots mean the secondary derivative. Let $\boldsymbol{p}_i = m_i \dot{\boldsymbol{r}}_i$ $(i = 1, 2, 3)$, then the Hamiltonian of the system reads

$$H = \frac{1}{2}\sum_{k=1}^{3}\frac{1}{m_k}\boldsymbol{p}_k \cdot \boldsymbol{p}_k - U(\boldsymbol{r}_1, \boldsymbol{r}_2, \boldsymbol{r}_3). \tag{1.7}$$

Equation (1.2) can be written in a Hamiltonian formulation as follows

$$\dot{\boldsymbol{r}}_i = \frac{1}{m_i}\boldsymbol{p}_i = \frac{\partial H}{\partial \boldsymbol{p}_i},$$
$$\dot{\boldsymbol{p}}_i = -\frac{\partial H}{\partial \boldsymbol{r}_i}. \qquad (i = 1, 2, 3), \tag{1.8}$$

Noting the relations between $\boldsymbol{r}_1, \boldsymbol{r}_2, \boldsymbol{r}_3$ and $\boldsymbol{r}, \boldsymbol{R}$:

$$\boldsymbol{r}_1 = -\frac{m_2}{m_1 + m_2}\boldsymbol{r} - \frac{m_3}{m_1 + m_2 + m_3}\boldsymbol{R},$$
$$\boldsymbol{r}_2 = \frac{m_1}{m_1 + m_2}\boldsymbol{r} - \frac{m_3}{m_1 + m_2 + m_3}\boldsymbol{R}, \tag{1.9}$$
$$\boldsymbol{r}_3 = \frac{m_1 + m_2}{m_1 + m_2 + m_3}\boldsymbol{R},$$

the total kinetic energy of the system in the Jacobi coordinate system can be written as

$$T = \frac{1}{2} \sum_{i=1}^{3} m_i \dot{\boldsymbol{r}}_i \cdot \dot{\boldsymbol{r}}_i = \frac{1}{2} \frac{m_1 m_2}{m_1 + m_2} v^2 + \frac{1}{2} \frac{(m_1 + m_2)m_3}{m_1 + m_2 + m_3} V^2, \qquad (1.10)$$

where $v = \|\dot{\boldsymbol{r}}\|, V = \|\dot{\boldsymbol{R}}\|$. Thus the Hamiltonian in the Jacobi coordinate system is

$$H = T - U, \quad U = U(\boldsymbol{r}, \boldsymbol{R}). \qquad (1.11)$$

1.1.4 *Integral invariants*

The 10 integral invariants (also known as "classical integrals") of the system are listed as below.

(1) Integral of center of mass

As $\boldsymbol{r}_{ij} = -\boldsymbol{r}_{ji}$, from Eq. (1.1) one gets the first identity

$$\frac{\mathrm{d}^2 (m_1 \boldsymbol{r}_1 + m_2 \boldsymbol{r}_2 + m_3 \boldsymbol{r}_3)}{\mathrm{d}t^2} = 0. \qquad (1.12)$$

Denoting

$$\boldsymbol{r}_0 = \frac{m_1 \boldsymbol{r}_1 + m_2 \boldsymbol{r}_2 + m_3 \boldsymbol{r}_3}{M}, \quad M = m_1 + m_2 + m_3.$$

Apparently $\boldsymbol{r}_0$ is the radius-vector of the mass center. By integrating Eq. (1.12), one obtains

$$\dot{\boldsymbol{r}}_0 = \boldsymbol{A}, \quad \boldsymbol{r}_0 = \boldsymbol{A}t + \boldsymbol{B} \qquad (1.13)$$

with $\boldsymbol{A}, \boldsymbol{B}$ being constant vectors. Surely, $\boldsymbol{A} = \boldsymbol{B} = 0$ means that the center of mass is at the origin, and the system is in an inertial barycentric system. Thus, Eq. (1.13) is called the "integral of center of mass". In Jacobi formulation of equations of motion, it refers only to the relative motion of three bodies. In practice, the Jacobi formulation is obtained through utilizing the integral of center of mass.

(2) Integral of angular momentum.

From Eq. (1.1) one obtains the second identity:

$$\sum_{j=1}^{3} m_j \boldsymbol{r}_j \times \left(\frac{\mathrm{d}^2 \boldsymbol{r}_j}{\mathrm{d}t^2} \right) = 0. \qquad (1.14)$$

Integrate this equation and denote $\boldsymbol{v}_j = \dot{\boldsymbol{r}}_j \quad (j = 1, 2, 3)$, one has

$$\sum_{j=1}^{3} m_j \boldsymbol{r}_j \times \boldsymbol{v}_j = \boldsymbol{C}, \qquad (1.15)$$

with C being a constant vector. This is the integral of angular momentum. In Jacobi coordinate system, this integral can be written as

$$C = \mu_1 r \times v + \mu_2 R \times V. \tag{1.16}$$

If $C = 0$, but $\mu_1\mu_2 \neq 0$, from Eq. (1.16) one knows that the four vectors r, v, R and V are on the same plane, and the motion of three bodies will maintain on the very same plane. Such a plane perpendicular to the vector C is called the "invariant plane" of the motion.

(3) Energy integral.

As the Hamiltonian in Eq. (1.7) is independent of time, from Eq. (1.2) one has immediately the energy integral $H = h$ with h being a constant, i.e.:

$$\frac{1}{2} \sum_{i=1}^{3} m_i \dot{r}_i \cdot \dot{r}_i - U(r_1, r_2, r_3) = h. \tag{1.17}$$

In Jacobi coordinate system, Eq. (1.17) becomes

$$H = T - U = h, \tag{1.18}$$

where

$$T = \frac{1}{2}\frac{m_1 m_2}{m_1 + m_2} v^2 + \frac{1}{2}\frac{(m_1 + m_2)m_3}{m_1 + m_2 + m_3} V^2, \tag{1.19}$$
$$U = U(r, R).$$

(4) Lagrange-Jacobi identity.

The semi-moment of inertia I of the system is defined as

$$I = \frac{1}{2}\sum_{j=1}^{3} m_j r_j^2. \tag{1.20}$$

Using the integral of center of mass, I can be written as

$$I = \frac{1}{2}\frac{m_1 m_2 m_3}{M}\left(\frac{r_{12}^2}{m_3} + \frac{r_{13}^2}{m_2} + \frac{r_{23}^2}{m_1}\right) = \frac{1}{2}(\mu_1 r \cdot r + \mu_2 R \cdot R). \tag{1.21}$$

The first and second derivatives of I are respectively

$$\frac{dI}{dt} = m_1 r_1 \cdot v_1 + m_2 r_2 \cdot v_2 + m_3 r_3 \cdot v_3, \tag{1.22}$$

and

$$\frac{d^2 I}{dt^2} = \sum_{j=1}^{3} m_j \left(v_j^2 + r_j \cdot \frac{d^2 r_j}{dt^2}\right) = 2T - U. \tag{1.23}$$

Thus the Lagrange-Jacobi identity can be obtained from Eq. (1.23) immediately

$$\frac{\mathrm{d}^2 I}{\mathrm{d}t^2} = T + h = U + 2h. \tag{1.24}$$

Now, some preliminary qualitative results of three bodies motion can be derived.

(a) For total energy $h < 0$.

As $U = T - h \geq -h > 0$, from Eq. (1.3) one knows that

$$\min(r_{12}, r_{13}, r_{23}) \leq -\frac{G}{h}(m_1 m_2 + m_1 m_3 + m_2 m_3). \tag{1.25}$$

It implies that if the total energy of the 3-body system is negative, the minimum of the distances between any two bodies will be bounded, and the minimum of the mutual distance of three bodies does not excess $-\frac{G}{h}(m_1 m_2 + m_1 m_3 + m_2 m_3)$.

(b) For total energy $h \geq 0$.

As $U > 0$, from Eq. (1.24) one gets

$$\frac{\mathrm{d}^2 I}{\mathrm{d}t^2} = U + 2h > 0, \tag{1.26}$$

i.e. $I = I(t)$ is a concave upward function. It has only one minimum. When $t \to \infty, I(t) \to \infty$, so the 3-body system is unstable, and the maximum of the mutual distances is always unbounded.

Great efforts have been made, by many famous scientists such as Bruns, Poincaré, Siegel, Painlevé etc. to find other integrals than these 10 known integrals. But only negative conclusions were obtained. In fact, Poincaré showed that the 3-body problem is not integrable.

1.2   Condition of permissible motion

For three bodies P_1, P_2, P_3 with masses m_1, m_2, m_3 and position vectors $\boldsymbol{r}_1, \boldsymbol{r}_2, \boldsymbol{r}_3$ in an inertial coordinate system, their equations of motion have been given in Eqs. (1.1)–(1.3). And in the Jacobi coordinate system, the equations of motion become Eq. (1.6). These equations possess the integral of angular momentum Eq. (1.16) and the energy integral Eq. (1.18), and they are rewritten as follows.

$$C = \mu_1 r \times \dot{r} + \mu_2 R \times \dot{R}, \tag{1.27}$$

$$T = \frac{1}{2}\left(\mu_1 \dot{r} \cdot \dot{r} + \mu_2 \dot{R} \cdot \dot{R}\right) = U(r, R) + h. \tag{1.28}$$

Take the invariant plane as the (x, y)-plane and C as the direction of z-axis. Let $r = (x_1, y_1, z_1)$, $\dot{r} = (\dot{x}_1, \dot{y}_1, \dot{z}_1)$, $R = (x_2, y_2, z_2)$, $\dot{R} = (\dot{x}_2, \dot{y}_2, \dot{z}_2)$. Thus $(r, R) = (x_1, y_1, z_1, x_2, y_2, z_2)$ comprise a 6-dimensional Euclidean space $\mathbb{R}^{(6)}$ and it is called the position space. Similarly, $\dot{\mathbb{R}}^{(6)}$ that indicates the six-dimensional space $(\dot{r}, \dot{R}) = (\dot{x}_1, \dot{y}_1, \dot{z}_1, \dot{x}_2, \dot{y}_2, \dot{z}_2)$ is called the velocity space. Let $\bar{\mathbb{R}}^{(6)}$ represent the set $\mathbb{R}^{(6)} \backslash \{O\}$ with $\{O\}$ being a set in which at least one of the three distances r_{ij} is zero. For a point $(r, R) \in \bar{\mathbb{R}}^{(6)}$, if there exist the solutions to Eqs. (1.27) and (1.28) for $(\dot{r}, \dot{R})$, this point (r, R) is then called a permissible point. All these points comprise the permissible set, denoted by $\mathbb{L}_6$.

If $(r, R) \in \mathbb{L}_6$, for a certain $\dot{R}$ Eqs. (1.27) and (1.28) are solvable for $\dot{r}$,

$$\mu_1 r \times \dot{r} = C - \mu_2 R \times \dot{R}, \tag{1.29}$$

$$\mu_1 |\dot{r}|^2 = 2\left[U(r, R) + h\right] - \mu_2 |\dot{R}|^2. \tag{1.30}$$

And the necessary and sufficient condition for the existence of a solution for Eq. (1.29) is

$$r \cdot \left(C - \mu_2 R \times \dot{R}\right) = 0. \tag{1.31}$$

When Eq. (1.31) is satisfied, Eq. (1.29) determines a line in three-dimensional space $(\dot{x}_1, \dot{y}_1, \dot{z}_1)$. Its distance to the origin is $\frac{|C - \mu_2 R \times \dot{R}|}{\mu_1 r}$, and Eq. (1.30) represents a sphere in this space. Then the necessary and sufficient condition for a solution of Eqs. (1.29) and (1.30) is that, this distance is smaller than the radius of the sphere defined by Eq. (1.30). So the necessary and sufficient condition for a solution of $\dot{r}$ with respect to r, R and a certain $\dot{R}$ will be Eq. (1.31) and

$$\frac{\left|C - \mu_2 R \times \dot{R}\right|^2}{\mu_1 r^2} \leq 2\left[U(r, R) + h\right] - \mu_2 |\dot{R}|^2,$$

i.e. Eq. (1.31) and

$$\bar{D}(r, R, \dot{R}, C) = \frac{\left|C - \mu_2 R \times \dot{R}\right|^2}{\mu_1 r^2} + \mu_2 |\dot{R}|^2 \leq 2\left[U(r, R) + h\right]. \tag{1.32}$$

When $\bar{D}(\boldsymbol{r}, \boldsymbol{R}, \dot{\boldsymbol{R}}, \boldsymbol{C}) < 2[U(\boldsymbol{r}, \boldsymbol{R}) + h]$, Eqs. (1.29) and (1.30) for $\dot{\boldsymbol{r}}$ have two solutions. And when an equality is taken in Eq. (1.32), Eqs. (1.29) and (1.30) can determine the unique $\dot{\boldsymbol{r}}$.

Let $D^2(\boldsymbol{r}, \boldsymbol{R}, \boldsymbol{C})$ be the minimum of $\bar{D}(\boldsymbol{r}, \boldsymbol{R}, \dot{\boldsymbol{R}}, \boldsymbol{C})$ with condition Eq. (1.32) being satisfied. From Eq. (1.31) one can see that the three bodies must be collinear in the invariant plane, i.e. when $\boldsymbol{r} \times \boldsymbol{R} = \boldsymbol{0}$, one must have $z_1 = z_2 = 0$. Therefore, when $\boldsymbol{r} \times \boldsymbol{R} = \boldsymbol{0}$, the point whose $z_1 \neq 0$ or $z_2 \neq 0$ (denoted by $\boldsymbol{Q} = (\boldsymbol{r}, \boldsymbol{R})$) does not exist. Let $\bar{\bar{\mathbb{R}}}^{(6)} = \bar{\mathbb{R}}^{(6)} \backslash \{\boldsymbol{Q}\}$, then $\mathbb{L}_6$ is the set in $\bar{\bar{\mathbb{R}}}^{(6)}$ that satisfies

$$D^2(\boldsymbol{r}, \boldsymbol{R}, \boldsymbol{C}) \leq 2[U(\boldsymbol{r}, \boldsymbol{R}) + h]. \tag{1.33}$$

If $\bar{D}(\boldsymbol{r}, \boldsymbol{R}, \dot{\boldsymbol{R}}, \boldsymbol{C})$ can attain its minimum $D^2(\boldsymbol{r}, \boldsymbol{R}, \boldsymbol{C})$, Eq. (1.33) is the necessary and sufficient condition for permissible region of motion. When $\mu_2 \boldsymbol{R} \times \dot{\boldsymbol{R}}$ keeps invariant, the minimum of $|\dot{\boldsymbol{R}}|$ is $\frac{|\mu_2 \boldsymbol{R} \times \dot{\boldsymbol{R}}|}{\mu_2 R}$, then $D^2(\boldsymbol{r}, \boldsymbol{R}, \boldsymbol{C})$ is the minimum of

$$\tilde{D}(\boldsymbol{r}, \boldsymbol{R}, \boldsymbol{P}, \boldsymbol{C}) = \frac{1}{\mu_1 r^2}\, |\boldsymbol{C} - \boldsymbol{P}|^2 + \frac{1}{\mu_2 R^2}\, |\boldsymbol{P}|^2\,, \tag{1.34}$$

with condition

$$\boldsymbol{r} \cdot (\boldsymbol{C} - \boldsymbol{P}) = 0, \quad \boldsymbol{R} \cdot \boldsymbol{P} = 0, \quad \boldsymbol{P} = \mu_2 \boldsymbol{R} \times \dot{\boldsymbol{R}}. \tag{1.35}$$

Take $(\boldsymbol{r}, \boldsymbol{R})$ as the new $(\bar{x}, \bar{y})$-plane, on which $\boldsymbol{r}$ is in the direction of the $\bar{x}$-axis, and the direction of $\boldsymbol{r} \times \boldsymbol{R}$ coincides with the $\bar{z}$-axis. In the original coordinate system, $\boldsymbol{C} = (0, 0, C)$. In this new coordinate system, let $\boldsymbol{P} = (\bar{x}, \bar{y}, \bar{z})$, $\boldsymbol{C} = (C_1, C_2, C_3)$, $\boldsymbol{r} = (\bar{x}_1, \bar{y}_1, \bar{z}_1)$, $\boldsymbol{R} = (\bar{x}_2, \bar{y}_2, \bar{z}_2)$. Obviously $\bar{y}_1 = \bar{z}_1 = 0$, $\bar{z}_2 = 0$, therefore, $\tilde{D}(\boldsymbol{r}, \boldsymbol{R}, \boldsymbol{P}, \boldsymbol{C})$ in the condition of Eq. (1.35) reads

$$\tilde{D}(\boldsymbol{r}, \boldsymbol{R}, \boldsymbol{P}, \boldsymbol{C}) = \frac{1}{\mu_1 r^2} \left[\left(C_2 + \frac{\bar{x}_2}{\bar{y}_2} C_1 \right)^2 + (C_3 - \bar{z})^2 \right]$$
$$+ \frac{1}{\mu_2 R^2} \left(C_1^2 + \frac{\bar{x}_2^2}{\bar{y}_2^2} C_1^2 + \bar{z}^2 \right), \tag{1.36}$$

with

$$C_1 = \frac{z_1}{r} C, \quad C_2 = \frac{r^2 z_2 - (\boldsymbol{r} \cdot \boldsymbol{R}) z_1}{r |\boldsymbol{R} \times \boldsymbol{r}|} C, \quad C_3 = \frac{x_1 y_2 - x_2 y_1}{|\boldsymbol{r} \times \boldsymbol{R}|} C,$$
$$C_1^2 + C_2^2 + C_3^2 = C^2. \tag{1.37}$$

When $\bar{z} = \frac{\mu_2 R^2}{\mu_1 r^2 + \mu_2 R^2} C_3$, $\tilde{D}(\boldsymbol{r}, \boldsymbol{R}, \boldsymbol{P}, \boldsymbol{C})$ takes the minimum as

$$D^2(\boldsymbol{r}, \boldsymbol{R}, \boldsymbol{C}) = \frac{1}{\mu_1 r^2}\left(C_2 + \frac{\bar{x}_2}{\bar{y}_2}C_1\right)^2 + \frac{C_1^2}{\mu_2 \bar{y}_2^2} + \frac{C_3^2}{\mu_1 r^2 + \mu_2 R^2}. \qquad (1.38)$$

Finally, the conditions for permissible motion in the phase space and configuration space are obtained, as follows.

$$\frac{1}{\mu_1 r^2}\left[\left(C_2 + \frac{\bar{x}_2}{\bar{y}_2}C_1\right)^2 + (C_3 - \bar{z})^2\right] + \frac{1}{\mu_2 R^2}\left(C_1^2 + \frac{\bar{x}_2^2}{\bar{y}_2^2}C_1^2 + \bar{z}^2\right)$$
$$\leq 2[U(\boldsymbol{r}, \boldsymbol{R}) + h], \qquad (1.39)$$

and

$$\frac{1}{\mu_1 r^2}\left(C_2 + \frac{\bar{x}_2}{\bar{y}_2}C_1\right)^2 + \frac{C_1^2}{\mu_2 \bar{y}_2^2} + \frac{C_3^2}{\mu_1 r^2 + \mu_2 R^2} \leq 2[U(\boldsymbol{r}, \boldsymbol{R}) + h]. \qquad (1.40)$$

From the above analysis, conditions in Eqs. (1.39) and (1.40) are the necessary and sufficient conditions for permissible motion. They are the basic inequalities for the variations of orbits.

1.3 Variations of configuration and position

After the conditions of permissible motion Eqs. (1.39) and (1.40) are known, the variation of orbit of three bodies can be discussed.

1.3.1 *Change of inclination of orbits*

Let i_0 be the angle between the normal direction of the orbit of m_1, m_2 and vector $\boldsymbol{C}$, then

$$\cos^2 i_0 = \frac{\left[\boldsymbol{C} \cdot (\boldsymbol{C} - \mu_2 \boldsymbol{R} \times \dot{\boldsymbol{R}})\right]^2}{C^2 \left|\boldsymbol{C} - \mu_2 \boldsymbol{R} \times \dot{\boldsymbol{R}}\right|^2}. \qquad (1.41)$$

Take the plane consisting of $\boldsymbol{r}, \boldsymbol{R}$ as $(\bar{x}, \bar{y})$-plane, the $\bar{x}$-axis along the direction of $\boldsymbol{r}$, and the $\bar{z}$-axis along the direction of $\boldsymbol{r} \times \boldsymbol{R}$, then

$$\boldsymbol{r} = (r, 0, 0), \quad \boldsymbol{R} = (\bar{x}_2, \bar{y}_2, 0), \quad \boldsymbol{C} = (C_1, C_2, C_3),$$

where C_1, C_2, C_3 are given by Eq. (1.37), and $\boldsymbol{P} = (\bar{x}, \bar{y}, \bar{z})$. From Eq. (1.35) one knows $C_1 - \bar{x} = 0$, $\bar{x}_2 \bar{x} + \bar{y}_2 \bar{y} = 0$, i.e. $\bar{x} = C_1$, $\bar{y} = -\frac{\bar{x}_2}{\bar{y}_2}C_1$. Then

Eq. (1.41) can be rewritten as

$$\cos^2 i_0 = \frac{\left[C_2\left(C_2 + \frac{\bar{x}_2}{\bar{y}_2}C_1\right) + C_3\left(C_3 - \bar{z}\right)\right]^2}{C^2\left[\left(C_2 + \frac{\bar{x}_2}{\bar{y}_2}C_1\right)^2 + (C_3 - \bar{z})^2\right]}. \tag{1.42}$$

Let $C_3 - \bar{z} = \xi, C_2 + \frac{\bar{x}_2}{\bar{y}_2}C_1 = b$, then

$$\cos^2 i_0 = f(\xi) = \frac{(C_2 b + C_3 \xi)^2}{C^2(b^2 + \xi^2)}.$$

Since the problem is trivial if $C_3 = 0$, hereafter, it is supposed that $C_3 \neq 0$. Calculations show that when $C_2 \neq 0$, the function $f(\xi)$ has two extreme points at

$$\xi_1 = -\frac{C_2 b}{C_3}, \qquad \xi_2 = \frac{C_3 b}{C_2}. \tag{1.43}$$

And the corresponding minimum and maximum are

$$f(\xi_1) = 0, \qquad f(\xi_2) = \frac{C_2^2 + C_3^2}{C^2}, \tag{1.44}$$

respectively. When $\xi \to \pm\infty$, $f(\xi) \to C_3^2/C^2$. Accordingly for $\xi_1 < \xi_2$ and $\xi_2 < \xi_1$, the diagram of $\eta = f(\xi)$ is shown in Fig. 1.2.

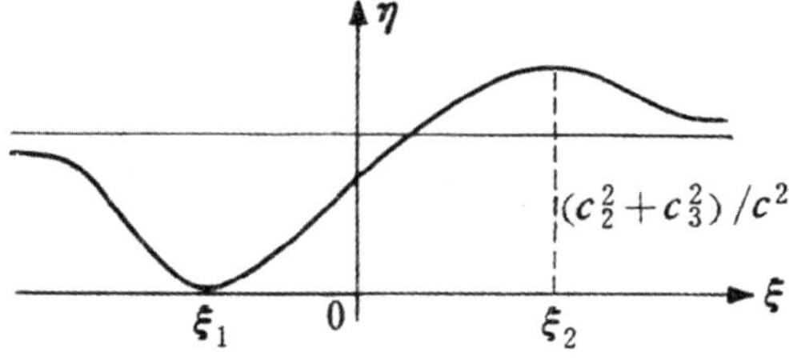

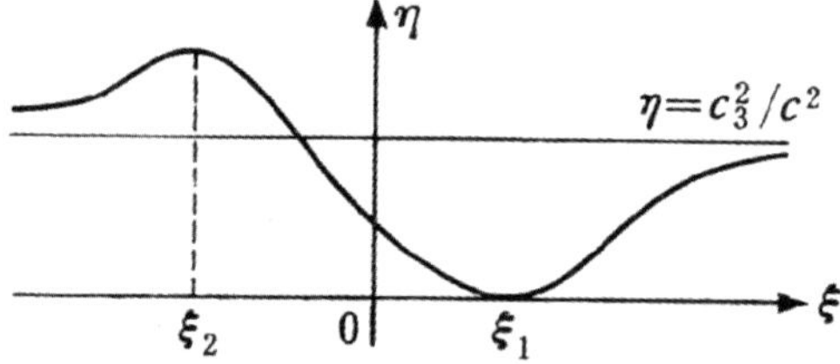

Fig. 1.2   The function $f(\xi)$. Taken from Sun & Luo (1980b).

Moreover, $\xi = C_3 - \bar{z}$ will be restricted by condition in Eq. (1.39). Let

$$g(\xi) = \frac{1}{\mu_1 r^2}\left[\left(C_2 + \frac{\bar{x}_2}{\bar{y}_2}C_1\right)^2 + \xi^2\right] + \frac{1}{\mu_2 R^2}\left[C_1^2 + \frac{\bar{x}_2^2}{\bar{y}_2^2}C_1^2 + (C_3 - \xi)^2\right],$$

then the function $g(\xi)$ has only one minimum point at $\bar{\xi} = \mu_1 C_3 r^2/(\mu_1 r^2 + \mu_2 R^2)$. When $\xi > \bar{\xi}$, $g(\xi)$ increases monotonically, and *vice versa*. The condition in Eq. (1.39) now can be written as

$$g(\xi) \le 2[U(\boldsymbol{r}, \boldsymbol{R}) + h].$$

Therefore, the permissible range of ξ is an abscissa interval between two intersection points of $\eta = g(\xi)$ with the line $\eta = 2\left[U(\boldsymbol{r}, \boldsymbol{R}) + h\right]$. Denote the interval by $[\xi^{(1)}, \xi^{(2)}]$. Then the maximum or minimum of inclination of orbit of P_2 will be the maximum or minimum of $f(\xi)$ in $[\xi^{(1)}, \xi^{(2)}]$.

When $C_2 = 0$, the function

$$f(\xi) = \frac{C_3^2 \xi^2}{\left[\left(\frac{\bar{x}_2}{\bar{y}_2}\right)^2 C_1^2 + \xi^2\right]C^2},$$

has only one minimum point at $\bar{\xi} = 0$, and correspondingly $f(\bar{\xi}) = 0$, while as $\xi \to \pm\infty$, $f(\xi) \to C_3^2/C^2$. Similarly, one can get the maximum and minimum of $f(\xi)$ (i.e. $\cos^2 i_0$) in this case.

Only $\cos^2 i_0$ was discussed here, consequently two values of $\cos i_0$, one positive and one negative, were obtained. Both of them define the varying range of the inclination of orbit, with the positive and negative values for the direct and retrograde motion respectively.

1.3.2 *Change of angle between orbits*

Let I be the angle between orbits, then

$$\cos I = \frac{(\boldsymbol{C} - \boldsymbol{P}) \cdot \boldsymbol{P}}{|\boldsymbol{C} - \boldsymbol{P}||\boldsymbol{P}|}, \tag{1.45}$$

where $\boldsymbol{P} = \mu_2 \boldsymbol{R} \times \dot{\boldsymbol{R}}$ and $\boldsymbol{C}$ is the vector of angular momentum. In the new coordinate system introduced in Sec. 1.2, since $\bar{x} = C_1$, $\bar{y} = -C_1\bar{x}_2/\bar{y}_2$, when $\boldsymbol{C}, \boldsymbol{r}, \boldsymbol{R}$ have been chosen, only $\bar{z}$ in $\boldsymbol{P}$ is changeable. The angle I can be calculated from Eq. (1.45) as

$$\cos I = \frac{-\left(C_2 + \frac{\bar{x}_2}{\bar{y}_2}C_1\right)\frac{\bar{x}_2}{\bar{y}_2}C_1 + (C_3 - \bar{z})\bar{z}}{\sqrt{\left(C_2 + \frac{\bar{x}_2}{\bar{y}_2}C_1\right)^2 + (C_3 - \bar{z})^2}\sqrt{C_1^2\left(1 + \frac{\bar{x}_2^2}{\bar{y}_2^2}\right) + \bar{z}^2}}.$$

After denoting

$$C_2 + \frac{\bar{x}_2}{\bar{y}_2}C_1 = b, \quad C_1^2\left(1 + \frac{\bar{x}_2^2}{\bar{y}_2^2}\right) = B^2, \quad -\frac{\bar{x}_2}{\bar{y}_2}bC_1 = A, \quad (1.46)$$

for simplicity, $\cos I$ becomes

$$\cos I = \frac{-\bar{z}^2 + C_3\bar{z} + A}{\sqrt{\bar{z}^2 - 2C_3\bar{z} + b^2 + C_3^2}\sqrt{\bar{z}^2 + B^2}}. \quad (1.47)$$

When the right side of Eq. (1.47) is positive, I will vary in $(0, \frac{\pi}{2})$, implying that P_2 and P_3 have the same directions of motion, i.e. the relative motion of P_1 and P_2 is the direct motion. When the right side of Eq. (1.47) is negative, I varies in $(\frac{\pi}{2}, \pi)$ and the motions of P_2, P_3 are in opposite directions, i.e. the relative motion of P_1, P_2 is retrograde. For $I = \frac{\pi}{2}$ ($\cos I = 0$), it is the critical case, in which the orbits are perpendicular to each other. For convenience, this critical case is also regarded as the retrograde one.

The discriminant of $-\bar{z}^2 + C_3\bar{z} + A$ in Eq. (1.47) is $\Delta = C_3^2 + 4A$. If $\Delta \leq 0$ the motion is retrograde, while if $\Delta > 0$, the mode of motion depends on the relative positions of $\bar{z}$ of permissible motion and the zeros of function $-\bar{z}^2 + C_3\bar{z} + A$.

In order to investigate the change of $\cos I$, the derivative of $\cos I$ with respect to $\bar{z}$ is calculated

$$\frac{\mathrm{d}(\cos I)}{\mathrm{d}\bar{z}} = \frac{-(B^2 + b^2 + 2A)\bar{z}^3 + 3(B^2 + A)C_3\bar{z}^2}{(\bar{z}^2 - 2C_3\bar{z} + b^2 + C_3^2)^{3/2}(\bar{z}^2 + B^2)^{3/2}}$$
$$+ \frac{-B^2\left[2b^2 + 3C_3^2 + A(b^2 + B^2 + C_3^2)\right]\bar{z} + B^2C_3(b^2 + C_3^2 + A)}{(\bar{z}^2 - 2C_3\bar{z} + b^2 + C_3^2)^{3/2}(\bar{z}^2 + B^2)^{3/2}}.$$
$$(1.48)$$

The coefficient of $\bar{z}^3$ in numerator of Eq. (1.48) is

$$-(B^2 + b^2 + 2A) = -\left[\left(1 + \frac{\bar{x}_2^2}{\bar{y}_2^2}\right)C_1^2 + b^2 - \frac{2\bar{x}_2}{\bar{y}_2}bC_1\right] = -(C_1^2 + C_2^2).$$

It is identically negative, indicating that when $\bar{z}$ increases from $-\infty$, $\cos I$ begins to decrease from $+1$, and when $\bar{z} \to +\infty$, $\cos I$ decreases to -1. To

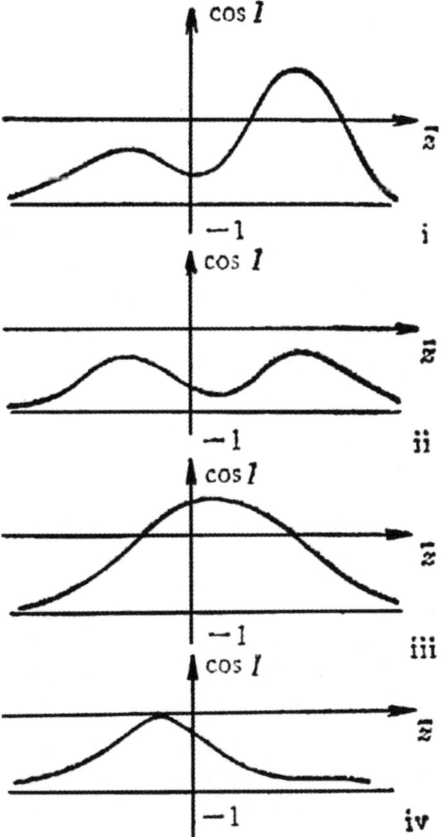

Fig. 1.3   The variation of $\cos I$. From Sun & Luo (1980a).

determine the zero points of numerator in Eq. (1.48), a translation of $\bar{z}$-axis is made and the coefficient of $\bar{z}^3$ is reduced to one. After such an operation, the numerator in Eq. (1.48) can be reduced to $\bar{z}^3 + p\bar{z} + q$ (in fact here $\bar{z}$ is a summation of the original $\bar{z}$ and a constant). Obviously, when $p \geq 0$ or, when $p > 0$ but with $q^2/4 + p^3/27 < 0$, the numerator has only one zero point. For the situation $p > 0$ and $q^2/4 + p^3/27 \geq 0$, there are three zero points. The sketch picture of the variation of $\cos I$ with respect to $\bar{z}$ is shown in Fig. 1.3. There are four different cases as follows.

(i) $p < 0$, $\frac{q^2}{4} + \frac{p^3}{27} \geq 0$, $\Delta > 0$. $\frac{\mathrm{d}\cos I}{\mathrm{d}\bar{z}}$ has three zeros, and $\cos I$ has three extreme points, with the maximum being positive.

(ii) $p < 0$, $\frac{q^2}{4} + \frac{p^3}{27} \geq 0$, $\Delta \leq 0$. $\cos I$ has three extreme points, with the maximum being nonpositive.

(iii) $p < 0$, $\frac{q^2}{4} + \frac{p^3}{27} < 0$ or $p \geq 0$, $\Delta > 0$. $\cos I$ has only one positive maximum.

(iv) $p < 0$, $\frac{q^2}{4} + \frac{p^3}{27} < 0$ or $p \geq 0$, $\Delta \leq 0$. $\cos I$ has one nonpositive maximum.

In Case (ii) and (iv), $\cos I \leq 0$, i.e. $I \geq \frac{\pi}{2}$, thus the relative motion of P_1, P_2 can only be retrograde. While in Cases (i) and (iii), the relative motion can be direct or retrograde. Thus further analysis about the restriction of $\bar{z}$ for the permissible motion is needed.

By using the notations introduced in Eq. (1.46), the condition Eq. (1.36) can be written as

$$\tilde{D} = \frac{1}{\mu_1 r^2} \left[b^2 + (C_3 - \bar{z})^2 \right] + \frac{1}{\mu_2 R^2} (B^2 + \bar{z}^2), \qquad (1.49)$$

and the necessary and sufficient condition for the permissible motion is

$$g(\bar{z}) \leq 2[U(\boldsymbol{r}, \boldsymbol{R}) + h]. \qquad (1.50)$$

The function $g(\bar{z})$ has only one minimum at $\bar{z}_0 = (\mu_2 R^2 C_3)/(\mu_1 r^2 + \mu_2 R^2)$. When $\bar{z} \to \pm\infty$, $g(\bar{z}) \to +\infty$. Therefore, the above inequality determines an interval $[\bar{z}^{(1)}, \bar{z}^{(2)}]$ of $\bar{z}$. To analyze the location of interval $[\bar{z}^{(1)}, \bar{z}^{(2)}]$ in Cases (i) or (iii), one can distinguish in which segment of $\bar{z}$-axis the motion is direct, and in which segment the motion is retrograde. For example, in Case (i), suppose the two zero points of $\cos I$ are $\bar{z}_1$ and $\bar{z}_2$, satisfying $\bar{z}_1 < \bar{z}^{(1)} < \bar{z}^{(2)} < \bar{z}_2$, then always $\cos I > 0$ for permissible motion, and the motion is direct. If $\bar{z}^{(1)} \leq \bar{z}_1$, or $\bar{z}_1 \leq \bar{z}^{(2)}$ in a segment, the motion is direct, while in the other segment the motion is retrograde. For other cases, analogue discussions also can be made.

1.3.3   *Changes of orbital semi-major axis and eccentricity*

For convenience, a new barycenter coordinate system $O - X'Y'Z'$ is chosen, in which OX' is along the direction of $\boldsymbol{R}$, the plane $X'OY'$ is the one hosting $\boldsymbol{R}$ and $\boldsymbol{r}$, thus the Z'-axis is determined by $\boldsymbol{R} \times \boldsymbol{r}$.

First of all, the relation between the original and new coordinate system is derived. In the original coordinate system, $\boldsymbol{R} = (x_1, y_1, z_1), \boldsymbol{r} = (x_2, y_2, z_2), \boldsymbol{C} = (0, 0, C)$, thus $\boldsymbol{R} \times \boldsymbol{r} = (y_1 z_2 - y_2 z_1, z_1 x_2 - z_2 x_1, x_1 y_2 - x_2 y_1)$. The directions of X', Y', Z' axes in the new coordinate system are determined by $\boldsymbol{R}, (\boldsymbol{R} \times \boldsymbol{r}) \times \boldsymbol{R}, \boldsymbol{R} \times \boldsymbol{r}$ respectively. Simple calculations show

$$(\boldsymbol{R} \times \boldsymbol{r}) \times \boldsymbol{R} = (\boldsymbol{R} \cdot \boldsymbol{R})\boldsymbol{r} - (\boldsymbol{R} \cdot \boldsymbol{r})\boldsymbol{R} = R^2 \boldsymbol{r} - (\boldsymbol{R} \cdot \boldsymbol{r})\boldsymbol{R},$$

$$|(\boldsymbol{R} \times \boldsymbol{r}) \times \boldsymbol{R}| = |\boldsymbol{R} \times \boldsymbol{r}| \cdot |\boldsymbol{R}| = R|\boldsymbol{R} \times \boldsymbol{r}|.$$

Denote the unit vector in the direction of a vector $\boldsymbol{a}$ by $(\boldsymbol{a})_0$, and suppose (l, m, n) is the direction cosine of the X-axis in the new coordinate system, then

$$l = (1,0,0) \cdot (\boldsymbol{R})_0 = \frac{x_1}{R},$$

$$m = (1,0,0) \cdot ((\boldsymbol{R} \times \boldsymbol{r}) \times \boldsymbol{R})_0 = \frac{R^2 x_2 - (\boldsymbol{R} \cdot \boldsymbol{r})x_1}{R|\boldsymbol{R} \times \boldsymbol{r}|}, \tag{1.51}$$

$$n = (1,0,0) \cdot (\boldsymbol{R} \times \boldsymbol{r})_0 = \frac{y_1 z_2 - y_2 z_1}{|\boldsymbol{R} \times \boldsymbol{r}|}.$$

Similarly, the direction cosine (l', m', n') of Y-axis in the new coordinate system is

$$l' = \frac{y_1}{R}, \quad m' = \frac{R^2 y_2 - (\boldsymbol{R} \cdot \boldsymbol{r})y_1}{R|\boldsymbol{R} \times \boldsymbol{r}|}, \quad n' = \frac{z_1 x_2 - z_2 x_1}{|\boldsymbol{R} \times \boldsymbol{r}|}. \tag{1.52}$$

In the new coordinate system, let $\boldsymbol{R} = (x_1', y_1', z_1'), \boldsymbol{r} = (x_2', y_2', z_2'), \boldsymbol{C} = (C_1, C_2, C_3)$. Since $\boldsymbol{R}$ is along the X'-axis and $\boldsymbol{r}$ is in the $X'OY'$ plane, $y_1' = z_1' = 0$, $z_2' = 0$.

Following the same calculations performed for Eqs. (1.51) and (1.52), C_1, C_2, C_3 are obtained

$$C_1 = \frac{z_1}{R}C, \quad C_2 = \frac{R^2 z_2 - (\boldsymbol{R} \cdot \boldsymbol{r})z_1}{R|\boldsymbol{R} \times \boldsymbol{r}|}C, \quad C_3 = \frac{x_1 y_2 - x_2 y_1}{|\boldsymbol{R} \times \boldsymbol{r}|}C. \tag{1.53}$$

Define $\boldsymbol{Q} = \mu_1 \boldsymbol{r} \times \dot{\boldsymbol{r}} = (x', y', z')$. From Eq. (1.29), $\boldsymbol{C} - \boldsymbol{Q} = \boldsymbol{P}$. Obviously, $\boldsymbol{Q}$ satisfies

$$\boldsymbol{r} \cdot \boldsymbol{Q} = 0, \quad \boldsymbol{R} \cdot (\boldsymbol{C} - \boldsymbol{Q}) = 0, \tag{1.54}$$

thus

$$x' = C_1, \quad y' = -\frac{x_2'}{y_2'}C_1. \tag{1.55}$$

Now the necessary and sufficient condition for the permissible motion can be expressed as

$$\frac{1}{\mu_2 R^2}\left[\left(C_2 + \frac{x_2'}{y_2'}C_1\right)^2 + (C_3 - z')^2\right] + \frac{1}{\mu_1 r^2}\left(C_1^2 + \frac{x_2'^2}{y_2'^2}C_1^2 + z'^2\right)$$

$$\leq 2\left[U(\boldsymbol{r}, \boldsymbol{R}) + h\right]. \tag{1.56}$$

Let a, e be the semi-major axis and eccentricity of the orbit of P_2 related to P_1. Since $|\mathbf{Q}| = |\mu_1 \mathbf{r} \times \dot{\mathbf{r}}|$, the solution of 2-body problem tells

$$|\mathbf{Q}| = \mu_1 \sqrt{\mu a(1 - e^2)}, \quad \mu = G(m_1 + m_2)$$

or alternatively,

$$\mu a(1 - e^2) = \frac{|\mathbf{Q}|^2}{\mu_1^2} = \frac{1}{\mu_1^2}\left(C_1^2 + \frac{x_2'}{y_2'}C_1^2 + z'^2\right). \tag{1.57}$$

Denote the right hand side of above equation by $f(z')$, and $\mu a, (1 - e^2)$ by ξ, η, this equation turns to

$$\xi\eta = f(z'). \tag{1.58}$$

For a permissible value $z' \in [z_1', z_2']$, Eq. (1.58) is a hyperbolic equation. Thus the z' taken over the range $[z_1', z_2']$ defines a family of hyperbolic curves. The outmost one among these curves corresponds to $f[\max(|z_1'|, |z_2'|)]$. But the one closest to the origin can be divided into two classes (see Fig. 1.4): (1) for $z_1' \cdot z_2' > 0$, it corresponds to $f[\min(|z_1'|, |z_2'|)]$; and (2) for $z_1' \cdot z_2' \leq 0$, it corresponds to $f(0)$. When ξ, η are in the first quadrant, the instantaneous orbit is an ellipse, and for ξ, η in the third quadrant the instantaneous orbit is a hyperbola. Here only the elliptic orbit will be discussed, while for the hyperbolic case, analogous discussion can be made.

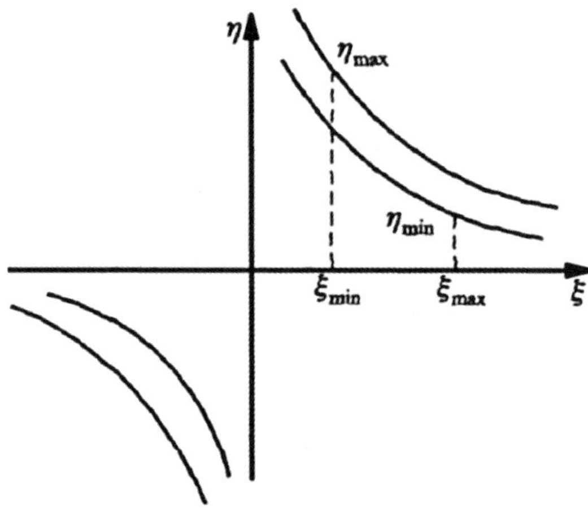

Fig. 1.4   The function $\xi\eta = f(z')$. Adapted from Sun & Luo (1980b).

According to the energy integral of the 2-body problem (*vis viva* equation)

$$|\dot{\boldsymbol{r}}|^2 = \mu \left(\frac{2}{r} - \frac{1}{a} \right), \tag{1.59}$$

the semi major axis a and $|\dot{\boldsymbol{r}}|$ for a fixed r reach their minimum and maximum simultaneously. Each time when $|\dot{\boldsymbol{r}}|$ attains its minimum or maximum, a attains the corresponding minimum or maximum. As $\boldsymbol{Q} = \mu_1 \boldsymbol{r} \times \dot{\boldsymbol{r}}$, for a fixed $\boldsymbol{r}$, when $\boldsymbol{Q}$ is given, all the permissible $\dot{\boldsymbol{r}}$ is situated in a plane, which includes $\boldsymbol{r}$ and is perpendicular to $\boldsymbol{Q}$. When $\dot{\boldsymbol{r}}$ is perpendicular to $\boldsymbol{r}$, $|\dot{\boldsymbol{r}}|$ reaches its minimum. And in this case, $\mu_1 |\boldsymbol{r}||\dot{\boldsymbol{r}}| = |\boldsymbol{Q}|$. Thus for a fixed $\boldsymbol{r}$, $|\dot{\boldsymbol{r}}|_{\min} = |\boldsymbol{Q}|_{\min}/(\mu_1 |\boldsymbol{r}|)$. Because $|\boldsymbol{Q}| = \mu_1 \sqrt{f(z')}$, one has

$$|\boldsymbol{Q}|_{\min} = \begin{cases} \mu_1 \sqrt{f(0)}, & \text{when } z_1' \cdot z_2' \le 0; \\ \mu_1 \sqrt{f[\min(|z_1'|, |z_2'|)]}, & \text{when } z_1' \cdot z_2' > 0. \end{cases} \tag{1.60}$$

This determines $|\dot{\boldsymbol{r}}|_{\min}$. Using Eq. (1.59), finally one obtains

$$a_{\min} = \left(\frac{2}{r} - \frac{|\dot{\boldsymbol{r}}|^2_{\min}}{\mu} \right)^{-1}.$$

And similarly, the $a_{\max}$ can be determined from $|\dot{\boldsymbol{r}}|_{\max}$.

When r, R, h, c as well as the angle α between $\boldsymbol{r}$ and $\boldsymbol{R}$ are fixed, $U(\boldsymbol{r}, \boldsymbol{R})$ and h are fixed too. Meanwhile, one knows that $|\dot{\boldsymbol{r}}|_{\max}$ corresponds to $|\dot{\boldsymbol{R}}|_{\min}$. Because $\mu_2 \boldsymbol{R} \times \dot{\boldsymbol{R}} = \boldsymbol{P} = \boldsymbol{C} - \boldsymbol{Q}$, in a similar way, one knows that $|\dot{\boldsymbol{R}}|_{\min}$ will be attained at $|\boldsymbol{C} - \boldsymbol{Q}|_{\min}$. In the new coordinate system, $\boldsymbol{C} - \boldsymbol{Q} = (0, C_2 + \frac{x_2'}{y_2'}C_1, C_3 - z')$, therefore,

$$|\dot{\boldsymbol{R}}|_{\min} = \frac{|\boldsymbol{C} - \boldsymbol{Q}|_{\min}}{\mu_2 R}$$

$$= \begin{cases} \dfrac{\left| C_2 + \frac{x_2'}{y_2'}C_1 \right|}{\mu_2 R}, & \text{when } (C_3 - z_1')(C_3 - z_2') \le 0; \\[6mm] \dfrac{\sqrt{\left(C_2 + \frac{x_2'}{y_2'}C_1 \right)^2 + \min\left[(C_3 - z_1')^2, (C_3 - z_2')^2 \right]}}{\mu_2 R}, \\[2mm] \qquad\qquad \text{when } (C_3 - z_1')(C_3 - z_2') > 0. \end{cases} \tag{1.61}$$

Thus, the $|\dot{\boldsymbol{r}}|_{\max}$ can be determined and finally $a_{\max}$ can be deduced.

Let $\mu a_{\min} = \xi_{\min}, \mu a_{\max} = \xi_{\max}$. If $\xi_{\min} \le \xi \le \xi_{\max}$, from Fig. 1.4, the η value at the intersection of the line $\xi = \xi_{\max}$ with the hyperbolic curve nearest to the origin will be the minimum, i.e. $\eta_{\min}$, thus $e_{\max} = \sqrt{1 - \eta_{\min}}$. And the η corresponding to the intersection point of $\xi = \xi_{\min}$ and the outmost hyperbolic curve will be the maximum $\eta_{\max}$, resulting in $e_{\min} = \sqrt{1 - \eta_{\max}}$. So, the maximum and minimum of a and e are determined.

From the above discussions, one knows that when r, R, h, C as well as the angle α between $\boldsymbol{r}$ and $\boldsymbol{R}$ are given, one obtains the change ranges of configuration and of position of orbits in configuration space. Because the condition is necessary and sufficient, the change range of configuration and position of orbits of 3-body in configuration space is completely solved.

1.4   Restricted 3-body problem and singular points of motion

The restricted 3-body problem is a special form of general 3-body problem, in which the two "primaries" have nonzero masses m_1 and m_2 while the third body has an infinitesimal mass. The two primaries move under their mutual attraction as in an ordinary 2-body problem, and the third body moves in the gravitational field of the primaries but without any influence over them.

The small bodies in the Solar System are mainly attracted by the Sun and the planet Jupiter, thus in many cases the model of restricted 3-body problem is often applied to study approximately the motions of small bodies in the Solar System.

1.4.1   *Equations of motion and Jacobi integral*

Denote the masses of three bodies by m_1, m_2, m_3, and their coordinates in the inertia system by $(x_1, y_1, z_1), (x_2, y_2, z_2)$ and (x_3, y_3, z_3), respectively. If $m_3 = 0$, the equations of motion for m_3 under the attractions of m_1, m_2 are:

$$\frac{\mathrm{d}^2 x_3}{\mathrm{d}t^2} = G\left(m_1 \frac{x_1 - x_3}{r_{31}^3} + m_2 \frac{x_2 - x_3}{r_{32}^3}\right),$$

$$\frac{\mathrm{d}^2 y_3}{\mathrm{d}t^2} = G\left(m_1 \frac{y_1 - y_3}{r_{31}^3} + m_2 \frac{y_2 - y_3}{r_{32}^3}\right), \qquad (1.62)$$

$$\frac{\mathrm{d}^2 z_3}{\mathrm{d}t^2} = G\left(m_1 \frac{z_1 - z_3}{r_{31}^3} + m_2 \frac{z_2 - z_3}{r_{32}^3}\right),$$

with G being the gravitation constant, and

$$r_{31}^2 = (x_1 - x_3)^2 + (y_1 - y_3)^2 + (z_1 - z_3)^2,$$
$$r_{32}^2 = (x_2 - x_3)^2 + (y_2 - y_3)^2 + (z_2 - z_3)^2. \tag{1.63}$$

The x_i, y_i, z_i $(i = 1, 2)$ in these equations are known functions of time t, since the motion of primaries m_1 and m_2 is described by a 2-body problem that is integrable. Based on the different types of motions of m_1, m_2, the above 3-body problem is called the elliptic (circular as a special case), parabolic and hyperbolic restricted 3-body problem, respectively. The motions of planets in the Solar System are elliptic orbits with small eccentricities. Therefore, for the motions of small bodies in the Solar System, the model of circular restricted 3-body problem is generally adopted. For convenience, the distance between m_1 and m_2 is regarded as the unit of length. Also, to define other dimensionless variables, set $m_1 + m_2 = 1$ and $m_2 = \mu$, $m_1 = 1 - \mu$. Without loss of generality, assume $m_2 \leq m_1$, thus $\mu \leq 1/2$. The angular velocity n of circular motions of m_1, m_2 is set equal to 1. Suppose the motions of m_1, m_2 take place on the (x, y)-plane, and the z-axis is perpendicular to the (x, y)-plane. The origin of the coordinate system is at the center of masses. Rotate the (x, y)-plane around the z-axis, such that m_1, m_2 are on the x-axis. In such a rotating coordinate system, the coordinates of m_1, m_2 are $(\mu, 0, 0)$, $(-1 + \mu, 0, 0)$ respectively. Now for simplicity, let the coordinates of m_3 be (x, y, z). Thus the equations of motion Eq. (1.62) now can be written as

$$\frac{\mathrm{d}^2 x}{\mathrm{d}t^2} - 2\frac{\mathrm{d}y}{\mathrm{d}t} - x = -\left[(1 - \mu)\frac{x - \mu}{r_1^3} + \mu\frac{x - \mu + 1}{r_2^3}\right],$$

$$\frac{\mathrm{d}^2 y}{\mathrm{d}t^2} + 2\frac{\mathrm{d}x}{\mathrm{d}t} - y = -\left[\frac{1 - \mu}{r_1^3} + \frac{\mu}{r_2^3}\right]y, \tag{1.64}$$

$$\frac{\mathrm{d}^2 z}{\mathrm{d}t^2} = -\left[\frac{1 - \mu}{r_1^3} + \frac{\mu}{r_2^3}\right]z,$$

where

$$r_1^2 = r_{31}^2 = (x - \mu)^2 + y^2 + z^2,$$
$$r_2^2 = r_{32}^2 = (x + 1 - \mu)^2 + y^2 + z^2.$$

The equations of motion Eq. (1.64) can be rewritten in an alternative way

$$\frac{\mathrm{d}^2 x}{\mathrm{d}t^2} - 2\frac{\mathrm{d}y}{\mathrm{d}t} = \frac{\partial \Omega}{\partial x},$$

$$\frac{\mathrm{d}^2 y}{\mathrm{d}t^2} + 2\frac{\mathrm{d}x}{\mathrm{d}t} = \frac{\partial \Omega}{\partial y}, \tag{1.65}$$

$$\frac{\mathrm{d}^2 z}{\mathrm{d}t^2} = \frac{\partial \Omega}{\partial z},$$

in which

$$\Omega = \Omega(x, y, z) = \frac{1}{2}(x^2 + y^2) + \frac{1-\mu}{r_1} + \frac{\mu}{r_2}. \tag{1.66}$$

An integral can be derived from Eq. (1.65),

$$2\Omega - v^2 = C, \tag{1.67}$$

where

$$v^2 = \left(\frac{\mathrm{d}x}{\mathrm{d}t}\right)^2 + \left(\frac{\mathrm{d}y}{\mathrm{d}t}\right)^2 + \left(\frac{\mathrm{d}z}{\mathrm{d}t}\right)^2.$$

Equation (1.67) is a manifold of motion state, called Jacobi integral.
Since $v^2 = 2\Omega - C$, the equation

$$2\Omega - C = (x^2 + y^2) + 2\left(\frac{1-\mu}{r_1} + \frac{\mu}{r_2}\right) = 0, \tag{1.68}$$

determines the location where $v = 0$. For a given Jacobi integral constant
C, Eq. (1.68) defines a set of surfaces, known as the *zero-velocity surfaces*.
Because the third body m_3 cannot penetrate these surfaces in its motion,
they serve as the boundaries of m_3's motion. On the (x, y)-plane, the intersection of zero-velocity surfaces with the (x, y)-plane produces a set of *zero-
velocity curves*. Figure 1.5 illustrates examples of these curves for $\mu = 0.2$.

In Fig. 1.5, the permissible region of motion changes with the integral
constant C. As C decreases from $C = 4$ to $C = 3.6$, the two separated
permissible regions around m_1 and m_2 come into contact with each other
first at a point L_1 on the x-axis, and they merge to one connected region.
Then this region grows larger and larger as C decreases, till this inner
permissible region connects to the outer permissible region starting from
another point (L_2) on the x-axis. And finally the forbidden region breaks
into two separated parts, and this begins to happen at the third point L_3
on x-axis. Below, it will be shown that these critical points L_1, L_2, L_3 are
in fact the singular points of the equations of motion.

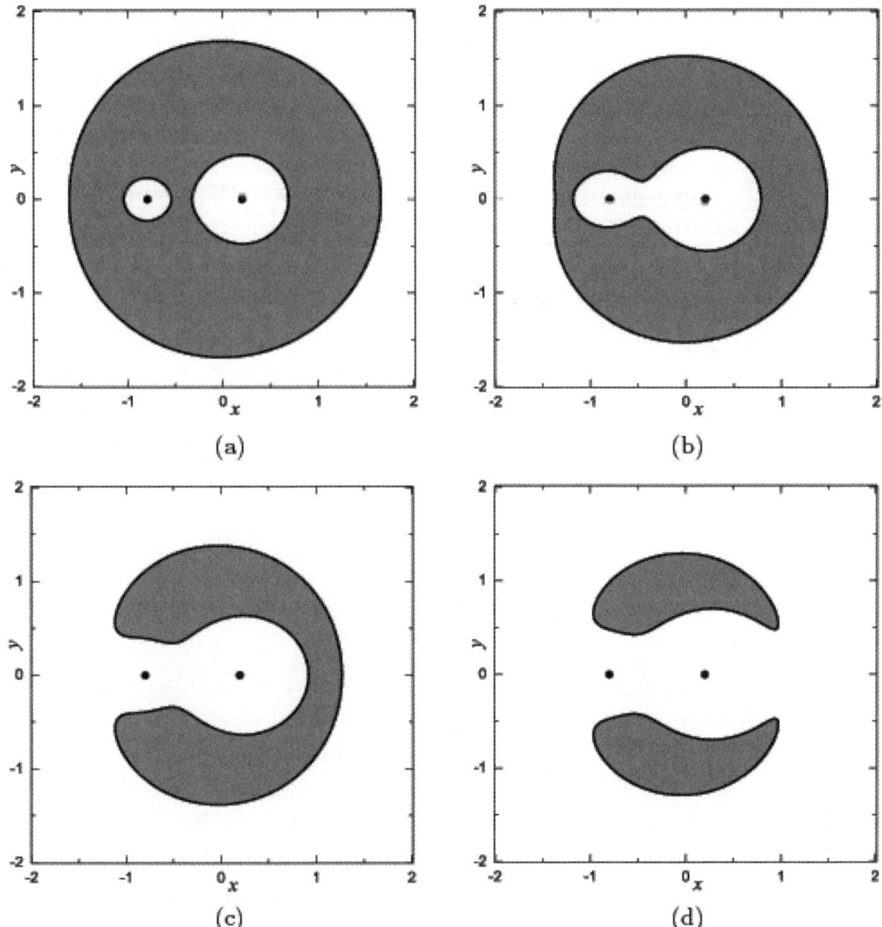

Fig. 1.5   The zero-velocity curves for $\mu = 0.2$. The Jacobi integral constant $C = 4$, 3.6, 3.3 and 3.15 in (a), (b), (c) and (d), respectively. Two dots stand for m_1 and m_2. Shaded regions are forbidden for m_3.

1.4.2   *Singular points of equations of motion*

The equations of singular points of state manifold read

$$
\frac{\mathrm{d}x}{\mathrm{d}t} = 0, \quad \frac{\mathrm{d}y}{\mathrm{d}t} = 0, \quad \frac{\mathrm{d}z}{\mathrm{d}t} = 0,
$$
$$
\frac{\partial\Omega}{\partial x} = 0, \quad \frac{\partial\Omega}{\partial y} = 0, \quad \frac{\partial\Omega}{\partial z} = 0. \tag{1.69}
$$

These are also the equations of equilibrium solutions. According to Eq. (1.69),

$$\frac{\partial \Omega}{\partial z} = -z \left(\frac{1-\mu}{r_1^3} + \frac{\mu}{r_2^3} \right) = 0,$$

thus $z = 0$, indicating that the singular points of state manifold or the equilibrium solutions (i.e. particular solutions) of the equations of motion are all on the (x, y)-plane.

From the above discussion, the equilibrium solutions should satisfy the equations

$$\frac{\partial \Omega}{\partial x} = 0, \qquad \frac{\partial \Omega}{\partial y} = 0,$$

that is,

$$x - \frac{(1-\mu)(x-\mu)}{r_1^3} - \frac{\mu(x+1-\mu)}{r_2^3} = 0, \tag{1.70}$$

$$y \left(1 - \frac{1-\mu}{r_1^3} - \frac{\mu}{r_2^3} \right) = 0. \tag{1.71}$$

Obviously, $r_1 = r_2 = 1$ satisfy Eqs. (1.70) and (1.71). In this case, each of the equilibrium solutions constitutes an equilateral triangle with m_1, m_2, thus they are called the equilateral triangle solutions. The coordinates of these solutions are $x = \mu - \frac{1}{2}, y = \pm \frac{\sqrt{3}}{2}$ (and surely $z = 0$). Moreover, from Eq. (1.71) one obtains $y = 0$. Substituting it into Eq. (1.70), one has

$$x - \frac{(1-\mu)(x-\mu)}{|x-\mu|^3} - \frac{\mu(x+1-\mu)}{|x+1-\mu|^3} = 0. \tag{1.72}$$

Divide the x-axis into three intervals $(-\infty, \mu - 1)$, $(\mu - 1, \mu)$ and $(\mu, +\infty)$, the solutions of Eq. (1.72) can be investigated in these intervals separately.

(a) $x \in (-\infty, \mu - 1)$. In this case, Eq. (1.72) is

$$x + \frac{1-\mu}{(x-\mu)^2} + \frac{\mu}{(x+1-\mu)^2} = 0. \tag{1.73}$$

Define an auxiliary value $\xi^{(1)}$ so that $x = \mu - 1 - \xi^{(1)}$. One must have $\xi^{(1)} > 0$ and

$$\xi^{(1)} + 1 - \mu + \frac{\mu - 1}{(1 + \xi^{(1)})^2} - \frac{\mu}{(\xi^{(1)})^2} = 0. \tag{1.74}$$

Therefore, $\xi^{(1)}$ is the root of the quintic equation

$$\xi^5 + (3-\mu)\xi^4 + (3-2\mu)\xi^3 - \mu\xi^2 - 2\mu\xi - \mu = 0. \tag{1.75}$$

According to the Cartesian discriminant for the positive roots of polynomial, when $0 < \mu \leq \frac{1}{2}$, Eq. (1.75) has a unique positive root, which can be found by applying an iterative method. Rewrite Eq. (1.75) as

$$\xi^3 = \frac{\mu(1+\xi)^2}{3 - 2\mu + \xi(3 - \mu + \xi)}. \tag{1.76}$$

Arbitrarily, take $\xi = 0$ as the initial guess of the solution. After the first iteration, one gets $\xi = [\mu/(3-2\mu)]^{\frac{1}{3}}$. In fact, one may take $[\mu/3(1-\mu)]^{\frac{1}{3}}$ as the initial value. Rewrite Eq. (1.75) as

$$\xi + (1-\mu)\left[1 - (1+\xi)^{-2}\right] + \mu\xi^{-2} = 0,$$

or

$$(1-\mu)\left[1 + \xi - (1+\xi)^{-2}\right] + \mu(\xi - \xi^{-2}) = 0,$$

then one has

$$\frac{\mu}{3(1-\mu)} = \frac{\xi^3(1 + \xi + \xi^2/3)}{(1+\xi)^2(1-\xi^3)}. \tag{1.77}$$

Finally a power series solution can be obtained from Eq. (1.77),

$$\xi^{(1)} = \nu\left(1 + \frac{1}{3}\nu - \frac{1}{9}\nu^2 - \frac{31}{81}\nu^3 - \frac{119}{243}\nu^4 - \frac{1}{9}\nu^5\right) + O(\nu^7), \tag{1.78}$$

where $\nu = [\mu/3(1-\mu)]^{\frac{1}{3}}$. This supports that the value $[\mu/3(1-\mu)]^{\frac{1}{3}}$ is a reasonable initial guess of the solution for the iterating method.

(b) $x \in (\mu - 1, \mu)$. In this case, Eq. (1.72) can be rewritten as

$$x + \frac{1-\mu}{(x-\mu)^2} - \frac{\mu}{(x+1-\mu)^2} = 0. \tag{1.79}$$

Define $\xi^{(2)}$ so that $x = \mu - 1 + \xi^{(2)}$. Then $\xi^{(2)} > 0$ and

$$\xi^{(2)} - 1 + \mu + \frac{1-\mu}{\left(\xi^{(2)} - 1\right)^2} - \frac{\mu}{\left(\xi^{(2)}\right)^2} = 0, \tag{1.80}$$

or

$$\xi^3 = \frac{\mu(1-\xi)^2}{3 - 2\mu - \xi(3 - \mu - \xi)}, \tag{1.81}$$

where $\xi = \xi^{(2)}$. One may obtain from Eq. (1.80) the following quintic equation

$$\xi^5 - (3-\mu)\xi^4 + (3-2\mu)\xi^3 - \mu\xi^2 + 2\mu\xi - \mu = 0. \tag{1.82}$$

Meanwhile, Eq. (1.81) has at least one positive root. Similar to the expansion in Eq. (1.78), for this case one has

$$\xi^{(2)} = \nu \left(1 - \frac{1}{3}\nu - \frac{1}{9}\nu^2 - \frac{23}{81}\nu^3 + \frac{151}{243}\nu^4 - \frac{1}{9}\nu^5 \right) + O(\nu^7), \quad (1.83)$$

with $\nu = [\mu/3(1-\mu)]^{\frac{1}{3}}$.

(c) $x \in (\mu, +\infty)$. Now Eq. (1.72) possesses the following form

$$x - \frac{1-\mu}{(x-\mu)^2} - \frac{\mu}{(x+1-\mu)^2} = 0. \qquad (1.84)$$

Set $x = \mu + \xi^{(3)}$, then $\xi^{(3)} > 0$ and

$$\xi^{(3)} + \mu - \frac{1-\mu}{\left(\xi^{(3)}\right)^2} - \frac{\mu}{\left(1+\xi^{(3)}\right)^2} = 0, \qquad (1.85)$$

or

$$\xi^3 = \frac{(1-\mu)(1+\xi)^2}{1+2\mu+\xi(2+\mu+\xi)}, \qquad (1.86)$$

where $\xi = \xi^{(3)}$. Again, from Eq. (1.86) another quintic equation is obtained

$$\xi^5 + (2+\mu)\xi^4 + (1+2\mu)\xi^3 - (1-\mu)\xi^2 - 2(1-\mu)\xi - (1-\mu) = 0. \quad (1.87)$$

The signs of coefficients have only one change, implying that this equation has only one unique positive root. It is obvious that the root is close to $+1$. Set $\eta = \xi - 1$, then η will satisfy

$$\eta^5 + (7+\mu)\eta^4 + (19+6\mu)\eta^3 + (24+13\mu)\eta^2 + 2(6+7\mu)\eta + 7\mu = 0. \quad (1.88)$$

From the last two terms of Eq. (1.88), the initial value of η can be taken as $\eta = -\frac{7}{12}\mu$. The expansion of the solution of Eq. (1.88) is

$$\eta = -\nu \left(1 + \frac{23}{84}\nu^2 + \frac{23}{84}\nu^3 + \frac{761}{2352}\nu^4 + \frac{3163}{7056}\nu^5 + \frac{30703}{49392}\nu^6 \right) + O(\nu^8).$$
$$(1.89)$$

The above three collinear solutions are called Euler particular solutions, and generally denoted by L_1, L_2, L_3. While the equilateral triangular particular solutions are called Lagrange solutions and denoted by L_4, L_5.

1.5 Stabilities of Lagrange and Euler solutions

In Sec. 1.4, the equations of motion of an infinitesimal body in the circular restricted 3-body problem have been introduced. Five particular solutions of the equations, i.e. Lagrange solutions and Euler solutions, have been revealed. They are all on the (x, y)-plane. In this section, the stability of these solutions is discussed.

1.5.1 *Planar problem*

At first, let us discuss the planar problem. The equations of motion are

$$\frac{\mathrm{d}^2 x}{\mathrm{d}t^2} - 2\frac{\mathrm{d}y}{\mathrm{d}t} = \frac{\partial \Omega}{\partial x},$$
$$\frac{\mathrm{d}^2 y}{\mathrm{d}t^2} + 2\frac{\mathrm{d}x}{\mathrm{d}t} = \frac{\partial \Omega}{\partial y},$$

(1.90)

where

$$\Omega = \frac{1}{2}(x^2 + y^2) + \frac{1 - \mu}{r_1} + \frac{\mu}{r_2},$$
$$r_1^2 = (x - \mu)^2 + y^2,$$
$$r_2^2 = (x + 1 - \mu)^2 + y^2.$$

Suppose $L(a, b)$ be the equilibrium points. Let

$$x = a + \xi, \quad y = b + \eta,$$

where ξ, η are the coordinates with respect to the equilibrium points L, and a, b are the coordinates of L. For the equilibrium solutions L_1, L_2, L_3, $a = x_1, x_2, x_3$ (i.e. $\xi^{(1)}, \xi^{(2)}, \xi^{(3)}$ in Sec. 1.4) and $b = 0$. For the triangular equilibrium solutions, they are $a = \mu - 1/2, b = \pm\sqrt{3}/2$.

$\Omega(x, y)$ can be expanded in L,

$$\Omega = \Omega(a, b) + \Omega_x(a, b)\xi + \Omega_y(a, b)\eta$$
$$+ \frac{1}{2}\Omega_{xx}(a, b)\xi^2 + \Omega_{xy}(a, b)\xi\eta + \frac{1}{2}\Omega_{yy}(a, b)y^2 + O(3),$$

(1.91)

and the equations in Eq. (1.90) becomes

$$\frac{\mathrm{d}^2 x}{\mathrm{d}t^2} - 2\frac{\mathrm{d}y}{\mathrm{d}t} = \Omega_{xx}(a, b)\xi + \Omega_{xy}(a, b)\eta + O(2),$$
$$\frac{\mathrm{d}^2 y}{\mathrm{d}t^2} + 2\frac{\mathrm{d}x}{\mathrm{d}t} = \Omega_{xy}(a, b)\xi + \Omega_{yy}(a, b)\eta + O(2),$$

(1.92)

where the subscript 'x' or 'y' denotes the partial differential calculation over 'x' or 'y', and $O(2), O(3)$ denote the terms with orders of and higher than 2 and 3 in ξ, η. Neglect the high order terms $O(2)$, Eq. (1.92) turns to be the variation equations, and the corresponding equation of eigenvalues reads

$$\begin{vmatrix} \lambda^2 - \Omega_{xx}(a,b) & -2\lambda - \Omega_{xy}(a,b) \\ 2\lambda - \Omega_{xy}(a,b) & \lambda^2 - \Omega_{yy}(a,b) \end{vmatrix} = 0. \tag{1.93}$$

This equation can be rewritten as

$$\lambda^4 + \left[4 - \Omega_{xx}(a,b) - \Omega_{yy}(a,b)\right]\lambda^2 + \Omega_{xx}(a,b)\Omega_{yy}(a,b) - \left[\Omega_{xy}(a,b)\right]^2 = 0. \tag{1.94}$$

For three collinear equilibrium solutions (when $0 < \mu < \frac{1}{2}$),

$$\Omega_{xy} = 0, \quad \Omega_{xx} > 0, \quad \Omega_{yy} < 0,$$

therefore,

$$\Omega_{xx}\Omega_{yy} - \Omega_{xy}^2 < 0.$$

Thus, Eq. (1.94) has the solutions as

$$\lambda_{1,2} = \pm \left[-\beta_1 + (\beta_1^2 + \beta_2^2)^{\frac{1}{2}}\right]^{\frac{1}{2}},$$

$$\lambda_{3,4} = \pm \left[-\beta_1 - (\beta_1^2 + \beta_2^2)^{\frac{1}{2}}\right]^{\frac{1}{2}}, \tag{1.95}$$

with

$$\beta_1 = 2 - \frac{\Omega_{xx}(a,b) + \Omega_{yy}(a,b)}{2}, \quad \beta_2^2 = -\Omega_{xx}(a,b)\Omega_{yy}(a,b) + \left[\Omega_{xy}(a,b)\right]^2.$$

Since

$$-\beta_1 + (\beta_1^2 + \beta_2^2)^{\frac{1}{2}} > 0, \quad -\beta_1 - (\beta_1^2 + \beta_2^2)^{\frac{1}{2}} < 0,$$

$\lambda_{1,2}$ is a pair of real roots with opposite signs, while $\lambda_{3,4}$ is a pair of conjugate complex roots. This indicates that three collinear equilibrium solutions L_1, L_2, L_3 are linearly unstable. According to the Lyapunov theorem of stability, they are unstable.

For the triangular equilibrium solutions L_4, L_5,

$$\Omega_{xx}(L_{4,5}) = \frac{3}{4}, \quad \Omega_{xy}(L_4) = \frac{3(3)^{\frac{1}{2}}}{2}\left(\mu - \frac{1}{2}\right),$$

$$\Omega_{yy}(L_{4,5}) = \frac{9}{4}, \quad \Omega_{xy}(L_5) = -\frac{3(3)^{\frac{1}{2}}}{2}\left(\mu - \frac{1}{2}\right).$$

Thus the equation of the eigenvalues is

$$\lambda^4 + \lambda^2 + \frac{27}{4}\mu(1-\mu) = 0. \tag{1.96}$$

Define

$$\Lambda_1 = \frac{1}{2}\left\{-1 + [1 - 27\mu(1-\mu)]^{\frac{1}{2}}\right\},$$

$$\Lambda_2 = \frac{1}{2}\left\{-1 - [1 - 27\mu(1-\mu)]^{\frac{1}{2}}\right\}, \tag{1.97}$$

The four roots of Eq. (1.96) are

$$\lambda_1 = +\Lambda_1^{\frac{1}{2}}, \quad \lambda_2 = -\Lambda_1^{\frac{1}{2}}, \quad \lambda_3 = +\Lambda_2^{\frac{1}{2}}, \quad \lambda_4 = -\Lambda_2^{\frac{1}{2}}, \tag{1.98}$$

Let μ_0 be the root of equation

$$1 - 27\mu(1-\mu) = 0.$$

For $0 < \mu \le \frac{1}{2}$, μ_0 is

$$\mu_0 = \frac{1}{2}\left(1 - \sqrt{\frac{23}{27}}\right).$$

Obviously, when $0 < 1 - 27\mu(1-\mu) \le 1$, or $\mu_0 > \mu \ge 0$,

$$-\frac{1}{2} < \Lambda_1 \le 0, \qquad -\frac{1}{2} > \Lambda_2 \ge -1.$$

Because in this case, Λ_1, Λ_2 are negative, $\lambda_1, \lambda_2, \lambda_3, \lambda_4$ are pure imaginary numbers. Therefore, L_4, L_5 are linearly stable. The value μ_0 is called Routh critical value (Routh, 1875). And this is the critical case in the stability theory.

However, the nonlinear stability cannot be decided only by the linear stability. By using Kolmogorov–Arnold–Moser (KAM) theorem, the conditions for stability can be derived. In the interval $0 < \mu < \mu_0$, the Lagrange equilibrium solutions are stable (Siegel & Moser, 1971), only except for three μ values of $\mu_1 = 0.010913\cdots$, $\mu_2 = \frac{1}{2}(1 - \frac{1}{45}\sqrt{1833}) = 0.024293\cdots$, $\mu_3 = \frac{1}{2}(1 - \frac{1}{15}\sqrt{213}) = 0.013516\cdots$.

For $\mu_0 < \mu < \frac{1}{2}$,

$$-\frac{23}{4} < 1 - 27\mu(1-\mu) < 0,$$

and Eq. (1.97) can be rewritten as

$$\Lambda_1 = \frac{1}{2}(-1 + i\delta), \quad \Lambda_2 = \frac{1}{2}(-1 - i\delta),$$

where $i = \sqrt{-1}$ and $\delta = \sqrt{27\mu(1-\mu) - 1}$, $0 < \delta < \sqrt{23}/2$. Thus the equation of eigenvalues Eq. (1.96) has four roots as following.

$$
\begin{aligned}
\lambda_1 &= +\sqrt{\frac{-1+i\delta}{2}} = \alpha_1 + i\beta_1, \\[2mm]
\lambda_2 &= -\sqrt{\frac{-1+i\delta}{2}} = \alpha_2 + i\beta_2, \\[2mm]
\lambda_3 &= +\sqrt{\frac{-1-i\delta}{2}} = \alpha_3 + i\beta_3, \\[2mm]
\lambda_4 &= -\sqrt{\frac{-1-i\delta}{2}} = \alpha_4 + i\beta_4.
\end{aligned}
\tag{1.99}
$$

The module of these eigenvalues is

$$
|\lambda| = |\lambda_{1,2,3,4}| = \frac{1}{\sqrt{2}}\left[27\mu(1-\mu)\right]^{\frac{1}{4}}.
\tag{1.100}
$$

The principle argument of λ_1 is

$$
\theta = \theta_1 = \arctan\left(\frac{1 + (1+\delta^2)^{\frac{1}{2}}}{\delta}\right).
$$

For $\mu_0 < \mu < \frac{1}{2}$,

$$
\frac{1}{\sqrt{2}} < |\lambda| < \frac{27^{\frac{1}{4}}}{2} \approx 1.1397535285.
$$

Thus,

$$
\frac{\pi}{2} > \theta > \arctan\left(\frac{3\sqrt{3}+2}{\sqrt{23}}\right) \approx 56.3187717113^\circ.
\tag{1.101}
$$

Four arguments of eigenvalues $\theta_{1,2,3,4}$ hold the following relations

$$
\theta = \theta_1 = \theta_2 - \pi = 2\pi - \theta_3 = \pi - \theta_4
\tag{1.102}
$$

Correspondingly, the α_k, β_k $(k = 1, 2, 3, 4)$ are related to each other by

$$
\begin{aligned}
\alpha &= \alpha_1 = -\alpha_2 = \alpha_3 = -\alpha_4, \\[2mm]
\beta &= \beta_1 = -\beta_2 = -\beta_3 = \beta_4,
\end{aligned}
\tag{1.103}
$$

where

$$
\alpha = \frac{\delta}{2\left(1 + 2|\lambda|^2\right)^{\frac{1}{2}}}, \qquad \beta = \frac{\left(1 + 2|\lambda|^2\right)^{\frac{1}{2}}}{2},
\tag{1.104}
$$

or, alternatively

$$
\alpha = \frac{1}{2}\left\{\left[27\mu(1-\mu)\right]^{\frac{1}{2}} - 1\right\}^{\frac{1}{2}}, \qquad \beta = \frac{1}{2}\left\{\left[27\mu(1-\mu)\right]^{\frac{1}{2}} + 1\right\}^{\frac{1}{2}}.
\tag{1.105}
$$

Since two of the four eigenvalues have positive real parts, the equilibrium solution L_4, L_5 are linearly unstable when $\mu_0 < \mu < \frac{1}{2}$. According to the stability theory, they are unstable.

1.5.2 *Spatial problem*

To investigate the stabilities of the equilibrium solutions in the spatial restricted 3-body problem, the small deviation (ξ, η, ζ) from the equilibrium points is defined as follows. Set $x = a + \xi$, $y = b + \eta$, $z = c + \zeta$, where (a, b, c) is the coordinate of the equilibrium solution, and substitute this (x, y, z) into the equation of motion Eq. (1.65). After some simple calculations, the variational equations are obtained

$$\frac{\mathrm{d}^2\xi}{\mathrm{d}t^2} - 2\frac{\mathrm{d}\eta}{\mathrm{d}t} = \Omega^0_{xx}\xi + \Omega^0_{xy}\eta + \Omega^0_{xz}\zeta,$$

$$\frac{\mathrm{d}^2\eta}{\mathrm{d}t^2} + 2\frac{\mathrm{d}\xi}{\mathrm{d}t} = \Omega^0_{yx}\xi + \Omega^0_{yy}\eta + \Omega^0_{yz}\zeta, \tag{1.106}$$

$$\frac{\mathrm{d}^2\zeta}{\mathrm{d}t^2} = \Omega^0_{zx}\xi + \Omega^0_{zy}\eta + \Omega^0_{zz}\zeta,$$

where $\Omega^0 = \Omega(a, b, c)$ and

$$\Omega_{zx} = 3z\left[\frac{(1-\mu)(x-\mu)}{r_1^5} + \frac{\mu(1-\mu+x)}{r_2^5}\right],$$

$$\Omega_{zy} = 3zy\left(\frac{1-\mu}{r_1^5} + \frac{\mu}{r_2^5}\right), \tag{1.107}$$

$$\Omega_{zz} = 3z^2\left(\frac{1-\mu}{r_1^5} + \frac{\mu}{r_2^5}\right) - \left(\frac{1-\mu}{r_1^3} + \frac{\mu}{r_2^3}\right).$$

Five equilibrium solutions are all on the plane of $z = 0$, thus here $c = 0$, $\Omega^0_{zx} = \Omega^0_{zy} = 0$, and

$$\Omega_{zz}(x, y, 0) = -\left(\frac{1-\mu}{r_1^3} + \frac{\mu}{r_2^3}\right) < 0. \tag{1.108}$$

Moreover, on the x-axis, $y = 0$ and

$$\Omega_{xx}(x, 0, 0) = 1 + 2\left(\frac{1-\mu}{r_1^3} + \frac{\mu}{r_2^3}\right) = 1 - 2\Omega_{zz}(x, 0, 0),$$

$$\tag{1.109}$$

$$\Omega_{yy}(x, 0, 0) = 1 - \left(\frac{1-\mu}{r_1^3} + \frac{\mu}{r_2^3}\right) = 1 + \Omega_{zz}(x, 0, 0).$$

Therefore,

$$\Omega_{zz}(x,0,0) = \Omega_{yy}(x,0,0) - 1, \tag{1.110}$$

or alternatively,

$$\Omega_{zz}(x,0,0) = \frac{1}{2}\left[1 - \Omega_{xx}(x,0,0)\right].$$

On the equilibrium solutions L_4 and L_5, $\Omega_{zz}(L_{4,5}) = -1$, while on the equilibrium solutions L_1, L_2 and L_3, $\Omega_{zz}(L_{1,2,3}) = \Omega_{yy}(L_{1,2,3}) - 1$. Moreover, $\Omega_{xz}^0 = \Omega_{yz}^0 = 0$. Finally, the equations of Eq. (1.106) can be separated into two independent systems: One is the system of ξ and η, and the other is the system of ζ. Thus the conclusions about the stability of equilibrium solutions in the planar restricted 3-body problem are still available for the space case. Nevertheless in the space case, the stability in the extra z-axis dimension is described by the variational equation of ζ as

$$\frac{\mathrm{d}^2\zeta}{\mathrm{d}t^2} = \Omega_{zz}^0\zeta. \tag{1.111}$$

This is an oscillatory equation. The corresponding eigenvalues are pure imaginary, hence the five equilibrium solutions along the z-axis are linearly stable. According to the above discussions, three Euler collinear equilibrium solutions are unstable. For $\mu_0 > \mu \geq 0$, two Lagrange triangular equilibrium solutions are linearly stable. And due to the nonlinear terms in the equations of ξ, η, ζ, the problem of stability becomes very complex.

1.6　Elliptic restricted 3-body problem

For the restricted 3-body problem, when the "primaries" move in the elliptic, parabolic and hyperbolic orbits, they are called elliptic, parabolic and hyperbolic restricted 3-body problem, respectively. These kinds of restricted 3-body problems are much complex than the circular one, in which there exists a well-known Jacobi integral. By means of this integral, one can deduce the Hill surface that gives the permissible and forbidden regions. But in the elliptic, parabolic or hyperbolic case, there is no such an integral. Nevertheless, the permissible region in the elliptic restricted 3-body problem can be investigated by using the minimum energy surface as proposed by Luk'Yanov (Luk'Yanov, 2005).

1.6.1　*Analyses on equations of motion*

In the same way as in the circular case discussed in previous sections, the motion of the two "primaries" m_1 and m_2 is assumed to take place on the

(x, y)-plane. A rotating coordinate system is chosen so that m_1 and m_2 are always on the x-axis. Thus the rotation of coordinate system is not uniform, and the distance r between m_1 and m_2 oscillates periodically. According to the well-known solution of the 2-body problem, r is given by

$$r = \frac{a(1 - e^2)}{1 + e \cos v},$$

where a, e is the semi-major axis and eccentricity of the ellipse of the relative motion of m_1 and m_2, while v is the true anomaly and considered as the independent variable. To make all the distances in this system to be dimensionless, all distances are divided by r, hereafter. In the dimensionless coordinate system, the distance between m_1 and m_2 is always 1. The sum of masses of primaries is set to be 1, i.e. $m_1 + m_2 = 1$. Without loss of generality, set $m_2 = \mu$, thus $m_1 = 1 - \mu$ and $0 < \mu < 1$. The center of masses is chosen as the origin of the coordinate system, so that the coordinates of m_1 and m_2 are $(\mu, 0, 0)$ and $(-1 + \mu, 0, 0)$, respectively. The equation of motion of the third body, i.e. the infinitesimal mass m_3 reads

$$\frac{\mathrm{d}^2 x}{dv^2} - 2\frac{\mathrm{d}y}{dv} = \frac{\partial \Omega}{\partial x},$$

$$\frac{\mathrm{d}^2 y}{dv^2} + 2\frac{\mathrm{d}x}{dv} = \frac{\partial \Omega}{\partial y}, \tag{1.112}$$

$$\frac{\mathrm{d}^2 z}{dv^2} = \frac{\partial \Omega}{\partial z},$$

where

$$\Omega = \rho \left(\frac{x^2 + y^2 - ez^2 \cos v}{2} + U \right),$$

$$\rho = \frac{1}{1 + e \cos v},$$

$$U = \frac{1 - \mu}{r_1} + \frac{\mu}{r_2}, \tag{1.113}$$

$$r_1 = \left[(x - \mu)^2 + y^2 + z^2 \right]^{\frac{1}{2}},$$

$$r_2 = \left[(x - \mu + 1)^2 + y^2 + z^2 \right]^{\frac{1}{2}}.$$

If $e = 0$, Eq. (1.112) turns to be the equations of motion for circular restricted 3-body problem, and the Jacobi integral exists. Below for Eq. (1.112) the analogous integral invariant to Jacobi integral is looked for. Multiply the three equations by $\frac{\mathrm{d}x}{\mathrm{d}v}$, $\frac{\mathrm{d}y}{\mathrm{d}v}$ and $\frac{\mathrm{d}z}{\mathrm{d}v}$ respectively, then the calculations of summation and integration give

$$\frac{V^2}{2} - \frac{V_{v_0}^2}{2} = \int_{v_0}^{v} \left(\frac{\partial \Omega}{\partial x}\frac{\mathrm{d}x}{\mathrm{d}v} + \frac{\partial \Omega}{\partial y}\frac{\mathrm{d}y}{\mathrm{d}v} + \frac{\partial \Omega}{\partial z}\frac{\mathrm{d}z}{\mathrm{d}v} \right) \mathrm{d}v, \qquad (1.114)$$

where

$$V^2 = \left(\frac{\mathrm{d}x}{\mathrm{d}v} \right)^2 + \left(\frac{\mathrm{d}y}{\mathrm{d}v} \right)^2 + \left(\frac{\mathrm{d}z}{\mathrm{d}v} \right)^2, \quad V_{v_0}^2 = (V^2)_{v=v_0} = V_0^2.$$

If $e \neq 0$, Eq. (1.114) is not a first integral. It is an integral invariant relation, from which another integral invariant relation is derived

$$\frac{V^2}{2} - \Omega = \frac{V_0^2}{2} - \Omega_0 - \int_{v_0}^{v} \frac{\partial \Omega}{\partial v}\mathrm{d}v, \quad \Omega_0 = (\Omega)_{v=v_0}. \qquad (1.115)$$

Substitute

$$\frac{\partial \Omega}{\partial v} = \frac{e \sin v}{(1 + e \cos v)^2} \left[\frac{1}{2}(x^2 + y^2 + z^2) + U \right], \qquad (1.116)$$

into Eq. (1.115), it can be reformed as

$$\frac{V^2}{2} - \Omega = \frac{V_0^2}{2} - \Omega_0 - \int_{v_0}^{v} \frac{e \sin v}{(1 + e \cos v)^2} W \mathrm{d}v, \qquad (1.117)$$

with

$$W = \frac{1}{2}(x^2 + y^2 + z^2) + U.$$

Further, replace the variable v by $\rho = 1/(1 + e \cos v)$, Eq. (1.117) becomes

$$\frac{V^2}{2} - \Omega = \frac{V_0^2}{2} - \Omega_0 - \int_{\rho_0}^{\rho} W \mathrm{d}\rho. \qquad (1.118)$$

Because $W > 0$, the integration in Eq. (1.118) satisfies

$$\int_{\rho_0}^{\rho} W \mathrm{d}\rho > 0 \qquad \text{when } \rho > \rho_0,$$

$$\int_{\rho_0}^{\rho} W \mathrm{d}\rho < 0 \qquad \text{when } \rho < \rho_0. \qquad (1.119)$$

To calculate Eq. (1.115) (or Eq. (1.117)), the expression of v as the function of x, y, z, i.e. the general solution of the equations of motion, must be given explicitly. Suppose the general solution was known and the integration in Eq. (1.115) is

$$\int_{v_0}^{v} \frac{\partial \Omega}{\partial v} \mathrm{d}v = u(v) - u(v_0). \tag{1.120}$$

Then Eq. (1.115) can be rewritten as

$$\frac{V^2}{2} - \Omega + u(v) = h, \tag{1.121}$$

where $h = \frac{1}{2}V_0^2 - \Omega_0 + u(v_0)$. But in fact, $u(v)$ is an unknown function, because the general solution is not known.

In the invariant relation Eq. (1.121), the first two terms are similar to the Jacobi integral in the circular restricted 3-body problem. Sometimes such an integral is regarded as the energy of the infinitesimal body m_3, thus Eq. (1.121) reflects the energy conservation law, where $\frac{V^2}{2} - \Omega$ is the Jacobi energy, $u(v)$ the additional energy, and h the energy constant. Apparently, for every particular solution, h has a specific value. Set $v_0 = \pi$, and $v_0 = 0$, Eq. (1.121) gives

$$\frac{V_a^2}{2} - \Omega_a + u(\pi) = \frac{V^2}{2} - \Omega + u(v) = \frac{V_p^2}{2} - \Omega_p + u(0), \tag{1.122}$$

where the subscripts 'a' and 'p' denote the apastron and periastron, respectively. As $\rho(v + 2\pi) = \rho(v)$, $W > 0$ and

$$\int_{v_0}^{v} \frac{\partial \Omega}{\partial v} \mathrm{d}v = \int_{\rho_0}^{\rho} W \mathrm{d}\rho = u(v) - u(v_0),$$

$u(v)$ is a periodic function with a period of 2π. Moreover, when $\pi \geq v \geq 0$, Eq. (1.119) tells that

$$u(\pi) \geq u(v) \geq u(0), \tag{1.123}$$

i.e. $u(v)$ attains its maximum and minimum at apastron and periastron respectively. Combined with Eq. (1.122), this proves

$$\frac{V_a^2}{2} - \Omega_a \leq \frac{V^2}{2} - \Omega \leq \frac{V_p^2}{2} - \Omega_p. \tag{1.124}$$

1.6.2 *Permissible region of m_3*

With all the above results, the permissible regions of m_3 can be determined, if the minimum of the function W is known. At its minimum, W satisfies the following equations

$$\frac{\partial W}{\partial x} = x - \frac{1-\mu}{r_1^3}(x-\mu) - \frac{\mu}{r_2^3}(x-\mu+1) = 0,$$

$$\frac{\partial W}{\partial y} = y\left(1 - \frac{1-\mu}{r_1^3} - \frac{\mu}{r_2^3}\right) = 0, \tag{1.125}$$

$$\frac{\partial W}{\partial z} = z\left(1 - \frac{1-\mu}{r_1^3} - \frac{\mu}{r_2^3}\right) = 0.$$

Apparently, $r_1 = r_2 = 1$ is a solution to these equations. This solution is a circle in the (x, y, z) space with the equal distances to m_1 and m_2, and it is called *libration torus*.

Meanwhile, $y = z = 0$ fulfills the second and the third equations in Eq. (1.125), thus another solution may be determined if $y = z = 0$ is substituted into the first equation. By doing this, a quintic algebraic equation of x is obtained. According to the discussion in Sec. 1.4, such a quintic equation has three real roots. Finally, W meets its minima at three collinear libration points and a libration torus. The libration torus is determined by equation $x = \frac{1}{2}(2\mu - 1)$, $y^2 + z^2 = \frac{3}{4}$, thus

$$W_{\min} = \frac{1}{2}\left[1 - \mu(1-\mu)\right]. \tag{1.126}$$

Introduce

$$\tilde{W} = W - W_{\min} = \frac{1}{2}(x^2 + y^2 + z^2) + U - \frac{1}{2}[1 - \mu(1-\mu)],$$

$$\tilde{u}(v) = u(v) - \frac{W_{\min}}{1 + e\cos v},$$

Eq. (1.122) can be rewritten as

$$\frac{V_a^2}{2} - \Omega_a + \frac{W_{\min}}{1-e} + \tilde{u}(\pi) = \frac{V^2}{2} - \Omega + \frac{W_{\min}}{1+e\cos v} + \tilde{u}(v)$$

$$= \frac{V_p^2}{2} - \Omega_p + \frac{W_{\min}}{1+e} + \tilde{u}(0) = h. \tag{1.127}$$

Taking into account the inequalities as follows

$$\tilde{W} \geq 0, \quad \int_{\frac{1}{1-e}}^{\rho} \tilde{W}d\rho \leq 0, \quad \int_{\frac{1}{1+e}}^{\rho} \tilde{W}d\rho \geq 0, \quad \tilde{u}(\pi) \geq \tilde{u}(v) \geq \tilde{u}(0),$$

$$\tag{1.128}$$

an inequality can be obtained from Eq. (1.127)

$$\frac{V_a^2}{2} - \Omega_a + \frac{W_{\min}}{1-e} \leq \frac{V^2}{2} - \Omega + \frac{W_{\min}}{1+e\cos v} \leq \frac{V_p^2}{2} - \Omega_p + \frac{W_{\min}}{1+e}.$$

$$\tag{1.129}$$

This is a generalization of the inequality in Eq. (1.124). Combining Eq. (1.128) and Eq. (1.129), an inequality is given

$$\frac{V_p^2}{2} - \Omega_p - \left(\frac{V_a^2}{2} - \Omega_a\right) = u(\pi) - u(0) \geq \frac{\rho}{1-e^2}\left[1 - \mu(1-\mu)\right].$$

$$(1.130)$$

Because of $V^2 \geq 0$, Eq. (1.124) may define the permissible motion regions for m_3 as

$$\Omega \geq \Omega_p - \frac{1}{2}V_p^2. \tag{1.131}$$

Its boundary is

$$\Omega = \Omega_p - \frac{1}{2}V_p^2,$$

i.e.

$$x^2 + y^2 - ez^2\cos v + 2U = C_p(1 + e\cos v), \tag{1.132}$$

where $C_p = 2\Omega_p - V_p^2$ is the Jacobi energy when $v = 0$. Another inequality for the permissible regions may be derived from Eq. (1.129), and it reads

$$\Omega - \frac{W_{\min}}{1 + e\cos v} \geq \Omega_p - \frac{V_p^2}{2} - \frac{W_{\min}}{1+e}. \tag{1.133}$$

The boundary of the region defined above is called the *minimum energy surface*, which can also be described by

$$x^2 + y^2 - ez^2\cos v + 2U - 2W_{\min} = C_p'(1 + e\cos v), \tag{1.134}$$

where $C_p' = C_p - \frac{2W_{\min}}{1+e}$. For given μ, e, C_p' and v, the infinitesimal body m_3 cannot move outside this boundary. Apparently, the permissible region bounded by Eq. (1.134) is smaller than the region surrounded by the boundary Eq. (1.132). Therefore, the minimum energy surface defined by Eq. (1.134) is generally adopted as the boundary of the permissible region. It is worth noting that the minimum energy surface is a periodic function of v, and it is always bounded between the following two surfaces

$$x^2 + y^2 - ez^2 + 2U - 2W_{\min} = C_p'(1 + e), \tag{1.135}$$

$$x^2 + y^2 + ez^2 + 2U - 2W_{\min} = C_p'(1 - e). \tag{1.136}$$

For any definitive C'_p, μ and e, Eq. (1.135) and Eq. (1.136) determine two boundaries, and all the surfaces defined by Eq. (1.134) locate between them.

Last but not least, the singular points and the topological structure of the minimum energy surface in the elliptical restricted 3-body problem can also be investigated, like in the circular restricted 3-body problem introduced in Sec. 1.4 and Sec. 1.5.

1.7 Hill region in 3-body problem

In the restricted 3-body problem, the Hill surface gives the permissible and forbidden region for massless body (see Sec. 1.4 for details). When the parameter hC^2 (h and C are the total energy and angular momentum, respectively) varies, the topological structure of phase space including the connexity of permissible regions will change. In this section, the generalized Hill surface and the permissible and forbidden configurations are discussed.

1.7.1 *Hill-type stability in 3-body problem*

Some notations as listed below will be used in this section.

$$M = m_1 + m_2 + m_3,$$

$$M^* = m_1 m_2 + m_1 m_3 + m_2 m_3,$$

$$\mu = GM,$$

$$I = \frac{1}{2} \sum_{i=1}^{3} m_i r_i^2,$$

$$U = G \left(\frac{m_1 m_2}{r_{12}} + \frac{m_1 m_3}{r_{13}} + \frac{m_2 m_3}{r_{23}} \right),$$

where G is the gravitational constant, m_i ($i = 1, 2, 3$) the masses of three bodies, r_i ($i = 1, 2, 3$) the distances of body P_i to the center of masses (i.e. the origin of the coordinate system), and r_{ij} ($i, j = 1, 2, 3; i \neq j$) the distances between bodies P_i and P_j. The I here is the semi moment of inertia. Besides, two constant lengths, the generalized semi-major axis a and generalized semi-latus rectum p, are defined as

$$a = -\frac{GM^*}{2h},$$

$$p = \frac{MC^2}{GM^{*2}}. \tag{1.137}$$

Moreover, another two variables of length, the mean quadratic distance ρ and the mean harmonic distance ν, are also defined:

$$M^*\rho^2 = m_1 m_2 r_{12}^2 + m_1 m_3 r_{13}^2 + m_2 m_3 r_{23}^2, \tag{1.138}$$

$$\frac{M^*}{\nu} = \frac{m_1 m_2}{r_{12}} + \frac{m_1 m_3}{r_{13}} + \frac{m_2 m_3}{r_{23}}. \tag{1.139}$$

Apparently, ρ, ν and I, U are related by

$$M^*\rho^2 = 2MI, \quad \frac{GM^*}{\nu} = U. \tag{1.140}$$

Write GM^*/ν in an alternative way:

$$\frac{M^*}{\nu} = \frac{(m_1 m_2)^{\frac{3}{2}}}{(m_1 m_2 r_{12}^2)^{\frac{1}{2}}} + \frac{(m_1 m_3)^{\frac{3}{2}}}{(m_1 m_3 r_{13}^2)^{\frac{1}{2}}} + \frac{(m_2 m_3)^{\frac{3}{2}}}{(m_2 m_3 r_{23}^2)^{\frac{1}{2}}}.$$

Using the Hölder inequality, one gets

$$\frac{M^*}{\nu} \geq \left(m_1 m_2 r_{12}^2 + m_1 m_3 r_{13}^2 + m_2 m_3 r_{23}^2\right)^{\frac{1}{2}} \cdot \left(m_1 m_2 + m_1 m_3 + m_2 m_3\right)^{\frac{3}{2}}$$

$$= \left(M^*\rho^2\right)^{-\frac{1}{2}} \left(M^*\right)^{\frac{3}{2}} = \frac{M^*}{\rho}. \tag{1.141}$$

Immediately, one knows from Eq. (1.141) that $\rho \geq \nu$. Only when all the three distances r_{ij} $(i, j = 1, 2, 3; i \neq j)$ are equal to each other, i.e. when the three bodies constitute an equilateral triangle in space, one has $\rho = \nu$. It can be seen that $\rho \geq \nu \geq 0$ corresponds to

$$U > 0, \quad IU^2 \geq \frac{G^2 M^{*3}}{2M}. \tag{1.142}$$

While the Lagrange-Jacobi identity is equivalent to

$$\frac{\mathrm{d}^2 \rho^2}{\mathrm{d}t^2} = 2\mu \left(\frac{1}{\nu} - \frac{1}{a}\right). \tag{1.143}$$

Substituting Eq. (1.140) into the Sundman inequality

$$C^2 + \left(\frac{\mathrm{d}I}{\mathrm{d}t}\right)^2 \leq 4I(U + h),$$

and using Eq. (1.137), one obtains the Sundman inequality

$$\frac{\rho}{\nu} \geq \frac{\rho}{2a} + \frac{p}{2\rho} + \frac{\rho\dot{\rho}^2}{2\mu} \qquad \left(\dot{\rho} = \frac{\mathrm{d}\rho}{\mathrm{d}t}\right). \tag{1.144}$$

The right hand side of this inequality

$$j = \frac{\rho}{2a} + \frac{p}{2\rho} + \frac{\rho\dot{\rho}^2}{2\mu} \tag{1.145}$$

is called Sundman function.

It will be shown below that the necessary and sufficient condition for the permissible configuration of r_{12}, r_{23}, r_{13} is

$$\frac{2}{\nu} \geq \frac{1}{a} + \frac{p}{\rho^2},$$

or equivalently

$$\frac{\rho}{\nu} \geq \frac{\rho}{2a} + \frac{p}{2\rho}. \tag{1.146}$$

Here the "necessary and sufficient" means that any permissible configuration must satisfy this condition, and when this condition is satisfied, the permissible configuration will exist.

The necessity

If the condition in Eq. (1.146) is not fulfilled, neither the condition in Eq. (1.144) is. The necessity is proven.

The sufficiency

Assume that three bodies are on the invariant plane that is determined by the vector of the total angular moment, and their velocities are

$$\boldsymbol{v}_i = A\boldsymbol{C} \times \boldsymbol{r}_i + B\boldsymbol{r}_i \qquad (i = 1, 2, 3), \tag{1.147}$$

where $\boldsymbol{C}$ is the vector of angular moment and

$$A = \frac{1}{2I} = \frac{M}{M^* \rho^2},$$

$$B = \pm \sqrt{\frac{\mu}{\rho^2} \left(\frac{2}{\nu} - \frac{1}{a} - \frac{p}{\rho^2} \right)}.$$

A simple calculation shows that the $\boldsymbol{r}_i, \boldsymbol{v}_i$ given by Eq. (1.147) are allowed by the integral of energy and the integral of angular moment, suggesting that this motion is permissible.

Below, all the permissible configuration of the three bodies will be searched with ρ/ν being fixed as a constant. Since ρ/ν is independent of the scale change of the configurations, all similar triangles composed by the three bodies correspond to the same value of ρ/ν. If the total energy $h \geq 0$ and $1/a \leq 0$, the condition in Eq. (1.146) will always be satisfied for large enough ρ. However, this is the case of unstable system, i.e. the system of three bodies will definitely disintegrate.

The more interesting case, obviously is the one with negative h and positive a. The right-hand side of Eq. (1.146) attains its minimum $\sqrt{p/a}$

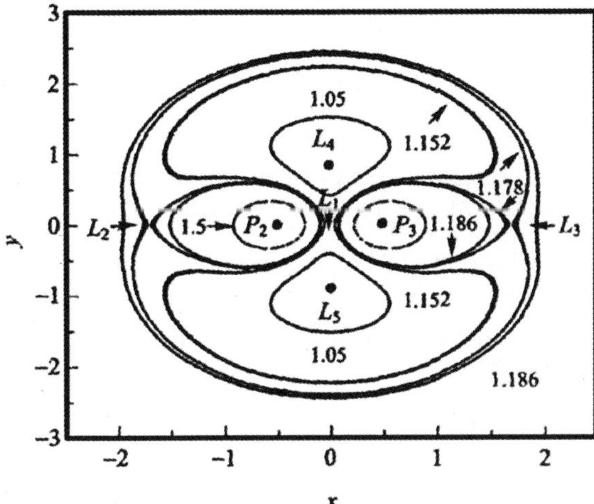

Fig. 1.6 The contours of ρ/ν, in which the masses are assumed to be $m_1 = 2m_2 = 2m_3 = 1$ and the distance between P_2 and P_3 are fixed. Adapted from Sun & Zhou (2008).

when $\rho = \sqrt{ap}$, therefore, for both of the above cases, the necessary and sufficient condition for the permissible configuration is

$$\left(\frac{\rho}{\nu}\right)^2 \geq \frac{p}{a}. \tag{1.148}$$

Figure 1.6 displays the generalized Hill curves for the case of $m_1 = 2m_2 = 2m_3 = 1$. The value of ρ/ν is determined by the relative distances among the three bodies, which are free of the specific scale of the system. Without loss of generality, the distance between P_2 and P_3 can be fixed and the value of ρ/ν can be determined by the distances of P_1 to P_2 and P_1 to P_3. When all the three distances among three bodies are equal to each other, i.e. the configuration of three bodies is an equilateral triangle, ρ/ν attains its minimum $\rho/\nu = 1$. This is indicated by two points L_4, L_5 in Fig. 1.6.

In fact, other three minima of $(\rho/\nu)^2$ can be found along the straight line connecting P_2 and P_3, as indicated by the points L_1, L_2 and L_3 in Fig. 1.6. The body P_1 resting on these points comprise with P_2 and P_3 three different collinear central configurations respectively. Without loss of generality, the three bodies are set on the abscissa, with P_2 being at the origin and between P_1 and P_3. Meanwhile the coordinates of P_1 and P_3 are supposed to be $(-1, 0)$ and $(x, 0)$, respectively. Obviously, $x > 0$ and the center of mass of this system locates at C^* (the abscissa)

$$C^* = \frac{(m_3 x - m_1)}{M}.$$

If the three masses constitute a central configuration, the acceleration of each body is proportional to their position vector with respect to the center of mass. Therefore, x will satisfy the following equation

$$\frac{m_2 + m_3/(1+x)^2}{1 + C^*} = \frac{m_3/x^2 - m_1}{C^*} = \frac{m_2/x^2 + m_1/(1+x)^2}{x - C^*}. \tag{1.149}$$

A quintic algebraic equation can be derived from the above equation, and it reads

$$(m_1 + m_2)x^5 + (3m_1 + 2m_2)x^4 + (3m_1 + m_2)x^3$$
$$- (m_2 + 3m_3)x^2 - (2m_2 + 3m_3)x - (m_2 + m_3) = 0. \tag{1.150}$$

A simple calculation may verify that this equation is consistent with the equation for the collinear central configuration. Now, it will be shown that the positions of these collinear central configuration are just the minimum points of $(\rho/\nu)^2$ on the line. The mutual distances among P_1, P_2, P_3, which can be calculated from their coordinates, are $r_{12} = 1, r_{23} = x, r_{13} = 1 + x$. Thus,

$$M^* \left(\frac{\rho}{\nu}\right)^2 = \left[m_1 m_2 + m_1 m_3 (1+x)^2 + m_2 m_3 x^2\right]$$
$$\times \left(m_1 m_2 + \frac{m_1 m_3}{1+x} + \frac{m_2 m_3}{x}\right)^2. \tag{1.151}$$

And then,

$$\frac{(1+x)^2 x^2}{2m_3} \frac{\mathrm{d}}{\mathrm{d}x} \left[M^{*3} \left(\frac{\rho}{\nu}\right)^2\right]$$
$$= m_1 m_2 \left[(m_1 + m_2)x^5 + (3m_1 + 2m_2)x^4 + (3m_1 + m_2)x^3\right.$$
$$\left. -(m_2 + 3m_3)x^2 - (2m_2 + 3m_3)x - (m_2 + m_3)\right]. \tag{1.152}$$

The extreme points can be determined by finding the solution to the following equation

$$F(x) = (m_1 + m_2)x^5 + (3m_1 + 2m_2)x^4 + (3m_1 + m_2)x^3$$
$$- (m_2 + 3m_3)x^2 - (2m_2 + 3m_3)x - (m_2 + m_3) = 0. \tag{1.153}$$

Apparently, this equation coincides completely with Eq. (1.150). Although this is only for the situation where P_1 locates between P_2 and P_3, similar calculations can be made and results can be obtained for other two sequences of P_1, P_2 and P_3 on a line, implying that the extreme points of $(\rho/\nu)^2$ in the collinear cases are just the positions of the collinear central configurations of three bodies.

According to the analysis about the equilibrium solution to the 3-body problem, $(\rho/\nu)^2$ attains its local minima at L_i $(i = 1, 2, 3)$ when the three bodies are collinear, but in space, these points are neither minima nor maxima.

Since $\rho/\nu \geq 1$, the inequality in Eq. (1.148) does not supply any useful information if $p/a \leq 1$ or equivalently $hC^2 \geq -G^2 M^{*3}/2M$. On the contrast, some conclusions about the motion region of three bodies may be drawn from Eq. (1.148) when $p/a > 1$. As $\rho/\nu = 1$ happens only at the L_4, L_5 points, the close vicinity of these two points is the forbidden region. If $(p/a)^{1/2} \geq \max\{(\rho/\nu)_{L_i} \ (i = 1, 2, 3)\}$, the permissible configuration region of three bodies will be divided into three disconnected regions. Because ρ/ν reaches infinity when P_1 is close to one of P_2 and P_3 or is close to infinity, P_1 will be either near one of P_2 and P_3 or at a large distance from both of them, i.e. P_1 will be in one of the isolated regions (Fig. 1.6). And the motion remains for all time confined to one of these regions.

The above results are very similar to the Hill surface in the restricted 3-body problem, hence the stability is called "Hill-type stability". Because ρ/ν is independent of scale change of configuration, Hill-type stability can't deny the escape phenomenon. This is quite different from the Hill stability in the restricted 3-body problem.

1.7.2 *Generalized Hill-type stability in N-body problem*

The above results can be generalized to broader applications in the problem of N-body's motion in a field of general attraction, i.e. the system with potential function as:

$$U = G \sum_{1 \leq i < j \leq N} \frac{m_i m_j}{r_{ij}^{k-1}} \quad (1 < k < 3). \tag{1.154}$$

In such a system,

$$I = \frac{1}{2} \sum_{i=1}^{N} m_i r_i^2 = \frac{1}{2} \sum_{i=1}^{N} m_i (x_i^2 + y_i^2 + z_i^2),$$

$$\frac{\mathrm{d}^2 I}{\mathrm{d}t^2} = \sum_{i=1}^{N} m_i (\dot{x}_i^2 + \dot{y}_i^2 + \dot{z}_i^2)$$

$$+ \sum_{i=1}^{N} m_i (x_i \ddot{x}_i + y_i \ddot{y}_i + z_i \ddot{z}_i) \quad (i = 1, 2, \ldots, N)$$

$$= 2T + \sum_{i=1}^{N} m_i(x_i U_{x_i} + y_i U_{y_i} + z_i U_{z_i}),$$

where $\dot{x} = \frac{\mathrm{d}x}{\mathrm{d}t}, \ddot{x} = \frac{\mathrm{d}^2 x}{\mathrm{d}t^2}, U_x = \frac{\partial U}{\partial x}$. Since

$$\sum_{i=1}^{N} m_i \left(x_i U_{x_i} + y_i U_{y_i} + z_i U_{z_i}\right) = (1 - k)U,$$

one knows

$$\frac{\mathrm{d}^2 \rho^2}{\mathrm{d}t^2} = \frac{2M}{M^*} \frac{\mathrm{d}^2 I}{\mathrm{d}t^2} = \frac{2M}{M^*} \left[(3 - k)U + 2h\right].$$

In virtue of

$$U = \frac{GM^*}{\nu}, \quad 2h = -\frac{GM^*}{a},$$

one obtains the Lagrange-Jacobi identity in the system of the general potential function (Eq. (1.154)) as below

$$\frac{\mathrm{d}^2 \rho^2}{\mathrm{d}t^2} = 2\mu \left[\frac{3 - k}{\nu} - \frac{1}{a}\right]. \tag{1.155}$$

The Sundman inequality Eq. (1.144) is still valid in this case, thus the necessary and sufficient condition for the permissible configuration is the same as the one listed in Eq. (1.148). But in this case, for different scales of configuration, the configurations are different even for the same value of p/a. When the scale changes, for example, r_{ij} becomes sr_{ij}, the potential U turns to $s^{1-k}U$, ν to $s^{k-1}\nu$, and ρ^2 to $s^2\rho^2$. Hence, for a fixed p/a, after scaling one has

$$\bar{\rho}^2 = s^2 \rho^2,$$
$$\bar{\nu} = s^{k-1}\nu,$$
$$\left(\frac{\bar{\rho}}{\bar{\nu}}\right)^2 = \frac{s^2 \rho^2}{s^{2(k-1)}\nu^2} = s^{4-2k} \left(\frac{\rho}{\nu}\right)^2.$$

Therefore, after the scaling transformation, $(\rho/\nu)^2 \geq p/a$ should be replaced by $(\bar{\rho}/\bar{\nu})^2 \geq s^{2k-4}p/a$. Equivalently, the scaling operation changes p/a to $s^{2k-4}p/a$.

1.7.3 *Hill stability in hierarchical triple system*

Because the parameter hC^2 controls the topology of the system, it plays an important role in the general 3-body system as the Jacobi constant does in the circular restricted 3-body problem. There exists in the triple stellar systems a special case: The hierarchical triple system, which comprises a close binary and a distant component. The asteroidal binary systems in the Kuiper belt of the Solar System can be regarded as the hierarchical triple systems as well, in which the binary's masses are relatively low with respect to the third body (the Sun) at large distance. The parameter hC^2 can be used as a criterion of the Hill-type stability in such a special hierarchical triple system.

A hierarchical triple system can be approximately regarded as two 2-body systems: One consists of m_1 and m_2, and the other one consists m_3 and the barycenter of m_1, m_2. Consider such a hierarchical triple system in which two components m_1 and m_2 form a close binary with semi-major axis a_1 and eccentricity e_1, while the third body m_3 moves around the barycenter of the binary on a much wider orbit with semi-major axis a_2 ($\gg a_1$), eccentricity e_2 and inclination i (with respect to the orbital plane of the binary). In the special case of asteroidal binary in the Solar System, the mass of the third body is much larger than the binaries' mass, i.e. $m_3 \gg m_1 + m_2$. Without loss of generality, suppose $m_1 \geq m_2$.

With a 2-body approximation, the total kinetic energy of the triple system can be written as

$$h = -\frac{G}{2}\left(\frac{m_1 m_2}{a_1} + \frac{\mu m_3}{a_2}\right), \tag{1.156}$$

with $\mu = m_1 + m_2$. The squared total angular momentum reads

$$\begin{aligned}
C^2 = G\Bigg[& \frac{m_1^2 m_2^2}{\mu}a_1(1 - e_1^2) + \frac{\mu^2 m_3^2}{M}a_2(1 - e_2^2) \\
& + \frac{2m_1 m_2 m_3 \mu^{1/2}}{M^{1/2}}\sqrt{a_1 a_2(1 - e_1^2)(1 - e_2^2)}\cos i \Bigg],
\end{aligned} \tag{1.157}$$

where $M = m_1 + m_2 + m_3$. As for the inclination i in the system, when $0 \leq i \leq 90°$, the motion of m_3 around the binary is said to be on a "direct" orbit, and the motion is "retrograde" if $90° < i \leq 180°$.

Define $S = -hC^2/G^2$. It is known that this quantity can be expressed in terms of the mutual distances among this 3-body system r_{12}, r_{13}, r_{23}, but is independent of the scale of the system configuration. As Szebehely and Zare did (Szebehely & Zare, 1977), the integral manifold of the 3-body problem can be parameterized by the scale-independent quantity

$$S = \frac{m_1^3 m_2^3}{2\mu}(1 - e_1^2) + \frac{m_1^2 m_2^2 m_3 \mu^{\frac{1}{2}}}{M^{\frac{1}{2}}}\left(\frac{a_2}{a_1}\right)^{\frac{1}{2}}\sqrt{(1 - e_1^2)(1 - e_2^2)}\cos i$$

$$+ \frac{m_1 m_2 m_3^2 \mu^2}{2M}\left(\frac{a_2}{a_1}\right)(1 - e_2^2) + \frac{m_1^2 m_2^2 m_3}{2}\left(\frac{a_1}{a_2}\right)(1 - e_1^2)$$

$$+ \frac{m_1 m_2 m_3^2 \mu^{\frac{3}{2}}}{M^{\frac{1}{2}}}\left(\frac{a_1}{a_2}\right)^{\frac{1}{2}}\sqrt{(1 - e_1^2)(1 - e_2^2)}\cos i + \frac{m_3^3 \mu^3}{2M}(1 - e_2^2).$$

$$(1.158)$$

Note that the right-hand side of this equation has been arranged in increasing powers of m_3 together with $M \sim m_3$.

The topology of the integral manifold of the 3-body problem will be changed when S passes through its critical values (Chen, 1979). Fix the distance r_{13} as the unit for length and use the scale-independent parameter S, the critical values of S can be determined in terms of the relative position of m_2 with respect to m_1 and m_3. The minimum critical S is attained at the triangular Lagrangian equilibrium points L_4 and L_5, when the three bodies comprise an equilateral triangle. Furthermore, the quantity S has three local minima at the collinear Lagrangian equilibrium points L_1, L_2 and L_3, which are specified by the same subscripts as the middle body in the collinear 3-body configuration. If S is larger than the highest value (denoted by S_{cr}) of the critical S at these three saddle points, the region of possible motion of m_2 is divided into three disconnected parts: Two around m_1 and m_3 respectively, and the rest one at a large distance. Then the body m_2 (originally near m_1) can never move near the third body m_3, since to do so it would have to cross the forbidden region of motion. This persistence of hierarchy type is called the Hill stability.

The condition $m_2 \leq m_1 \ll m_3$ implies that the critical value S_{cr} is achieved at L_2. Denote the distance ratio by x, i.e. $r_{12} : r_{23} = x \in (0, 1)$. The value x in this configuration will be determined by the well-known fifth order algebraic equation

$$(m_2 + m_3)x^5 + (2m_2 + 3m_3)x^4 + (m_2 + 3m_3)x^3$$
$$- (3m_1 + m_2)x^2 - (3m_1 + 2m_2)x - (m_1 + m_2) = 0.$$

$$(1.159)$$

Note the evident similarity between this equation and Eq. (1.150).

In terms of x, S_{cr} can be expressed in a convenient form (Zare, 1977)

$$S_{\mathrm{cr}} = \frac{f^2(x)g(x)}{2M}, \tag{1.160}$$

where f, g are functions as

$$\begin{aligned}
f(x) &= m_2 m_3 + m_1 m_3(1+x)^{-1} + m_1 m_2 x^{-1}, \\
g(x) &= m_2 m_3 + m_1 m_3(1+x)^2 + m_1 m_2 x^2.
\end{aligned} \tag{1.161}$$

The approximate value of x may be readily determined as

$$x = x_0 \left(1 + \frac{2m_1 + m_2}{3\mu} x_0 + \frac{2m_1 - m_2}{9\mu} x_0^2 + O(x_0^3) \right), \tag{1.162}$$

where the first approximation to x is

$$x_0 = \left(\frac{\mu}{3m_3} \right)^{\frac{1}{3}}. \tag{1.163}$$

Under the condition $m_3 \gg \mu$, x_0 is a small quantity.

Substituting the expression of x into Eq. (1.161) and expanding Eq. (1.160) in power of x_0, the critical parameter S_{cr} can be calculated by

$$S_{\mathrm{cr}} = \frac{m_3^3}{2M} \left[\mu^3 + 9m_1 m_2 \mu x_0^2 + 2m_1 m_2(m_1 - m_2)x_0^3 + O(x_0^4) \right]. \tag{1.164}$$

The sufficient condition for the triple system to be Hill stability is $S \geq S_{\mathrm{cr}}$, i.e. $hC^2 \leq \left(hC^2 \right)_{\mathrm{cr}}$. Since x_0 is a small quantity and $m_3/M = 1 - 3x_0^3 + O(x_0^6)$, S and S_{cr} can be expanded as a power series in x_0. Define the parameter $\beta = m_2^3 m_3^2/2$, the mass ratio $\lambda = m_1/m_2 \geq 1$, $\gamma = 1 + \lambda$, and the semimajor axes ratio $\alpha = a_2/a_1$, one obtains

$$\begin{aligned}
S = \beta \Big\{ &\gamma^3(1 - e_2^2) + 2\sqrt{3}\lambda\gamma\alpha^{-\frac{1}{2}} \sqrt{(1 - e_1^2)(1 - e_2^2)} \cos i \cdot x_0^{\frac{3}{2}} \\
&+ \left[3\lambda\gamma\alpha(1 - e_2^2) - 3\gamma^3(1 - e_2^2) + 3\lambda^2\gamma^{-1}\alpha^{-1}(1 - e_1^2) \right] x_0^3 \\
&+ 6\sqrt{3}\lambda^2\gamma^{-1}\alpha^{\frac{1}{2}} \sqrt{(1 - e_1^2)(1 - e_2^2)} \cos i \cdot x_0^{\frac{9}{2}} + o(x_0^{\frac{9}{2}}) \Big\},
\end{aligned} \tag{1.165}$$

and

$$S_{\mathrm{cr}} = \beta\{ \gamma^3 + 9\lambda\gamma x_0^2 + [2\lambda(\lambda - 1) - 3\gamma^3]x_0^3 + O(x_0^4) \}. \tag{1.166}$$

Because α is large in a hierarchical triple system, one may assume $\alpha \sim x_0^{-3}$. In this case, one may retain the terms up to $(\alpha^{\frac{1}{2}} \cdot x_0^{\frac{9}{2}})$ in Eq. (1.165), which is of order x_0^3. Furthermore, one may also omit the lowest order term in α

for the coefficients of terms x_0^3. Then, subtract Eq. (1.166) from Eq. (1.165) and retain terms up to the third order in x_0, one obtains the sufficient condition for the hierarchical triple systems to be Hill stable, as below:

$$S - S_{\text{cr}} = \beta \Bigg\{ -(1+\lambda)^3 e_2^2$$

$$+ 2\sqrt{3}\lambda(1+\lambda)\alpha^{-\frac{1}{2}}\sqrt{(1-e_1^2)(1-e_2^2)}\cos i \cdot x_0^{\frac{3}{2}} - 9\lambda(1+\lambda)x_0^2$$

$$+ \left[3\lambda(1+\lambda)\alpha(1-e_2^2) + 3(1+\lambda)^3 e_2^2 - 2\lambda(\lambda-1)\right]x_0^3$$

$$+ 6\sqrt{3}\lambda^2(1+\lambda)^{-1}\alpha^{\frac{1}{2}}\sqrt{(1-e_1^2)(1-e_2^2)}\cos i \cdot x_0^{\frac{9}{2}} \Bigg\} \geq 0.$$

$$(1.167)$$

One may derive from this inequality the critical value of Hill stability in terms of the semi-major axes ratio α. By introducing a new variable $X = \alpha^{\frac{1}{2}}$ and dividing $S = S_{\text{cr}}$ by $\beta\lambda(1+\lambda)\alpha^{\frac{1}{2}}$, one obtains the critical condition of Hill stability $S = S_{\text{cr}}$ as

$$AX^3 + BX^2 + CX + D = 0, \qquad (1.168)$$

where the coefficients A, B, C and D are as follows.

$$A = 3(1-e_2^2)x_0^3,$$

$$B = 6\sqrt{3}\lambda(1+\lambda)^{-2}\sqrt{(1-e_1^2)(1-e_2^2)}\cos i \cdot x_0^{\frac{9}{2}}$$

$$C = -\lambda^{-1}(1+\lambda)^2 e_2^2 - 9x_0^2 \qquad (1.169)$$

$$+ \left[3\lambda^{-1}(1+\lambda)^2 e_2^2 - 2(\lambda-1)(1+\lambda)^{-1}\right]x_0^3$$

$$D = 2\sqrt{3(1-e_1^2)(1-e_2^2)}\cos i \cdot x_0^{\frac{3}{2}}.$$

The cubic equation Eq. (1.168) can be solved following a standard procedure. Defining $Y = X + B/(3A)$, this cubic equation becomes

$$Y^3 + pY + q = 0, \qquad (1.170)$$

where

$$p = \frac{C}{A} - \frac{B^2}{3A^2}, \qquad q = \frac{2B^3}{27A^3} - \frac{BC}{3A^2} + \frac{D}{A}. \qquad (1.171)$$

The discriminant Δ of Eq. (1.170) reads

$$\Delta = \left(\frac{q}{2}\right)^2 + \left(\frac{p}{3}\right)^3 = -\frac{\left[9x_0^2 + \lambda^{-1}(1+\lambda)^2 e_2^2\right]^3}{729(1-e_2^2)^3} \cdot x_0^{-9} + O(x_0^{-6}) < 0,$$

$$(1.172)$$

which means that Eq. (1.170) has three real roots

$$Y_k = \frac{2}{3}\sqrt{\frac{\rho}{(1-e_2^2)x_0}} \cdot \cos\left\{\frac{1}{3}\arccos\left[\frac{3\sqrt{3}(3-4e_2^2)\sqrt{1-e_1^2}\cos i}{\rho^{\frac{3}{2}}}\right] + \theta_k\right\},$$

$$(k = 1, 2, 3)$$

$$(1.173)$$

with $\rho = 9 + \lambda^{-1}(1+\lambda)^2 e_2^2(x_0^{-2} - 3x_0) + 2(\lambda-1)(1+\lambda)^{-1}x_0$ and $\theta_1 = -\pi/3$, $\theta_2 = \pi/3$, $\theta_3 = \pi$. In the process of deriving these roots, the high order terms $\sim x_0^3$ and $\sim x_0^{1/2}$ have been removed in the coefficients p and q respectively, and such approximation is equivalent to the omission of terms much smaller than $O(x_0^3)$ in the inequality of Eq. (1.167) by considering the above transformation.

Since the quantity $B/(3A)$ $(\sim x_0^{3/2})$ is much smaller than Y_k (at least of order $x_0^{-1/2}$), the roots of Eq. (1.168) can be approximated by $X_k \approx Y_k$ ($k = 1, 2, 3$). Taking into account the fact that the Hill stability requires the semi-major axis ratio $(a_2/a_1) \geq (a_2/a_1)_{\mathrm{cr}} = X^2$, one obtains the solution $X \approx Y = \max\{Y_1, Y_2, Y_3\} > 0$ for the sufficient condition. According to the facts that $\rho \geq 9$ as a result of small x_0 and the quantity $\sqrt{1-e_1^2}\cos i \cdot (3 - 4e_2^2) \in [-3, 3]$, it is easy to determine that $X = X_1 > 0$. Finally, the critical semi-major axis ratio in a closed form is obtained:

$$\left(\frac{a_2}{a_1}\right)_{\mathrm{cr}} = \frac{4\rho}{9(1-e_2^2)x_0}$$

$$\times \cos^2\left\{\frac{1}{3}\arccos\left[\frac{3\sqrt{3}(3-4e_2^2)\sqrt{1-e_1^2}\cos i}{\rho^{3/2}}\right] - \frac{\pi}{3}\right\},$$

$$(1.174)$$

where $\rho = 9 + \lambda^{-1}(1+\lambda)^2 e_2^2(x_0^{-2} - 3x_0) + 2(\lambda-1)(1+\lambda)^{-1}x_0$ as in Eq. (1.173).

However, when the third body's eccentricity $e_2 \neq 0$, Eq. (1.174) contains too many variables and takes a very complex form, hence it is not easy to clarify the influence of each system parameter on the critical semi-major axis ratio $(a_2/a_1)_{\mathrm{cr}}$. Below two cases of $e_2 = 0$ and $e_2 \neq 0$ are considered separately.

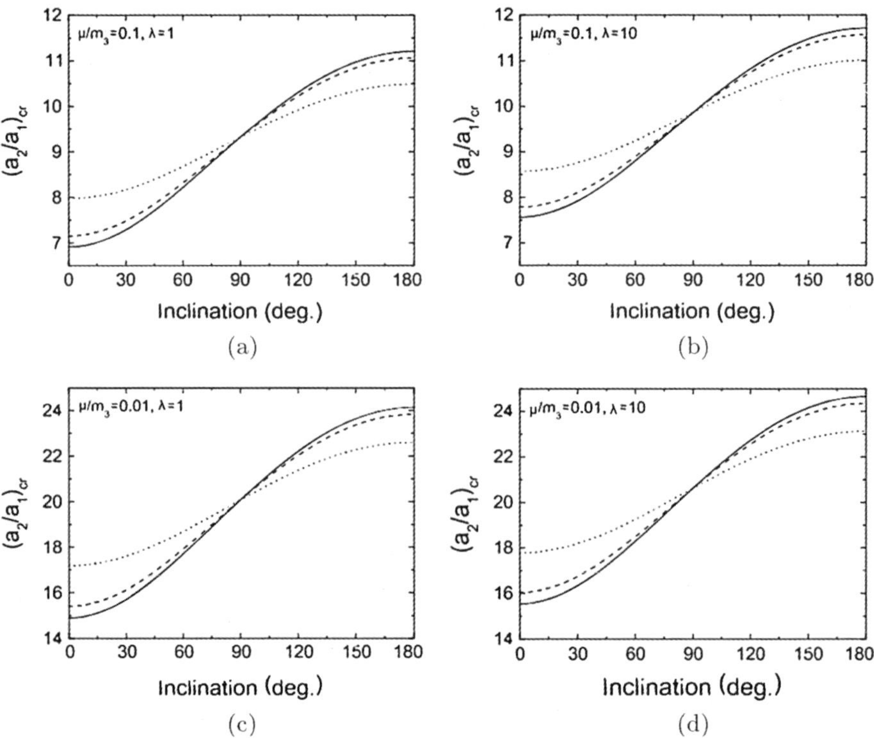

Fig. 1.7   The variation of critical semi-major axis ratio $(a_2/a_1)_{cr}$ in the case of $e_2 = 0$. The solid, dashed and dotted curves are for different eccentricities of $e_1 = 0, e_1 = 0.4$ and $e_1 = 0.8$ respectively. Four panels are for (a) $\mu/m_3 = 0.1, \lambda = 1$, (b) $\mu/m_3 = 0.1, \lambda = 10$, (c) $\mu/m_3 = 0.01, \lambda = 1$, and (d) $\mu/m_3 = 0.01, \lambda = 10$. Taken from Li *et al.* (2010).

Case $e_2 = 0$

In the case of $e_2 = 0$, Eq. (1.174) is simplified to

$$\left(\frac{a_2}{a_1}\right)_{cr} = \frac{4\rho'}{9x_0}\cos^2\left\{\frac{1}{3}\arccos\left[\frac{9\sqrt{3}\sqrt{1-e_1^2}\cos i}{\rho'^{3/2}}\right] - \frac{\pi}{3}\right\}, \qquad (1.175)$$

where $\rho' = 9 + 2(\lambda - 1)(1 + \lambda)^{-1}x_0$ and the λ-dependence is due to the inclusion of the term proportional to $(\alpha^0 \cdot x_0^3)$ in the inequality of Eq. (1.167).

Figure 1.7 shows the diagram of $(a_2/a_1)_{cr}$ against the inclination i for the representative eccentricities e_1 and different mass ratio pairs $(\mu/m_3, \lambda)$. The region of Hill stability lies to the upside of each curve, and has some apparent characteristics: (1) All curves intersect at $i = 90°$, which is an obvious consequence of Eq. (1.175), as $(a_2/a_1)_{cr} = 4[9 + \lambda^{-1} + 2(\lambda - 1)$

$(1+\lambda)^{-1}x_0]/(9x_0)$, independent of e_1, for $i = 90°$; (2) As e_1 increases, the Hill stable region contracts for the direct motion ($0° \leq i \leq 90°$), whilst it expands for the retrograde motion ($90° < i \leq 180°$). The different trends can be analytically checked by taking the derivative of the right-hand side of Eq. (1.175) with respect to e_1, which has the form $\frac{\partial (a_2/a_1)_{\mathrm{cr}}}{\partial e_1} = \tau \cdot \cos i$ ($\tau > 0$); (3) The critical semi-major axis ratio $(a_2/a_1)_{\mathrm{cr}}$ increases with i, since Eq. (1.175) gives the derivative $\frac{\partial (a_2/a_1)_{\mathrm{cr}}}{\partial i} \geq 0$. This characteristic suggests in particular that systems in direct motion are more stable than those in retrograde motion; (4) The region of stability decreases in size as the binary/third body mass ratio μ/m_3 decreases; (5) As λ increases from 1, i.e. equal-mass binary, to higher values, the region of stability decreases only slightly. Besides, such variation tends to zero as μ/m_3 ($\sim x_0^3$) tends to zero, as the right-hand side of Eq. (1.175) does not depend on λ anymore in the limit $x_0 = 0$.

Case $e_2 \neq 0$

In Eq. (1.174), e_1 and i appear in a combined form $\sqrt{1-e_1^2}\cos i$. Defining a new parameter $K = \sqrt{1-e_1^2}\cos i$ and taking derivative of Eq. (1.174) with respect to K, it is easy to see that $(a_2/a_1)_{\mathrm{cr}}$ increases monotonically with K decreasing when $(3 - 4e_2^2) > 0$, i.e. $e_2 < 0.86$. While for $e_2 > 0.86$, it increases monotonically with K increasing. According to Eq. (1.174), the critical semi-major axis ratio now can be computed for given $e_2, \mu/m_3$ and K. Several typical examples are illustrated in Fig. 1.8.

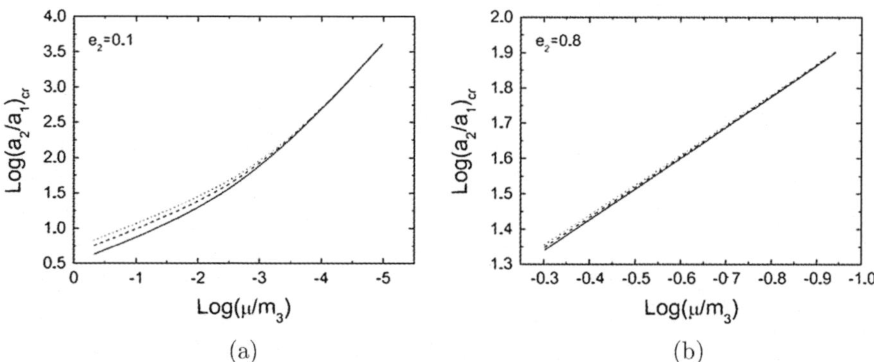

Fig. 1.8　The variation of critical semi-major axis ratio $(a_2/a_1)_{\mathrm{cr}}$ with respect to mass ratio μ/m_3. The solid, dashed and dotted curves are for $K = 1, K = 0$ and $K = -1$ respectively. Panels (a) and (b) refer to case $e_2 = 0.1$ and $e_2 = 0.8$, and the mass ratio $\lambda = 1$ in both panels. Taken from Li *et al.* (2010).

Table 1.1 The values of $(a_2/a_1)_{\mathrm{cr}}$ for various mass ratio μ/m_3 and orbital parameter K (see text), for the case $\lambda = 1$.

μ/m_3	10^{-1}	10^{-2}	10^{-3}	10^{-4}	10^{-5}
			$e_2 = 0.01$		
$K = 1$	6.92	14.95	32.59	73.97	195.64
$K = 0.5$	8.22	17.75	38.58	86.47	219.51
$K = 0$	9.33	20.12	43.67	97.23	240.85
$K = -0.5$	10.31	22.24	48.23	106.90	260.45
$K = -1$	11.21	24.19	52.41	115.81	278.74
			$e_2 = 0.1$		
$K = 1$	7.46	19.78	76.84	489.10	4233.48
$K = 0.5$	8.70	22.15	80.53	493.65	4238.35
$K = 0$	9.77	24.28	84.06	498.15	4243.22
$K = -0.5$	10.74	26.24	87.45	502.62	4248.08
$K = -1$	11.63	28.06	90.71	507.05	4252.93
			$e_2 = 0.5$		
$K = 1$	23.07	157.13	1387.94	13454.51	133597.99
$K = 0.5$	23.76	157.96	1388.81	13455.40	133598.88
$K = 0$	24.42	158.77	1389.68	13456.28	133599.77
$K = -0.5$	25.07	159.58	1390.56	13457.17	133600.66
$K = -1$	25.70	160.39	1391.43	13458.05	133601.54
			$e_2 = 0.8$		
$K = 1$	89.43	759.29	7223.68	71362.42	711661.35
$K = 0.5$	89.66	759.53	7223.93	71362.68	711661.60
$K = 0$	89.89	759.78	7224.18	71362.93	711661.86
$K = -0.5$	90.11	760.03	7224.44	71363.19	711662.11
$K = -1$	90.34	760.27	7224.69	71363.44	711662.36

As displayed in Fig. 1.8, the critical semi-major axis ratio $(a_2/a_1)_{\mathrm{cr}}$ increases as the eccentricity e_2 increases and as the mass ratio μ/m_3 decreases. For given e_2, the relative difference among the illustrated critical curves in Fig. 1.8 for Hill stability decreases down to negligible level with μ/m_3. A more intuitive idea about the variation of $(a_2/a_1)_{\mathrm{cr}}$ with respect to parameters μ/m_3 and K can be found in Table 1.1, in which the values of $(a_2/a_1)_{\mathrm{cr}}$ are listed for $\lambda = 1$ and $e_2 = 0.01, 0.1, 0.5, 0.8$.

When μ/m_3 is small, the weak dependence of $(a_2/a_1)_{\mathrm{cr}}$ on (e_1, i) is apparent in Table 1.1. In fact, for nonzero e_2, as x_0 in Eq. (1.174) gets smaller, the dominator of argument of the arccos function, which is of order x_0^{-2}, gets larger and the fraction will tend to zero. Thus the Hill stability limit would be virtually independent of e_1 and i for sufficiently small μ/m_3.

Such a weak dependence of $(a_2/a_1)_{\mathrm{cr}}$ on (e_1, i) implies that the Hill stability for some real systems can be effectively analyzed without knowing e_1 and i. Besides, the investigations on the general cases of $\lambda > 1$ show that they have nearly the same features as shown in Fig. 1.8. And for small μ/m_3, the negligible influence of the binary mass ratio λ on the stable region is analogous to the case $e_2 = 0$.

For systems with low mass ratio μ/m_3 and not extremely small e_2, as shown in Table 1.1, the critical semi-major axis ratio $(a_2/a_1)_{\mathrm{cr}}$ may be approximated to its value at $i = 90°$ (i.e. $K = 0$), which can be obtained from Eq. (1.174) and takes a very simple from

$$\left(\frac{a_2}{a_1}\right)_{\mathrm{cr}} = \frac{(1+\lambda)^2(1 - 3x_0^3)e_2^2 + 9\lambda x_0^2 + 2\lambda(\lambda - 1)(1 + \lambda)^{-1}x_0^3}{3\lambda(1 - e_2^2)x_0^3}.$$

$$(1.176)$$

Although this formula cannot be applied to the case of large μ/m_3 or very small e_2, it is still a very useful criterion due to its simple form and independence of e_1 and i. Because $(a_2/a_1)_{\mathrm{cr}}$ increases monotonically with a decreasing K when $e_2 < 0.86$ (this can be seen in Table 1.1 too), the Hill-type stability can be assured for a real system that is in direct motion $(0° \leq i \leq 90°)$ and has an (a_2/a_1) value larger than $(a_2/a_1)_{\mathrm{cr}}$ calculated from Eq. (1.176). On the contrary, if $(a_2/a_1) < (a_2/a_1)_{\mathrm{cr}}$, the system must be Hill unstable for all retrograde motions $(90° < i \leq 180°)$. Exactly speaking, the system in direct motion turns to be less stable when the quantity $(3 - 4e_2^2) < 0$, i.e. $e_2 > 0.86$ (derived from Eq. (1.174)). According to the changing rule of the derivative of $(a_2/a_1)_{\mathrm{cr}}$ with respect to K, the $(a_2/a_1)_{\mathrm{cr}}$ in retrograde motion would be larger than the $(a_2/a_1)_{\mathrm{cr}}$ in direct motion. This implies that the system in direct motion is less stable than that in retrograde motion. However, for even larger e_2, the difference between the direct and the retrograde motion for Hill stability would nearly disappear, and the arguments presented above would remain valid.

Moreover, a further approximation can be made to Eq. (1.176) for the case $x_0^2 \ll e_2^2$, which leads to

$$\left(\frac{a_2}{a_1}\right)_{\mathrm{cr}} = \frac{(1+\lambda)^2}{3\lambda x_0^3}e_2^2 + O(e_2^4). \qquad (1.177)$$

According to this formulation, the value of $(a_2/a_1)_{\mathrm{cr}}$ is proportional to e_2^2, implying that the Hill stable region shrinks rather quickly with increasing e_2. This formulation is more convenient for the analysis of sufficiently low mass binary systems, such as the Sun and Kuiper belt binary system.

1.8 Evolution of inertia momentum in N-body problem

It is well-known that the evolution of inertia moment plays an important role in the qualitative studies of N-body problem. For example, the bounded motions, escapes, the general collision, etc. can be characterized by the behavior of the inertia moment.

In Sec. 1.7, the ratio of the "mean quadratic distance" ρ over "mean harmonic distance" ν has been used to describe the generalized Hill curves. Because the ratio ρ/ν is independent of the scale change of configuration of three bodies, it cannot determine the real forbidden and attainable domains. In this section, the variations of the attainable and forbidden domains with respect to the evolution of inertia moment I will be discussed. As the square root of inertia moment I is proportional to the "mean quadratic distance" ρ, one may make use of ρ to study the evolution of I.

Some notations, as listed below, will be used in this section.

$$m_i \quad (i = 1, 2, \ldots, N), \qquad\qquad \text{mass of the } i\text{th body;}$$

$$M = \sum_{i=1}^{N} m_i, \qquad\qquad \text{total mass;}$$

$$M^* = \sum_{1 \leq i < j \leq N} m_i m_j,$$

$$\mu = GM, \qquad\qquad G \text{ is the gravitational constant;}$$

$$\boldsymbol{r}_i \quad (i = 1, 2, \ldots, N), \qquad\qquad \text{radius vector of the } i\text{th body;}$$

$$r_{ij}, \quad (i, j = 1, 2, \ldots, N; i \neq j), \qquad \text{mutual distances;}$$

$$h = \frac{1}{2} \sum_{i=1}^{N} m_i \left(\frac{\mathrm{d}\boldsymbol{r}_i}{\mathrm{d}t} \cdot \frac{\mathrm{d}\boldsymbol{r}_i}{\mathrm{d}t} \right)$$
$$- G \sum_{1 \leq i < j \leq N} \frac{m_i m_j}{r_{ij}}, \qquad\qquad \text{energy integral;}$$

$$\boldsymbol{C} = \sum_{i=1}^{N} m_i \boldsymbol{r}_i \times \frac{\mathrm{d}\boldsymbol{r}_i}{\mathrm{d}t}, \qquad\qquad \text{integral of angular momentum;}$$

$$a = -\frac{GM^*}{2h}, \qquad\qquad \text{generalized semi-major axis;}$$

$$p = \frac{MC^2}{GM^*}, \quad C = |\boldsymbol{C}|, \qquad\qquad \text{generalized semi-latus rectum.}$$

In a restricted 3-body problem and also in a 2-body problem, the fixed lengths a and p are actually the semi-major axis and semi-latus rectum of the relative motion of the primaries.

The variable lengths "mean quadratic distance" ρ and "mean harmonic distance" ν are defined as before (in Sec. 1.7),

$$M^*\rho^2 = \sum_{1 \le i < j \le N}{}^{\backprime} m_i m_j r_{ij}^2,$$

$$\frac{M^*}{\nu} = \sum_{1 \le i < j \le N} \frac{m_i m_j}{r_{ij}}.$$

Note that $M^*\rho^2/2M$ is the inertia moment I in the axes of the general center of mass and GM^*/ν is the potential U. In fact,

$$\frac{M^*\rho^2}{2M} = \frac{1}{2}\sum_{i=1}^{N} m_i \boldsymbol{r}_i \cdot \boldsymbol{r}_i = I,$$

$$\frac{GM^*}{\nu} = G \sum_{1 \le i < j \le N} \frac{m_i m_j}{r_{ij}} = U.$$

On the other hand, the harmonic mean of non-negative quantities is always smaller than or equal to their quadratic mean, as expressed by

$$\inf(r_{ij}) \le \nu \le \rho \le \sup(r_{ij}). \tag{1.178}$$

The equality $\nu = \rho$ can be attained either for a 2-body case, or for a restricted problem of 3-body when all r_{ij} are equal.

1.8.1 *Some relations and a theorem*

With the above notations, some usual relations can be found, as in Sec. 1.7.

$$\frac{\mathrm{d}^2\rho^2}{\mathrm{d}t^2} = 2\mu\left(\frac{1}{\nu} - \frac{1}{a}\right) \quad \text{(Lagrange-Jacobi equation)}, \tag{1.179}$$

$$\frac{\rho}{\nu} \ge j \quad \text{(Sundman inequality)}, \tag{1.180}$$

$$j = \frac{p}{2\rho} + \frac{\rho}{2a} + \frac{\rho\dot{\rho}^2}{2\mu} \quad \text{(Sundman function)}, \tag{1.181}$$

where $\dot{\rho} = \frac{\mathrm{d}\rho}{\mathrm{d}t}$. From Eqs. (1.178) and (1.180), one obtains

$$\frac{1}{\nu} \ge \frac{1}{\rho}, \quad \frac{1}{\nu} \ge \frac{j}{\rho}. \tag{1.182}$$

Hence, if ν is eliminated from Eqs. (1.179) and (1.182), a simple inequality containing the maximum information about the variables $\rho, \dot{\rho}$ is derived:

$$\frac{\mathrm{d}^2\rho^2}{\mathrm{d}t^2} \geq 2\mu\left[\frac{\sup(1;j)}{\rho} - \frac{1}{a}\right], \quad \text{i.e.:}$$

$$\ddot{\rho} \geq \sup\left[\left(\frac{\mu}{\rho^2} - \frac{\mu}{a\rho} - \frac{\dot{\rho}^2}{\rho}\right); \left(\frac{\mu p}{2\rho^3} - \frac{\mu}{2a\rho} - \frac{\dot{\rho}^2}{2\rho}\right)\right]. \tag{1.183}$$

Note first that the limit case (when "=" is attained) in Eq. (1.183) is integrable. In such a case,

$$\frac{\mathrm{d}^2\rho^2}{\mathrm{d}t^2} = 2\mu\left[\frac{\sup(1;j)}{\rho} - \frac{1}{a}\right]. \tag{1.184}$$

According to the value of j, the study on Eq. (1.184) can be divided into two cases as following.

Case I: $j \leq 1$. In such a case,

$$\frac{\mathrm{d}^2\rho^2}{\mathrm{d}t^2} = 2\mu\left(\frac{1}{\rho} - \frac{1}{a}\right). \tag{1.185}$$

This is the differential equation of mutual distance ρ in a 2-body problem. Consider

$$\frac{\mathrm{d}^2\rho^2}{\mathrm{d}t^2} = 2\dot{\rho}^2 + 2\rho\ddot{\rho},$$

rewrite $\ddot{\rho} = \dot{\rho}\frac{\mathrm{d}\dot{\rho}}{\mathrm{d}\rho}$, and multiply both sides of Eq. (1.185) by ρ, one has

$$\rho\dot{\rho}^2 + \rho^2\dot{\rho}\frac{\mathrm{d}\dot{\rho}}{\mathrm{d}\rho} = \mu - \frac{\mu\rho}{a}.$$

Because $\mathrm{d}(\rho^2\dot{\rho}^2) = 2\rho\dot{\rho}^2\mathrm{d}\rho + 2\rho^2\dot{\rho}\mathrm{d}\dot{\rho}$, the above equation can be integrated, and one obtains

$$\rho^2\dot{\rho}^2 - 2\mu\rho + \frac{\mu\rho^2}{a} = 2\mu\rho(j-1) - \mu p = \text{constant}. \tag{1.186}$$

In this integral, $\rho(j-1)$ is a constant ($j \leq 1$). The evolution of ρ is just the evolution of mutual distance in a 2-body problem, in which

$$\mu_1 = \text{gravitation constant} = \mu,$$
$$a_1 = \text{semi-major axis} = a, \tag{1.187}$$
$$p_1 = \text{semi-latus rectum} = p + 2\rho(1-j) \geq p.$$

Note that the left-hand side of Eq. (1.186) is equal to $\rho^2[\dot{\rho}^2 - \mu(\frac{2}{\rho} - \frac{1}{a})]$. From the integral of *force vive* in 2-body problem, $\dot{\rho}^2 = \mu_1(\frac{2}{\rho} - \frac{1}{a_1}) - \frac{\mu_1 p_1}{\rho^2}$, one gets $-\mu p_1 = 2\mu\rho(j-1) - \mu p$, i.e. $p_1 = p + 2\rho(1-j)$.

Case II: $j \geq 1$. In this case,

$$\frac{\mathrm{d}^2 \rho^2}{\mathrm{d}t^2} = 2\mu \left(\frac{j}{\rho} - \frac{1}{a} \right), \tag{1.188}$$

or, alternatively,

$$2\rho\ddot{\rho} + 2\dot{\rho}^2 = \frac{\mu p}{\rho^2} - \frac{\mu}{a} + \dot{\rho}^2.$$

Multiply both sides of above equation by $\dot{\rho}$, one has

$$2\rho\dot{\rho}\ddot{\rho} + \dot{\rho}^3 - \frac{\mu p}{\rho^2}\dot{\rho} + \frac{\mu}{a}\dot{\rho} = 0.$$

This can be integrated to obtain

$$\rho\dot{\rho}^2 + \frac{\mu p}{\rho} + \frac{\mu\rho}{a} = \text{ constant } = 2\mu j. \tag{1.189}$$

Therefore, in Eq. (1.189) the Sundman function j is a constant. Hence j remains larger than or equal to 1, and two cases described by Eq. (1.186) and Eq. (1.189) will never mix with each other. The evolution of ρ is now the evolution of mutual distance in another 2-body problem, in which

$$\begin{aligned}
\mu_2 &= \text{gravitation constant } = \mu j, \\
a_2 &= \text{semi-major axis } = aj, \\
p_2 &= \text{semi-latus rectum } = \frac{p}{j} \leq p.
\end{aligned} \tag{1.190}$$

Now, with the inequality Eq. (1.183), one may turn to the critical question as below: How much can we know about the future of ρ and $\dot{\rho}$ if at an initial time t_0 the initial values ρ_0 and $\dot{\rho}_0$ are given, and if all the possible trajectories compatible with the condition in Eq. (1.183) are considered? As an answer, a theorem is presented as follows.

Theorem. *At the final time t_f larger than t_0, the final conditions ρ_f, $\dot{\rho}_f$ can be attained if and only if there exist $\dot{\rho}_1, \dot{\rho}_2$ such that*

(a) *$\dot{\rho}_1 \geq \dot{\rho}_0, \quad \dot{\rho}_2 \leq \dot{\rho}_f$,*
(b) *as the limit of Eq. (1.183), the differential equation Eq. (1.184) leads $\rho_0, \dot{\rho}_1, t_0$ to $\rho_f, \dot{\rho}_2, t_f$.*

Proof. Assume $\rho_f, \dot{\rho}_f, t_f$ verify the above conditions (a) and (b). Because there is no upper limit on $\ddot{\rho}$ given in Eq. (1.183), one can approach as near as possible the positive discontinuity of $\dot{\rho}$ ($\dot{\rho}(t)$ has a positive discontinuity at $t = t_0$ if $\lim_{t \to t_{0-}} \dot{\rho}(t) = \dot{\rho}_0 < \dot{\rho}_0' = \lim_{t \to t_{0+}} \dot{\rho}(t)$). The discontinuity comes from the difference between Eq. (1.186) and Eq. (1.189). For instance, the

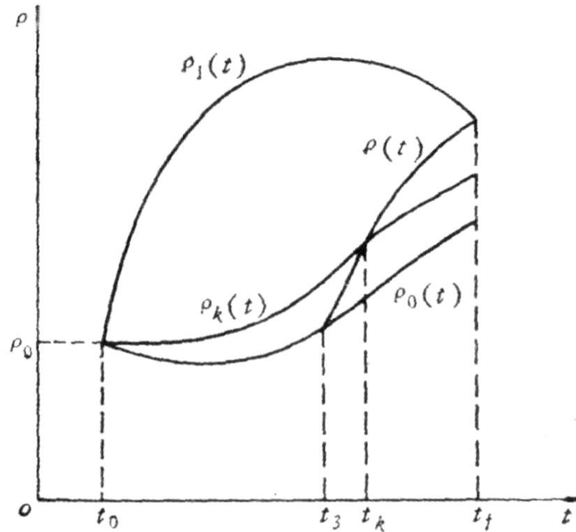

Fig. 1.9   The functions $\rho(t)$, $\rho_1(t)$, $\rho_0(t)$, $\rho_k(t)$. From Marchal & Sun (1985).

discontinuities $\dot\rho_0 \to \dot\rho_1$ and $\dot\rho_2 \to \dot\rho_f$. Since Eq. (1.184) is the limit of the differential inequality Eq. (1.183), the final state $\rho_f, \dot\rho_f, t_f$ is either attainable or at least the limit of attainable state. In addition, if the twice continuously differentiable function $\rho(t)$ verifies Eq. (1.183) and it leads $\rho_0, \dot\rho_0, t_0$ to $\rho_f, \dot\rho_f, t_f$, one may demonstrate the following: There exists a function $\rho_1(t)$ verifying Eq. (1.184) and leading $\rho_0, \dot\rho_1, t_0$ to $\rho_f, \dot\rho_2, t_f$, such that for all $t \in [t_0, t_f]$, $\rho_1(t) \geq \rho(t)$. This, of course, implies that $\dot\rho_1 \geq \dot\rho_0$ and $\dot\rho_2 \leq \dot\rho_f$, hence the function $\rho_1(t)$ will be the function used in conditions (a) and (b).

Consider first the function $\rho_0(t)$ starting at $\rho_0, \dot\rho_0, t_0$ and verifying Eq. (1.184). This function remains equal to $\rho(t)$ in some interval $[t_0, t_3]$ (see Fig. 1.9). Certainly it is possible that $t_3 = t_0$, but if $t_3 = t_f$ the function $\rho_0(t)$ is just the function $\rho_1(t)$ being looked for.

If $t_3 < t_f$, one has necessarily $\rho_0(t) < \rho(t)$ immediately after t_3 since at t_3 both functions have the same ρ and $\dot\rho$. And, because $\rho_0(t)$ takes the smallest possible $\ddot\rho$ given in Eq. (1.183), while for $\rho(t)$ the second derivative is larger.

Consider now the function $\rho_k(t)$ verifying Eq. (1.184) and starting at $\rho_0, \dot\rho_k, t_0$ with $\dot\rho_k > \dot\rho_0$. There are of course some intervals after t_0 during which $\rho_k(t) > \rho(t)$. Denote $t_k(\dot\rho_k)$ the time at which the difference $\rho_k(t) - \rho(t)$ comes back to zero for the first time after t_0. When $\dot\rho_k > \dot\rho_0$, the two curves $\rho_k(t)$ and $\rho(t)$ cannot be tangent to each other at the time $t_k(\dot\rho_k)$.

This is because $\rho_k(t) > \rho(t)$ just before $t_k(\dot\rho_k)$, and the second derivative of $\rho_k(t)$ is given by Eq. (1.184) while the second derivative of $\rho(t)$ satisfies Eq. (1.183), hence, $t_k(\dot\rho_k)$ is a continuous function of $\dot\rho_k$ as soon as $\dot\rho_k > \dot\rho_0$. Finally, for very large $\dot\rho_k$, the integration of Eq. (1.189) shows that $\rho_k(t)$ will remain very large for at least a long period and will thus be far above $\rho(t)$. Thus, for some intermediate $\dot\rho_k$ (that will be $\dot\rho_1$), $t_k(\dot\rho_k) = t_f$. $\qquad\square$

1.8.2 *Evolution of ρ*

The evolution of the "mean quadratic distance" ρ will be analyzed in this part. According to the definition of ρ, it is proportional to the square root of inertia moment I. Apparently the evolution of I will be directly reflected by ρ. Finally, the attainable and forbidden domain for the motion can be deduced.

With the theorem, to obtain in the future all $\rho_f, \dot\rho_f, t_f$ compatible with $\rho_0, \dot\rho_0, t_0$ and with the condition Eq. (1.183), one needs only to consider the integral of Eq. (1.184) starting at $\rho_0, \dot\rho_1, t_0$ with $\dot\rho_1 > \dot\rho_0$.

As in the previous demonstration, $\rho_k(t)$ is called the solution of Eq. (1.184) starting at $\rho_0, \dot\rho_k, t_0$ and it always leads to similar results as in Fig. 1.10 and Fig. 1.11. There is, on the (t, ρ) plane, a shaded attainable domain above the full curve $\rho = \phi(t)$, which is the envelope of the curves $\rho_k(t)$. This limit curve always begins with the part AB where $\rho = \rho_0(t)$.

Define some notations first as follows.

$$j_0 = \frac{p}{2\rho_0} + \frac{\rho_0}{2a} + \frac{\rho_0\dot\rho_0^2}{2\mu},$$

$$\mu_0 = \text{gravitational constant} = \mu \cdot \sup(1; j_0),$$

$$a_0 = \text{semi-major axis} = a \cdot \sup(1; j_0),$$

$$p_0 = \text{semi-latus rectum} = \begin{cases} p + 2\rho_0(1 - j_0) & \text{if } j_0 \le 1 \\ \dfrac{p}{j_0} & \text{if } j_0 \ge 1. \end{cases}$$

$$n_0 = \text{mean angular motion} = \frac{n}{\sup(1; j_0)},$$

$$e = \text{eccentricity} = \sqrt{1 - \frac{p_0}{a_0}},$$

$$T_0 = \text{period} = T \cdot \sup(1; j_0).$$

$$(1.191)$$

Now the evolution of ρ is divided into four main cases as follows.

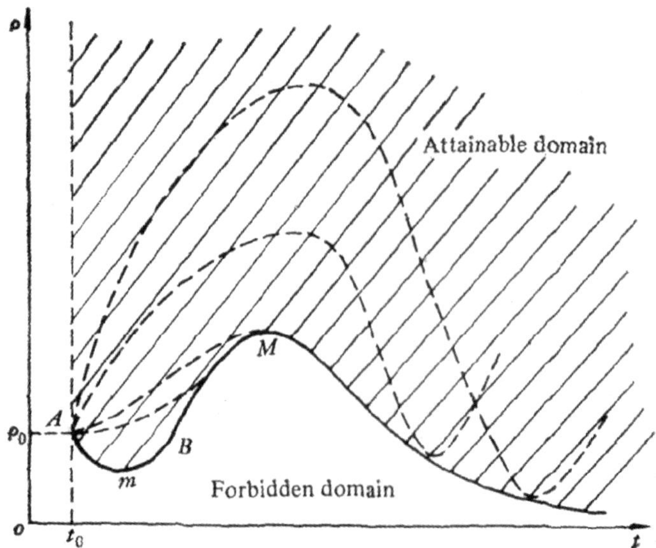

Fig. 1.10   The curve AB is the limit curve $\rho_0(t)$, the solution of Eq. (1.184) starting at $\rho_0, \dot\rho_0, t_0$. The dotted curves are the $\rho_k(t)$ solutions starting at $\rho_0, \dot\rho_k, t_0$ with $\dot\rho_k > \dot\rho_0$. The full curve $AmBM$ will be called $\phi(t)$ and is (after B) the envelope of the above $\rho_k(t)$ curves. From Marchal & Sun (1985).

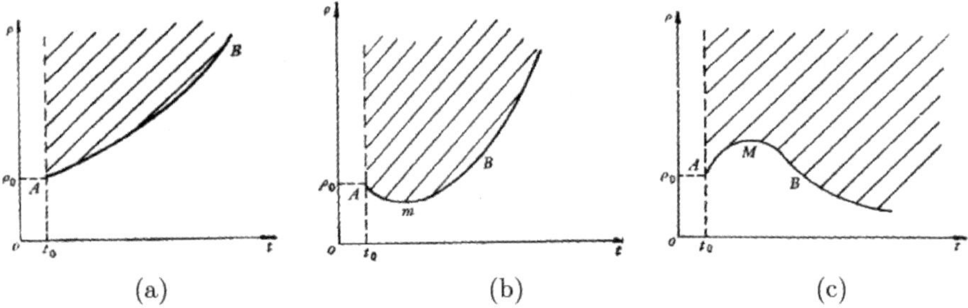

(a)   (b)   (c)

Fig. 1.11   The limit curves $\phi(t)$ for (a) $h \geq 0, \dot\rho_0 \geq 0$, (b) $h \geq 0, \dot\rho_0 < 0$, and (c) $h < 0, \dot\rho_0 \geq 0$, respectively. From Marchal & Sun (1985).

Case I. $h \geq 0, \dot\rho_0 \geq 0$

As shown in Fig. 1.11(a), $\phi(t)$ remains forever increasing and equal to $\rho_0(t)$. When t goes to infinity $\phi(t)$ also goes to infinity. If $h > 0$, it is the hyperbolic orbit, at infinity $\dot\rho \approx \sqrt{-\mu/a}$ thus $\rho \approx \sqrt{-\mu/a}\,t$. If $h = 0$, it is the parabolic orbit and $\rho = \frac{p_0}{1+\cos\nu} = \frac{p_0}{2}\sec^2\frac{\nu}{2}$. Define the "mean angular motion" $\bar{n}_0$

such that $\bar{n}_0^2 p_0^3 = \mu_0$, then

$$\frac{1}{2}\tan\frac{\nu}{2} + \frac{1}{6}\tan^3\frac{\nu}{2} = \bar{n}_0 t.$$

When the distance between two bodies goes to infinity, $\nu \to \pi$, hence $\tan\frac{\nu}{2} \ll \tan^3\frac{\nu}{2}$ and

$$\tan\frac{\nu}{2} \approx (6\bar{n}_0 t)^{\frac{1}{3}}.$$

And finally,

$$\rho \approx \frac{p_0}{2}(6\bar{n}_0 t)^{\frac{2}{3}} = \left(\frac{9}{2}\mu_0 t^2\right)^{\frac{1}{3}} = \left[\frac{9\mu t^2 \sup(1;j)}{2}\right]^{\frac{1}{3}}.$$

Case II. $h \geq 0$, $\dot{\rho}_0 < 0$

This case is shown in Fig. 1.11(b). Clearly, $\phi(t)$ has only one minimum at m and it remains equal to $\rho_0(t)$ until B is further than m. When $h > 0$, it is the hyperbolic orbit, and $\rho \approx \sqrt{-\mu/a}$ when $t \to \infty$. For the limit case $h = 0$, as $t \to \infty$ one has $\dot{\rho}_0 = 0$, thus $j_0 = \frac{p}{2\rho_0}$ and

$$\rho \approx \left[\frac{9\mu t^2 \sup(1;\frac{p}{2\rho_0})}{2}\right]^{\frac{1}{3}}.$$

Case III. $h < 0$, $\dot{\rho}_0 \geq 0$

As shown in Fig. 1.11(c), in this case $\phi(t)$ has only one maximum equal to $\rho_0(t)$ until B is further than M (if and only if $\dot{\rho}_0 = 0$ and $\rho_0 \leq \sup(a;\sqrt{ap})$, B and M are at the same place).

Case IV. $h < 0$, $\dot{\rho}_0 < 0$

This is the case shown in Fig. 1.10. $\phi(t)$ has one minimum at m and one maximum at M, and it remains equal to $\rho_0(t)$ before point B, which locates between m and M.

Some more conclusions about the limit curve can be obtained by further investigations. According to the formulae of 2-body problem, one has

$$\rho = \frac{p_0}{1 + e\cos\nu} = \begin{cases} a_0(1 - e\cos E) & \text{if } h < 0, \\ a_0(1 - e\cosh F) & \text{if } h > 0, \end{cases}$$

$$\dot{\rho} = e\sqrt{\frac{\mu_0}{p_0}}\sin\nu = \begin{cases} \dfrac{n_0 a_0^2 e\sin E}{\rho} & \text{if } h < 0, \\[3mm] \dfrac{e\sqrt{1 - \mu_0 a_0}\sinh F}{\rho} & \text{if } h > 0. \end{cases} \tag{1.192}$$

If t_m is the time of "perihelion passage" (i.e. the moment of passing through point m), for $h < 0$ (elliptic orbit) one has $E - e\sin E = n_0(t - t_m)$, thus $t - t_m = (E - e\sin E)/n_0$. Similar calculations can also be made for $h = 0$ and $h > 0$. The results are summarized as below.

$$t - t_m = \begin{cases} \dfrac{E - e\sin E}{n_0} & \text{if } h < 0, \\[3mm] \dfrac{p_0}{6}\left(\tan^3\dfrac{\nu}{2} + 3\tan\dfrac{\nu}{2}\right)\sqrt{\dfrac{p_0}{\mu_0}} & \text{if } h = 0, \\[3mm] (e\sinh F - F)\sqrt{-\dfrac{a_0^3}{\mu_0}} & \text{if } h > 0. \end{cases} \tag{1.193}$$

Moreover, if $h = 0$,

$$\begin{aligned} (t - t_m)^2 &= \frac{p_0^3}{36\mu_0}\left(\tan^3\frac{\nu}{2} + 3\tan\frac{\nu}{2}\right)^2 \\ &= \frac{p_0^3}{36\mu_0}\left(\tan^6\frac{\nu}{2} + 6\tan^4\frac{\nu}{2} + 9\tan^2\frac{\nu}{2}\right). \end{aligned} \tag{1.194}$$

Since

$$\rho = \frac{p_0}{1 + \cos\nu} = \frac{p_0}{2}\sec^2\frac{\nu}{2} = \frac{p_0}{2}\left(1 + \tan^2\frac{\nu}{2}\right),$$

one has

$$\tan^2\frac{\nu}{2} = \frac{2\rho}{p_0} - 1.$$

Substitute this into the equation of $(t - t_m)^2$ Eq. (1.194) and let $p_0 = 2\rho_m$ (ρ_m is the value of ρ at m), one gets

$$9\mu_0(t - t_m)^2 = 2(\rho - \rho_m)(\rho + 2\rho_m)^2. \tag{1.195}$$

In addition, when $h = 0$,

$$\dot\rho = \sqrt{\frac{\mu_0}{p_0}}\sin\nu, \qquad \sin\nu = \frac{2\tan\frac{\nu}{2}}{1 + \tan^2\frac{\nu}{2}},$$

hence

$$\sin^2\nu = \frac{4\tan^2\frac{\nu}{2}}{\left(1 + \tan^2\frac{\nu}{2}\right)^2} = \frac{4\left(2\rho/p_0 - 1\right)}{\left(2\rho/p_0\right)^2}.$$

Finally,

$$\begin{aligned} \rho^2\dot\rho^2 &= \rho^2\cdot\frac{\mu_0}{p_0}\cdot\frac{4\left(2\rho/p_0 - 1\right)}{\left(2\rho/p_0\right)^2} = \mu_0(2\rho - p_0) \\ &= \mu_0(2\rho - 2\rho_m) = 2\mu_0(\rho - \rho_m), \end{aligned} \tag{1.196}$$

where, $\rho_m = p_0/2$. These expressions are also valid for other $\rho_k(t)$ and $\rho_1(t)$ curves, with only modifications on the value of t_m and on the expression of j (e.g. for a departure at $\rho_0, \dot{\rho}_1, t_0$, $j = j_1 = \frac{p}{2\rho_0} + \frac{\rho_0}{2a} + \frac{\rho_0 \dot{\rho}_1^2}{2\mu}$).

1.9 Motion of isolated body in 3-body problem

Concerning the Hill-type stability in 3-body problem (see Sec. 1.7), when the energy and angular momentum constants satisfy certain conditions, the motion of three bodies is confined in three disconnected regions. The motion region of the third celestial object P_3 may be not connected with the regions of other two objects P_1 and P_2, and this will be investigated further in this section.

As in previous sections, denote the masses of P_1, P_2, P_3 by m_1, m_2, m_3, and suppose $0 < m_2 \le m_1$. In the Jacobi coordinate system, O denotes the center of mass of three bodies, O_{12} the center of mass of P_1 and P_2, $\boldsymbol{r}_1, \boldsymbol{r}_2, \boldsymbol{r}_3$ the positional vectors of three bodies from O, while $\boldsymbol{r}$ and $\boldsymbol{R}$ are the vectors from P_1 to P_2 and from O_{12} to P_3 respectively, as shown in Fig. 1.12.

Apparently,

$$\boldsymbol{r} = \boldsymbol{r}_2 - \boldsymbol{r}_1 = \boldsymbol{r}_{12}, \tag{1.197}$$

$$\boldsymbol{R} = \frac{M\boldsymbol{r}_3}{m_1 + m_2}, \quad M = m_1 + m_2 + m_3. \tag{1.198}$$

And the equation of motion of P_3 reads,

$$\frac{\mathrm{d}^2 \boldsymbol{R}}{\mathrm{d}t^2} = -\mu \left(\alpha \frac{\boldsymbol{r}_{13}}{r_{13}^3} + \beta \frac{\boldsymbol{r}_{23}}{r_{23}^3} \right), \tag{1.199}$$

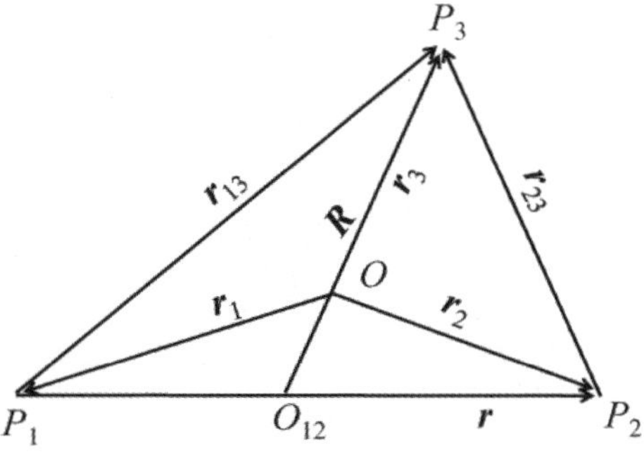

Fig. 1.12   The Jacobi coordinate system.

where $\mu = GM$, r_{ij} $(i, j = 1, 2, 3)$ is the distance between P_i and P_j, and

$$\alpha = \frac{m_1}{m_1 + m_2}, \quad \beta = \frac{m_2}{m_1 + m_2} = 1 - \alpha, \quad 0 \leq \beta \leq \frac{1}{2}. \tag{1.200}$$

In Jacobi coordinate system, the integral of angular momentum and the integral of energy are

$$\boldsymbol{C} = \mu_1 \boldsymbol{r} \times \boldsymbol{v} + \mu_2 \boldsymbol{R} \times \boldsymbol{V}, \tag{1.201}$$

$$h = \frac{1}{2} \left(\mu_1 v^2 + \mu_2 V^2 \right) - G \left(\frac{m_1 m_2}{r} + \frac{m_1 m_3}{r_{13}} + \frac{m_2 m_3}{r_{23}} \right), \tag{1.202}$$

where,

$$\mu_1 = \frac{m_1 m_2}{m_1 + m_2}, \quad \mu_2 = \frac{m_3(m_1 + m_2)}{M}, \tag{1.203}$$

$$\boldsymbol{v} = \frac{\mathrm{d}\boldsymbol{r}}{\mathrm{d}t}, \quad \boldsymbol{V} = \frac{\mathrm{d}\boldsymbol{R}}{\mathrm{d}t}. \tag{1.204}$$

Thus $\boldsymbol{C}$ and h are the constant vector of angular momentum and the constant of energy, respectively. Also, as in Sec. 1.7 let

$$M^* = m_1 m_2 + m_1 m_3 + m_2 m_3,$$
$$a = -\frac{GM^*}{2h}, \quad p = \frac{MC^2}{GM^{*2}}. \tag{1.205}$$

Then a is the generalized semi-major axis and p the generalized semi-latus rectum. Obviously, in the restricted 3-body problem and 2-body problem, they are just the semi-major axis and semi-latus rectum, since $m_3 = 0$. And again, the mean quadratic distance ρ and the mean harmonic distance ν are introduced as

$$M^* \rho^2 = \sum_{1 \leq i < j \leq 3} m_i m_j r_{ij}^2,$$
$$\frac{M^*}{\nu} = \sum_{1 \leq i < j \leq 3} \frac{m_i m_j}{r_{ij}}. \tag{1.206}$$

Obviously, ρ and ν satisfy the inequalities as below

$$\min\{r_{12}, r_{13}, r_{23}\} \leq \nu \leq \rho \leq \max\{r_{12}, r_{13}, r_{23}\}.$$

And these two distances ρ and ν are related to the potential function U and the momentum of inertia I by

$$2I = \sum_{i=1}^{3} m_i r_i^2 = \frac{M^* \rho^2}{M}, \tag{1.207}$$

and

$$U = G \sum_{1 \leq i < j \leq 3} \frac{m_i m_j}{r_{ij}} = \frac{GM^*}{\nu}. \tag{1.208}$$

Note that notations like a, p, ρ, ν, etc. have been introduced in previous sections.

Denote

$$\dot{\rho} = \frac{\mathrm{d}\rho}{\mathrm{d}t}, \quad j = \frac{\rho}{2a} + \frac{p}{2\rho} + \frac{\rho\dot{\rho}^2}{2\mu}, \tag{1.209}$$

and surely j is the Sundman function. The Lagrange-Jacobi equality $\mathrm{d}^2 I/\mathrm{d}t^2 = U + 2h$ and Sundman's inequality $4I(U + h) \geq C^2 + (\mathrm{d}I/\mathrm{d}t)^2$ now can be written as

$$\frac{\mathrm{d}^2 \rho^2}{\mathrm{d}t^2} = 2\mu \left(\frac{1}{\nu} - \frac{1}{a} \right) \tag{1.210}$$

and

$$\frac{\rho}{\nu} \geq j. \tag{1.211}$$

For the system with negative energy, one has

$$\frac{\rho}{\nu} \geq \sqrt{\frac{p}{a}}. \tag{1.212}$$

1.9.1 *Isolated body*

When the motion region of P_3 is disconnected with the motion regions of P_1 and P_2, and if $r \leq kR \quad (0 < k < 1/\alpha \leq 2)$, the third body P_3 will be called the third isolated body. In this section, the motion of the third isolated body P_3 will be studied carefully under the assumptions that the total energy of the system $h < 0$ and the total angular momentum $C \neq 0$.

Without loss of generality, suppose $r = r_{12} = \min\{r_{12}, r_{13}, r_{23}\}$, according to the results in Chen *et al.* (1978), the necessary and sufficient condition of permissible motion is

$$\frac{C^2}{\mu_1 r^2 + \mu_2 R^2} \leq 2(U + h), \tag{1.213}$$

i.e.

$$\frac{C^2}{\mu_1 r^2 + \mu_2 R^2} \leq 2\frac{\bar{M}^*}{r} + 2h, \quad \bar{M}^* = GM^*. \tag{1.214}$$

Or, alternatively

$$C^2 r \leq 2\bar{M}^* \left(\mu_1 r^2 + \mu_2 R^2 \right) + 2hr(\mu_1 r^2 + \mu_2 R^2). \tag{1.215}$$

Define

$$f(r, R) \equiv 2\mu_2 \left(\bar{M}^* + hr \right) R^2 + 2\mu_1 hr^3 + 2\mu_1 \bar{M}^* r^2 - C^2 r,$$

then $f(r, R) = 0$ determines a curve in the plane (r, R), which can be alternatively written as

$$R^2 = -\frac{2\mu_1 hr^3 + 2\mu_1 \bar{M}^* r^2 - C^2 r}{2\mu_2 \left(\bar{M}^* + hr\right)}. \tag{1.216}$$

Obviously, this curve is symmetric in R. According to the definition of "isolated body", the intersection of this curve with the line $r = kR$ gives the condition under which P_3 is an isolated body. The intersection is determined by the following equation

$$2\mu_2(\bar{M}^* + hkR)R^2 = -(2\mu_1 hk^3 R^3 + 2\mu_1 \bar{M}^* k^2 R^2 - C^2 kR). \tag{1.217}$$

From Eq. (1.217), one knows

$$R = 0,$$

or

$$R^2 + \frac{\bar{M}^*}{kh}R - \frac{C^2}{2h(\mu_2 + \mu_1 k^2)} = 0. \tag{1.218}$$

Apparently, if

$$\left(\frac{\bar{M}^*}{kh}\right)^2 + \frac{2C^2}{h(\mu_2 + \mu_1 k^2)} < 0, \tag{1.219}$$

except for the origin $(r, R) = (0, 0)$ no other intersection point of the curve $f(r, R) = 0$ and line $r = kR$ exists. The derivative along the curve described by Eq. (1.216) is

$$\frac{dR}{dr} = -\frac{6\mu_1 hr^2 + 4\mu_1 \bar{M}^* r + 2\mu_2 hR^2 - C^2}{4\mu_2 \left(\bar{M}^* + hr\right) R}. \tag{1.220}$$

Because $C \neq 0$, $\frac{dR}{dr}\big|_{(0,0)} = \infty$, hence the curve Eq. (1.216) is tangent to OR axis at the origin $(0, 0)$. Moreover, since r is the smallest among the mutual distances of three bodies and $h < 0$, one has $r \leq -\bar{M}^*/h$, i.e. $\bar{M}^* + hr \geq 0$. From Eqs. (1.216) and (1.220), both R and $\frac{dR}{dr}$ tend to $\pm\infty$ when $r \to -\bar{M}^*/h$. Therefore, only the situation $\bar{M}^* + hr > 0$ will be discussed below. From $f(r, R) > 0$, one has

$$R^2 > -\frac{2\mu_1 hr^3 + 2\mu_1 \bar{M}^* r^2 - C^2 r}{2\mu_2(\bar{M}^* + hr)}. \tag{1.221}$$

Thus, under the condition Eq. (1.219) the permissible motion will be confined in a zone D, which is illustrated in Fig. 1.13 (shaded zone), and in this zone $r < kR$ forever. Besides, $f(r, R) = 0$ determines a curve in the (r, R) plane. In the right-hand side of Eq. (1.216), the denominator

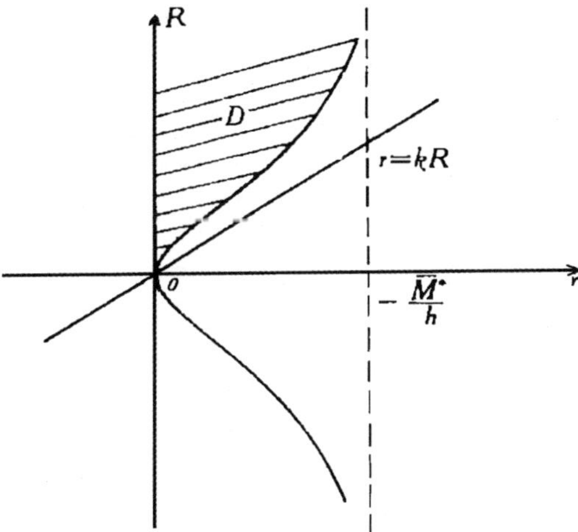

Fig. 1.13 The permissible zone D on (r, R) plane. Taken from Sun & Marchal (1985).

$2\mu_2(\bar{M}^* + hr) > 0$, and when

$$\left(\frac{\bar{M}^*}{h}\right)^2 + \frac{2C^2}{\mu_1 h} < 0, \tag{1.222}$$

the numerator $2\mu_1 hr^3 + 2\mu_1 \bar{M}^* r^2 - C^2 r$ has no real zero point except for $r = 0$. Therefore one always has $2\mu_1 hr^3 + 2\mu_1 \bar{M}^* r^2 - C^2 r < 0$ under the condition Eq. (1.222), which can be easily deduced from Eq. (1.219).

Below, the condition that guarantees $r = r_{12}$ is the smallest mutual distance, and another condition of P_3 being the isolated body, will be given.

Let

$$r_{12} = \xi r_{13}, \quad r_{23} = \eta r_{13},$$

$$W(\xi, \eta) = G\left(\frac{m_1 m_2 m_3}{M}\right)^{\frac{1}{2}} \left(\frac{\xi^2}{m_3} + \frac{\eta^2}{m_1} + \frac{1}{m_2}\right)^{\frac{1}{2}} \tag{1.223}$$

$$\times \left(\frac{m_1 m_2}{\xi} + \frac{m_2 m_3}{\eta} + m_1 m_3\right).$$

The necessary and sufficient condition for permissible motion now reads

$$W(\xi, \eta) \geq -2hC^2. \tag{1.224}$$

Taking into account the property of function $W(\xi, \eta)$, if

$$-2hC^2 \geq \max\left\{W^*, W(1, 2), W\left(\frac{1}{2}, \frac{1}{2}\right)\right\}, \tag{1.225}$$

where W^* is one of the values of $W(\xi, \eta)$ corresponding to three collinear Euler solutions, the motion region will be divided into two or three

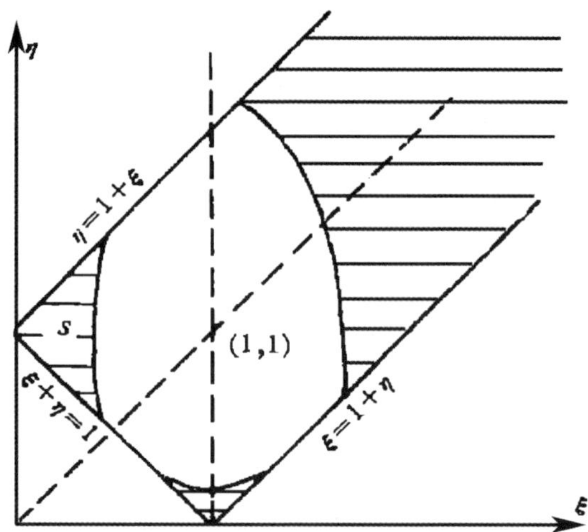

Fig. 1.14   The permissible zone on (ξ, η) plane. From Sun & Marchal (1985).

disconnected zones (Chen, 1979), as shown by the shadowed parts in Fig. 1.14. Consequently, if the (ξ, η) corresponding to a certain configuration of three bodies at a certain time resides in the region indicated by S in Fig. 1.14, one has $\xi < \eta, \xi < 1$ forever, that is, $r = r_{12}$ will continue to be the smallest one among the three mutual distances forever, and obviously the other condition of P_3 being an isolated body is satisfied too.

1.9.2   *Acceleration of isolated body*

The acceleration of the isolated body P_3 is

$$\frac{\mathrm{d}^2 \boldsymbol{R}}{\mathrm{d}t^2} = -\mu \left(\alpha \frac{\boldsymbol{r}_{13}}{r_{13}^3} + \beta \frac{\boldsymbol{r}_{23}}{r_{23}^3} \right),$$

or, alternatively

$$\frac{\mathrm{d}^2 \boldsymbol{R}}{\mathrm{d}t^2} = -\mu \left[\frac{\alpha(\boldsymbol{R} + \beta \boldsymbol{r})}{r_{13}^3} + \frac{\beta(\boldsymbol{R} - \alpha \boldsymbol{r})}{r_{23}^3} \right]. \tag{1.226}$$

Suppose the radial and transversal components of $\frac{\mathrm{d}^2 \boldsymbol{R}}{\mathrm{d}t^2}$ are X and Y, then

$$X = \frac{\mathrm{d}^2 \boldsymbol{R}}{\mathrm{d}t^2} \cdot \frac{\boldsymbol{R}}{R} = -\mu R \left(\frac{\alpha}{r_{13}^3} + \frac{\beta}{r_{23}^3} \right) - \mu \alpha \beta \frac{\boldsymbol{r} \cdot \boldsymbol{R}}{R} \left(\frac{1}{r_{13}^3} - \frac{1}{r_{23}^3} \right),$$

$$Y = \frac{\mathrm{d}^2 \boldsymbol{R}}{\mathrm{d}t^2} \cdot \boldsymbol{\tau} = -\mu \alpha \beta \boldsymbol{\tau} \cdot \boldsymbol{r} \left(\frac{1}{r_{13}^3} - \frac{1}{r_{23}^3} \right),$$

$$\tag{1.227}$$

where $\boldsymbol{\tau}$ is the transversal unit vector. From Eq. (1.227), one obtains

$$\left[X + \mu R \left(\frac{\alpha}{r_{13}^3} + \frac{\beta}{r_{23}^3} \right) \right]^2 + Y^2 = \mu^2 \alpha^2 \beta^2 r^2 \left(\frac{1}{r_{13}^3} - \frac{1}{r_{23}^3} \right)^2$$

$$\leq \frac{\mu^2 \alpha^2 \beta^2 k^2}{R^4} \max \left\{ \left[\frac{1}{(1 - \beta k)^3} \quad \frac{1}{(1 + \alpha k)^3} \right]^2 ; \right. \tag{1.228}$$

$$\left. \left[\frac{1}{(1 - \alpha k)^3} - \frac{1}{(1 + \beta k)^3} \right]^2 \right\},$$

and

$$\frac{\mu R}{(R^2 + \alpha \beta r^2)^{3/2}} \leq \mu R \left(\frac{\alpha}{r_{13}^3} + \frac{\beta}{r_{23}^3} \right)$$

$$\leq \mu R \max \left\{ \left[\frac{\alpha}{(R + \beta r)^3} + \frac{\beta}{(R - \alpha r)^3} \right] ; \left[\frac{\alpha}{(R - \beta r)^3} + \frac{\beta}{(R + \alpha r)^3} \right] \right\}. \tag{1.229}$$

Obviously, $\frac{1}{(R^2 + \alpha \beta r^2)^{3/2}}$ decreases monotonically as r increases, while $\frac{\alpha}{(R+\beta r)^3} + \frac{\beta}{(R-\alpha r)^3}$ and $\frac{\alpha}{(R-\beta r)^3} + \frac{\beta}{(R+\alpha r)^3}$ increase monotonically as r increases when $r \in (0, kR]$. Therefore,

$$\frac{\mu}{(1 + \alpha \beta k^2)^{3/2}} \cdot \frac{1}{R^2} \leq \mu R \left(\frac{\alpha}{r_{13}^3} + \frac{\beta}{r_{23}^3} \right)$$

$$\leq \frac{\mu}{R^2} \max \left\{ \left[\frac{\alpha}{(1 + \beta k)^3} + \frac{\beta}{(1 - \alpha k)^3} \right] ; \left[\frac{\alpha}{(1 - \beta k)^3} + \frac{\beta}{(1 + \alpha k)^3} \right] \right\}. \tag{1.230}$$

Let

$$X_1 = \frac{\mu}{R^2} \frac{1}{(1 + \alpha \beta k^2)^{3/2}},$$

$$X_2 = \frac{\mu}{R^2} \max \left\{ \left[\frac{\alpha}{(1 + \beta k)^3} + \frac{\beta}{(1 - \alpha k)^3} \right] ; \left[\frac{\alpha}{(1 - \beta k)^3} + \frac{\beta}{(1 + \alpha k)^3} \right] \right\},$$

$$d^2 = \frac{\mu^2 \alpha^2 \beta^2 k^2}{R^4} \max \left\{ \left[\frac{1}{(1 - \beta k)^3} - \frac{1}{(1 + \alpha k)^3} \right]^2 ; \right.$$

$$\left. \left[\frac{1}{(1 - \alpha k)^3} - \frac{1}{(1 + \beta k)^3} \right]^2 \right\}. \tag{1.231}$$

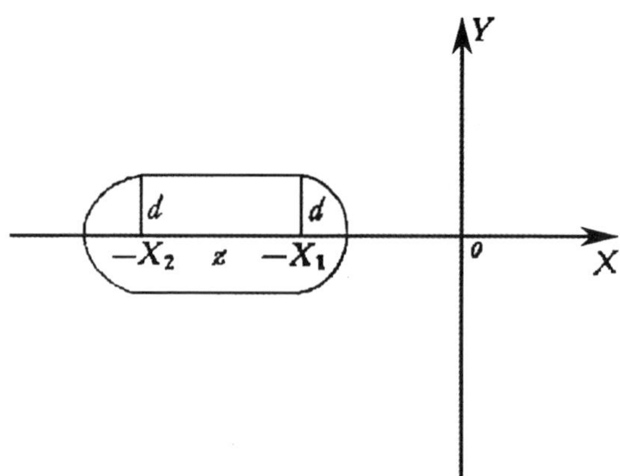

Fig. 1.15 The acceleration of isolated P_3 in (X, Y) plane. Adopted from Sun & Marchal (1985).

Then the acceleration $\frac{\mathrm{d}^2 \boldsymbol{R}}{\mathrm{d}t^2}$ of the isolated body P_3 is confined in a zone of the (X, Y)-plane, as shown in Fig. 1.15.

Moreover, under the condition $r \leq kR$, the radial component of P_3's acceleration is always negative. Thus, only the case $X_1 \geq d$ (e.g. when k is small) will be discussed below, while the case of $X_1 < d$ is omitted, being aware of that analogue discussions can be made too.

1.9.3 *Position and velocity of isolated body*

First the permissible range of the distance from P_3 to O_{12}, i.e. R, will be investigated. The system with negative energy retains the inequality in Eq. (1.212):

$$\frac{\rho}{\nu} \geq \sqrt{\frac{p}{a}}.$$

And from the Lagrange-Jacobi equality Eq. (1.210), one has

$$\frac{\mathrm{d}^2 \rho^2}{\mathrm{d}t^2} = 2\mu \left(\frac{1}{\nu} - \frac{1}{a} \right) \geq 2\mu \left(\frac{1}{\rho} \sqrt{\frac{p}{a}} - \frac{1}{a} \right). \tag{1.232}$$

Consider the equation

$$\frac{\mathrm{d}^2 \rho^2}{\mathrm{d}t^2} = 2\mu \left(\frac{1}{\rho} \sqrt{\frac{p}{a}} - \frac{1}{a} \right), \tag{1.233}$$

and integrate it, one obtains

$$\rho^2 \dot\rho^2 = \mu \left(2\sqrt{\frac{p}{a}}\rho - \frac{1}{a}\rho^2 \right) + c_1, \tag{1.234}$$

$$\pm t = \int \frac{\rho d\rho}{\sqrt{c_1 + 2\mu\sqrt{\frac{p}{a}}\rho - \frac{\mu}{a}\rho^2}} + c_2, \tag{1.235}$$

where c_1, c_2 are integral constants. Since c_2 is only related to the initial time, without loss of generality, one can take $c_2 = 0$. In addition, assume $\rho = \rho_0$ and $\dot\rho = \dot\rho_0$ at $t = t_0$. If these initial conditions satisfy the following inequality

$$2\mu\sqrt{\frac{p}{a}}\frac{1}{\rho_0} - \frac{\mu}{a} - \mu p\frac{1}{\rho_0^2} < \dot\rho_0^2 < 2\mu\sqrt{\frac{p}{a}}\frac{1}{\rho_0} - \frac{\mu}{a}, \tag{1.236}$$

one obtains from Eq. (1.235)

$$\pm t = -\frac{a}{\mu}\sqrt{c_1 + 2\mu\sqrt{\frac{p}{a}}\rho - \frac{\mu}{a}\rho^2} + a\sqrt{\frac{p}{\mu}}\arccos\left[\frac{\sqrt{\mu p}\left(1 - \frac{1}{\sqrt{pa}}\rho\right)}{\sqrt{c_1 + \mu p}} \right], \tag{1.237}$$

with $c_1 = \rho^2\dot\rho^2 - \mu\left(2\sqrt{\frac{p}{a}}\rho - \frac{1}{a}\rho^2\right)$. Introduce an auxiliary variable τ so that

$$\rho = \sqrt{pa}\left(1 - \sqrt{1 + \frac{c_1}{\mu p}}\cos\tau\right). \tag{1.238}$$

Under the condition Eq. (1.236), $\rho > 0$ and $1 + \frac{c_1}{\mu p} > 0$, one has

$$\pm t = \sqrt{\frac{p}{\mu}}a\left(\tau - \sqrt{1 + \frac{c_1}{\mu p}}\sin\tau\right). \tag{1.239}$$

On the (ρ, t)-plane, Eq. (1.237) represents an epicycloid, which has been plotted in Fig. 1.16. According to the comparison theorem of differential equation, the region D (shadowed part in Fig. 1.16) is the permissible region of motion. Therefore the permissible range of R can be obtained from

$$\rho^2 = \frac{M}{M^*}\left(\mu_1 r^2 + \mu_2 R^2\right) \le \frac{M}{M^*}\left(\mu_1 k^2 + \mu_2\right)R^2. \tag{1.240}$$

Surely, a more accurate inequality than Eq. (1.232) would improve the above estimate of the permissible region.

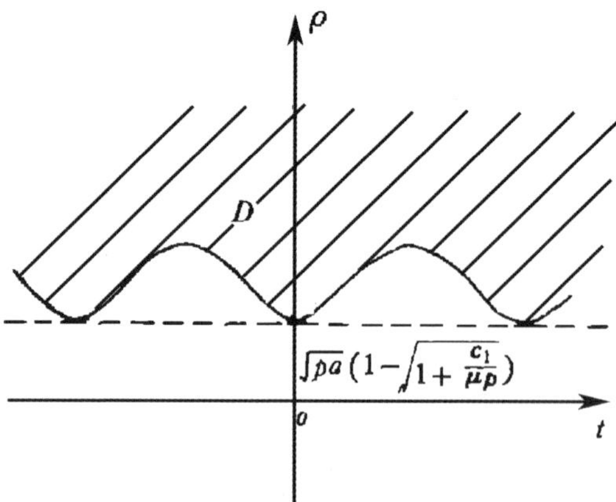

Fig. 1.16   The curve of Eq. (1.237). The shadowed region is the permissible region of motion. Adopted from Sun & Marchal (1985).

As for the velocity $\boldsymbol{V} = \frac{\mathrm{d}\boldsymbol{R}}{\mathrm{d}t}$ of P_3, denote the radial and transversal components of $\boldsymbol{V}$ by x and y, then the equations of motion of P_3 can be written as

$$
\begin{aligned}
\frac{\mathrm{d}R}{\mathrm{d}t} &= x, \\
\frac{\mathrm{d}x}{\mathrm{d}t} &= \frac{y^2}{R} + X, \\
\frac{\mathrm{d}y}{\mathrm{d}t} &= -\frac{xy}{R} + Y.
\end{aligned}
\tag{1.241}
$$

After introducing auxiliary variables ϕ and ψ

$$
\phi = x\sqrt{R}, \quad \psi = y\sqrt{R},
\tag{1.242}
$$

the equations of motion become

$$
\begin{aligned}
\frac{\mathrm{d}R}{\mathrm{d}t} &= \frac{\phi}{\sqrt{R}}, \\
\frac{\mathrm{d}\phi}{\mathrm{d}t} &= \frac{1}{R\sqrt{R}}\left(\frac{\phi^2}{2} + \psi^2 + R^2 X\right), \\
\frac{\mathrm{d}\psi}{\mathrm{d}t} &= \frac{1}{R\sqrt{R}}\left(-\frac{\phi\psi}{2} + R^2 Y\right).
\end{aligned}
\tag{1.243}
$$

Therefore,

$$
\frac{\mathrm{d}\psi}{\mathrm{d}\phi} = \frac{-\frac{\phi\psi}{2} + R^2 Y}{\frac{\phi^2}{2} + \psi^2 + R^2 X}.
\tag{1.244}
$$

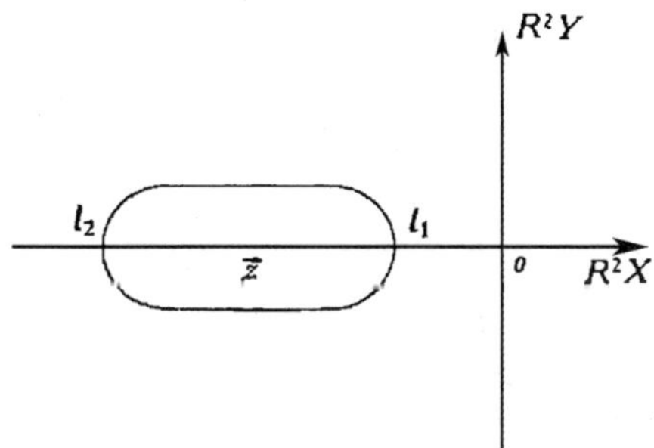

Fig. 1.17   The variation range of acceleration on the (R^2X, R^2Y) plane corresponding to the Z zone in Fig. 1.15. Adopted from Sun & Marchal (1985).

From Eq. (1.231), one knows that the change of R^2X and R^2Y is independent of R, thus the right-hand side of Eq. (1.244) is independent of R, and the variation region of (R^2X, R^2Y) in Z zone of Fig. 1.15 will correspond to the zone $\bar{Z}$ in Fig. 1.17. It is easy to verify that the extreme values will attain at the two semi-circles l_1 and l_2 of the boundary $\partial\bar{Z}$ of $\bar{Z}$.

Write down l_1, l_2 in the parametric form as follows

$$l_1 : \begin{cases} R^2X + R^2X_1 = R^2 d\cos u, \\ R^2Y = R^2 d\sin u, \end{cases} \quad -\frac{\pi}{2} \le u \le \frac{\pi}{2},$$

$$l_2 : \begin{cases} R^2X + R^2X_2 = R^2 d\cos u, \\ R^2Y = R^2 d\sin u, \end{cases} \quad \frac{\pi}{2} \le u \le \frac{3\pi}{2}.$$

Then the extreme values of $\frac{d\psi}{d\phi}$ will attain at certain value $\bar{u}$ of u.

In the (ϕ, ψ) plane, two ellipses, E_1 and E_2, and two hyperbolas H_1 and H_2 are defined by

$$E_1 : \quad \frac{\phi^2}{2} + \psi^2 = R^2X_1 - R^2d,$$

$$E_2 : \quad \frac{\phi^2}{2} + \psi^2 = R^2X_2 - R^2d,$$

$$H_1 : \quad -\frac{\phi\psi}{2} + R^2d = 0,$$

$$H_2 : \quad -\frac{\phi\psi}{2} - R^2d = 0,$$

and they are illustrated in Fig. 1.18.

In the shadowed region I in the first quadrant of the (ϕ, ψ) plane shown in Fig. 1.18, the denominator of the right hand side of Eq. (1.244) is negative

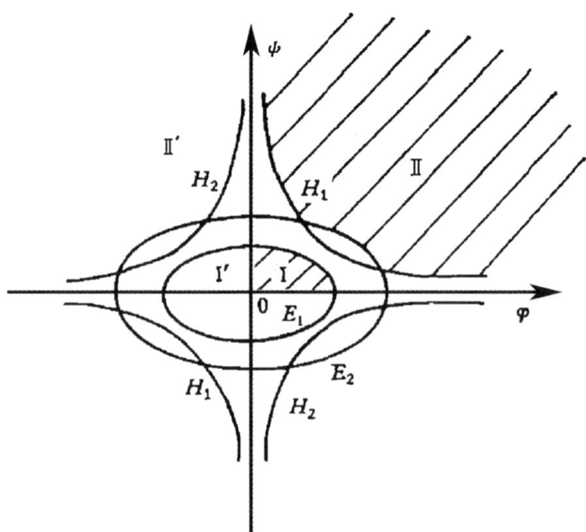

Fig. 1.18   Ellipses E_1, E_2 and hyperbolas H_1, H_2 (see text). Adopted from Sun & Marchal (1985).

while the sign of the numerator is definite. Therefore, the minimum of $\frac{\mathrm{d}\psi}{\mathrm{d}\phi}$ should be attained on the semi-circle l_1, that is,

$$\left(\frac{\mathrm{d}\psi}{\mathrm{d}\phi}\right)_{\max}^{(\mathrm{I})} = \frac{-\frac{\phi\psi}{2} + R^2 d\sin\bar{u}}{\frac{\phi^2}{2} + \psi^2 + R^2 d\cos\bar{u} - R^2 X_1} \geq 0, \quad -\frac{\pi}{2} \leq \bar{u} \leq 0.$$

$$(1.245)$$

In the shadowed region II, the denominator is positive and the numerator is negative, hence the minimum of $\frac{\mathrm{d}\psi}{\mathrm{d}\phi}$ is attained on the semi-circle l_2, i.e.

$$\left(\frac{\mathrm{d}\psi}{\mathrm{d}\phi}\right)_{\min}^{(\mathrm{II})} = \frac{-\frac{\phi\psi}{2} + R^2 d\sin\bar{u}}{\frac{\phi^2}{2} + \psi^2 + R^2 d\cos\bar{u} - R^2 X_2} \leq 0, \quad \pi \leq \bar{u} \leq \frac{3\pi}{2}.$$

$$(1.246)$$

For the regions I′, II′ that are symmetric to I, II with respect to the ψ-axis in the second quadrant, correspondingly, one has

$$\left(\frac{\mathrm{d}\psi}{\mathrm{d}\phi}\right)_{\min}^{(\mathrm{I}')} = \frac{-\frac{\phi\psi}{2} + R^2 d\sin\bar{u}}{\frac{\phi^2}{2} + \psi^2 + R^2 d\cos\bar{u} - R^2 X_1} \leq 0, \quad 0 \leq \bar{u} \leq \frac{\pi}{2}, \quad (1.247)$$

and

$$\left(\frac{\mathrm{d}\psi}{\mathrm{d}\phi}\right)_{\max}^{(\mathrm{II}')} = \frac{-\frac{\phi\psi}{2} + R^2 d\sin\bar{u}}{\frac{\phi^2}{2} + \psi^2 + R^2 d\cos\bar{u} - R^2 X_2} \geq 0, \quad \frac{\pi}{2} \leq \bar{u} \leq \pi. \quad (1.248)$$

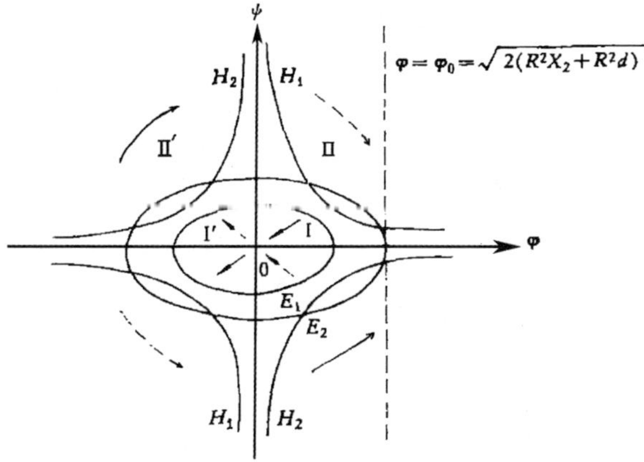

Fig. 1.19   The velocity variation of P_3. Adopted from Sun & Marchal (1985).

For the third and fourth quadrant in the (ϕ, ψ) plane, similar discussions can be made. Finally, the velocity variation of P_3 in the plane (ϕ, ψ) is shown in Fig. 1.19. The solid line represents the maximum curve of $\frac{d\psi}{d\phi}$ and the dashed line is the minimum curve. The arrows on them indicate the motion direction of points on these lines. The $\bar{u}$ in Eqs. (1.245), (1.246), (1.247) and (1.248) correspond to the extreme points. For the points on l_1, it should satisfy the following equation

$$\left(\frac{\phi^2}{2} + \psi^2 - R^2 X_1\right) R^2 d \cos \bar{u} - \frac{\phi\psi}{2} R^2 d \sin \bar{u} + (R^2 d)^2 = 0, \qquad (1.249)$$

or, alternatively

$$\sin(\bar{u} + \theta_0) = -\frac{R^2 d}{\sqrt{\left(\frac{\phi^2}{2} + \psi^2 - R^2 X_1\right)^2 + \frac{\phi^2 \psi^2}{4}}}, \qquad (1.250)$$

where

$$\sin \theta_0 = \frac{\frac{\phi^2}{2} + \psi^2 - R^2 X_1}{\sqrt{\left(\frac{\phi^2}{2} + \psi^2 - R^2 X_1\right)^2 + \frac{\phi^2 \psi^2}{4}}},$$

$$\cos \theta_0 = \frac{-\frac{\phi\psi}{2}}{\sqrt{\left(\frac{\phi^2}{2} + \psi^2 - R^2 X_1\right)^2 + \frac{\phi^2 \psi^2}{4}}}. \qquad (1.251)$$

For the points on l_2, the results are analog. Consequently, in Regions I and I', for the points that are symmetric with respect to the $O\psi$ axis, the absolute values of $(\frac{\mathrm{d}\psi}{\mathrm{d}\phi})_{\max}^{(\mathrm{I})}$ and $(\frac{\mathrm{d}\psi}{\mathrm{d}\phi})_{\min}^{(\mathrm{I}')}$ are equal but they have opposite signs. For points that are symmetric to Regions I and I' with respect to the $O\phi$ axis, the results are similar. Moreover, in the region defined by $\frac{\phi^2}{2} + \psi^2 \le R^2 X_1 - R^2 d$, the points on the plane (ϕ, ψ) always move from right to left, thus they will never stay in this region forever. In the region $\frac{\phi^2}{2} + \psi^2 \ge R^2 X_2 + R^2 d$, the points however move from left to right. As soon as the points pass through the line $\phi = \phi_0 = \sqrt{2(R^2 X_2 + R^2 d)}$, one must have $\frac{\mathrm{d}\phi}{\mathrm{d}t} > 0$, but $\frac{\mathrm{d}R}{\mathrm{d}t} = \frac{\phi}{\sqrt{R}} \ge \frac{\phi_0}{\sqrt{R}}$, hence P_3 will escape.

1.10 Sitnikov motion and its generalization

Sitnikov motion, sometimes called Sitnikov problem, is a special case of the spatial restricted 3-body problem, in which two primary bodies m_1 and m_2 of equal mass move around each other on elliptic Keplerian orbits while the third infinitesimally small body m_3 is confined to the line perpendicular to the primaries' orbit plane through the center of mass. Sitnikov has proven the existence of oscillatory motion in the 3-body problem of the above model. In addition, as a nonintegrable dynamical system, the Sitnikov problem often serves as a perfect example of the coexistence of regularity and chaoticity in a dynamical system.

Since long time ago, there have been many literatures discussing this problem. But in this section, a mapping model will be derived and used to study the Sitnikov problem. (The advantages of mapping method and the construction method of mapping will be discussed in detail in Chapter 2.)

1.10.1 *Model and mappings*

Brief astronomical background

The schematic diagram of Sitnikov problem is shown in Fig. 1.20. Two bodies of masses $m_1 = m_2 > 0$ move on the elliptic orbits and the center of mass is at rest. The third body m_3 with an infinitesimal mass moves on the vertical line that is perpendicular to the orbital plane of m_1 and m_2. This is a rather simple restricted 3-body problem.

The time is normalized in such a way that the motion period of primaries m_1, m_2 is 2π, and the mass unit is taken as the total mass. Thus, both the semi-major axis and mean motion of the 2-body orbit of m_1 and m_2 are 1, and the gravitational constant is 1 as well. Let z be the coordinate describing the position of m_3 and $z = 0$ corresponds to the center of mass.

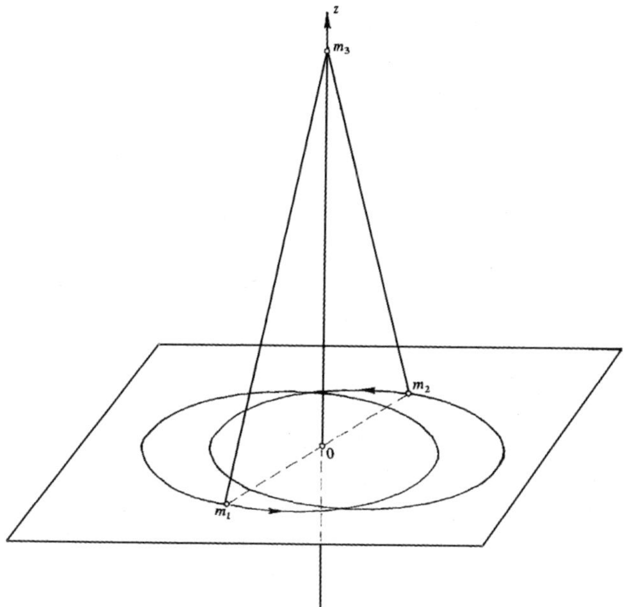

Fig. 1.20   Schematic configuration of Sitnikov problem. Adapted from Liu & Sun (1990).

The equation of motion of m_3 is

$$\frac{\mathrm{d}^2 z}{\mathrm{d}t^2} = -\frac{z}{[z^2 + r^2(t)]^{3/2}}, \tag{1.252}$$

where,

$$r(t) = a(1 - \epsilon \cos t) + o(\epsilon^2), \tag{1.253}$$

describes the elliptic motion of the primaries with respect to the center of mass, with ϵ being the eccentricity of elliptic orbit and a the semi-major axis ($a = 1/2$ since the relative orbit of primaries has a semi-major axis of 1 and $m_1 = m_2$). To the ϵ^2 order,

$$\frac{\mathrm{d}^2 z}{\mathrm{d}t^2} = -\frac{z}{(z^2 + a^2)^{3/2}} - \frac{3\epsilon a^2 z \cos t}{(z^2 + a^2)^{5/2}}. \tag{1.254}$$

Let $v = \frac{\mathrm{d}z}{\mathrm{d}t}$, then the equation of motion can be written as

$$\frac{\mathrm{d}z}{\mathrm{d}t} = v,$$

$$\frac{\mathrm{d}v}{\mathrm{d}t} = -\frac{z}{(z^2 + a^2)^{3/2}} - \frac{3\epsilon a^2 z \cos t}{(z^2 + a^2)^{5/2}}. \tag{1.255}$$

When $\epsilon = 0$, the unperturbed equation of motion

$$\frac{\mathrm{d}z}{\mathrm{d}t} = v,$$
$$\frac{\mathrm{d}v}{\mathrm{d}t} = -\frac{z}{(z^2 + a^2)^{3/2}},$$

(1.256)

is integrable, and it has an integral

$$v^2 - \frac{2}{(z^2 + a^2)^{1/2}} = c.$$

(1.257)

When $z = 0$, let $v = v_c$, so that

$$v_c^2 = c + \frac{2}{a}.$$

(1.258)

It is easy to derive from Eqs. (1.257) and (1.258) that: When $v_c^2 < 2/a$ (i.e. $c < 0$), the motion of m_3 is bounded; while for the case $v_c^2 > 2/a$ (i.e. $c > 0$) the motion is unbounded.

Mappings

Equation (1.255) is nonintegrable when $\epsilon \neq 0$, and the integral constant c in Eq. (1.257) varies with time. In virtue of Eq. (1.255), one obtains

$$\frac{\mathrm{d}c}{\mathrm{d}t} = -\frac{6\epsilon a^2 vz \cos t}{(z^2 + a^2)^{5/2}}.$$

(1.259)

This formula gives the change of energy of m_3 with time t. For $\cos t > 0$, when m_3 moves towards the plane $z = 0$, that is $z > 0$ and $v < 0$, m_3 gets energy from the primaries; conversely, when m_3 goes away from the plane $z = 0$, that is $z < 0, v < 0$, m_3 loses energy. Apparently, for $\cos t < 0$, opposite conclusions may be drawn. Because the strongest interaction between m_3 and the primaries m_1, m_2 is attained near $z = 0$, it is reasonable to suppose that the main change in energy c happens near the plane $z = 0$. This is an intuitive approximation, under which the energy c of m_3 experiences a jump Δc only when crossing the plane $z = 0$, and m_3 then keeps this energy of $c + \Delta c$ until next crossing through the $z = 0$ plane. The energy change Δc can be estimated in the following way.

Near $z = 0$, consider an approximate solution

$$z = \mathrm{sgn}(v_c)\sqrt{c + 2/a}(t - \tau_c),$$
$$v = \mathrm{sgn}(v_c)\sqrt{c + 2/a},$$

(1.260)

where τ_c is the time when a crossing with the $z = 0$ plane takes place. Substitute Eq. (1.260) into Eq. (1.259),

$$\frac{dc}{dt} = -\frac{6\epsilon a^2(c + 2/a)(t - \tau_c)\cos t}{[(c + 2/a)(t - \tau_c)^2 + a^2]^{5/2}}. \tag{1.261}$$

Integrate this from $-\infty$ to $+\infty$

$$\Delta c = -6\epsilon a^2\left(c + \frac{2}{a}\right)\int_{-\infty}^{+\infty}\frac{(t - \tau_c)\cos t}{[(c + 2/a)(t - \tau_c)^2 + a^2]^{5/2}}dt, \tag{1.262}$$

one finally obtains

$$\Delta c = \frac{4\epsilon a}{v_c^2}K_1\left[\frac{a}{(c + 2/a)^{1/2}}\right]\sin\tau_c, \tag{1.263}$$

where $K_1(x)$ is the Bessell function.

From Eq. (1.257), $v_c^2 = c + 2/a$, $(v_c + \Delta v_c)^2 = c + \Delta c + 2/a$, then one has $\Delta v_c \approx \Delta c/(2v_c)$. Combined with Eq. (1.263), this leads to the jump of velocity at the crossing of $z = 0$ plane

$$\Delta v_c = \text{sgn}(v_c)\frac{2\epsilon a}{|v_c^3|}K_1\left(\frac{a}{|v_c|}\right)\sin\tau_c. \tag{1.264}$$

The time interval T between two consecutive crossings can be estimated from Eqs. (1.256) and (1.257). Define an auxiliary variable s as

$$s = \frac{2}{v^2 - v_c^2 + 2/a},$$

then the time interval T can be calculated as

$$T = \int_{-v_c}^{v_c}\frac{s^3}{(s^2 - a^2)^{1/2}}dv = 2\int_0^{v_c}\frac{s^3}{(s^2 - a^2)^{1/2}}dv. \tag{1.265}$$

Combine Eqs. (1.264) and (1.265), one obtains the fundamental mapping

$$\begin{cases} v_c' = v_c + \text{sgn}(v_c)\dfrac{2\epsilon a}{|v_c^3|}K_1\left(\dfrac{a}{|v_c|}\right)\sin\tau_c, \\[3mm] \tau_c' = \tau_c + T(v_c'), \end{cases} \tag{1.266}$$

where the time increment T is a function of v_c as described in Eq. (1.265).

The original differential equations describe a continuous dynamical system of three dimensions, but now it is approximated by a mapping, i.e. a discrete dynamical system of only two dimensions.

Since the equations in Eq. (1.255) are invariant under the reflection transform $z \to -z$, it is not necessary to distinguish the cases of positive and negative values of z. Thus, to simplify the discussion, set $v_c = |v_c|$, $a = 1/2$, and replace the mapping in Eq. (1.266) by

$$\begin{cases} v_c' = v_c + \dfrac{\epsilon}{v_c^3} K_1 \left(\dfrac{1}{2v_c} \right) \sin \tau_c, \\[2mm] \tau_c' = \tau_c + T(v_c') \quad \mod (2\pi). \end{cases} \tag{1.267}$$

In general, celestial mechanics deals with conservative dynamical systems, which generally can be described by Hamiltonian systems or measure-preserving mappings. Write down the equations of motion of the Sitnikov problem Eq. (1.255) in a Hamiltonian form:

$$\dot{z} = H_v, \quad \dot{v} = -H_z, \tag{1.268}$$

where the subscript indicates partial derivative, and the Hamiltonian function is

$$H = \frac{1}{2} v^2 - \frac{1}{[z^2 + r^2(t)]^{1/2}}. \tag{1.269}$$

The Cartan differential form

$$\mathrm{d}v \wedge \mathrm{d}z - \mathrm{d}H \wedge \mathrm{d}t, \tag{1.270}$$

is preserved under the flow of Hamiltonian system. Consequently the mapping derived afore should preserve the restriction of this form to $z = 0, \mathrm{d}z = 0$. Because

$$\mathrm{d}H = v\mathrm{d}v - \frac{\partial}{\partial z} \left(\frac{1}{[z^2 + r^2(t)]^{1/2}} \right) \mathrm{d}z - \frac{\partial}{\partial t} \left(\frac{1}{[z^2 + r^2(t)]^{1/2}} \right) \mathrm{d}t,$$

$$\mathrm{d}z = 0, \quad \mathrm{d}t \wedge \mathrm{d}t = 0,$$

one has

$$\mathrm{d}v \wedge \mathrm{d}z - \mathrm{d}H \wedge \mathrm{d}t = -\mathrm{d}H \wedge \mathrm{d}t = -\mathrm{d}\left(\frac{v^2}{2} \right) \wedge \mathrm{d}t = -v\mathrm{d}v \wedge \mathrm{d}t. \tag{1.271}$$

Thus the fundamental mapping in Eq. (1.266) should preserve the area element $v\mathrm{d}v\,\mathrm{d}t$. For this reason, a modification should be made to the mapping in Eq. (1.267) correspondingly and it turns to

$$M : \begin{cases} v_c' = v_c + \epsilon q(v_c) \sin \tau_c, \\[2mm] \tau_c' = \tau_c + T(v_c') + \epsilon g(v_c, \tau_c), \end{cases} \tag{1.272}$$

where

$$q(v_c) = \frac{1}{v_c^3} K_1 \left(\frac{1}{2v_c} \right), \tag{1.273}$$

and

$$g(v_c, \tau_c) = \left(\frac{\partial q}{\partial v_c} + \frac{q}{v_c} \right) \cos \tau_c. \tag{1.274}$$

Obviously, the mapping M in Eq. (1.272) will be area-preserving, that is, it preserves the area element $v \, dv \, dt$. It is called the fundamental mapping now.

To get the inverse form of M, one may replace t by $-t$ in the differential equation Eq. (1.255) and get

$$\frac{dz}{dt} = -v,$$

$$\frac{dv}{dt} = -\frac{z}{(z^2 + a^2)^{3/2}} - \frac{3\epsilon a^2 z \cos t}{(z^2 + a^2)^{5/2}}, \tag{1.275}$$

and

$$\frac{dc}{dt} = \frac{6\epsilon a^2 v z \cos t}{(z^2 + a^2)^{5/2}}. \tag{1.276}$$

Then, following the same calculating procedure as above, one may obtain the inverse mapping of the fundamental mapping:

$$M^{-1} : \begin{cases} v_c = v_c' - \epsilon q(v_c') \sin \tau_c', \\ \tau_c = \tau_c' - T(v_c) - \epsilon g(v_c', \tau_c'), \end{cases} \tag{1.277}$$

The energy change Δc can be computed either from Eq. (1.263) or by numerically integrating the original differential equations Eq. (1.255). A comparing of these values verifies the validity of the above approximation.

1.10.2 *Topological structure of phase space*

The time interval $T(v_c)$ between two consecutive crossings of m_3 through the $z = 0$ plane, Eq. (1.265), and its partial derivative $\partial T / \partial v_c$ increase monotonically with v_c. For $\tau_c' = \tau_c + T(v_c') + O(\epsilon)$, so the sign $\Delta v_c' = v_c'' - v_c'$ $(\sim 2 \sin \tau_c')$ depends only on τ_c' according to Eq. (1.266). If $\pi > \tau_c' > 0$, $\Delta v_c'$ is positive; additionally, if $2\pi > \tau_c' > \pi$, $\Delta v_c'$ is negative. The sketch of the regions, in which $\Delta v_c'$ is positive or negative, are plotted in Fig. 1.21. These regions are called the positive and negative region for short, and they have the following properties:

(1) The positive and negative regions are symmetric to each other with respect to the origin ($v_c = 0$).
(2) The positive region and negative region entangle with each other. As v_c goes to $\sqrt{2/a}$ the regions become increasingly narrow, and their boundary curves tend to be circular.

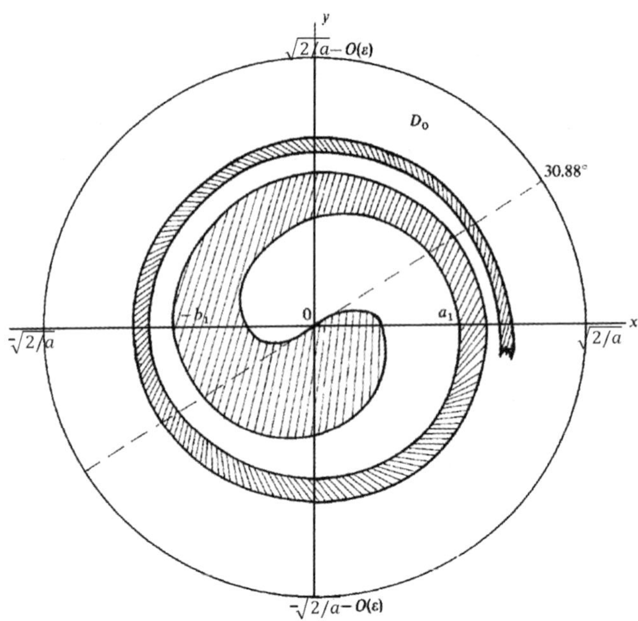

Fig. 1.21   The region of positive and negative $\Delta v'_c$ in a polar coordinate system. The polar axis is v_c and the argument is τ_c. The shaded and unshaded region is the positive and negative region of the mapping M, respectively. Adapted from Liu & Sun (1990).

For the inverse mapping M^{-1}, same discussions can be made and the positive and negative regions are sketched in Fig. 1.22.

The topological structure of phase space is closely related to the locations and characteristics of fixed points. According to Eq. (1.272), the fixed points of the mapping M satisfy the following equations

$$\epsilon q(v_c)\sin\tau_c = 0,$$
$$T(v'_c) + \epsilon g(v_c, \tau_c) = 0,$$

(1.278)

that is

$$\bar\tau = 0$$

$$T(\bar v_c) + \epsilon\left(\frac{\partial q}{\partial v_c} + \frac{q}{v_c}\right)\bigg|_{v_c=\bar v_c} = 2k\pi,$$

(1.279)

or

$$\bar\tau = \pi$$

$$T(\bar v_c) - \epsilon\left(\frac{\partial q}{\partial v_c} + \frac{q}{v_c}\right)\bigg|_{v_c=\bar v_c} = 2k\pi,$$

(1.280)

with k being any integer.

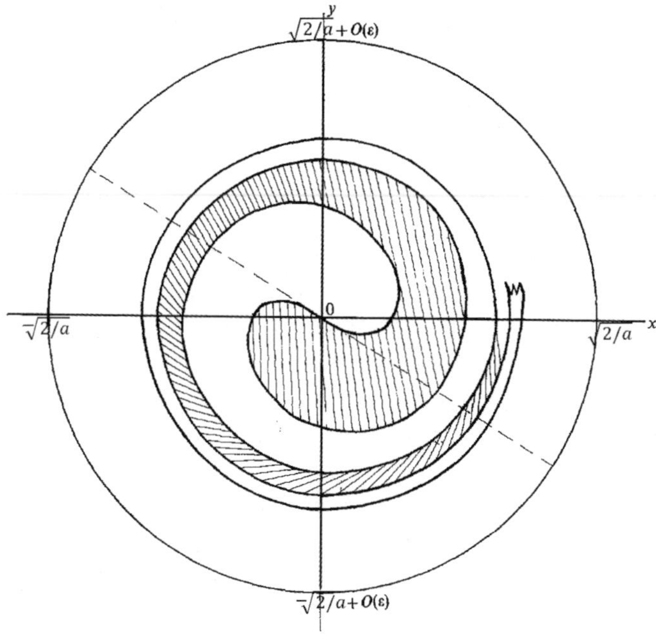

Fig. 1.22　The positive and negative region of the inverse mapping M^{-1}. Adapted from Liu & Sun (1990).

For the unperturbed system described by Eq. (1.256), its fundamental period is $T(v_c)$. The period of the perturbation term is 2π, so the solutions of these fixed points correspond to the resonance phenomena.

For further discussions, a set of Cartesian coordinates is introduced

$$x = v_c \cos \tau_c, \quad y = v_c \sin \tau_c, \tag{1.281}$$

then the mapping Eq. (1.272) is transformed to the following form

$$\bar{M} : \begin{cases} x' = F(x, y), \\ y' = G(x, y). \end{cases} \tag{1.282}$$

This new mapping preserves the area element $dxdy$. The first class of fixed points of this mapping is

$$(a_1, 0), (a_2, 0), \ldots, (a_k, 0), \ldots, \tag{1.283}$$

where $a_k > 0$ and satisfies the following equation

$$T(a_k) + \epsilon \left(\frac{\partial q}{\partial v_c} + \frac{q}{v_c} \right) \Bigg|_{v_c = a_k} = 2k\pi. \tag{1.284}$$

The second class of fixed points is

$$(-b_1, 0), (-b_2, 0), \ldots, (-b_k, 0), \ldots, \tag{1.285}$$

where $b_k > 0$ and satisfies

$$T(b_k) - \epsilon \left(\frac{\partial q}{\partial v_c} + \frac{q}{v_c} \right) \Bigg|_{v_c = b_k} = 2k\pi. \tag{1.286}$$

The fixed points in Eqs. (1.283) and (1.285) can be denoted as (x_k, y_k) with $x_k = a_k$ (or $x_k = -b_k$) and $y_k = 0$ for $k = 1, 2, \ldots$.

To determine the properties of the mapping $\bar{M}$ near the fixed points, one may linearize the mapping at these points and consider the tangent mapping

$$D\bar{M}_{(x_k, y_k)} = \begin{pmatrix} \frac{\partial F}{\partial x} & \frac{\partial F}{\partial y} \\ \frac{\partial G}{\partial x} & \frac{\partial G}{\partial y} \end{pmatrix}_{(x_k, y_k)}. \tag{1.287}$$

The characteristic equation of the first class of fixed points is

$$(\lambda - 1)^2 - \left[\epsilon q(v_c) \frac{\partial T}{\partial v_c} (\lambda - 1) + \epsilon q(v_c) \frac{\partial T}{\partial v_c} \right]_{v_c = a_k} = 0. \tag{1.288}$$

The eigenvalues are

$$\lambda_{1,2} = 1 + \frac{1}{2} \left[\epsilon q \frac{\partial T}{\partial v_c} \pm \sqrt{\left(\epsilon q \frac{\partial T}{\partial v_c} \right)^2 + 4\epsilon q \frac{\partial T}{\partial v_c}} \right]_{v_c = a_k}. \tag{1.289}$$

In the same way, the characteristic equations and the corresponding eigenvalues of the second class of fixed points are obtained as follows.

$$(\lambda + 1)^2 + \left[\epsilon q(v_c) \frac{\partial T}{\partial v_c} (\lambda + 1) + \epsilon q(v_c) \frac{\partial T}{\partial v_c} \right]_{v_c = b_k} = 0. \tag{1.290}$$

$$\lambda_{1,2} = -1 + \frac{1}{2} \left[-\epsilon q \frac{\partial T}{\partial v_c} \pm \sqrt{\left(\epsilon q \frac{\partial T}{\partial v_c} \right)^2 - 4\epsilon q \frac{\partial T}{\partial v_c}} \right]_{v_c = b_k}. \tag{1.291}$$

As ϵ is small and $\partial T / \partial v_c$ is not too large, one may conclude that the first class of fixed points are hyperbolic points and the second ones are elliptical points.

1.10.3 *Chaotic domain*

After a crossing of the $z = 0$ plane, a jump Δv_c of v_c occurs, and $\Delta v_c = \epsilon K_1(\frac{1}{2v_c})/v_c^3$ according to the mapping Eq. (1.272). As discussed in Sec. 1.10.2, when $v_c \to \sqrt{2/a}$, the positive region and negative regions become more and more narrow. They entangle with each other and the boundary curves between them tend to be circular (see Fig. 1.21). The width of the positive (or negative) region can be estimated as follows.

From the mapping M in Eq. (1.272), one has

$$\tau_c' = \tau_c + T(v_c') + \epsilon g(v_c, \tau_c),$$

$$\tau_c'' = \tau_c' + T'(v_c'') + \epsilon g(v_c', \tau_c'),$$

thus approximately

$$\tau_c'' - \tau_c' = \tau_c' - \tau_c + T'(v_c'') - T(v_c').$$

For the positive region (or negative region), $\pi > \tau_c' > 0$ and $\tau_c'' > \tau_c' > \tau_c$, therefore $\Delta T = T'(v_c'') - T(v_c')$ will attain its maximum when $\tau_c'' - \tau_c'$ attains its maximum π (while $\tau_c' - \tau_c$ reaches its minimum 0 simultaneously).

Let Δl be the width of the positive region. Since there is an approximated relation

$$\Delta T \approx \left(\frac{\partial T}{\partial v_c}\right) \Delta l = \pi, \tag{1.292}$$

one has

$$\Delta l \approx \left(\frac{\partial T}{\partial v_c}\right)^{-1} \pi. \tag{1.293}$$

If the change on v_c at each crossing is larger than Δl, it means that the orbits starting from there are sensitive to the initial conditions and the motion would be chaotic. Therefore, there is a criterion for chaotic motion as

$$\Delta v_c > \Delta l, \tag{1.294}$$

that is

$$\frac{\epsilon}{v_c^3} K_1\left(\frac{1}{2v_c}\right) > \left(\frac{\partial T}{\partial v_c}\right)^{-1} \pi. \tag{1.295}$$

The value of $\partial T/\partial v_c$ can be calculated from the expression of T in Eq. (1.265), then v_c corresponding to chaotic motion can be estimated by substituting $\partial T/\partial v_c$ into Eq. (1.295).

The topological structure of phase space, including the fixed points and the chaotic region, is illustrated in Fig. 1.23.

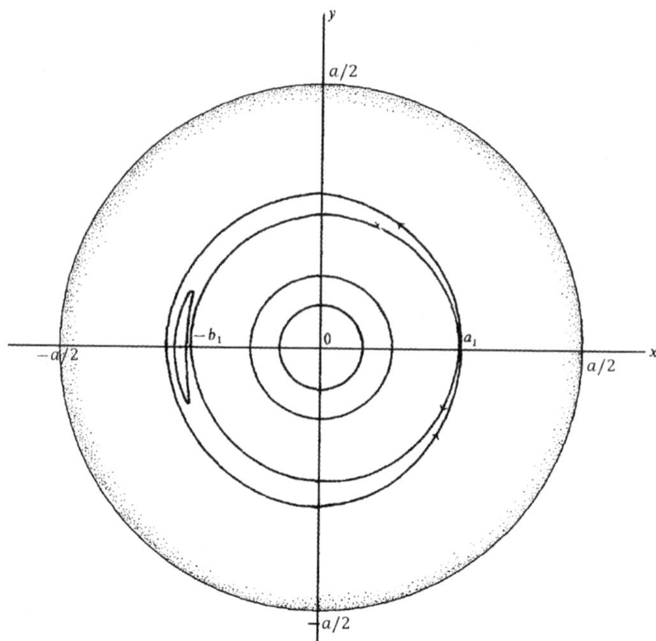

Fig. 1.23   Phase space of the mapping M. The fixed point at $(a_1, 0)$ is a hyperbolic fixed point, and $(-b_1, 0)$ is an elliptic fixed point. Near the boundary curve, there are a chaotic region. Adapted from Liu & Sun (1990).

1.10.4   *Extended Sitnikov problem*

If the third body in Sitnikov problem is massless, two primaries are always on fixed Keplerian orbits and they are always equally far away from the barycenter of the system and from the third body. If the third body has the same mass as the primaries, i.e. $m_1 = m_2 = m_3$, the two primaries will not move on Keplerian orbits and their orbital plane are not fixed anymore. Such a system is called the "Extended Sitnikov Problem", and it is in fact a special kind of general 3-body problem, with much more complex dynamics than the usual Sitnikov problem.

In a Cartesian coordinate system with the origin at the center of mass, denote the coordinates of m_1, m_2 and m_3 by (x_i, y_i, z_i), $i = 1, 2, 3$ respectively. Thanks to the special configuration of Sitnikov problem, the following symmetries hold all the time: $x_1 = -x_2$, $y_1 = -y_2$, $x_3 = y_3 = 0$ and $z_1 = z_2 = -\frac{1}{2}z_3$. Thus the initial conditions of the system can be chosen as $x_1^0 = -x_2^0$, $y_1^0 = -y_2^0$, $x_3^0 = y_3^0 = 0$ and $z_1^0 = z_2^0 = -\frac{1}{2}z_3^0$.

Taking into account this symmetry, the extended Sitnikov problem is reduced to a dynamical system of three degrees of freedom. Define $x = x_1$, $y = y_1$, $z = z_3$ and $m = m_1 = m_2 = m_3$, the equations of motion reads

$$\frac{\mathrm{d}^2 x}{\mathrm{d}t^2} = -\frac{2mx}{(4x^2 + 4y^2)^{3/2}} - \frac{mx}{(x^2 + y^2 + 9z^2)^{3/2}},$$

$$\frac{\mathrm{d}^2 y}{\mathrm{d}t^2} = -\frac{2my}{(4x^2 + 4y^2)^{3/2}} - \frac{my}{(x^2 + y^2 + 9z^2)^{3/2}}, \qquad (1.296)$$

$$\frac{\mathrm{d}^2 z}{\mathrm{d}t^2} = -\frac{3mz}{(x^2 + y^2 + 9z^2)^{3/2}}.$$

Through a coordinate transformation $x = r\cos\theta, y = r\sin\theta$ and taking into account the angular momentum integral $r^2\dot{\theta} = C/2m$ (C being the angular momentum constant of this system), Eq. (1.296) becomes

$$\frac{\mathrm{d}^2 r}{\mathrm{d}t^2} = \frac{C^2}{4m^2 r^3} - \frac{mr}{(r^2 + 9z^2)^{3/2}} - \frac{m}{4r^2},$$

$$\frac{\mathrm{d}^2 z}{\mathrm{d}t^2} = -\frac{3mz}{(r^2 + 9z^2)^{3/2}}. \qquad (1.297)$$

Apparently, the system has been reduced to a dynamical system of 2 degrees of freedom having the energy integral

$$m\left(\dot{r}^2 + \frac{C^2}{4m^2 r^2} + \frac{3}{4}\dot{z}^2\right) - \frac{m^2}{2r} - \frac{2m^2}{\sqrt{r^2 + 9z^2}} = h = \text{const.} \qquad (1.298)$$

First, some qualitative results about the motion of these three bodies will be presented below.

The smallest value of the energy h can be determined for a given C. An inequality can be derived easily from Eq. (1.298):

$$h \leq \frac{C^2}{4mr^2} - \frac{5m^2}{2r} \equiv f(r). \qquad (1.299)$$

Since the minimum of $f(r)$ is attained at

$$r_0 = \frac{C^2}{5m^3}, \qquad (1.300)$$

the minimum of h for a given C is

$$h_{\min} = f(r_0) = -\frac{25m^5}{4C^2} \equiv h_0. \qquad (1.301)$$

The values of h for bounded motion with a given C can also be computed. The minimum of $\frac{C^2}{4mr^2} - \frac{m^2}{2r}$ is

$$\left(\frac{C^2}{4mr^2} - \frac{m^2}{2r} \right)\Bigg|_{r=C^2/m^3} = -\frac{m^5}{4C^2} \equiv h_1. \qquad (1.302)$$

If

$$h_0 \le h < h_1, \qquad (1.303)$$

the following inequality holds

$$h \ge h_1 - \frac{2m^2}{\sqrt{r^2 + 9z^2}}, \qquad (1.304)$$

or, alternatively

$$\sqrt{r^2 + 9z^2} \le \frac{2m^2}{h_1 - h}. \qquad (1.305)$$

In this case the motions of m_1, m_2 and m_3 are bounded, and the inequality in Eq. (1.303) is a sufficient condition of bounded motion for the value of h.

Also, the region of possible motion in the plane $(r, \dot{r})$ can be determined. Define

$$g(r, \dot{r}) \equiv -m\dot{r}^2 - \frac{C^2}{4mr^2} + \frac{m^2}{2r},$$

$$q(r, \dot{r}, z) \equiv \frac{3}{4}m\dot{z}^2 - \frac{2m^2}{\sqrt{r^2 + 9z^2}}, \qquad (1.306)$$

the energy integral in Eq. (1.298) can be rewritten as

$$q(r, \dot{r}, z) = h + g(r, \dot{r}). \qquad (1.307)$$

It is easy to verify that $g(r, \dot{r})$ has the only equilibrium point $(r_0, \dot{r}_0)$ at $(C^2/m^3, 0)$, where it reaches its maximum

$$g_{\max} = \frac{m^5}{4C^2}. \qquad (1.308)$$

Let

$$\Delta \equiv g_{\max} + h. \qquad (1.309)$$

If $\Delta < 0$, $h < -g_{\max} = -\frac{m^5}{4C^2} = h_1$, and Eqs. (1.303)–(1.305) give the bounded motion. When $\Delta > 0$, a closed curve L on the plane $(r, \dot{r})$ can be defined as follows

$$g(r, \dot{r}) + h = 0. \qquad (1.310)$$

Inside the curve L, $g(r, \dot{r}) + h < 0$ but $\Delta > 0$; and outside the curve L, $g(r, \dot{r}) + h > 0$ and there is no contradiction with $\Delta > 0$. Therefore, roughly speaking, the curve L can be regarded as the boundary between the admissible region (outside L) and the forbidden region (inside L) for bounded motion.

Since $\dot{z}^2 \geq 0$, the energy integral Eq. (1.298) gives

$$\frac{h}{m} - r^2 - \frac{C^2}{4m^2r^2} + \frac{5m}{2r} \geq 0, \tag{1.311}$$

or in an alternative way

$$\dot{r}^2 + \frac{C^2}{4m^2r^2} - \frac{5m}{2r} \leq \frac{h}{m}. \tag{1.312}$$

These equations normally define a bounded region in the plane $(r, \dot{r})$. For example, a negative h will exclude the regions near and far from the origin in the $(r, \dot{r})$ plane.

The extended Sitnikov problem is a special case of general 3-body problem and virtually no theoretical result about this problem is known (in contrary, the normal Sitnikov problem is a special case of restricted 3-body problem). Under such circumstance, the numerical computation is useful for the investigation of this problem. Meanwhile, the surface of section is often used to display the topological structure of phase space of a dynamical system. For the extended Sitnikov problem, a section defined on the $(r, \dot{r})$ plane with $z = 0$ may serve as a proper surface of section.

As will be seen in the following figures, the character of the phase space for bounded motions is dominated by the 1:2 resonance, in which m_3 oscillates along the z-axis once while m_1 and m_2 orbit around each other exactly twice. In the surface of section, this resonance is represented by a fixed point, which is more or less always in the middle of the region of motion for all negative energy where bounded motion happens. According to the definition, all points on the surface of section are plotted whenever m_3 crosses the motion plane of m_1 and m_2 that is identical with the moment of m_3 crossing the barycenter of the system.

As Eqs. (1.311) and (1.312) show, the admissible and forbidden region in the surface of section depend on the total energy h. Below, some typical examples of surface of section with different h will be presented.

(1) $h = -0.26$

Figure 1.24 displays the temporal variations of r (mutual distance between m_1 and m_2) and z of an orbit close to the 1:2 resonance and with $h = -0.26$.

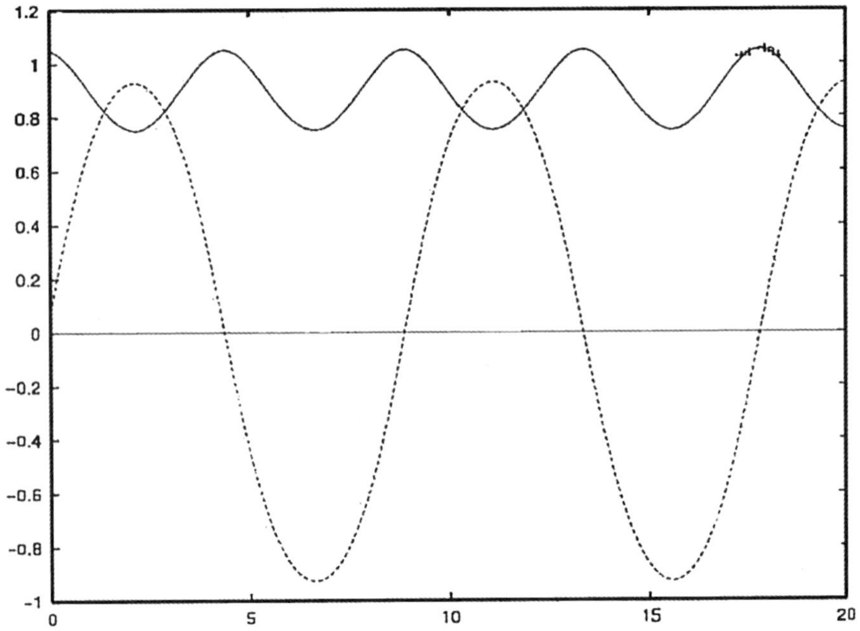

Fig. 1.24   An orbit close to the 1:2 resonance with $h = -0.26$. The temporal evolutions of z (dotted line) and r (the distance between primaries, solid line) are plotted. Adapted from Dvorak & Sun (1997).

The resonant behavior can be seen clearly, and judging from the figure, the motion is very regular. This regularity can be verified more clearly by the surface of section in Fig. 1.25.

In the surface of section, among the invariant tori around the central fixed point, three large islands lying on the invariant tori around the 3:4 resonant periodic orbits can be recognized immediately. Small chaotic region can be found around the separatrix "connecting" the hyperbolic fixed points in between the 3:4 islands.

(2) $h = -0.20$

When the energy h increases from $h = -0.26$ to $h = -0.20$, the central elliptic fixed point is bifurcated into two elliptic fixed points (stable fixed points), while the original elliptic point is turned into a hyperbolic one (unstable fixed point) that still corresponds to the 1:2 resonance, as shown in Fig. 1.26.

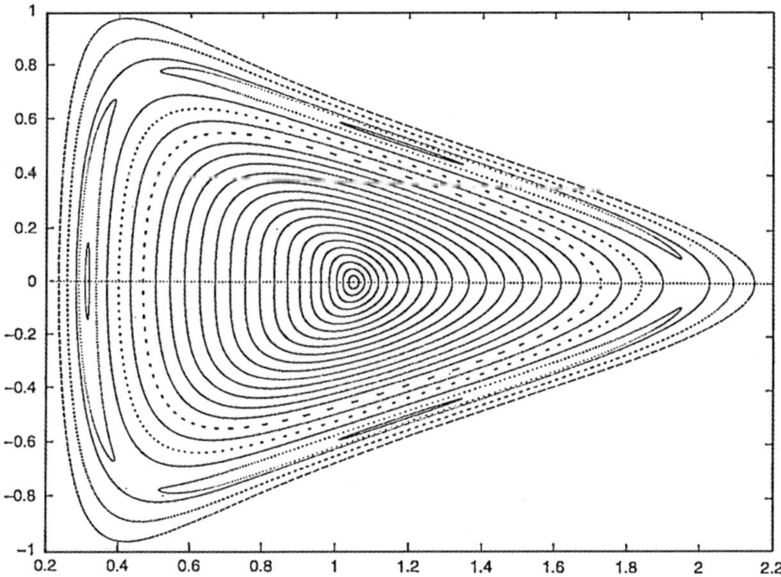

Fig. 1.25   Surface of section $(r, \dot{r})$ for $h = -0.26$. Adapted from Dvorak & Sun (1997).

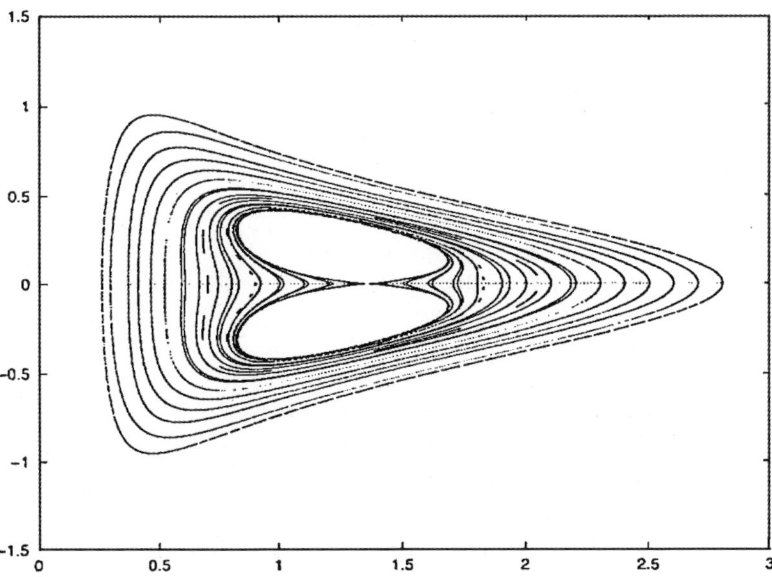

Fig. 1.26   Surface of section $(r, \dot{r})$ for $h = -0.20$. Adapted from Dvorak & Sun (1997).

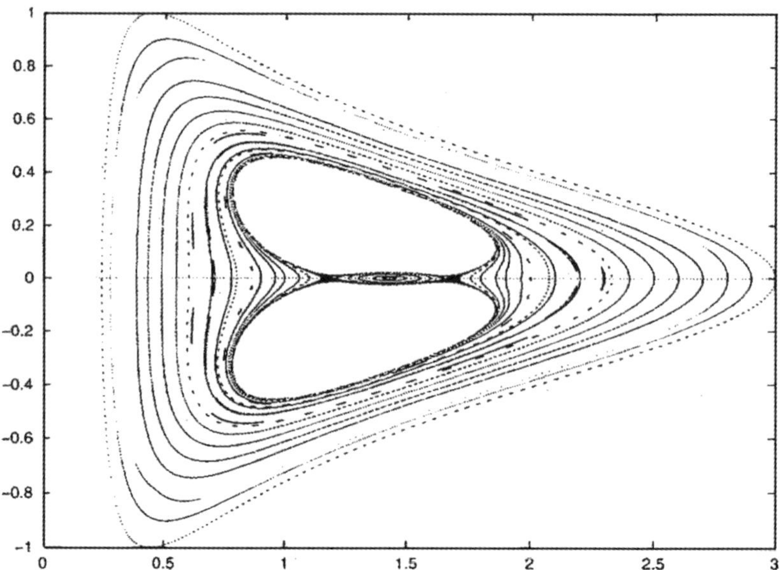

Fig. 1.27   Surface of section $(r, \dot{r})$ for $h = -0.19$. Adapted from Dvorak & Sun (1997).

(3) $h = -0.19$

As shown in Fig. 1.27, when h increases to $h = -0.19$, the central fixed point splits into one stable and two unstable points on the r-axis corresponds to the 1:2 resonance. The central point now becomes stable again.

(4) $h = -0.16$

At $h = -0.16$, chaotic region arises around the separatrix "connecting" the two unstable 1:2 resonant orbits, as illustrated in Fig. 1.28.

(5) $h = -0.04$

The regular region decreases and the chaotic region increases as h increases to $h = -0.04$. The central fixed point is now surrounded by invariant curves, indicating that the 1:2 resonance is still a stable fixed point. A large chaotic region is visible in Fig. 1.29.

In summary, the phase space structure depends on the energy h. When h increases, the chaotic region appears and grows gradually, as a result, the originally ordered phase space changes to a mixture of ordered region coexisting with chaotic regions. As h increases, the chaotic region increases

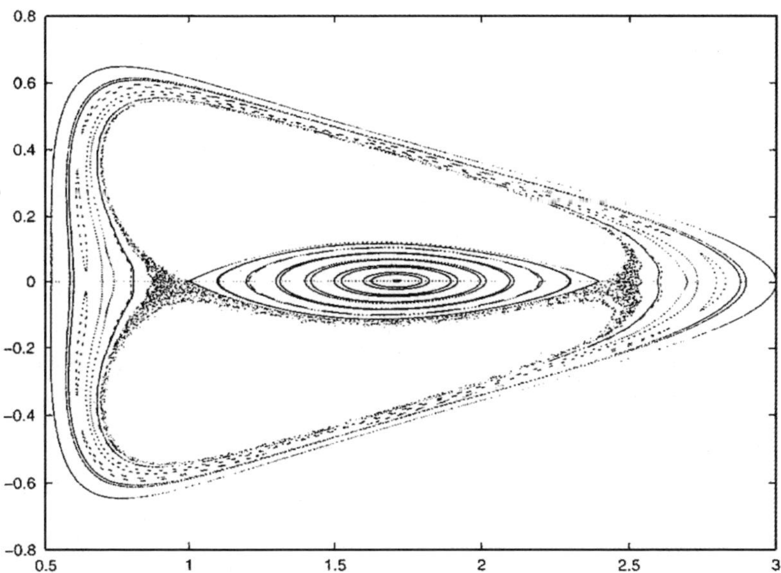

Fig. 1.28　Surface of section $(r, \dot{r})$ for $h = -0.16$. Adapted from Dvorak & Sun (1997).

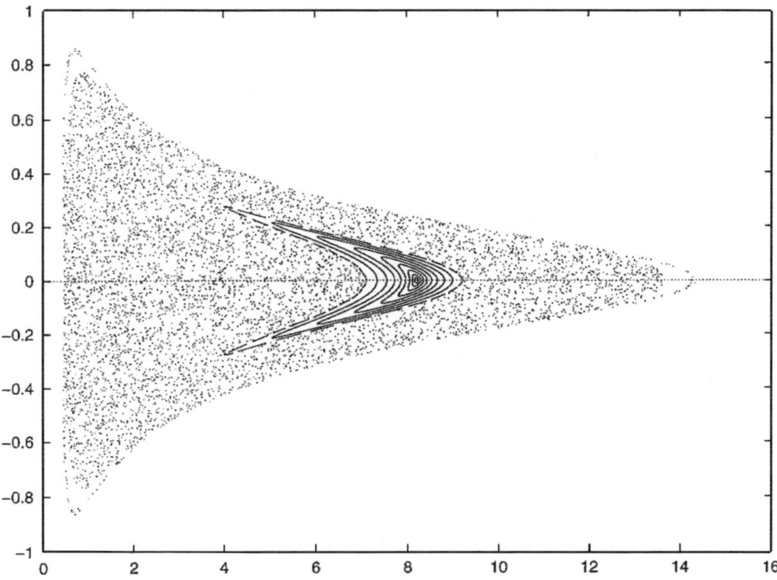

Fig. 1.29　Surface of section $(r, \dot{r})$ for $h = -0.04$. Adapted from Dvorak & Sun (1997).

while the ordered region shrinks. Moreover, other typical phenomena of nonlinear dynamics, such as the bifurcation phenomenon, can be found too in the extended Sitnikov problem.

1.11 Central configuration of 4-body problem

It is well-known that the general N-body problem is nonintegrable for $N \geq 3$, but some particular solutions do exist. For $N = 3$, there are five particular solutions, i.e. three Euler collinear solutions and two Lagrange equilateral triangle solutions. These five solutions have such a property that the acceleration vector of each body is equal to the product of a common scalar and its respective coordinate vector relative to the center of mass of the system. Such a configuration that holds the aforementioned property is called "central configuration". The adjective "central" here refers, on one hand, to the geometric property of the gravitation vectors, on the other hand, to the important role played by these configurations in the analysis of particular orbits, of collision orbits, of the expanding gravitational systems, and of the topological structures of phase space in a gravitational system. For the general N-body problem, if N' ($\leq N$) bodies collide or expand, the subsystem will line up in a central configuration as time $t \to \infty$. Similar to the case of the 3-body system, it is at the central configurations where restrictions to configurations begin to occur and surfaces meet. Nevertheless, little is known about the basic properties of central configurations when $N > 3$, even whether the number of central configurations is finite or not, is not clear yet.

Based on the equations of central configurations, such a coordinate system will be chosen that the equation of central configuration can be reduced to a simple form. Then the problem of whether the number of central configurations in the general 4-body problem is finite or not, will be studied. In addition, a global numerical survey through the solution-scanning method will be made to find the central configurations and to confirm numerically some theoretical results. However, some questions will be raised meanwhile.

1.11.1 *Elementary equations and theorem*

Consider N bodies with masses $m_i > 0$ and position vectors $\boldsymbol{r}_i$ ($i = 1, 2, \ldots, N$) relative to the center of mass of the system. Start at first with the problem of inverse $q-$force law where $1 < q < 3$.

Let

$$U = \sum_{1 \le i < j \le N} \frac{G m_i m_j}{r_{ij}^{q-1}}, \tag{1.313}$$

where G is the gravitational constant and $r_{ij} = |\boldsymbol{r}_i - \boldsymbol{r}_j|$ is the distance between m_i and m_j. Also let

$$I = \frac{1}{2} \sum_{i=1}^{N} m_i |\boldsymbol{r}_i|^2 = \frac{1}{2M} \sum_{1 \le i < j \le N} m_i m_j r_{ij}^2, \quad M = \sum_{i=1}^{N} m_i. \tag{1.314}$$

With the definition of central configuration, it is easy to prove that $\boldsymbol{r} = (\boldsymbol{r}_1, \boldsymbol{r}_2, \ldots, \boldsymbol{r}_N)$ consists a central configuration if and only if

$$\nabla \left(I U^{\frac{2}{q-1}} \right) = 0, \tag{1.315}$$

at $\boldsymbol{r}$. In fact, the definition of central configuration in N-body problem requires that a central configuration must satisfy the following equations

$$\lambda m_i \boldsymbol{r}_i = \frac{\partial U}{\partial \boldsymbol{r}_i}, \quad i = 1, 2, \ldots, N, \tag{1.316}$$

where λ is a constant. Make inner product of both sides of Eq. (1.316) with $\boldsymbol{r}_i$, sum over all indices i, and then use the fact that U is homogeneous of degree $(1 - q)$ and Euler's theorem, one obtains

$$2\lambda I = (1 - q)U,$$

or

$$\lambda = \frac{(1 - q)U}{2I}.$$

Adopting this value of λ, it follows that the central configurations are determined by the zero sets of

$$\frac{(1 - q)U}{2I} \frac{\partial I}{\partial \boldsymbol{r}_i} - \frac{\partial U}{\partial \boldsymbol{r}_i} \quad (i = 1, 2, \ldots, N),$$

which is the same as the set of critical points of $I U^{2/(q-1)}$. Therefore, the necessary and sufficient condition Eq. (1.315) is proven.

Let P_0' be the coordinate set in the original inertial coordinate system, i.e. $P_0' = \{x_i, y_i, z_i; \ i = 1, 2, \ldots, N\}$ and P_j be the coordinate set in such a coordinate system that the origin is at m_1, bodies $m_1, m_2, \ldots, m_j$ on the x-axis, m_2 at $x = 1$, m_{j+1} on the upper-half-plane and $m_{j+2}, m_{j+3}, \ldots, m_N$ at arbitrary positions in the space but not collide with each other. It is evident that $P_j = \{x_3, \ldots, x_j, x_{j+1}, y_{j+1}, x_{j+2}, y_{j+2}, z_{j+2}, \ldots, x_N, y_N, z_N\}$ has a total of $3(N - 1) - 2j$ elements (coordinate components), and among them $y_{j+1} > 0$. Let $\mathrm{root}[\nabla_{P_j}(IU^{2/(q-1)})]$ be the solution set of $\nabla_{P_j}(IU^{2/(q-1)}) = 0$, where ∇_{P_j} denotes the gradient relative to the coordinates in set P_j. A basic theorem is stated first as below.

Theorem (Basic Theorem). *Under rotation, translation and scale change of coordinates, the following relation holds.*

$$\mathrm{root}\left[\nabla_{P_0'}\left(IU^{\frac{2}{q-1}}\right)\right] = \bigcup_{j=2}^{N} \mathrm{root}\left[\nabla_{P_j}\left(IU^{\frac{2}{q-1}}\right)\right].$$

Proof. $P_0' \to P_j$ can be accomplished by a series of rotations, translations and scale-changes of coordinate system. When the ranks of the Jacobi matrices of these transformations are equal to the number of new variables respectively at any permissible positions, it is easy to prove that the new equations still have the form of Eq. (1.315). Moreover, the number of new equations is smaller. $\qquad\square$

From this basic theorem, the study on the central configuration can be divided into several problems in the space with fewer dimensions.

1.11.2   *General 4-body problem*

According to the basic theorem, the studies on central configurations of general 4-body problem ($N = 4, q = 2$) can be led to solve the equations $\nabla_{P_2}(IU^2) = 0$, $\nabla_{P_3}(IU^2) = 0$ and $\nabla_{P_4}(IU^2) = 0$, where

$$I = 1 + ar_{13}^2 + br_{23}^2 + acr_{14}^2 + bcr_{24}^2 + abcr_{34}^2,$$

$$U = 1 + \frac{a}{r_{13}} + \frac{b}{r_{23}} + \frac{ac}{r_{14}} + \frac{bc}{r_{24}} + \frac{abc}{r_{34}}, \tag{1.317}$$

$$a = \frac{m_3}{m_2}, \quad b = \frac{m_3}{m_1}, \quad c = \frac{m_4}{m_3}.$$

Here I and U differ from those in Eqs. (1.313) and (1.314) only by a constant factor, but the form of equations does not change. Through appropriate arrangement of m_1, m_2, m_3 and m_4, one now has three different cases: (1) $m_1, \ldots, m_4$ are all on the x-axis; (2) m_1, m_2, m_3 are on the x-axis while m_4 on the upper-half-plane (thus $z_4 = 0$); and (3) m_1, m_2 are on the x-axis, $z_3 = 0$.

The aforementioned equations will be considered separately as follows.

Case $\nabla_{P_4}(IU^2) = 0$

Some good results about this collinear 4-body problem have been obtained. For example, there are twelve central configurations for each set of mass. For details, one can refer to relative literatures (e.g. Saari (1980)).

Case $\nabla_{P_3}(IU^2) = 0$

This is the equation for constructing the point-line type central configuration, in which three (m_1, m_2, m_3) of the four mass-points lie on a line while the fourth does not.

Theorem. *There is not any point-line type central configuration in the general 4-body problem.*

Proof. An explanation will be given below. $\qquad\square$

Case $\nabla_{P_2}(IU^2) = 0$

The set of coordinate variables must be $P_2 = \{x_3, y_3, x_4, y_4, z_4\}$ where $y_3 > 0$ and the subscripts correspond to the bodies. Then $r_{13}^2 = x_3^2 + y_3^2$, $r_{23}^2 = (x_3 - 1)^2 + y_3^2$, $r_{14}^2 = x_4^2 + y_4^2 + z_4^2$, $r_{24}^2 = (x_4 - 1)^2 + y_4^2 + z_4^2$, $r_{34}^2 = (x_4 - x_3)^2 + (y_4 - y_3)^2 + z_4^2$. Because IU^2 is the function of P_2 and also $P_2' = \{r_{13}^2, r_{23}^2, r_{14}^2, r_{24}^2, r_{34}^2\}$, while P_2' and P_2 have the above relations $f : P_2 \to P_2'$ with the same number of variables. One has $\nabla_{P_2}(IU^2) = Df \cdot \nabla_{P_2'}(IU^2)$, where

$$Df = 2 \cdot \begin{pmatrix} x_3 & x_3 - 1 & 0 & 0 & x_3 - x_4 \\ y_3 & y_3 & 0 & 0 & y_3 - y_4 \\ 0 & 0 & x_4 & x_4 - 1 & x_4 - x_3 \\ 0 & 0 & y_4 & y_4 & y_4 - y_3 \\ 0 & 0 & z_4 & z_4 & z_4 \end{pmatrix}. \tag{1.318}$$

The determinant of it is $2z_4 y_3^2$.

In this case, two situations, divided according to the value of z_4, should be discussed separately.

(1) $z_4 \neq 0$. The central configuration can be obtained by solving equation $\nabla_{P_2'}(IU^2) = 0$ instead of $\nabla_{P_2}(IU^2) = 0$. The former equation reads

$$\frac{1}{r_{ij}} \frac{\partial(IU^2)}{\partial r_{ij}} = 0,$$

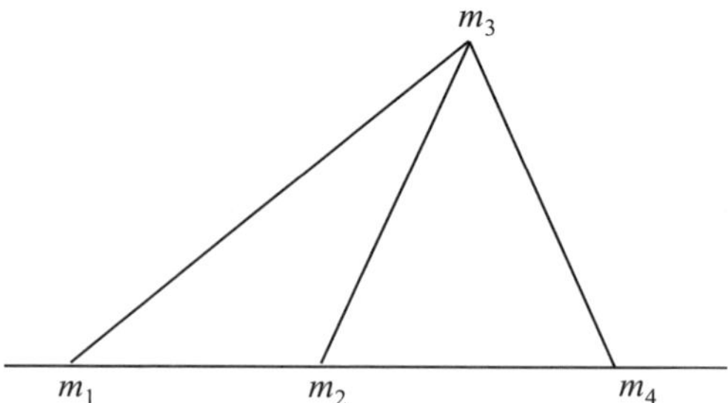

Fig. 1.30   Point-line type central configuration with m_1, m_2, m_4 on the same line.

and the solution can be found easily by substituting Eq. (1.317) into it. The only one positive real solution is a configuration of regular tetrahedron.

(2) $z_4 = 0$. Some elementary row transformations, whose feasibilities are evident, can be performed and useful results can be obtained. When $y_4 = 0$ ($x_4 \neq 0$ otherwise m_1 will collide with m_4), Df can be transformed to

$$Df = \begin{pmatrix} 1 & 0 & 0 & 0 & 0 \\ 0 & 1 & 0 & 0 & 0 \\ 0 & 0 & x_4 & x_4 - 1 & 0 \\ 0 & 0 & 0 & 0 & 1 \\ 0 & 0 & 0 & 0 & 0 \end{pmatrix}. \tag{1.319}$$

Now the central configurations of point-line type as shown in Fig. 1.30 will be searched.

From the first, the second and the fourth rows of the matrix Df in Eq. (1.319),

$$\frac{\partial(IU^2)}{\partial r_{13}} = \frac{\partial(IU^2)}{\partial r_{23}} = \frac{\partial(IU^2)}{\partial r_{34}} = 0,$$

thus $r_{13} = r_{23} = r_{34} = \sqrt[3]{I/U}$. Apparently, this is impossible since the three bodies on a line cannot keep the same distance from one point off the line. Further, using the above results, it is easy to prove the theorem in the case $\nabla_{P_3}(IU^2) = 0$.

Up to now, only two cases are left, namely $x_4 = 0, y_4 \neq 0, x_3 \neq 0$, and $x_4 \neq 0, y_4 \neq 0, x_3/y_3 \neq x_4/y_4$. The former one is a special case of the latter one. If we exchange m_3 with m_4 and let $x_3 = 0$ (it is permissible for the latter case), the latter case turns into the former case.

Then the problem of solving $\nabla_{P_2}(IU^2) = 0$ is led to the solution of the equation $Df \cdot \nabla_{P_2'}(IU^2) = 0$, in which $x_4 \neq 0, y_4 \neq 0, x_3/y_3 \neq x_4/y_4$. And the equations are:

$$I = \tau^3 U,$$

$$S_{123}\left(\frac{1}{\tau^3} - \frac{1}{r_{13}^3}\right) = -cS_{124}\left(\frac{1}{\tau^3} - \frac{1}{r_{14}^3}\right),$$

$$S_{123}\left(\frac{1}{\tau^3} - \frac{1}{r_{23}^3}\right) = -cS_{124}\left(\frac{1}{\tau^3} - \frac{1}{r_{24}^3}\right), \qquad (1.320)$$

$$S_{124}\left(\frac{1}{\tau^3} - \frac{1}{r_{14}^3}\right) = -bS_{243}\left(\frac{1}{\tau^3} - \frac{1}{r_{34}^3}\right),$$

$$S_{124}\left(\frac{1}{\tau^3} - \frac{1}{r_{24}^3}\right) = -aS_{134}\left(\frac{1}{\tau^3} - \frac{1}{r_{34}^3}\right),$$

where S_{ijk} is the area of triangle formed by m_i, m_j, m_k. A positive value will be given to S_{ijk} if m_i, m_j, m_k are arranged in a counter-clockwise order, and otherwise it will have a negative value.

Equations in Eq. (1.320) are the elementary equations of central configuration for the planar 4-body problem and may serve as the basis for further investigations. They are irrational equations so that their solutions cannot be expressed by elementary functions, and even whether the number of their solutions is finite or not remains unknown. Compared with other forms of equations concerned, Eq. (1.320) have advantages of a simpler form and a smaller number of equations.

Subtracting the third equation from the second one in Eq. (1.320), one gets

$$S_{123}\left(\frac{1}{r_{23}^3} - \frac{1}{r_{13}^3}\right) = cS_{124}\left(\frac{1}{r_{14}^3} - \frac{1}{r_{24}^3}\right). \qquad (1.321)$$

Obviously, if $r_{13} = r_{23}$ then $r_{14} = r_{24}$. Therefore, a theorem is obtained as below.

Theorem. *For the central configuration of any 4-body problem, if there is a body that keeps the same distance from two of other three bodies, the fourth body must keep the same distance from these two bodies too. Such a central configuration is symmetric.*

Since $y_3 > 0$, when $y_4 = 2S_{124} > 0$, Eq. (1.321) indicates that m_3 and m_4 lie on different sides of vertical bisector of segment $\overline{m_1 m_2}$, and when

$y_4 < 0$ they lie on the same side. Moreover, none of $r_{13}, r_{23}, r_{14}, r_{24}, r_{34}$ can be equal to τ.

Now, using Eq. (1.318) the equations of central configurations can be written in the forms of several polynomial equations. By substituting $x_3 = (r_{13}^2 - r_{23}^2 + 1)/2$ and $x_4 = (r_{14}^2 - r_{24}^2 + 1)/2$ into the first and the third expressions in $Df \cdot \nabla_{P_2'}(IU^2) = 0$ where Df is defined in Eq. (1.318), one obtains two expressions, denoted by (A) and (B) respectively. A third expression called (C) can be obtained by making some simple algebraic operations over the four equations, namely the first row multiplying by x_3 plus the second row multiplying by y_3 and then subtract the third row multiplying by x_4 and the fourth row multiplying by y_4. The fourth expression called (D) can be obtained by squaring the fourth equation of Eq. (1.320) and substituting the triangle area formula into it. Finally, when four bodies stay in the same plane, a geometrical relation must be obeyed. Such an expression, called (E) here, comes directly from the reference (formula (24), page 249 in Hagihara (1970)).

So now, $I = \tau^3 U$ and expressions (A)–(E) are the equations of $\tau, r_{13}, r_{23}, r_{14}, r_{24}, r_{34}$ and these equations can be easily proven to be independent. Because there is no radical expression in them, these equations can be changed into a set of polynomial equations. Whether the number of the solutions is finite or not depends on whether the polynomial equations are relatively prime. The number of solutions will be finite and will be smaller than or equal to the product of the highest degree of polynomials, otherwise it is infinite.

The above polynomial equations are obtained following the principle of letting the degrees of the polynomials be as low as possible without decreasing the independent polynomial equations. Thus it will be easier to determine whether the polynomials are relatively prime.

1.11.3 *Numerical exploration*

In this part, all central configurations for some sets of masses of general 4-body problem will be explored numerically.

Let m_1 be at the origin, m_2 at $x = 1$ and m_3 on the upper half-plane. Then m_3 can be on both sides of $x = 1/2$, i.e. either $x_3 \leq 1/2$ or $x_3 > 1/2$. Only one of this configurations, namely the former one, will be considered, since the latter one becomes the former one when m_1 and m_2 are exchanged. According to the non-collision conditions in the chosen coordinate system and discussions in previous subsection (Sec. 1.11.2), m_4 can be in the regions I, II, III, IV and V that are shown in Fig. 1.31 and Fig. 1.32.

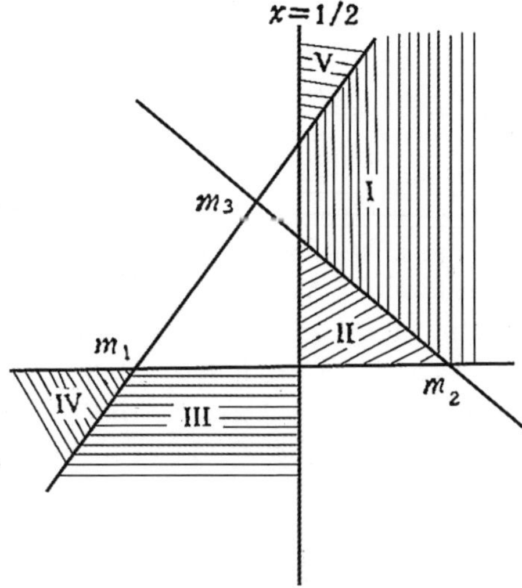

Fig. 1.31　Regions of m_4 when $x_3 > 0$. Adapted from Yan & Sun (1988).

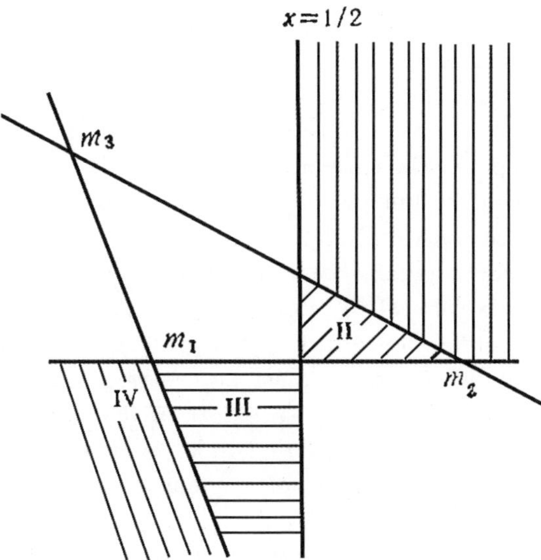

Fig. 1.32　Regions of m_4 when $x_3 \leq 0$. Adapted from Yan & Sun (1988).

Through a careful analyzing on these regions, the problem of finding central configurations is turned into the problem of determining the solutions (central configurations) in some compact sets. All the solutions in the above compact sets can be found by the scanning method that is widely applied in finding solutions of high degree polynomial equations with one variable. Consequently, all central configurations can be obtained. A "net" covering the above compact sets can be constructed by dividing the unit segment of each variable into H equal parts. At each "knot" of this net, evaluate the square sum of the right-hand sides of the equations. If the square sum is smaller than a given number ε, which is small but not too small, the corresponding knot is selected as the initial value for further searching for solution in the neighbourhood with the linear method. After working on all the knots, some solutions are achieved. Generally, the larger H is, the more solutions can be found. However, if H reaches such a value that no more solutions can be obtained by increasing it further, with a certainty, all the solutions have been found numerically.

In calculating some sets of masses, the results indicate that, except for the case of three or more than three equal masses, the total number of central configurations is 35 (including a regular tetrahedron).

For the case of three or more than three equal masses, letting $m_1 = m_2 = m_3 = 1$ and increasing m_4 from 0.05 to 1 by a step of 0.05, the numerical calculations reveal that every central configuration is symmetric and the total number of central configurations is 39 (including the regular tetrahedron) except for the case of $m_4 = 1$.

An analyzing on 26 out of the total 39 planar central configurations indicates that these central configurations can be "produced" by five symmetric ones (called "original central configurations"), which have a symmetric axis, i.e. the vertical bisector of $\overline{m_1 m_2}$. For example, if there is a central configuration as shown in Fig. 1.33(a) where m_3 and m_4 lie on the vertical bisector of $\overline{m_1 m_2}$, there must be central configurations that can be obtained by turning $\overline{m_2 m_3}$ or $\overline{m_3 m_1}$ into $\overline{m_1 m_2}$ and unitizing them, as shown in Figs. 1.33(b),(c). Thus these two latter central configurations can be regarded as derivatives of the former one, and this is what the word "produce" means here.

The five original central configurations appear in the forms as shown in Fig. 1.34. In the configuration (e), m_1, m_2 and m_3 forms an equilateral triangle while m_4 resides in the center. This central configuration is independent of m_4, hence it is the only one unchangeable original central configuration. It produces two other central configurations. While the forms (a), (b), (c)

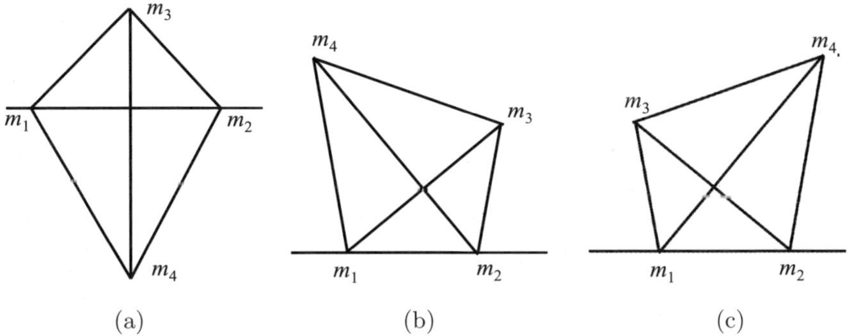

Fig. 1.33   Symmetric central configuration (a), in which m_3 and m_4 lie on the vertical bisector of $\overline{m_1 m_2}$, and two central configurations (b), (c) produced by this symmetric one (a). Adapted from Yan & Sun (1988).

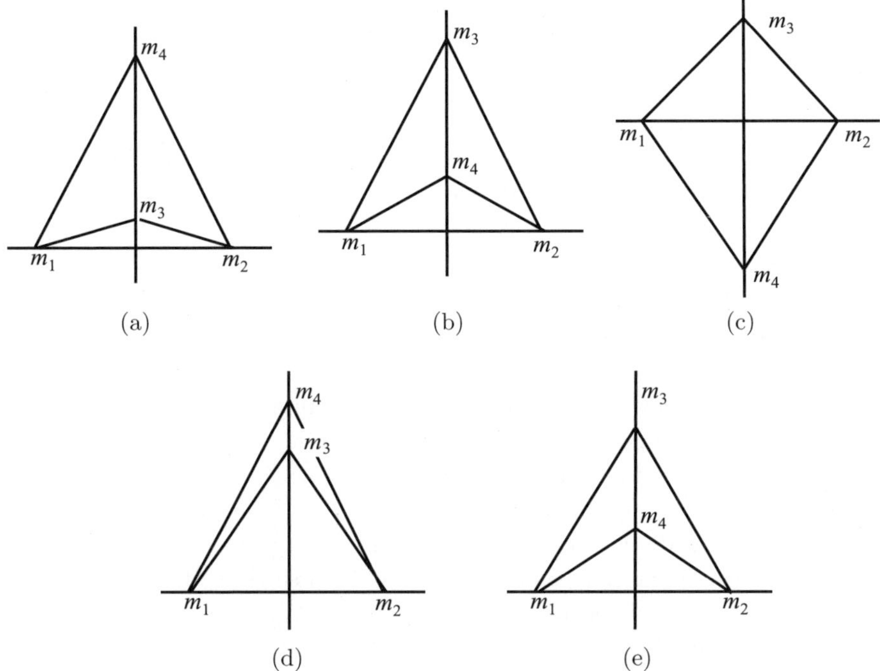

Fig. 1.34   Forms of five original central configurations. Adapted from Yan & Sun (1988).

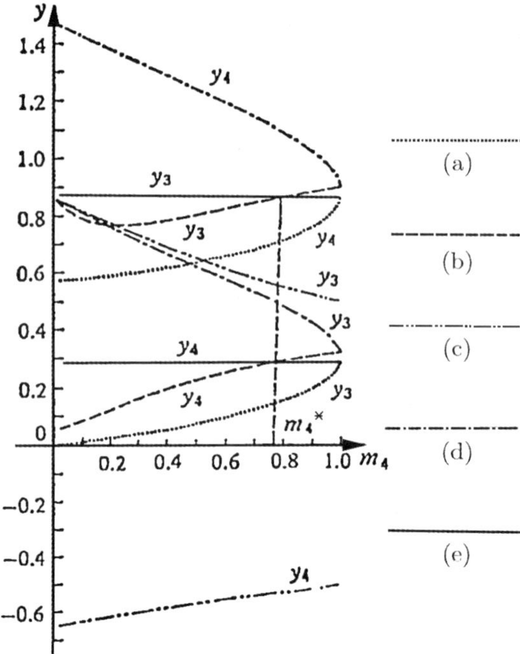

Fig. 1.35   Changes of central configurations of forms (a),(b) (c), (d) and (e) in Fig. 1.34. Adapted from Yan & Sun (1988).

and (d) all depend on m_4. By taking the original central configurations, which have been obtained through scanning directly, with $m_4 = 0.02$ as the initial value, 0.001 as the increment of m_4, and doing the scanning process in a similar way, the variations of the central configurations are calculated, with the results being plotted in Fig. 1.35.

In Fig. 1.35, sudden changes occur at $m_4 = 1$ or $m_4 = m_4^*$ where $m_4^* \approx 0.771$. The numerical results indicate that when $m_4 = m_4^*$, forms (b) and (e) coincide with each other and (b) will produce no more central configurations. When m_4 passes m_4^*, (b) and (e) separate from each other and the case comes back to that of $m_4 < m_4^*$. When $m_4 \to 1$, form (a) approaches to (e) (but m_4, m_3 being exchanged); (b) and (d) tend to be the same type, in which m_1, m_2 and m_3 lie at the vertexes of an isosceles but not equilateral triangle with m_4 on the segment of the symmetric axis; and form (c) tends to a square. Thus the original central configurations have three forms when $m_4 = 1$, i.e. equilateral triangle, isosceles but not equilateral triangle and square. These three kinds of original central configurations can produce all the central configurations for four-equi-masses case and obtain the same results as that obtained by direct scanning.

Therefore, in the case of at least three-equi-masses, when the fourth mass varies from a small value to 1, the total number of central configurations is $4 \times 6 + 1 \times 2 + 12 + 1 = 39$ in $0 < m_4 < m_4^*$ or $m_4^* < m_4 < 1$, $3 \times 6 + 1 \times 2 + 12 + 1 = 33$ at $m_4 = m_4^*$ and $8 + 20 + 6 + 12 + 1 = 47$ at $m_4 = 1$. Because the original central configurations can produce more central configurations in the four-equi-masses cases, it is no surprising that the number of central configurations increases suddenly when $m_4 = 1$.

1.12 Central configurations of N-body problem with general attraction and homographic solutions

The problem of central configurations of 4-body under Newton attraction is discussed in Sec. 1.11. In this section, the investigations on central configurations and the homographic solutions under the general attraction will be presented.

Let $m_i, \boldsymbol{r}_i$ $(i = 1, 2, \ldots, N)$ be the mass of the ith body and its position vector in the inertial coordinate system with origin at the center of mass. Denote the distance between m_i and m_j $(j \neq i, j = 1, 2, \ldots, N)$ by r_{ij}. The potential function of the N-body system is

$$U = G \sum_{1 \leq i < j \leq N} m_i m_j r_{ij}^{-k+1}, \quad (k \neq 1). \tag{1.322}$$

and the attraction on m_i reads

$$\boldsymbol{F}_i = (-k + 1)G \sum_{j=1, j \neq i}^{N} m_i m_j r_{ij}^{-k-1} (\boldsymbol{r}_i - \boldsymbol{r}_j), \tag{1.323}$$

where G $(\neq 0)$ is the proportional constant. Obviously, when $(-k+1)G < 0$ the bodies attract each other, otherwise they exclude each other.

The configurations of N bodies can be expressed by a set of vectors $(\boldsymbol{r}_1, \boldsymbol{r}_2, \ldots, \boldsymbol{r}_N)$, and the central configuration is defined below. If the configuration of an N-body system satisfies

$$\boldsymbol{F}_i = \sigma m_i \boldsymbol{r}_i, \quad (i = 1, 2, \ldots, N), \tag{1.324}$$

where σ is a constant, then the configuration $(\boldsymbol{r}_1, \boldsymbol{r}_2, \ldots, \boldsymbol{r}_N)$ is called a central configuration. As indicated in Sec. 1.11, one has

$$\sigma = \frac{(-k + 1)U}{2I}, \tag{1.325}$$

where I is the semi-moment of inertia of the system defined as

$$I = \frac{1}{2}\sum_i^N m_i\,|\boldsymbol{r}_i|^2 = \frac{1}{2M}\sum_{1\le i<j\le N} m_i m_j r_{ij}^2, \quad M = \sum_{i=1}^N m_i. \tag{1.326}$$

Substituting Eqs. (1.323) and (1.325) into Eq. (1.324), one gets

$$(-k+1)G\sum_{j=1,j\ne i}^N m_i m_j r_{ij}^{-k-1}(\boldsymbol{r}_i-\boldsymbol{r}_j) = \frac{(-k+1)U}{2I}m_i\boldsymbol{r}_i, \quad (i=1,2,\ldots,N),$$

i.e.

$$G\sum_{j=1,j\ne i}^N m_j r_{ij}^{-k-1}(\boldsymbol{r}_i-\boldsymbol{r}_j) = \frac{U}{2I}\boldsymbol{r}_i. \quad (i=1,2,\ldots,N).$$

Using the expressions of U and I in Eqs. (1.322) and (1.326), Eq. (1.324) can be transformed into an equivalent form as

$$\sum_{j=1,j\ne i}^N \frac{m_j}{r_{ij}^{k+1}}(\boldsymbol{r}_i-\boldsymbol{r}_j) = \frac{\sum_{1\le i<j\le N} m_i m_j r_{ij}^{-k+1}}{\sum_{1\le i<j\le N} m_i m_j r_{ij}^2}M\boldsymbol{r}_i, \tag{1.327}$$

$$(i=1,2,\ldots,N).$$

Considering the characteristics of $U(\boldsymbol{r}_i)$, $I(\boldsymbol{r}_i)$ and Eq. (1.325), it is easy to prove that a central configuration is still a central configuration after a similarity transformation. Naturally, if two or more central configurations can be transformed to each other, they are considered as the same central configuration.

1.12.1 *Some theoretical analysis*

According to Eq. (1.327), whether $\{m_1, m_2, \ldots, m_N\}$ constitutes a central configuration is determined by the parameter k and $N-1$ mass ratios $\{m_j/m_{i_0} : j \in \{1,2,\ldots,N\}, j \ne i_0\}$, where i_0 is an arbitrarily given index and $i_0 \in \{1,2,\ldots,N\}$.

From above discussions, the value of G, the units of mass and distance can be chosen independently. For convenience, $G = 1$ is taken hereafter. When $k = -1$, Eq. (1.327) will be satisfied for any configurations of N bodies, thus any N-body system trivially constitutes a central configuration. From now on it is supposed that $k \ne -1$.

If the N bodies reside all in an Euclidean space of dimension D, but not in a subspace of $(D-1)$-dimension, their configuration is called a

D-dimensional configuration. For example, collinear and coplanar configurations are one-dimensional and two-dimensional configurations, respectively. Because the central configurations depend on $N - 1$ parameters (mass ratios) for a fixed k, only when $N - 1 \geq D$, i.e. $N \geq D + 1$, the N bodies can constitute D-dimensional central configurations.

In a Cartesian coordinate system $\{X_1, X_2, \ldots, X_D\}$ in D-dimensional space, let $\{\boldsymbol{r}_1, \boldsymbol{r}_2, \ldots, \boldsymbol{r}_N\} = \{(x_{11}, \ldots, x_{D1}), \ldots, (x_{1N}, \ldots, x_{DN})\}$. Taking into account $G = 1$, Eq. (1.327) can be written in the components of $\boldsymbol{r}_i$ such that a scalar form as

$$\sum_{j=1, j \neq i}^{N} \frac{m_j}{r_{ij}^{k+1}} (x_{di} - x_{dj}) - \frac{U}{2I} x_{di} = 0. \tag{1.328}$$

$$(d = 1, 2, \ldots, D; \ i = 1, 2, \ldots, N).$$

A theorem can be derived from this equation.

Theorem 1. *When* $\{m_1, m_2, \ldots, m_N\}$ *($N \geq 3$) constitute a D-dimensional central configuration, for all possible (l, i, j, α, β), the following equations are satisfied.*

$$\sum_{j=1, j \neq i, l}^{N} m_j S(l, i, j, \alpha, \beta) \left(r_{lj}^{-k-1} - r_{ij}^{-k-1} \right) = 0, \tag{1.329}$$

where $1 \leq l \neq i \leq N$, $1 \leq \alpha \neq \beta \leq D$, *and*

$$S(l, i, j, \alpha, \beta) = \begin{vmatrix} x_{\alpha l} & x_{\beta l} & 1 \\ x_{\alpha i} & x_{\beta i} & 1 \\ x_{\alpha j} & x_{\beta j} & 1 \end{vmatrix}.$$

Proof. Simple calculations show that

$$\sum_{j=1, j \neq i, l}^{N} m_j (r_{lj}^{-k-1} - r_{ij}^{-k-1}) \begin{vmatrix} x_{\alpha l} & x_{\beta l} & 1 \\ x_{\alpha i} & x_{\beta i} & 1 \\ x_{\alpha j} & x_{\beta j} & 1 \end{vmatrix}$$

$$= (x_{\beta l} - x_{\beta i}) X_{\alpha l i} - (x_{\alpha l} - x_{\alpha i}) X_{\beta l i},$$

where

$$X_{dli} = \sum_{j=1, j \neq i, l}^{N} m_j \left(r_{lj}^{-k-1} - r_{ij}^{-k-1} \right) \left[x_{dj} - \frac{1}{2} (x_{di} + x_{dl}) \right], \quad d = \alpha, \beta.$$

Now, the theorem will be proved if for any given α, β and m_l, m_i,

$$(x_{\beta l} - x_{\beta i}) X_{\alpha l i} - (x_{\alpha l} - x_{\alpha i}) X_{\beta l i} = 0.$$

In fact, for body m_l, the dth dimensional equality in Eq. (1.328) reads

$$\sum_{j=1,j\neq l}^{N} \frac{m_j}{r_{lj}^{k+1}}(x_{dl} - x_{dj}) - \frac{U}{2I}x_{dl} = 0.$$

This equation can be written as

$$\sum_{j=1,j\neq l}^{N} \frac{m_j}{r_{lj}^{k+1}}(x_{dl} - x_{dj}) - \frac{U}{2I}x_{dl}$$

$$= -\sum_{j=1,j\neq l}^{N} \frac{m_j}{r_{lj}^{k+1}}\left[x_{dj} - \frac{1}{2}(x_{di} + x_{dl})\right] \tag{1.330}$$

$$+ \frac{1}{2}(x_{dl} - x_{di})\sum_{j=1,j\neq l}^{N} \frac{m_j}{r_{lj}^{k+1}} - \frac{U}{2I}x_{dl} = 0.$$

In the same way, for body m_i and the dth dimensional equality,

$$-\sum_{j=1,j\neq i}^{N} \frac{m_j}{r_{ij}^{k+1}}\left[x_{dj} - \frac{1}{2}(x_{di} + x_{dl})\right]$$

$$\tag{1.331}$$

$$+ \frac{1}{2}(x_{di} - x_{dl})\sum_{j=1,j\neq i}^{N} \frac{m_j}{r_{ij}^{k+1}} - \frac{U}{2I}x_{di} = 0.$$

Subtract Eq. (1.331) from Eq. (1.330), one obtains

$$\sum_{j=1,j\neq l,i}^{N} m_j\left(r_{lj}^{-k-1} - r_{ij}^{-k-1}\right)\left[x_{dj} - \frac{1}{2}(x_{di} + x_{dl})\right]$$

$$+ \frac{1}{2}(x_{di} - x_{dl})\sum_{j=1,j\neq l,i}^{N} m_j\left(r_{lj}^{-k-1} + r_{ij}^{-k-1}\right)$$

$$+ (x_{di} - x_{dl})(m_i + m_l)r_{li}^{-k-1} + \frac{U}{2I}(x_{dl} - x_{di}) = 0,$$

i.e.

$$X_{dli} = \sum_{j=1,j\neq l,i}^{N} m_j\left(r_{lj}^{-k-1} - r_{ij}^{-k-1}\right)\left[x_{dj} - \frac{1}{2}(x_{di} + x_{dl})\right] = (x_{dl} - x_{di})W$$

$$W = \frac{1}{2}\sum_{j=1,j\neq l,i}^{N} m_j\left(r_{lj}^{-k-1} + r_{ij}^{-k-1}\right) + (m_i + m_l)r_{li}^{-k-1} - \frac{U}{2I}.$$

Because W is independent of dimensions, one has

$$(x_{\beta l} - x_{\beta i})X_{\alpha l i} - (x_{\alpha l} - x_{\alpha i})X_{\beta l i}$$

$$= (x_{\beta l} - x_{\beta i})W(x_{\alpha l} - x_{\alpha i}) - (x_{\alpha l} - x_{\alpha i})W(x_{\beta l} - x_{\beta i}) = 0.$$

This completes the proof of the theorem. $\square$

The geometrical meaning of $S(l, i, j, \alpha, \beta)$ in Theorem 1 is the double area of triangle that consists of the projections of m_l, m_i, m_j on the two-dimensional plane spanned by two different axes X_α, X_β. With this, a theorem can be derived.

Theorem 2. *For a D-dimensional $(D \geq 2)$ central configuration of N bodies $\{m_1, \ldots, m_N\}$ $(N \geq 3)$, if there exists such a two-dimensional plane in the D-dimensional space of central configuration that the projections of $N-1$ bodies (suppose $m_1, \ldots, m_{N-1}$) on it are on a line but not overlapping on a point while the projection of m_N is not on this line, then for any two bodies whose projects are not overlapping, their distances to m_N are equal to each other.*

Proof. Consider a central configuration of N bodies satisfying the above conditions. Without loss of generality, take X_1 and X_2 as the unit vectors of axes on the above plane in Theorem 2, i.e. $\alpha = 1, \beta = 2$, and choose arbitrarily two bodies m_l and m_i among the $N-1$ bodies whose projections are not overlapping. Then, there would be the only one term of $j = N$ in the summation Eq. (1.329), i.e. $m_N S(l, i, N, 1, 2)(r_{lN}^{-k-1} - r_{iN}^{-k-1})$, that is not zero, since only m_N is not on the line connecting the projections of m_l and m_i. In the meantime, this configuration (m_l, m_i on the line but not overlap with m_N off the line) ensures that the triangle area $S(l, i, N, 1, 2) \neq 0$, and obviously $m_N \neq 0$. However, according to Theorem 1, the summation must be zero, indicating that this term must be zero, thus $r_{lN}^{-k-1} - r_{iN}^{-k-1} = 0$, that is, $r_{lN} = r_{iN}$. $\square$

From Theorem 2, two corollaries can be drawn.

Corollary 1. *If the projections of any two bodies of $m_1, m_2, \ldots, m_{N-1}$ in Theorem 2 do not overlap each other, these bodies must all reside on a $(D-1)$-dimension sphere centered at m_N.*

Corollary 2. *The central configuration, in which all but one of the $N(\geq 4)$ bodies are collinear, does not exist.*

In fact, another theorem as below can be derived from Theorem 2 too.

Theorem 3. *If a two-dimensional central configuration of N $(N \geq 4)$ bodies $\{m_1, \ldots, m_N\}$ satisfies the following conditions:*

(1) $m_1, \ldots, m_s$ $(2 \leq s \leq N - 1)$ are all on a line L_1;
(2) $m_{s+1}, \ldots, m_{N-1}$ are all on the perpendicular bisector L_2 of the line segment connecting bodies m_a and m_b $(1 \leq a \neq b \leq s)$.

then

$$m_N \in L_1 \cup L_2.$$

1.12.2 *Equivalent conditions of central configuration*

Define an $N \times N$ matrix Γ as

$$\Gamma = (\gamma_{ij}),$$
$$\gamma_{ij} = m_j r_{ij}^{-k-1}, \quad (1 \leq i \neq j \leq N),$$
$$\gamma_{ii} = \frac{U}{2I} - \sum_{j=1, j\neq i}^{N} m_j r_{ij}^{-k-1}, \quad (i = 1, 2, \ldots, N),$$

and an $N \times D$ matrix Q as

$$Q = \begin{pmatrix} x_{11} & \cdots & x_{D1} \\ x_{12} & \cdots & x_{D2} \\ \vdots & \vdots & \vdots \\ x_{1N} & \cdots & x_{DN} \end{pmatrix}.$$

Theorem 4. *The necessary and sufficient condition of a D-dimensional configuration of $\{m_1, m_2, \ldots, m_N\}$ being a central configuration is*

$$\Gamma Q = 0.$$

Proof. The ΓQ is an $N \times D$ matrix, whose elements can be calculated as:

$$(\Gamma Q)_{il} = \sum_{j=1}^{N} \gamma_{ij} Q_{jl} = \sum_{j=1, j\neq i}^{N} \gamma_{ij} Q_{jl} + \gamma_{ii} Q_{il}$$

$$= \sum_{j=1, j\neq i}^{N} m_j r_{ij}^{-k-1} x_{jl} + \left(\frac{U}{2I} - \sum_{j=1, j\neq i}^{N} m_j r_{ij}^{-k-1} \right) x_{il}$$

$$= \sum_{j=1, j\neq i}^{N} m_j (x_{jl} - x_{il}) r_{ij}^{-k-1} + \frac{U}{2I} x_{il}.$$

For a given i, $\Gamma Q = 0$ corresponds to D equalities

$$(\Gamma Q)_{il} = 0. \quad (l = 1, 2, \ldots, D).$$

These equalities may be written in a vector form, i.e.

$$\sum_{j=1, j \neq i}^{N} m_j r_{ij}^{-k-1}(\boldsymbol{r}_j - \boldsymbol{r}_i) + \frac{U}{2I}\boldsymbol{r}_i = 0 \quad (i = 1, 2, \ldots, N).$$

Obviously, this is equivalent to Eq. (1.327). This completes the proof. $\square$

This theorem leads to a corollary.

Corollary 3. *If N bodies $\{m_1, m_2, \ldots, m_N\}$ constitute a D-dimensional central configuration, then*

$$\det \Gamma = 0.$$

From Eq. (1.325), it can be proven that Eq. (1.324) is equivalent to

$$\frac{\partial\left(IU^{\frac{2}{k-1}}\right)}{\partial \boldsymbol{r}_i} = 0. \quad (i = 1, 2, \ldots, N). \tag{1.332}$$

Therefore, there exists a theorem as follows.

Theorem 5. *N bodies $\{m_1, m_2, \ldots, m_N\}$ constitute a central configuration if and only if Eq. (1.332) is satisfied.*

Proof. Eq. (1.332) is equivalent to

$$0 = \frac{\partial I}{\partial \boldsymbol{r}_i}U^{\frac{2}{k-1}} + I\frac{\partial}{\partial \boldsymbol{r}_i}\left(U^{\frac{2}{k-1}}\right) = \frac{\partial I}{\partial \boldsymbol{r}_i}U^{\frac{2}{k-1}} + I\frac{2}{k-1}U^{\frac{3-k}{k-1}}\frac{\partial U}{\partial \boldsymbol{r}_i}$$

$$= \frac{\partial I}{\partial \boldsymbol{r}_i}U^{\frac{2}{k-1}} + \frac{2}{k-1}IU^{\frac{3-k}{k-1}}\boldsymbol{F}_i,$$

that is

$$\boldsymbol{F}_i = -\left(\frac{\partial I}{\partial \boldsymbol{r}_i}U^{\frac{2}{k-1}}\right)\Bigg/\left(\frac{2}{k-1}IU^{\frac{3-k}{k-1}}\right) = (-k+1)\frac{U}{2I}\frac{\partial I}{\partial \boldsymbol{r}_i}$$

$$= \sigma\frac{\partial I}{\partial \boldsymbol{r}_i} = \sigma\frac{\partial}{\partial \boldsymbol{r}_i}\left(\frac{1}{2}\sum_{j=1}^{N} m_j\boldsymbol{r}_j \cdot \boldsymbol{r}_j\right) = \sigma m_i\boldsymbol{r}_i.$$

This is equivalent to Eq. (1.324), the definition of central configuration. $\square$

1.12.3 *Homographic solutions of general N-body problem*

Generally, the N-body problem is nonintegrable for $N \geq 3$, i.e. the general solution of this problem does not exist, but there are five particular solutions (Euler and Lagrange particular solutions) in 3-body problem with Newtonian potential. These solutions constitute central configurations. In N-body problem with general attraction, similar solutions, *homographic solutions*, may exist.

In the inertial system with origin at the center of mass, denote the position vectors of body m_i at initial time t_0 and moment t by $\boldsymbol{r}_{i0}$ and $\boldsymbol{r}_i(t)$, respectively. If a particular solution of N-body problem satisfies

$$\boldsymbol{r}_i(t) = a(t)\Omega(t)\boldsymbol{r}_{i0}, \quad (i = 1, 2, \ldots, N), \tag{1.333}$$

where $a(t)$ is a positive scalar function and $\Omega(t)$ is the rotation matrix (i.e. orthogonal matrix equal to 1), this solution is called the "homographic solution". Apparently,

$$a(0) = 1, \quad \Omega(0) = E,$$

where E is the unit matrix.

According to Eq. (1.333), for the homographic solutions, one has

$$I = a^2 I_0, \tag{1.334}$$

$$U = a^{-k+1} U_0, \tag{1.335}$$

$$\boldsymbol{F}_i = a^{-k}\Omega\boldsymbol{F}_{i0}, \tag{1.336}$$

where the subscript "0" denotes the quantities at initial time t_0. In addition, the derivative of $\boldsymbol{r}_i(t)$ with respect to time t can be calculated as

$$\frac{\mathrm{d}\boldsymbol{r}_i}{\mathrm{d}t} = \frac{\mathrm{d}a}{\mathrm{d}t}\Omega\boldsymbol{r}_{i0} + a\frac{\mathrm{d}\Omega}{\mathrm{d}t}\boldsymbol{r}_{i0} = \Omega\left(\frac{\mathrm{d}a}{\mathrm{d}t} + a\Omega^{-1}\frac{\mathrm{d}\Omega}{\mathrm{d}t}\right)\boldsymbol{r}_{i0}$$

$$= \Omega\left(\frac{\mathrm{d}a}{\mathrm{d}t}E + a\Omega^T\frac{\mathrm{d}\Omega}{\mathrm{d}t}\right)\boldsymbol{r}_{i0} \tag{1.337}$$

$$= \Omega\left(\frac{\mathrm{d}a}{\mathrm{d}t}E + aA\right)\boldsymbol{r}_{i0}, \quad (i = 1, 2, \ldots, N),$$

where Ω^T is the transpose of Ω, $A = \Omega^T\frac{\mathrm{d}\Omega}{\mathrm{d}t}$, and the orthogonality of Ω (i.e. $\Omega^T\Omega = E$) has been taken into account. Also from the orthogonality, one has

$$\frac{\mathrm{d}\Omega^T}{\mathrm{d}t}\Omega + \Omega^T\frac{\mathrm{d}\Omega}{\mathrm{d}t} = \frac{\mathrm{d}E}{\mathrm{d}t} = 0.$$

A transposition operation to this equation leads to

$$\Omega^T \frac{d\Omega}{dt} + \left(\Omega^T \frac{d\Omega}{dt}\right)^T = 0,$$

that is, $A^T = -A$, indicating that A is an anti-symmetric matrix. Denote A as

$$A = \begin{pmatrix} 0 & -e_3 & e_2 \\ e_3 & 0 & -e_1 \\ -e_2 & e_1 & 0 \end{pmatrix}. \tag{1.338}$$

By means of Eq. (1.337), the kinetic energy T can be written as

$$\begin{aligned}
T &= \frac{1}{2} \sum_{i=1}^{N} m_i \frac{d\boldsymbol{r}_i}{dt} \cdot \frac{d\boldsymbol{r}_i}{dt} \\
&= \frac{1}{2} \sum_{i=1}^{N} m_i \left[\Omega \left(\frac{da}{dt} E + aA \right) \boldsymbol{r}_{i0} \right] \cdot \left[\Omega \left(\frac{da}{dt} E + aA \right) \boldsymbol{r}_{i0} \right] \\
&= \frac{1}{2} \sum_{i=1}^{N} m_i \left\{ \left(\frac{da}{dt} \right)^2 (\Omega \boldsymbol{r}_{i0}) \cdot (\Omega \boldsymbol{r}_{i0}) + a^2 (\Omega A \boldsymbol{r}_{i0}) \cdot (\Omega A \boldsymbol{r}_{i0}) \right. \\
&\quad + a \left(\frac{da}{dt} \right) \left[(\Omega A \boldsymbol{r}_{i0}) \cdot (\Omega \boldsymbol{r}_{i0}) + (\Omega \boldsymbol{r}_{i0}) \cdot (\Omega A \boldsymbol{r}_{i0}) \right] \Big\} \\
&= \frac{1}{2} \sum_{i=1}^{N} m_i \left\{ \left(\frac{da}{dt} \right)^2 \boldsymbol{r}_{i0} \cdot \boldsymbol{r}_{i0} + a^2 (A \boldsymbol{r}_{i0}) \cdot (A \boldsymbol{r}_{i0}) + 2a \left(\frac{da}{dt} \right) (A \boldsymbol{r}_{i0}) \cdot \boldsymbol{r}_{i0} \right\}.
\end{aligned}$$

In the above equations, the dot "$\cdot$" indicates inner product of vectors, as usual. Since A is an anti-symmetric matrix, one may calculate

$$(A\boldsymbol{r}_{i0}) \cdot \boldsymbol{r}_{i0} = (A\boldsymbol{r}_{i0})^T \boldsymbol{r}_{i0} = \boldsymbol{r}_{i0}^T A^T \boldsymbol{r}_{i0} = -\boldsymbol{r}_{i0}^T A \boldsymbol{r}_{i0}.$$

Further, since $(A\boldsymbol{r}_{i0}) \cdot \boldsymbol{r}_{i0}$ is a scalar,

$$(A\boldsymbol{r}_{i0}) \cdot \boldsymbol{r}_{i0} = -\left(\boldsymbol{r}_{i0}^T A \boldsymbol{r}_{i0} \right)^T = -\boldsymbol{r}_{i0}^T A^T \boldsymbol{r}_{i0} = \boldsymbol{r}_{i0}^T A \boldsymbol{r}_{i0}.$$

These above two equations mean $(A\boldsymbol{r}_{i0}) \cdot \boldsymbol{r}_{i0} = -(A\boldsymbol{r}_{i0}) \cdot \boldsymbol{r}_{i0}$, therefore

$$(A\boldsymbol{r}_{i0}) \cdot \boldsymbol{r}_{i0} = 0$$

Finally, the kinetic energy T can be computed as

$$
\begin{aligned}
T &= \frac{1}{2} \sum_{i=1}^{N} m_i \left\{ \left(\frac{\mathrm{d}a}{\mathrm{d}t} \right)^2 \boldsymbol{r}_{i0} \cdot \boldsymbol{r}_{i0} + a^2 \left(A\boldsymbol{r}_{i0} \right) \cdot \left(A\boldsymbol{r}_{i0} \right) \right\} \\
&= \frac{1}{2} \sum_{i=1}^{N} m_i \left\{ \left(\frac{\mathrm{d}a}{\mathrm{d}t} \right)^2 \boldsymbol{r}_{i0} \cdot \boldsymbol{r}_{i0} + a^2 \left(\Omega^T \frac{\mathrm{d}\Omega}{\mathrm{d}t} \boldsymbol{r}_{i0} \right) \cdot \left(\Omega^T \frac{\mathrm{d}\Omega}{\mathrm{d}t} \boldsymbol{r}_{i0} \right) \right\} \\
&= \frac{1}{2} \sum_{i=1}^{N} m_i \left\{ \left(\frac{\mathrm{d}a}{\mathrm{d}t} \right)^2 \boldsymbol{r}_{i0} \cdot \boldsymbol{r}_{i0} + a^2 \left(\frac{\mathrm{d}\Omega}{\mathrm{d}t} \boldsymbol{r}_{i0} \right) \cdot \left(\frac{\mathrm{d}\Omega}{\mathrm{d}t} \boldsymbol{r}_{i0} \right) \right\} \\
&= \left(\frac{\mathrm{d}a}{\mathrm{d}t} \right)^2 I_0 + \frac{1}{2} a^2 \sum_{i=1}^{N} m_i \left| \frac{\mathrm{d}\Omega}{\mathrm{d}t} \boldsymbol{r}_{i0} \right|^2 .
\end{aligned}
\tag{1.339}
$$

The angular momentum is

$$
\begin{aligned}
\boldsymbol{J} &= \sum_{i=1}^{N} m_i \boldsymbol{r}_i \times \frac{\mathrm{d}\boldsymbol{r}_i}{\mathrm{d}t} \\
&= \sum_{i=1}^{N} m_i \left[a\Omega\boldsymbol{r}_{i0} \times \left(\frac{\mathrm{d}a}{\mathrm{d}t} \Omega\boldsymbol{r}_{i0} + a \frac{\mathrm{d}\Omega}{\mathrm{d}t} \boldsymbol{r}_{i0} \right) \right] \\
&= \sum_{i=1}^{N} m_i \left(a\Omega\boldsymbol{r}_{i0} \right) \times \left(a \frac{\mathrm{d}\Omega}{\mathrm{d}t} \boldsymbol{r}_{i0} \right) \\
&= a^2 \sum_{i=1}^{N} m_i \left(\Omega\boldsymbol{r}_{i0} \right) \times \left(\frac{\mathrm{d}\Omega}{\mathrm{d}t} \boldsymbol{r}_{i0} \right) .
\end{aligned}
\tag{1.340}
$$

The acceleration vector, can be derived from Eq. (1.337)

$$
\begin{aligned}
\frac{\mathrm{d}^2 \boldsymbol{r}_i}{\mathrm{d}t^2} &= \frac{\mathrm{d}\Omega}{\mathrm{d}t} \left(\frac{\mathrm{d}a}{\mathrm{d}t} E + aA \right) \boldsymbol{r}_{i0} + \Omega \left(\frac{\mathrm{d}^2 a}{\mathrm{d}t^2} E + \frac{\mathrm{d}a}{\mathrm{d}t} A + a \frac{\mathrm{d}A}{\mathrm{d}t} \right) \boldsymbol{r}_{i0} \\
&= \Omega \left[A \left(\frac{\mathrm{d}a}{\mathrm{d}t} E + aA \right) + \left(\frac{\mathrm{d}^2 a}{\mathrm{d}t^2} E + \frac{\mathrm{d}a}{\mathrm{d}t} A + a \frac{\mathrm{d}A}{\mathrm{d}t} \right) \right] \boldsymbol{r}_{i0} \\
&= \Omega \left[\frac{\mathrm{d}^2 a}{\mathrm{d}t^2} E + 2 \frac{\mathrm{d}a}{\mathrm{d}t} A + a \left(A^2 + \frac{\mathrm{d}A}{\mathrm{d}t} \right) \right] \boldsymbol{r}_{i0} \\
&= a^{-k} \left(\Omega B \right) \boldsymbol{r}_{i0},
\end{aligned}
\tag{1.341}
$$

where

$$
B = a^k \left[\frac{\mathrm{d}^2 a}{\mathrm{d}t^2} E + 2 \frac{\mathrm{d}a}{\mathrm{d}t} A + a \left(A^2 + \frac{\mathrm{d}A}{\mathrm{d}t} \right) \right] .
\tag{1.342}
$$

Because $\boldsymbol{F}_i = m_i \frac{\mathrm{d}^2 \boldsymbol{r}_i}{\mathrm{d}t^2}$, from Eqs. (1.336) and (1.341), one knows

$$a^{-k}\Omega\boldsymbol{F}_{i0} = m_i a^{-k}\Omega B\boldsymbol{r}_{i0}.$$

Recalling that $\boldsymbol{F}_{i0}$ is the attraction suffered by m_i at initial time t_0,

$$\boldsymbol{F}_{i0} = m_i \frac{\mathrm{d}^2 \boldsymbol{r}_i(t_0)}{\mathrm{d}t^2} \equiv m_i \frac{\mathrm{d}^2 \boldsymbol{r}_{i0}}{\mathrm{d}t^2},$$

one obtains the equation of motion for the homographic solution:

$$\Omega B\boldsymbol{r}_{i0} = \frac{\mathrm{d}^2 \boldsymbol{r}_{i0}}{\mathrm{d}t^2}. \tag{1.343}$$

In such a system, the Lagrange–Jacobi identity reads

$$\frac{\mathrm{d}^2 I}{\mathrm{d}t^2} = (-k + 3)U + 2h,$$

where h is the constant of energy integral. Substituting Eqs. (1.334) and (1.335) into it, this equation becomes

$$\frac{\mathrm{d}(a^2 I_0)}{\mathrm{d}t^2} = (-k + 3)\left(a^{-k+1}U_0\right) + 2h,$$

or, in an alternative way

$$2\left[\left(\frac{\mathrm{d}a}{\mathrm{d}t}\right)^2 + a\left(\frac{\mathrm{d}^2 a}{\mathrm{d}t^2}\right)\right]I_0 - (3 - k)a^{-k+1}U_0 = 2h. \tag{1.344}$$

This is in fact the Lagrange–Jacobi identity along the homographic solution.

For the planar N-body problem, take the plane of motion as the (X, Y)-plane of inertia coordinate system $O - XYZ$, and the Z-axis is chosen such that

$$\boldsymbol{J} = J\hat{z},$$

where $\hat{z}$ is the unit vector of Z-axis, $J = |\boldsymbol{J}|$. In this coplanar case, Ω is reduced to a simple form as

$$\Omega = \begin{pmatrix} \cos\varphi & -\sin\varphi & 0 \\ \sin\varphi & \cos\varphi & 0 \\ 0 & 0 & 1 \end{pmatrix}. \tag{1.345}$$

Then from Eq. (1.340), the value J can be calculated as

$$J = |\boldsymbol{J}| = \left|a^2 \sum_{i=1}^{N} m_i\left(\Omega\boldsymbol{r}_{i0}\right) \times \left(\frac{\mathrm{d}\Omega}{\mathrm{d}t}\boldsymbol{r}_{i0}\right)\right| = \left|a^2 \sum_{i=1}^{N} m_i \frac{\mathrm{d}\varphi}{\mathrm{d}t}|\boldsymbol{r}_{i0}|^2\right|$$

$$= \left|a^2 \frac{\mathrm{d}\varphi}{\mathrm{d}t}(2I_0)\right| = 2a^2 \frac{\mathrm{d}\varphi}{\mathrm{d}t}I_0. \qquad \left(\frac{\mathrm{d}\varphi}{\mathrm{d}t} > 0\right). \tag{1.346}$$

And for the energy integral, one has

$$h = \left(\frac{\mathrm{d}a}{\mathrm{d}t}\right)^2 I_0 + \frac{1}{2}a^2 \sum_{i=1}^{N} m_i \left|\frac{\mathrm{d}\Omega}{\mathrm{d}t}\boldsymbol{r}_{i0}\right|^2 - a^{-k+1}U_0$$

$$= \left[\left(\frac{\mathrm{d}a}{\mathrm{d}t}\right)^2 + a^2 \left(\frac{\mathrm{d}\varphi}{\mathrm{d}t}\right)^2\right] I_0 - a^{-k+1}U_0.$$

Combined with Eq. (1.344), this gives

$$a^k \left[\frac{\mathrm{d}^2 a}{\mathrm{d}t^2} - a\left(\frac{\mathrm{d}\varphi}{\mathrm{d}t}\right)^2\right] = \frac{(1-k)U_0}{2I_0}. \qquad (1.347)$$

This equation describes an equation for the planar homographic solution.

1.12.4 *Homographic solutions and central configurations*

According to Eq. (1.345), the matrix A in the planar case is

$$A = \Omega^T \frac{\mathrm{d}\Omega}{\mathrm{d}t} = \begin{pmatrix} 0 & \frac{\mathrm{d}\varphi}{\mathrm{d}t} & 0 \\ \frac{\mathrm{d}\varphi}{\mathrm{d}t} & 0 & 0 \\ 0 & 0 & 0 \end{pmatrix},$$

thus in this case the matrix B as in Eq. (1.342) reads

$$B = \begin{pmatrix} a^k\left[\frac{\mathrm{d}^2 a}{\mathrm{d}t^2} - a\left(\frac{\mathrm{d}\varphi}{\mathrm{d}t}\right)^2\right] & -a^{k-1}\frac{\mathrm{d}}{\mathrm{d}t}\left(a^2\frac{\mathrm{d}\varphi}{\mathrm{d}t}\right) & 0 \\ a^{k-1}\frac{\mathrm{d}}{\mathrm{d}t}\left(a^2\frac{\mathrm{d}\varphi}{\mathrm{d}t}\right) & a^k\left[\frac{\mathrm{d}^2 a}{\mathrm{d}t^2} - a\left(\frac{\mathrm{d}\varphi}{\mathrm{d}t}\right)^2\right] & 0 \\ 0 & 0 & a^k\frac{\mathrm{d}^2 a}{\mathrm{d}t^2} \end{pmatrix}.$$

From Eq. (1.346), $a^2\frac{\mathrm{d}\varphi}{\mathrm{d}t} = \frac{J}{2I_0}$ is a constant, therefore $\frac{\mathrm{d}}{\mathrm{d}t}\left(a^2\frac{\mathrm{d}\varphi}{\mathrm{d}t}\right) = 0$. Taking into account Eq. (1.347), B can be reduced to

$$B = \begin{pmatrix} \frac{(-k+1)U_0}{2I_0} & 0 & 0 \\ 0 & \frac{(-k+1)U_0}{2I_0} & 0 \\ 0 & 0 & \frac{(-k+1)U_0}{2I_0} + a^{k+1}\left(\frac{\mathrm{d}\varphi}{\mathrm{d}t}\right)^2 \end{pmatrix}.$$

It has been shown in literature Fu & Sun (1992), Fu & Sun (1993) that when $k \neq 3$, the homographic solutions are either planar solutions or non-rotational solutions (i.e. $\Omega = E$). Consequently, the configurations corresponding to all the homographic solutions will satisfy

$$\frac{\mathrm{d}^2 \boldsymbol{r}_{i0}}{\mathrm{d}t^2} = \frac{(-k+1)U_0}{2I_0}\boldsymbol{r}_{i0}. \quad (i = 1, 2, \ldots, N).$$

By means of Eqs. (1.324) and (1.325), it is easy to see that when $k \neq 3$ the configurations corresponding to homographic solutions are central configurations.

Moreover, it can be shown that when the motions of N bodies are along the homographic solutions, each body m_i moves on a Kepler orbit as in 2-body problem. Let $\{M_{1i}, M_{2i}\}$ be a 2-body system and satisfy

$$M_{1i}M_{2i}\left(\frac{M_{2i}}{M_{1i}+M_{2i}}\right)^{-k} = m_i \frac{U_0 \left|\boldsymbol{r}_{i0}\right|^{k+1}}{GI_0}, \tag{1.348}$$

then, according to the properties of central configuration and homographic solution, the resulting force acting on m_i is

$$\boldsymbol{F}_i = (-k+1)\frac{m_i U_0}{I_0}\left(\frac{\left|\boldsymbol{r}_{i0}\right|}{\left|\boldsymbol{r}_i\right|}\right)^{k+1}\boldsymbol{r}_i,$$

and in the 2-body system the resulting force acting on M_{1i} is

$$\boldsymbol{f}_{1i} = (-k+1)GM_{1i}M_{2i}\left(\frac{M_{1i}+M_{2i}}{M_{2i}}\right)^{-k}R_{1i}^{-k-1}\boldsymbol{R}_{1i},$$

where $\boldsymbol{R}_{1i}$ is the position vector of M_{1i} with respect to the center of mass of the system and $R_{1i} = |\boldsymbol{R}_{1i}|$. Obviously if the condition in Eq. (1.348) is satisfied, $\boldsymbol{F}_i$ and $\boldsymbol{f}_{1i}$ will possess the same form, implying the conclusion mentioned above.

Chapter 2

Motion of small bodies in the planetary system

One of the most important issues in celestial mechanics is the motion of celestial bodies in the Solar System. The stability, the origin and evolution of the system is the "holy grail" for researchers in the related area. The planets are in the center of interests in these studies. But the small celestial bodies may also bear very important information about the Solar System. Their dynamical behaviour, orbital stability, space distribution, etc. are now the main problems in the dynamics of the Solar System and may serve as the key to our understanding of many problems in this planetary system. In this chapter, we will present some results of our investigations on the dynamics of small bodies including the main belt asteroids, Trojans, Kuiper belt objects, comets and planetary rings. Many times, the mapping method was applied in these (and also in many other) studies, so a brief review of this method will be given first.

2.1 Mapping method in Hamiltonian system

The long term behavior of celestial bodies is a very important topic. Traditionally, there are three methods for the study of long-term behavior: The qualitative method, the analytic method and the numerical method. All of these methods have some insufficiencies. The qualitative method cannot show the details of motion. For the analytical method, since the N-body (for $N \geq 3$) problem is nonintegrable, it is impossible to make a rigorous analytic study in such a system. On the other hand, if an expansion in series for the solution of the equations of motion is adopted, the problem of convergence arises. For the numerical method, the accumulation of numerical error makes the results for long time intervals unreliable.

Like the technique of *Poincaré Surface of Section* (see Sec. 3.3), the mapping on a specific surface can be used to study the motion in space. But generally such an analytical mapping cannot be obtained, if the solution of the equation of motion is not known. Hadjidemetriou proposed a method to construct the mappings for near-integrable Hamiltonian systems (Hadjidemetriou, 1991), and three criteria were raised to make sure that the constructed mapping is a good approximation of the original Hamiltonian system.

(i) The mapping must be area-preserving, as the Hamiltonian flow is volume-preserving and the mapping on the Poincaré surface of section is area-preserving.

(ii) The mapping should have the same fixed points as the Hamiltonian system.

(iii) The fixed points have the correct stability index, i.e. both of the mapping system and the actual system possess the same topology.

The procedure to construct an appropriate mapping for a Hamiltonian system includes mainly several steps as follows. First the averaging system is derived from the original one, then a generating function is constructed using the averaging system, and finally a mapping is produced through the canonical transformation defined by the generating function. Such a mapping generated in this way will satisfy the criteria listed above. Generally speaking, the mapping method has two main advantages. First, the theoretical analysis on a mapping is much easier than on the equations of motion, and second, the numerical computing of a mapping is much faster than the numerical integration of the differential equations. Of course, it is only an approximation of equations of motion.

In celestial mechanics and astrodynamics, the gravitational attractions between bodies are the main force, hence a dynamical system in this field in general is a conservative system. But in some special cases, there exist the dissipative forces. Thus in this section, the mapping method proposed by Hadjidemetriou will be generalized to include the dissipative system. Because the dissipative forces are generally smaller than the gravitational attractions, only the near-conservative system is discussed below.

2.1.1 *Hamiltonian equation of near-conservative system*

A near-conservative system has the following general form of equations of motion

$$\frac{\mathrm{d}J_i}{\mathrm{d}t} = -\frac{\partial H}{\partial \theta_i} + \mu F_i, \qquad \frac{\mathrm{d}\theta_i}{\mathrm{d}t} = \frac{\partial H}{\partial J_i} - \mu G_i, \qquad (i = 1, 2, \ldots, n), \qquad (2.1)$$

where $H = H_0 + \epsilon H_1$ is the Hamiltonian of the conservative part, J_i, θ_i are the action-angle variables with respect to the Hamiltonian H_0, $\mu F_i, -\mu G_i$ are the dissipative terms, and ϵ, μ are small parameters. In this section, only the situation where $\mu \sim O(\epsilon^k)$, $(k \geq 2)$ will be discussed.

Assume the Hamiltonian in an autonomous near-conservative system Eq. (2.1) of two degrees of freedom has the form

$$H = H_0(J_1, J_2) + \epsilon H_1(J_1, J_2, \theta_1, \theta_2). \qquad (2.2)$$

For the conservative case ($\mu = 0$), the Hamiltonian system corresponds to a perturbed twist mapping on a two-dimensional surface of section ($H = $ const., $\theta_2 = 0$). Hadjidemetriou defined on the surface a perturbed twist mapping through a canonical transformation

$$J_{1,n} = \frac{\partial W}{\partial \theta_{1,n}}, \qquad \theta_{1,n+1} = \frac{\partial W}{\partial J_{1,n+1}},$$

where W is a generating function

$$W = J_{1,n+1}\theta_{1,n} + S_0(J_{1,n+1}) + \epsilon S_1(J_{1,n+1}, \theta_{1,n}). \qquad (2.3)$$

The perturbed mapping obtained in this way reads

$$\begin{aligned}
J_{1,n+1} &= J_{1,n} - \epsilon\frac{\partial S_1}{\partial \theta_{1,n}}, \\[2ex]
\theta_{1,n+1} &= \theta_{1,n} + \frac{\partial S_0}{\partial J_{1,n+1}} + \epsilon\frac{\partial S_1}{\partial J_{1,n+1}}.
\end{aligned} \qquad (2.4)$$

The functions S_0, S_1 are determined suitably so that the mapping preserves the topology of the original system.

While for the near-conservative situation ($\mu \neq 0$), the equations of motion Eq. (2.1) are a perturbed extension of the corresponding conservative system. Without the energy conservation, the Hamiltonian system in this case corresponds to a three-dimensional mapping in the surface of section ($\theta_2 = \theta_{2,0}$). Such a three-dimensional perturbed extension of the mapping Eq. (2.4) reads,

$$J_{1,n+1} = J_{1,n} - \epsilon\frac{\partial S_1}{\partial \theta_{1,n}} + \mu F_1',$$

$$\theta_{1,n+1} = \theta_{1,n} + \frac{\partial S_0}{\partial J_{1,n+1}} + \epsilon\frac{\partial S_1}{\partial J_{1,n+1}} - \mu G_1', \tag{2.5}$$

$$J_{2,n+1} = J_{2,n} + \mu F_2'.$$

In this mapping, F_1', G_1', F_2' are functions to be determined. Now the main problem is to find suitable functions so that the mapping model keeps all the main features of the actual system.

According to the construction method of mapping mentioned above, one should get the corresponding averaging system first.

Suppose the Hamiltonian in Eq. (2.1) near a resonance has the following form

$$H = H_0(J_1, J_2) + \epsilon H_{1,\text{res}}(J_1, J_2, r\theta_1 - s\theta_2) + \epsilon H_{1,\text{non}}(J_1, J_2, \theta_1, \theta_2), \tag{2.6}$$

where r, s are integers and $\frac{\partial H_0}{\partial J_1}/\frac{\partial H_0}{\partial J_2} = n_1/n_2 = s/r$, $H_{1,\text{res}}$ stands for the resonant part (including the secular part for simplicity) of H_1, while $H_{1,\text{non}}$ is the nonresonant part. The dissipative terms in Eq. (2.1) will be divided into two parts also:

$$F_i(J, \theta) = F_{i,\text{res}}(J_1, J_2, r\theta_1 - s\theta_2) + F_{i,\text{non}}(J_1, J_2, \theta_1, \theta_2),$$

$$G_i(J, \theta) = G_{i,\text{res}}(J_1, J_2, r\theta_1 - s\theta_2) + G_{i,\text{non}}(J_1, J_2, \theta_1, \theta_2). \tag{2.7}$$

Near the s/r resonance, a canonical transformation of variables from J, θ to $\hat{J}, \hat{\theta}$ through a generating function $W = (\theta_1 - \frac{s}{r}\theta_2)\hat{J}_1 + \theta_2\hat{J}_2$ is defined as follow,

$$\hat{J}_1 = J_1, \qquad\qquad \hat{J}_2 = -\frac{s}{r}J_1 + J_2,$$

$$\hat{\theta}_1 = \theta_1 - \frac{s}{r}\theta_2, \quad \hat{\theta}_2 = \theta_2, \tag{2.8}$$

where $\hat{\theta}_1$ and $\hat{\theta}_2$ are the slow and the fast angles respectively. The new equations of motion have the form

$$\frac{\mathrm{d}\hat{J}_i}{\mathrm{d}t} = -\frac{\partial \hat{H}}{\partial \hat{\theta}_i} + \mu\tilde{F}_i,$$

$$\qquad\qquad (i = 1, 2), \tag{2.9}$$

$$\frac{\mathrm{d}\hat{\theta}_i}{\mathrm{d}t} = \frac{\partial \hat{H}}{\partial \hat{J}_i} - \mu\tilde{G}_i,$$

where $\hat{H} = H(\hat{J}_1, \hat{J}_2 - \frac{s}{r}\hat{J}_1, \hat{\theta}_1 + \frac{s}{r}\hat{\theta}_2, \hat{\theta}_2)$ and

$$\tilde{F}_1 = \hat{F}_1, \quad \tilde{F}_2 = \frac{s}{r}\hat{F}_1 + \hat{F}_2, \quad \tilde{G}_1 = \hat{G}_1 - \frac{s}{r}\hat{G}_2, \quad \tilde{G}_2 = \hat{G}_2 \qquad (2.10)$$

with $\hat{F}_1 = F_1(\hat{J}_1, \hat{J}_2 - \frac{s}{r}\hat{J}_1, \hat{\theta}_1 + \frac{s}{r}\hat{\theta}_2, \hat{\theta}_2)$, etc.

2.1.2 *Averaging system and fixed points*

To get the averaging system corresponding to Eq. (2.9), one may first calculate the average of the Hamiltonian part. With a suitably defined function S, one can obtain a generating function $W = \bar{J}_1\hat{\theta}_1 + \bar{J}_2\hat{\theta}_2 + \epsilon S(\bar{J}_1, \bar{J}_2, \hat{\theta}_1, \hat{\theta}_2)$ so that the canonical transformation reads

$$\begin{aligned}
\hat{J}_i &= \frac{\partial W}{\partial \hat{\theta}_i} = \bar{J}_i + \epsilon\frac{\partial S}{\partial \hat{\theta}_i}, \\
\bar{\theta}_i &= \frac{\partial W}{\partial \bar{J}_i} = \hat{\theta}_i + \epsilon\frac{\partial S}{\partial \bar{J}_i},
\end{aligned} \qquad (i = 1, 2),$$

i.e.

$$\begin{aligned}
\hat{J}_i &= \bar{J}_i + \epsilon\frac{\partial S}{\partial \hat{\theta}_i}, \\
\hat{\theta}_i &= \bar{\theta}_i - \epsilon\frac{\partial S}{\partial \bar{J}_i}.
\end{aligned} \qquad (i = 1, 2), \qquad (2.11)$$

Such a canonical transformation changes the Hamiltonian into a new one $\bar{H} = \hat{H}_0(\bar{J}) + \epsilon\hat{H}_{\text{res}}(\bar{J}_1, \bar{J}_2, \bar{\theta}_1)$, which to the order of $O(\epsilon)$, does not contain the fast angle $\bar{\theta}_2$. Meanwhile, $\bar{\theta}_2$ still appear in $\tilde{F}_i, \tilde{G}_i$ $(i = 1, 2)$ after the transformation. A further transformation can be made as follows,

$$\begin{aligned}
\bar{J}_i &= \tilde{J}_i + \mu f_i(\tilde{J}_1, \tilde{J}_2, \tilde{\theta}_1, \tilde{\theta}_2), \\
\bar{\theta}_i &= \tilde{\theta}_i + \mu g_i(\tilde{J}_1, \tilde{J}_2, \tilde{\theta}_1, \tilde{\theta}_2),
\end{aligned} \qquad (i = 1, 2). \qquad (2.12)$$

with functions f_i, g_i to be determined. This is a noncanonical transformation, and to the order of $O(\mu)$ the equations of motion after this transformation become

$$\frac{d\tilde{J}_1}{dt} = -\frac{\partial\bar{H}(\tilde{J}, \tilde{\theta})}{\partial\tilde{\theta}_1} + \mu\tilde{F}_{1,\text{res}} + \left[-\mu\frac{\partial\hat{H}_0(\tilde{J})}{\partial\tilde{J}_2}\frac{\partial f_1}{\partial\tilde{\theta}_2} + \mu\tilde{F}_{1,\text{non}}\right],$$

$$\frac{d\tilde{\theta}_1}{dt} = \frac{\partial\bar{H}(\tilde{J}, \tilde{\theta})}{\partial\tilde{J}_1} - \mu\tilde{G}_{1,\text{res}} + \left[-\mu\frac{\partial\hat{H}_0(\tilde{J})}{\partial\tilde{J}_2}\frac{\partial g_1}{\partial\tilde{\theta}_2} - \mu\tilde{G}_{1,\text{non}}\right],$$

$$\frac{\mathrm{d}\tilde{J}_2}{\mathrm{d}t} = \mu\tilde{F}_{2,\mathrm{res}} + \left[-\mu\frac{\partial\hat{H}_0(\tilde{J})}{\partial\tilde{J}_2}\frac{\partial f_2}{\partial\tilde{\theta}_2} + \mu\tilde{F}_{2,\mathrm{non}} \right],$$

$$\frac{\mathrm{d}\tilde{\theta}_2}{\mathrm{d}t} = \frac{\partial\bar{H}(\tilde{J},\tilde{\theta})}{\partial\tilde{J}_2} - \mu\tilde{G}_{2,\mathrm{res}} + \left[-\mu\frac{\partial\hat{H}_0(\tilde{J})}{\partial\tilde{J}_2}\frac{\partial g_2}{\partial\tilde{\theta}_2} - \mu\tilde{G}_{2,\mathrm{non}} \right],$$

$$(2.13)$$

where $\bar{H} = \hat{H}_0(\tilde{J}) + \epsilon\hat{H}_{1,\mathrm{res}}(\tilde{J}_1, \tilde{J}_2, \tilde{\theta}_1)$. The functions f_i, g_i $(i = 1, 2)$ are so determined that the terms in the square brackets of Eq. (2.13) can be eliminated. Since $\frac{\partial\hat{H}_0(\tilde{J})}{\partial\tilde{J}_2} \approx n_2 \neq 0$, many f_i, g_i can be chosen. Thus the right hand side of the equations of motion Eq. (2.13) will not contain the fast angle $\tilde{\theta}_2$ and the equations become

$$\frac{\mathrm{d}\tilde{J}_1}{\mathrm{d}t} = -\frac{\partial\bar{H}}{\partial\tilde{\theta}_1} + \mu\bar{F}_1, \qquad \frac{\mathrm{d}\tilde{\theta}_1}{\mathrm{d}t} = \frac{\partial\bar{H}}{\partial\tilde{J}_1} - \mu\bar{G}_1,$$

$$\frac{\mathrm{d}\tilde{J}_2}{\mathrm{d}t} = \mu\bar{F}_2, \qquad \frac{\mathrm{d}\tilde{\theta}_2}{\mathrm{d}t} = \frac{\partial\bar{H}}{\partial\tilde{J}_2} - \mu\bar{G}_2.$$

$$(2.14)$$

In above equations, for simplicity $\tilde{F}_{i,\mathrm{res}}$ and $\tilde{G}_{i,\mathrm{res}}$ were denoted by $\bar{F}_i$ and $\bar{G}_i$ $(i = 1, 2)$, and the tildes over the averaged variables $\tilde{J}, \tilde{\theta}$ will be omitted below throughout this section.

To construct a mapping, the topology of the dynamical system should be clarified at first. The phase flow defined by the set of equations Eq. (2.14) is no longer volume-preserving, and the divergence of the phase flow is $A(J, \theta)$, the Jacobian of the differential equations. The instantaneous volume-changing rate is then $\chi = \mathrm{Trace}A(J, \theta)$. For the system described by Eq. (2.14), the Jacobian reads

$$A = \begin{pmatrix} -\alpha_{21} + \beta_{11} & -\alpha_{22} + \beta_{12} & -\alpha_{23} + \beta_{13} & 0 \\ \alpha_{11} - \gamma_{11} & \alpha_{12} - \gamma_{12} & \alpha_{13} - \gamma_{13} & 0 \\ \beta_{21} & \beta_{22} & \beta_{23} & 0 \\ \alpha_{31} - \gamma_{21} & \alpha_{32} - \gamma_{22} & \alpha_{33} - \gamma_{23} & 0 \end{pmatrix}, \qquad (2.15)$$

where

$$\alpha_{ij} = \frac{\partial^2\bar{H}}{\partial\sigma_i\partial\sigma_j}, \quad \beta_{ij} = \mu\frac{\partial\bar{F}_i}{\partial\sigma_j}, \quad \gamma_{ij} = \mu\frac{\partial\bar{G}_i}{\partial\sigma_j}, \quad (i = 1, 2; \; j = 1, 2, 3, 4),$$

$$\boldsymbol{\sigma} = (J_1, \theta_1, J_2, \theta_2)^T = (\sigma_1, \sigma_2, \sigma_3, \sigma_4)^T.$$

$$(2.16)$$

Therefore,

$$\chi(J, \theta) = \beta_{11}(J, \theta) - \gamma_{12}(J, \theta) + \beta_{23}(J, \theta). \qquad (2.17)$$

The fixed points of Eq. (2.14) in the surface of section $\theta_2 = \theta_{2,0}$ can be obtained by setting $\frac{\mathrm{d}J_1}{\mathrm{d}t} = 0, \frac{\mathrm{d}\theta_1}{\mathrm{d}t} = 0, \frac{\mathrm{d}J_2}{\mathrm{d}t} = 0$. The corresponding solution of Eq. (2.14) is,

$$J_1 = J_{1,0}, \quad \theta_1 = \theta_{1,0}, \quad J_2 = J_{2,0}, \quad \frac{\mathrm{d}\theta_2}{\mathrm{d}t} = \tilde{n}_{2,0} = \text{const.}, \tag{2.18}$$

where

$$\frac{\mathrm{d}\theta_2}{\mathrm{d}t} = \frac{\partial \bar{H}(J_{1,0}, J_{2,0}, \theta_{1,0})}{\partial J_2} - \mu \bar{G}_2(J_{1,0}, J_{2,0}, \theta_{1,0}). \tag{2.19}$$

Converting the averaged variables to the original ones through the transformations Eqs. (2.8, 2.11, 2.12), one may see that these fixed points are the r-multiple periodic solutions of the original system Eq. (2.1) at the s/r resonance with a period

$$T = \frac{2\pi}{\tilde{n}_{2,0}} r. \tag{2.20}$$

The variational equations of Eq. (2.14) can be obtained, with the Jacobian Eq. (2.15) being evaluated at the periodic solution Eq. (2.18). Then the stability of the T-periodic solution can be determined through the corresponding monodromy matrix defined as $M = \mathrm{e}^{AT}$. As $\alpha_{ij} \sim O(\epsilon), \beta_{ij}, r_{ij} \sim O(\mu) \sim O(\epsilon^k) \quad (k \geq 2)$, one has, to $O(\mu)$,

$$M = \begin{pmatrix} 1 + (-\alpha_{21} + \beta_{11})T + c_{11} & (-\alpha_{22} + \beta_{12})T & (-\alpha_{23} + \beta_{13})T + c_{13} & 0 \\ (\alpha_{11} - \gamma_{11})T & 1 + (\alpha_{12} - \gamma_{12})T + c_{22} & (\alpha_{13} - \gamma_{13})T + c_{23} & 0 \\ \beta_{21}T & \beta_{22}T & 1 + \beta_{23}T & 0 \\ (\alpha_{31} - \gamma_{21}) + c_{41} & (\alpha_{32} - \gamma_{22})T + c_{42} & (\alpha_{33} - \gamma_{23})T & 1 \end{pmatrix}. \tag{2.21}$$

In the above matrix, $c_{11} = c_{22} = (\alpha_{21}^2 - \alpha_{11}\alpha_{22})T^2/2 + o(\epsilon^2)$, $c_{13} = c_{42} = (\alpha_{21}\alpha_{23} - \alpha_{22}\alpha_{13})T^2/2 + o(\epsilon^2)$, $c_{23} = -c_{41} = (\alpha_{12}\alpha_{13} - \alpha_{11}\alpha_{23})T^2/2 + o(\epsilon^2)$. The three nontrivial eigenvalues of M are, to $O(\mu)$,

$$\lambda_{1,2} = 1 + \frac{1}{2}(\alpha + \beta - \nu) \pm \mathrm{i}\sqrt{-\alpha(1 + \frac{\alpha}{4})}, \quad \lambda_3 = 1 + \nu, \quad (\mathrm{i} = \sqrt{-1}), \tag{2.22}$$

where

$$\alpha = (\alpha_{12}^2 - \alpha_{11}\alpha_{22})T^2 + o(\epsilon^2),$$

$$\beta = (\beta_{11} - \gamma_{12} + \beta_{23})T,$$

$$\nu = \left[\frac{\beta_{21}(\alpha_{12}\alpha_{23} - \alpha_{13}\alpha_{22}) + \beta_{22}(\alpha_{11}\alpha_{23} - \alpha_{13}\alpha_{21})}{\alpha_{12}^2 - \alpha_{11}\alpha_{22}} + \beta_{23} \right] T,$$

$$(2.23)$$

and surely it is assumed that $\alpha_{12}^2 - \alpha_{11}\alpha_{22} \neq 0$.

2.1.3   *Construction of mapping*

Below the mapping like Eq. (2.5) but for the averaged system Eq. (2.14) will be derived. Compared with the averaged system, the mapping has: (a) Approximately, the same volume-changing rate in the phase space; (b) the same fixed points; and (c) the same stabilities of the fixed points.

According to Hadjidemetriou (1991), the averaged Hamiltonian is used to produce a generating function $S = J_{1,n+1}\theta_{1,n} + \tau\bar{H}(J_{n+1}, \theta_n)$. For the near-conservative system, from Eq. (2.5) a mapping is obtained:

$$J_{1,n+1} = J_{1,n} - \tau\frac{\partial\bar{H}}{\partial\theta_{1,n}}(J_{n+1}, \theta_n) + \tau\mu\bar{F}_1(J_{n+1}, \theta_n),$$

$$\theta_{1,n+1} = \theta_{1,n} + \tau\frac{\partial\bar{H}}{\partial J_{1,n+1}}(J_{n+1}, \theta_n) - \tau\mu\bar{G}_1(J_{n+1}, \theta_n), \qquad (2.24)$$

$$J_{2,n+1} = J_{2,n} + \tau\mu\bar{F}_2(J_{n+1}, \theta_n),$$

with $\bar{F}_1(J_{n+1}, \theta_n), \bar{G}_1(J_{n+1}, \theta_n)$ and $\bar{F}_2(J_{n+1}, \theta_n)$ stand for F_1', G_1', F_2' in the mapping in Eq. (2.5), respectively. To $O(\mu)$, these variables can be replaced by $\bar{F}_1(J_n, \theta_n), \bar{G}_1(J_n, \theta_n)$ and $\bar{F}_2(J_n, \theta_n)$ in the mapping Eq. (2.24).

Suppose a linear system $\mathrm{d}\boldsymbol{X}/\mathrm{d}t = \boldsymbol{A}\boldsymbol{X}$, where $\boldsymbol{X}$ is an n-vector, $\boldsymbol{A}$ is an $n \times n$ constant matrix, and its eigenvalues are λ_i ($i = 1, 2, \ldots, n$). The corresponding discretized mapping of the system with stepsize T is $\boldsymbol{X}_{n+1} = \boldsymbol{X}_n e^{\boldsymbol{A}T}$, and the eigenvalues of the Jacobian of the mapping, i.e. $e^{\boldsymbol{A}T} = \boldsymbol{A}'$, are $\mu_i = e^{\lambda_i T}$. Thus, $\mathrm{Trace}\boldsymbol{A}' = \sum_{i=1}^{n}\lambda_i = \frac{1}{T}\ln\Pi_{i=1}^{n}\mu_i = \frac{1}{T}\ln|\det\boldsymbol{A}'|$. Based on this fact, the instantaneous volume-changing rate for a mapping with a stepsize of τ as in (Lichtenberg & Lieberman, 1983) is

$$\chi(J_n, \theta_n) = \frac{1}{T}\ln|\det\boldsymbol{A}'(J_n, \theta_n)|, \qquad (2.25)$$

where $A'(J_n, \theta_n)$ is the Jacobian of the mapping. For the mapping Eq. (2.24), its Jacobian, to $O(\mu)$, is

$$A' = \begin{pmatrix} 1 + (-\alpha_{21} + \beta_{11})\tau + d_{11} & (-\alpha_{22} + \beta_{12})\tau + d_{12} & (-\alpha_{23} + \beta_{13})\tau + d_{13} \\ (\alpha_{11} - \gamma_{11})\tau + d_{21} & 1 + (\alpha_{12} - \gamma_{12})\tau + d_{22} & (\alpha_{13} - \gamma_{13})\tau + d_{23} \\ \beta_{21}\tau & \beta_{22}\tau & 1 + \beta_{23}\tau \end{pmatrix}$$

$$(2.26)$$

where

$$d_{11} = \alpha_{21}^2 \tau^2 + o(\epsilon^2), \qquad d_{12} = \alpha_{21}\alpha_{22}\tau^2 + o(\epsilon^2),$$

$$d_{13} = \alpha_{21}\alpha_{23}\tau^2 + o(\epsilon^2), \qquad d_{21} = -\alpha_{11}\alpha_{21}\tau^2 + o(\epsilon^2),$$

$$d_{22} = -\alpha_{11}\alpha_{22}\tau^2 + o(\epsilon^2), \quad d_{23} = -\alpha_{11}\alpha_{23}\tau^2 + o(\epsilon^2),$$

and $\alpha_{ij}, \beta_{ij}, \gamma_{ij}$ $(i,j = 1,2,3)$ are defined as in Eq. (2.16). Therefore, the volume-changing rate is

$$\chi(J_n, \theta_n) = \beta_{11}(J_n, \theta_n) - \gamma_{12}(J_n, \theta_n) + \beta_{23}(J_n, \theta_n) + O(\mu). \qquad (2.27)$$

It coincides, to $O(\mu)$, with that of the actual system, as one can see in Eq. (2.17).

The fixed points of mapping Eq. (2.24) are $J_{1,n+1} = J_{1,n} = J_{1,0}$, $\theta_{1,n+1} = \theta_{1,n} = \theta_{1,0}$, $J_{2,n+1} = J_{2,n} = J_{2,0}$, and from Eq. (2.24) one can see that they satisfy the same equations as those of system Eq. (2.14). Linearizing the mapping near the fixed points, the Jacobian of the variational equations is as Eq. (2.26), with $\alpha_{ij}, \beta_{ij}, \gamma_{ij}$ $(i,j = 1,2,3)$ being evaluated at the fixed points. It can be shown that the eigenvalues of A' have, to $O(\mu)$, the same expressions as in Eq. (2.22), with T being replaced by τ in Eq. (2.23). Thus, if $\tau = T$ in Eq. (2.24), the fixed points will have the same stability with those of the averaged system. Particularly, for the case of $\mu \sim O(\epsilon^2)$, this stability coincidence is hold not only qualitatively, but also quantitatively. As a consequence, the perturbed mapping Eq. (2.24) is such a proper mapping model that has all the required properties.

It is clear that the method introduced above can be generalized to the cases of three or more degrees of freedom, or to the cases where the Hamiltonian and the dissipative terms are time-dependent. For the latter ones, the time-dependent terms in fact can be eliminated by the method of averaging.

There is a theorem on the relation between the mapping Eq. (2.24) and the differential equations, as follows.

Theorem. *The mapping Eq. (2.24) exactly describes the time-τ evolution of a near-conservative system*

$$\frac{\mathrm{d}J_1}{\mathrm{d}t} = -\frac{\partial \tilde{H}}{\partial \theta_1} + \mu \tilde{F}_1,$$

$$\frac{\mathrm{d}\theta_1}{\mathrm{d}t} = \frac{\partial \tilde{H}}{\partial J_1} - \mu \tilde{G}_1, \tag{2.28}$$

$$\frac{\mathrm{d}J_2}{\mathrm{d}t} = \mu \tilde{F}_2,$$

with $\tilde{H}, \tilde{F}_1, \tilde{G}_1, \tilde{F}_2$ having the expressions of power series in τ, as

$$\tilde{H} = \tilde{H}^{(1)} + \tau \tilde{H}^{(2)} + \frac{\tau^2}{2} \tilde{H}^{(3)} + \cdots,$$

$$\tilde{G}_1 = \tilde{G}_1^{(1)} + \tau \tilde{G}_1^{(2)} + \frac{\tau^2}{2} \tilde{G}_1^{(3)} + \cdots, \tag{2.29}$$

$$\tilde{F}_i = \tilde{F}_i^{(1)} + \tau \tilde{F}_i^{(2)} + \frac{\tau^2}{2} \tilde{F}_i^{(3)} + \cdots,$$

where

$$\tilde{H}^{(1)} = \bar{H}, \quad \tilde{F}_1^{(1)} = \bar{F}_1, \quad \tilde{G}_1^{(1)} = \bar{G}_1, \quad \tilde{F}_2^{(1)} = \bar{F}_2,$$

$$\tilde{H}^{(2)} = -\frac{1}{2} \bar{H}_{J_1} \bar{H}_{\theta_1},$$

$$\tilde{F}_1^{(2)} = \frac{1}{2} \left(-\bar{H}_{\theta_1 J_1} \bar{F}_1 - \bar{H}_{\theta_1 \theta_1} \bar{G}_1 - \bar{H}_{J_1} \bar{F}_{1\theta_1} - \bar{H}_{\theta_1} \bar{F}_{1 J_1} \right.$$

$$\left. - \bar{H}_{\theta_1 J_1} \bar{F}_2 + \mu \bar{G}_1 \bar{F}_{1\theta_1} - \mu \bar{F}_1 \bar{F}_{1 J_1} - \mu \bar{F}_2 \bar{F}_{1 J_2} \right),$$

$$\tilde{G}_1^{(2)} = \frac{1}{2} \left(-\bar{H}_{J_1 J_1} \bar{F}_1 - \bar{H}_{J_1 \theta_1} \bar{G}_1 - \bar{H}_{J_1} \bar{G}_{1\theta_1} + \bar{H}_{\theta_1} \bar{G}_{1 J_1} \right. \tag{2.30}$$

$$\left. - \bar{H}_{J_1 J_2} \bar{F}_2 + \mu \bar{G}_1 \bar{G}_{1\theta_1} - \mu \bar{F}_1 \bar{G}_{1 J_1} - \mu \bar{F}_2 \bar{G}_{1 J_2} \right),$$

$$\tilde{F}_2^{(2)} = \frac{1}{2} \left(-\bar{H}_{J_1} \bar{F}_{2\theta_1} + \bar{H}_{\theta_1} \bar{F}_{2 J_1} + \mu \bar{G}_1 \bar{F}_{2\theta_1} \right.$$

$$\left. - \mu \bar{F}_1 \bar{F}_{2 J_1} - \mu \bar{F}_2 \bar{F}_{2 J_2} \right),$$

$$\ldots \ldots,$$

and the variables in the subscripts indicate the partial derivatives, e.g. $\bar{H}_{J_1} = \frac{\partial \bar{H}}{\partial J_1}$.

Proof. Define a linear differential operator $\Lambda_{\tilde{H},(\tilde{F},\tilde{G})}$ in the n-dimensional phase space $(\boldsymbol{J}, \boldsymbol{\theta})$ as,

$$\Lambda_{\tilde{H},(\tilde{F},\tilde{G})} = \sum_{i=1}^{n} \left[\left(\frac{\partial \tilde{H}}{\partial J_i} - \mu \tilde{G}_i \right) \frac{\partial}{\partial \theta_i} - \left(\frac{\partial \tilde{H}}{\partial \theta_i} - \mu \tilde{F}_i \right) \frac{\partial}{\partial J_i} \right], \qquad (2.31)$$

so that the system Eq. (2.28) can be written as

$$\frac{\mathrm{d}Z_i}{\mathrm{d}t} = \Lambda_{\tilde{H},(\tilde{F},\tilde{G})} Z_i,$$

with $\boldsymbol{Z} = (J_1, \theta_1, J_2)$. Therefore the exact time evolution of $\boldsymbol{Z}(t)$ from $t = 0$ to $t = \tau$ is given by

$$\boldsymbol{Z}(\tau) = \exp\left(\tau \Lambda_{\tilde{H},(\tilde{F},\tilde{G})} \right) \boldsymbol{Z}(0). \qquad (2.32)$$

Suppose that there exist $\tilde{H}, \tilde{F}_1, \tilde{G}_1, \tilde{F}_2$ so that the evolution Eq. (2.32) is just the mapping Eq. (2.24), with τ in Eq. (2.32) given by Eq. (2.20). For any analytical function $f(\boldsymbol{J}, \boldsymbol{\theta})$, one has

$$f(\boldsymbol{J}_{n+1}, \boldsymbol{\theta}_{n+1}) = \exp\left(\tau \Lambda_{\tilde{H},(\tilde{F},\tilde{G})} \right) f(\boldsymbol{J}_n, \boldsymbol{\theta}_n), \qquad (2.33)$$

for the evolution along Eq. (2.32). With the help of the mapping model Eq. (2.24), one can expand $f(\boldsymbol{J}_{n+1}, \boldsymbol{\theta}_{n+1})$ into a Taylor series in the neighborhood of $(\boldsymbol{J}_n, \boldsymbol{\theta}_n)$, while the right-hand side of Eq. (2.33) can be expanded into power series of τ. Then by comparing terms with the same power of τ in both sides of Eq. (2.33), one obtains Eq. (2.29). This proves the theorem.

$\square$

From the theorem, one can see that the mapping constructed is, in a sense, "close" to the averaged system Eq. (2.14). If $\tau = 0$, the mapping coincides with the averaged system.

For the case of $\mu = 0$, a corollary is obtained from the theorem.

Corollary. *The mapping studied in (Hadjidemetriou, 1991) exactly describes the time-τ evolution of an associated Hamiltonian system $\tilde{H}$ given by Eq. 2.29, which is close to the averaged Hamiltonian system $\bar{H}$. In particular, the mapping preserves $\tilde{H}$ exactly.*

In our own research, we have applied the above mapping method to the motion near the 3:1 mean motion resonance with the presence of Poynting–Robertson drag in the Solar System. The results showed that the mapping method is valid for very small drags, i.e. $\mu \sim O(\epsilon^k)$ $(k \geq 2)$. Thus it is suitable to apply this mapping method to the study of evolution of small dust grains with drags in the Solar System, such as the particles in the rings of planets.

2.2 Structure of phase space near Lagrange solutions

With the mapping method, this section presents the investigation on the dynamics of asteroids in the vicinity of the triangular Lagrange points in a circular restricted 3-body problem.

Five equilibrium solutions in the circular restricted 3-body problem have been discussed in Secs. 1.4 and 1.5. The triangular equilibrium solutions L_4, L_5 are stable for the Sun-planet-asteroid system for all planets in the Solar System. Meanwhile, except for Mercury, all planets in the Solar System evolve around the Sun on near circular orbits. Thus the circular restricted 3-body problem serves as a good dynamical model for studying the motion of asteroids moving around the vicinity of these fixed points. In fact, asteroids have been found around planets' L_4 and L_5 points, first around Jupiter, later Mars and Neptune, and recently the Earth (see for example a list of Trojans in the International Astronomical Union Minor Planet Center webpage http://www.minorplanetcenter.net/iau/lists/Trojans.html). These asteroids are named "Trojan asteroids" or just "Trojans". Below in this section, the topology of the phase space of the asteroids' motion near the L_4, L_5 points will be studied using a mapping method. Such investigations help people understand the orbital stability of Trojans.

2.2.1 *Hamiltonian equations*

Assume that a planet orbits the Sun on a circular orbit and a massless test particle moves in the plane defined by the planetary orbit under the influence of gravitations from the Sun and the planet. As usual, to simplify the equations of motion, a set of units are defined such that:

$$M_\odot + m = 1, \quad a_{S-P} = 1, \quad n = \left[G(M_\odot + m)/a_{S-P}^3\right]^{1/2} = 1, \quad (2.34)$$

where $M_\odot$ and m are masses of the Sun and the planet, a_{S-P} and n are the semi-major axis and mean motion of the planet–Sun orbit, and G is the gravitational constant. Obviously in such units $G = 1$. Denoting $\mu = \frac{m}{M_\odot + m}$, then the masses of the planet and the Sun are μ and $\mu' = 1 - \mu$, respectively. In a coordinate system centered on the barycenter and rotating with an angular velocity $n = 1$, the planet and the Sun are stationary points. Without any loss of generality, the abscissa is assumed to be the line pointing from the Sun to the planet, and their positions are $(-\mu, 0)$ and $(\mu', 0)$.

In this coordinate system, the Hamiltonian function describing the motion of the test particle is

$$H = H_0 + \mu H_1, \tag{2.35}$$

where

$$H_0 = \frac{1}{2}\left(P_r^2 + \frac{P_\psi^2}{r^2}\right) - \frac{\mu'}{r} - P_\phi, \tag{2.36}$$

$$\mu H_1 = \frac{\mu'}{r} - \frac{\mu'}{r_1} - \frac{\mu}{r_2}$$

$$= \frac{\mu'}{r} - \frac{\mu'}{\sqrt{r^2 + \mu^2 + 2\mu r \cos\phi}} - \frac{\mu}{\sqrt{r^2 + \mu'^2 + 2\mu'r \cos\phi}} \tag{2.37}$$

$$= \mu\left(\frac{\cos\phi}{r^2} - \frac{1}{\sqrt{r^2 + 1 - 2r \cos\phi}}\right) + O(\mu^2).$$

In above equations, r, r_1 and r_2 are the distances from the test particle to the barycenter (the origin of the coordinate system), to the Sun and to the planet, respectively. ϕ is the azimuth angle of the test particle, and $\phi = \omega + f - nt$ with ω and f representing the pericenter argument and the true anomaly of the test particle. Therefore, r, ϕ are the coordinates and P_r, P_ϕ are the conjugate momentums. H_0 is the unperturbed Hamiltonian describing the Keplerian motion of the test particle moving around the barycenter and μH_1 is regarded as the perturbation with μ being a small parameter (the biggest μ in our Solar System is of the order 10^{-3} for planet Jupiter).

As mentioned above, instead of solving the Hamiltonian equations directly, a mapping will be constructed to study the motion. For this sake, the action-angle variables $(I_r, I_\phi, \theta_r, \theta_\phi)$ of the unperturbed Hamiltonian are introduced first. Denoting the separable Hamiltonian characteristic function of the unperturbed Hamiltonian by $W(r, \phi, P_r, P_\phi) = W_1(r, P_r, P_\phi) + W_2(\phi, P_r, P_\phi)$, one has

$$\frac{1}{2}\left(\frac{\partial W}{\partial r}\right)^2 + \frac{1}{2r^2}\left(\frac{\partial W}{\partial \phi}\right)^2 - \frac{\partial W}{\partial \phi} - \frac{\mu'}{r} = h,$$

$$\frac{dW_2}{d\phi} = c_1, \tag{2.38}$$

$$\frac{dW_1}{dr} = \sqrt{2(c_1 + h) + \frac{2\mu'}{r} - \frac{c_1^2}{r^2}},$$

where h and c_1 are constants. Therefore,

$$W = c_1\phi + \int \sqrt{2(c_1 + h) + \frac{2\mu'}{r} - \frac{c_1^2}{r^2}}\,\mathrm{d}r, \tag{2.39}$$

and accordingly,

$$P_r = \frac{\partial W}{\partial r} = \sqrt{2(c_1 + h) + \frac{2\mu'}{r} - \frac{c_1^2}{r^2}},$$

$$P_\phi = \frac{\partial W}{\partial \phi} = c_1. \tag{2.40}$$

The action variables are defined as:

$$I_r = \frac{1}{2\pi} \oint P_r \mathrm{d}r = -c_1 + \frac{\mu'}{\sqrt{-2(c_1 + h)}},$$

$$I_\phi = \frac{1}{2\pi} \oint P_\phi \mathrm{d}\phi = c_1. \tag{2.41}$$

Substitute the c_1 and h in Eq. (2.39) by I_r, I_ϕ, one obtains

$$W(r, \phi, I_r, I_\phi) = \phi I_\phi + \sqrt{-I_\phi^2 + 2\mu'r - \frac{\mu'^2 r^2}{I_{\rm s}^2}}$$

$$- I_\phi \arcsin \frac{I_{\rm s}(\mu'r - I_\phi^2)}{\mu'r\sqrt{I_{\rm s}^2 - I_\phi^2}} - I_{\rm s} \arcsin \frac{I_{\rm s}^2 - \mu'r}{I_{\rm s}\sqrt{I_{\rm s}^2 - I_\phi^2}}, \tag{2.42}$$

where $I_{\rm s} = I_r + I_\phi$ has been used to abbreviate the formula. Finally, adopting $W(r, \phi, I_r, I_\phi)$ as the generating function, one obtains the canonical transformation from variables (P_r, P_ϕ, r, ϕ) to $(I_r, I_\phi, \theta_r, \theta_\phi)$ as:

$$I_r = -P_\phi + \frac{\mu'}{\sqrt{-P_r^2 + 2\mu'/r - P_\phi^2/r^2}},$$

$$I_\phi = P_\phi,$$

$$\theta_r = -\frac{1}{I_{\rm s}}\sqrt{-I_\phi^2 + 2\mu'r - \frac{\mu'^2 r^2}{I_{\rm s}^2}} - \arcsin\left[\frac{I_{\rm s}^2 - \mu'r}{I_{\rm s}\sqrt{I_{\rm s}^2 - I_\phi^2}}\right], \tag{2.43}$$

$$\theta_\phi = \phi + \theta_r - \arcsin\left[\frac{I_{\rm s}(\mu'r - I_\phi^2)}{\mu'r\sqrt{I_{\rm s}^2 - I_\phi^2}}\right].$$

The relation between (P_r, P_ϕ, r, ϕ) and the orbital elements of the test particle is well-known, through which the action-angle variables can be written as functions of orbital elements. They read

$$I_r = \sqrt{\mu' a}\left(1 - \sqrt{1 - e^2}\right),$$

$$I_\phi = \sqrt{\mu' a \left(1 - e^2\right)},$$

$$\theta_r - M \quad \frac{\pi}{2},$$

$$\theta_\phi = \phi + M - f = \omega + M - t,$$

(2.44)

where a, e, M and ω are the semi-major axis, eccentricity, mean anomaly and argument of pericenter, respectively. Particularly, the mean longitude for a planar orbit is $\lambda = \omega + M$. In the rotating coordinate system $\theta_\phi = \lambda - t$ is the difference between the mean longitudes of the test particle and the planet, because the planet is fixed at the abscissa and the system is rotating with an angular velocity $n = 1$. Thus, as the mean longitude difference between the test particle and the planet, θ_ϕ is in fact the resonant angle (or critical angle) of the 1:1 mean motion resonance. Inversely, the orbital elements a and e are given by

$$a = I_s^2/\mu',$$

$$e = \sqrt{1 - I_\phi^2/I_s^2}.$$

(2.45)

Finally, ignoring those terms of $O(\mu^2)$ or $O(\mu e)$, the new Hamiltonian written in action-angle variables reads

$$\bar{H} \approx -I_\phi - \frac{\mu'^2}{2I_s^2} + \mu \left[\frac{\cos\theta_\phi}{I_s^2} - \frac{1}{\sqrt{I_s^2 + 1 - 2I_s^2\cos\theta_\phi}} \right],$$

(2.46)

where the first two terms in the right-hand side comprise the unperturbed part $(\bar{H}_0)$ and the rest is the perturbation $(\mu\bar{H}_1)$. And the canonical equation of motion is

$$\frac{dI_r}{dt} = -\frac{\partial\bar{H}}{\partial\theta_r} = 0,$$

$$\frac{d\theta_r}{dt} = \frac{\partial\bar{H}}{\partial I_r} = \frac{\mu'^3}{I_s^4} - \frac{4\mu\cos\theta_\phi}{I_s^5} + \frac{2\mu(I_s^3 - I_s\cos\theta_\phi)}{\sqrt{(I_s^4 - 2I_s^2\cos\theta_\phi + 1)^3}},$$

$$\frac{dI_\phi}{dt} = -\frac{\partial\bar{H}}{\partial\theta_\phi} = \frac{\mu\sin\theta_\phi}{I_s^4} - \frac{\mu I_s^2\sin\theta_\phi}{\sqrt{(I_s^4 - 2I_s^2\cos\theta_\phi + 1)^3}},$$

$$\frac{d\theta_\phi}{dt} = \frac{\partial\bar{H}}{\partial I_\phi} = \frac{d\theta_r}{dt} - 1.$$

(2.47)

The motion of asteroids around the triangular equilibrium points is approximately described by these differential equations. Generally, it is not an easy job to solve the differential equations analytically, while the numerical method is expensive both in computing time and in computational machines. If the dynamics described by the differential equations can be transferred to a discrete mapping, the analysis and computation would be significantly simplified.

2.2.2 *Mapping model*

If a mapping is constructed to represent the principle dynamical property of a Hamiltonian system, the phase space of the mapping must have the same topological structure as the phase space of the Hamiltonian system. To determine whether a mapping is a good discretized approximation of the Hamiltonian system, three criteria and a systematic method of constructing such a mapping has been given in the previous section (Sec. 2.1).

The fixed points (equilibrium solutions) of the circular restricted 3-body problem has been introduced in previous section of this book (Sec. 1.4). They are three collinear fixed points ($L_{1,2,3}$) and two triangular points ($L_{4,5}$). Before constructing the mapping, first recover the fixed points from Eq. (2.47). Let $\dot{I}_\phi = \dot{\theta}_\phi = 0$, one gets

(i) $\sin\theta_\phi = 0$, $\cos\theta_\phi = c_0 = \pm 1$, $\mu'^2/I_{\mathrm{s}}^3 - 4\mu c_0/I_{\mathrm{s}}^5 + 2\mu I_{\mathrm{s}}/(I_{\mathrm{s}}^2 - c_0)^2 - 1 = 0$;
(ii) $\sin\theta_\phi \neq 0$, $\cos\theta_\phi = \mu'/2I_{\mathrm{s}}^2$, $I_{\mathrm{s}}^7 - (1 - \mu^2)I_{\mathrm{s}}^4 + 3\mu(1 - \mu)^3 = 0$.

The solution in case (i) is collinear points while case (ii) is for triangular points L_4, L_5. Note that the solution in case (ii) is not exactly the same as the triangular equilibrium solution of the circular restricted 3-body problem because the Hamiltonian is truncated in our model. For example, for a planet with mass $\mu = 0.01$, the solution found by numerical calculations is $I_{\mathrm{s}} = 0.98975$ (correspondingly $a = 0.99975$) and $\theta_\phi = 59.648°$, just a little different from the exact value ($I_{\mathrm{s}} = 1, \theta_\phi = 60°$). Because the collinear fixed points are hyperbolic, the structure of phase space in their vicinities is rather simple. Below in this section, only case (ii) will be studied.

For the Hamiltonian system of 2-degree-of-freedom, a Poincaré section can be made. On the three-dimensional hypersurface defined by a constant Hamiltonian, $\theta_r = c$ (a constant) is a Poincaré section, on which (I_ϕ, θ_ϕ) depict a section of the phase space. Here, following the procedure given in the previous section, a mapping $(I_\phi, \theta_\phi) \rightarrow (I'_\phi, \theta'_\phi)$ preserving the main features of the Hamiltonian system can be constructed through a generating function. The generating function is chosen as:

$$F = I'_\phi \theta_\phi + \tau[\bar{H}_0(I_r^0, I'_\phi) + \mu \bar{H}_1(I_r^0, I'_\phi, \theta_\phi)]. \tag{2.48}$$

In this generating function, $\tau = 2\pi/\dot{\theta}_r^0$, and $I_r^0, \dot{\theta}_r^0$ take the values at the fixed point. The mapping then is defined by

$$I_\phi = \frac{\partial F}{\partial \theta_\phi}, \quad \theta'_\phi = \frac{\partial F}{\partial I'_\phi}. \tag{2.49}$$

Denote $I'_s = I_r^0 + I'_\phi$, the mapping reads

$$I'_\phi = I_\phi + \tau \mu \sin \theta_\phi \left[\frac{1}{I'^4_s} - \frac{I'^2_s}{(I'^4_s - 2I'^2_s \cos \theta_\phi + 1)^{3/2}} \right],$$

$$\theta'_\phi = \theta_\phi + \tau \left[\frac{\mu'^2}{I'^3_s} - 1 - \frac{4\mu \cos \theta_\phi}{I'^5_s} + \frac{2\mu(I'^3_s - I'_s \cos \theta_\phi)}{(I'^4_s - 2I'^2_s \cos \theta_\phi + 1)^{3/2}} \right]. \tag{2.50}$$

At the triangular Lagrange points, $\dot{\theta}_r^0 = 1$ so that $\tau = 2\pi$.

2.2.3 *Structure of phase space*

Using the mapping Eq. (2.50), the surface of section can be computed. First of all, adopt $\mu = 9.53 \times 10^{-4}$, i.e. it is the Sun–Jupiter–Trojan system. Suppose that the test particles start from circular orbits with $e^0 = 0$, so that $I_r^0 = 0$. The phase space of the mapping is shown in Fig. 2.1. It is worth noting that only half of the phase space with $\theta_\phi \in [0, \pi]$ has been plotted, the other half with $\theta_\phi \in (\pi, 2\pi)$ is absolutely a symmetry of Fig. 2.1 with respect to the axis $\theta_\phi = \pi$. Surely this one represents the motion around the leading Lagrange point L_4, while the other one is for the trailing point L_5.

Apparently, the fixed point at $(I_\phi, \theta_\phi) \sim (1, \pi/3)$ is deeply embed in invariant curves, indicating the stability of this equilibrium point. The "island" of invariant curves extends from $\theta_\phi \sim 0.5$ to $\theta_\phi \sim 2.1$ in ordinate and from $I_\phi \sim 0.98$ to $I_\phi \sim 1.02$ in abscissa, implying the fact that the region is largely occupied by stable "tadpole orbits". The extent of such orbits, both in the resonant angle (mean longitude difference between the asteroid and Jupiter) and in the asteroid's semi-major axis, can be derived from the relation between the orbital elements and the action-angle variables Eq. (2.45). Meanwhile, within the island, secondary island chains can be seen, implying the complexity of motion around the triangular Lagrange points and the existence of plenty periodic orbits. Two groups of asteroids nicknamed Trojans and Greeks have been found around these points.

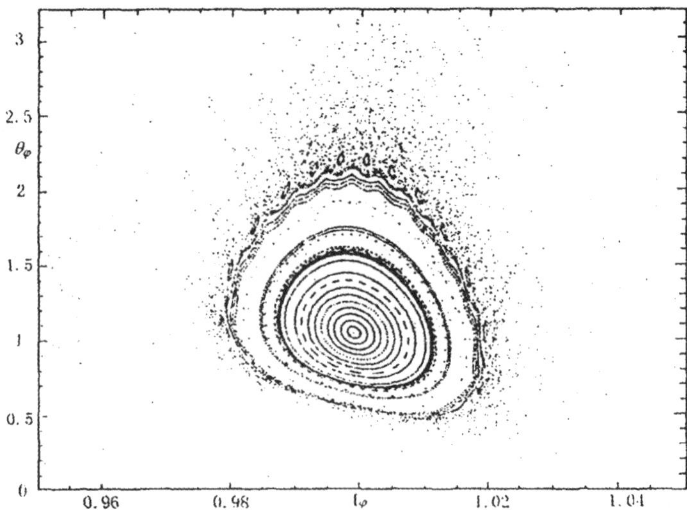

Fig. 2.1   Phase space of the mapping for $\mu = 9.53 \times 10^{-4}$, i.e. for the Jovian Trojans. Taken from Zhou, Sun & Zhou (2000).

Jupiter is the most massive planet in the Solar System and all other planets have masses smaller than it, that is, $\mu \leq 9.53 \times 10^{-4}$ in the Solar System. The linear stability analysis shows that when $\mu < \mu_1 = 0.03852$ the triangular Lagrange points are linearly stable, where μ_1 is the Routh critical value as mentioned above. And calculations show that the phase space around the L_4, L_5 points of other planets have nearly the same feature as in Fig. 2.1.

The application of KAM theorem has revealed two other critical μ values $\mu_2 = 0.02429$ and $\mu_3 = 0.01351$, both smaller than μ_1, at which the motion in the vicinities of L_4, L_5 points is in fact unstable (Siegel & Moser, 1971). Although μ_1, μ_2, μ_3 are pretty large for planets in our Solar System, such a mass ratio can be found in a satellite system, or probably in an extra-solar planetary system with a massive Jupiter-like planet and a brown dwarf host star. Therefore, it is reasonable to extend the investigation beyond $\mu = 9.53 \times 10^{-4}$. Some results are illustrated in Fig. 2.2.

The pictures in Fig. 2.2 show that as the value of μ approaches to μ_3, the structure of the phase space is nearly the same, with only little variation in the position of the fixed point and a continuous shrinking in the area surrounded by the invariant curves. Eventually, the island turns into a single point and the fixed point is no longer stable. A careful calculation gave the

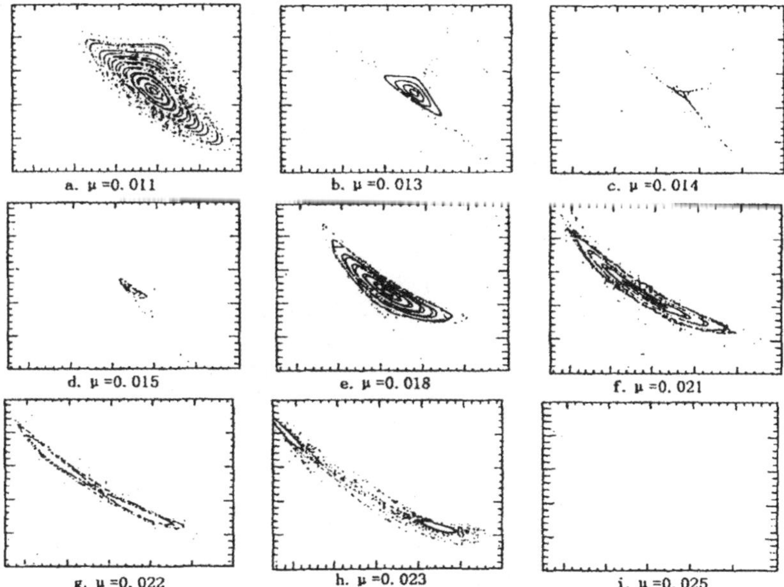

Fig. 2.2   Phase space for different μ. In all panels, the abscissa is I_ϕ from 0.955 to 1.01, and the ordinate is θ_ϕ from 0.8 to 1.3. Taken from Zhou, Sun & Zhou (2000).

critical value of $\mu_3' \approx 0.01440$. This is a little different from the theoretical value of μ_3, due to the fact that the mapping is an approximation of the original system. When μ goes beyond μ_3', the invariant curves reappear, and the area enclosed by them increases as μ increases. But the stable island now changes to a "banana-shaped" profile. The fixed point becomes unstable again at the value of $\mu_2' \approx 0.02165$, where the elliptic fixed point turns into a hyperbolic one. But this time, not like $\mu = \mu_3'$, two new fixed points, also surrounded by invariant curves, emerge. More details of the phase space at μ_2' is shown Fig. 2.3. Although the fixed point now is a hyperbolic point implying the unstable motion around it, the two new elliptic fixed points and the circumambient invariant curves indicate that stable orbits in the vicinity of this Lagrange point exist at this mass ratio.

Beyond μ_2', the motion becomes more and more chaotic, and the invariant curves surrounding the newly born fixed points begin to break as μ increases, until μ reaches $\mu_1' \approx 0.024$, when the phase space becomes completely chaotic.

Due to the approximating procedure employed in deducing the mapping, the critical values μ_1', μ_2', μ_3' obtained from the mapping by numerical

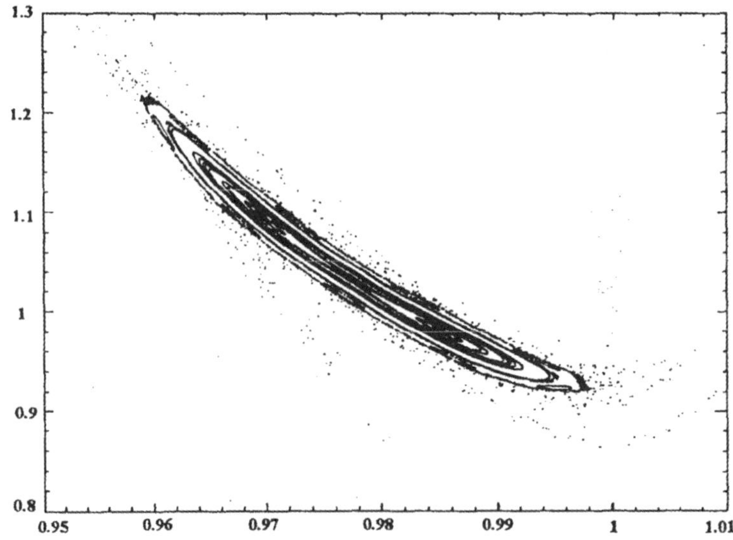

Fig. 2.3   Phase space at $\mu = 0.02165$. Taken from Zhou, Sun & Zhou (2000).

calculations are not exactly the same as the values μ_1, μ_2, μ_3 derived from the theoretical analysis. But the topological structure of the phase space has been revealed correctly in a qualitative way. It implies the structure stability of the system.

Last but not least, the phase space shown here are all for an initial condition of $I_r^0 = 0$ that corresponds to the situation where the test particle starts off in a circular orbit. For $I_r^0 > 0$ corresponding to initial orbits of nonzero eccentricity, our numerical computations show that the structure of the phase space and its variation with μ are very similar to the case of $I_r^0 = 0$, provided $I_r^0 < 0.019$ (corresponding to the initial eccentricity < 0.14).

2.3   Stability of asteroid orbits in resonances

As one of the most notable and important dynamical features in physics as well as in astronomy, the resonance problem has attracted much attention. A resonance may happen when two frequencies or their multiples in a dynamical system match to each other. In the Solar System, quite many objects had been or currently are deeply involved in resonances. For planets, dwarf planets and asteroids, the most significant resonant effect comes from the Mean Motion Resonance (MMR) where the mean motions of two objects are locked in a commensurability (note this is not an exact definition). The Trojans discussed in Sec. 2.2 have the same mean motion as the host planet so that they are in a 1:1 MMR with the planet.

The Main Belt Asteroids (MBAs) orbit the Sun in between the orbits of Mars and Jupiter. The spatial distribution of MBAs is known to be strongly influenced by MMRs with Jupiter (and probably with other planets too). Gaps in the semi-major axis distribution of MBAs, known as "Kirkwood gaps", have been noticed as early as in the mid 19th century. One of the most distinguishable gaps, with a distance of 2.5 AU from the Sun, is in fact in a 3:1 MMR with Jupiter. The mechanism of depletion of asteroids in this gap was first explained via a mapping method (Wisdom, 1982; Wisdom, 1983). Using the mapping, the motion of an MBA in this resonance was simulated for the first time to a timescale larger than million years, within the capability of computers by then. Except the 3:1 MMR, other gaps are observed near the 4:1, 5:2, 7:3 and 2:1 resonances. However, asteroids may also gather in some resonances. For example, an asteroid family named "Hilda group" locates in the 3:2 MMR while the "Thule group" resides in the 4:3 MMR with Jupiter. The complicated dynamics of MMR is always an interesting topic in celestial mechanics and astronomy.

In this section, using a mapping model the orbital stability of asteroids in different MMRs of order-one is investigated. A gap will be formed if the asteroids formerly residing in a region escape more quickly than in the neighboring region. Such an escape generally may be described as a "diffusion" process in the phase space of the corresponding dynamical system. Thus the diffusion phenomenon of the mapping describing the motion of asteroids in several MMRs will be analyzed.

2.3.1 *Mapping model*

A planar circular restricted 3-body problem is adopted as the dynamic model. In this model, the asteroid is a zero-mass test particle orbiting in a combined gravitation field of the Sun and a planet with semi-major axis a_p, eccentricity $e_p = 0$ and mass $m_p \ll M_\odot$, where $M_\odot$ is the mass of the Sun.

One-planet mapping

For the problem of two small masses orbiting around a massive central body and gravitationally interacting with each other, Hénon & Petit gave a series expansion for encounter-type solution (Hénon & Petit, 1986), in which the two small bodies approach each other from a faraway distance, interact for a while and then leave each other. In the restricted 3-body problem model consisting of the Sun, a planet and an asteroid, since the planet mass $m_p \ll M_\odot$, such an encounter-type solution can also be applied. Using this series solution, Duncan *et al.* constructed a mapping (Duncan *et al.*, 1989) describing the changes of the test particle's orbital elements during a conjunction with the planet. This mapping is valid if the test particle's initial eccentricity e and semi-major axis a satisfy $e \ll |a - a_p|/a_p \ll 1$. If further assuming that the perturbing effect of the planet away from the conjunction configuration is negligible, i.e. the motion of the test particle is just Keplerian, the motion of the test particle can be described by this mapping.

Before writing down explicitly the mapping, notations ϵ and z that will be used in the mapping are introduced as follows

$$\epsilon = \frac{a - a_p}{a_p}, \qquad z = e \exp(\mathrm{i}\varpi), \qquad (2.51)$$

where ϖ is the azimuth of perihelion of the test particle, and $\mathrm{i} = \sqrt{-1}$ is the imaginary unit. Note that the Jacobi integral of this restricted 3-body problem is

$$\gamma = \frac{3}{4}\epsilon^2 - |z|^2. \qquad (2.52)$$

Denoting the longitude of the test particle at the nth conjunction by λ_n, the mapping reads

$$z_{n+1} = z_n + \mathrm{sgn}(\epsilon_n) \frac{\mathrm{i}g \exp(\mathrm{i}\lambda_n)}{\epsilon_n^2} \frac{m_p}{M_\odot},$$

$$\epsilon_{n+1} = \epsilon_n \sqrt{1 + \frac{4\left(|z_{n+1}|^2 - |z_n|^2\right)}{3\epsilon_n^2}}, \qquad (2.53)$$

$$\lambda_{n+1} = \lambda_n + 2\pi f(\epsilon_{n+1}),$$

where $\mathrm{sgn}(\epsilon_n)$ is the sign of ϵ_n, $g = 2.23956667$ is a constant and $f(\epsilon) \equiv |(1 + \epsilon)^{-3/2} - 1|^{-1}$. Note that one iteration of this mapping stands for a conjunction between the test particle and the planet, thus the time leap in every iteration varies when different test particles and/or planets are investigated. Below, instead of the iteration number, the Julian year is adopted to measure the time. The variables z_n and λ_n in the mapping can be consolidated into one variable y_n as

$$y_n \equiv z_n \exp\{-\mathrm{i}[\lambda_n - \pi f(\epsilon_n) + \pi]\}, \qquad (2.54)$$

and then the iterating formula becomes

$$y_{n+1} = \left\{ y_n \exp\left[-i\pi f(\epsilon_n)\right] - \mathrm{sgn}(\epsilon_n)\frac{ig}{\epsilon_n^2}\frac{m}{M_\odot} \right\} \exp[-i\pi f(\epsilon_{n+1})]. \qquad (2.55)$$

Taking into account the Jacobi integral in Eq. (2.52), ϵ_n can also be eliminated from the mapping. For a given γ, ϵ_n can be replaced by

$$\epsilon_n = \mathrm{sgn}(\epsilon_1)\sqrt{\frac{4}{3}(\gamma + |y_n|^2)}. \qquad (2.56)$$

Substitute Eq. (2.56) into Eq. (2.55), three formulas in Eq. (2.53) are now combined into one iteration equation of y_n.

According to the perturbation theory, the MMR does not change very much the semi-major axes of objects involved in, thus the value of ϵ_n does not differ a lot from ϵ_1 if the test particle and the planet are in an MMR. Then one can replace approximately ϵ_n by ϵ_1 in the first formula of Eq. (2.53) without degrading the accuracy of the mapping. But this replacement makes the mapping symplectic as will be seen below, which is an essential character of a proper mapping that can correctly describe the original conservative dynamical system (Duncan *et al.*, 1989). A corresponding modification should be made to Eq. (2.55), and finally the mapping can be written as

$$y_{n+1} = \left\{ y_n \exp\left[-i\pi f\left(\mathrm{sgn}(\epsilon_1)\sqrt{\frac{4}{3}(\gamma + |y_n|^2)}\right)\right] - \mathrm{sgn}(\epsilon_1)\frac{ig}{\epsilon_1^2}\frac{m}{M_\odot} \right\}$$
$$\times \exp\left[-i\pi f\left(\mathrm{sgn}(\epsilon_1)\sqrt{\frac{4}{3}(\gamma + |y_{n+1}|^2)}\right)\right]. \qquad (2.57)$$

Since y is complex, the real part and imaginary part of the mapping can be separated and the mapping can be reformed as a two-dimensional mapping in the complex y plane as:

$$y_{n+1}^{\mathrm{r}} = y_n^{\mathrm{i}} \sin(\psi_n + \psi_{n+1}) + y_n^{\mathrm{r}} \cos(\psi_n + \psi_{n+1}) - C \sin\psi_{n+1},$$
$$y_{n+1}^{\mathrm{i}} = y_n^{\mathrm{i}} \cos(\psi_n + \psi_{n+1}) - y_n^{\mathrm{r}} \sin(\psi_n + \psi_{n+1}) - C \cos\psi_{n+1}, \qquad (2.58)$$

where y_n^{r} and y_n^{i} represent the real and imaginary part of y_n respectively, $\psi_k \equiv \psi_k(y_k^{\mathrm{r}}, y_k^{\mathrm{i}}, \gamma)$ $(k = n, n+1)$ is a function of $y_k^{\mathrm{r}}, y_k^{\mathrm{i}}$ and parameter γ, and $C \equiv \mathrm{sgn}(\epsilon_1)gm_p/(M_\odot\epsilon_1^2)$ is a constant. Note that the mapping in Eq. (2.58) is symmetric about the real axis in the complex y plane.

Fixed points of the mapping

As being stated in Sec. 2.1, the positions of fixed points and their linear stabilities are important topological properties of a mapping. The fixed points can be easily found by letting $y_{n+1}^r = y_n^r$ and $y_{n+1}^i = y_n^i$ in Eq. (2.58). The solution is

$$y_{\mathrm{fp}}^i = 0,$$

$$y_{\mathrm{fp}}^r = -\frac{C}{2\sin\left[\psi\left(y_{\mathrm{fp}}^r, y_{\mathrm{fp}}^i = 0, \gamma\right)\right]}. \tag{2.59}$$

Here the subscript "fp" indicates the fixed point. Obviously, all the fixed points lie on the real axis since $y_{\mathrm{fp}}^i = 0$. To find the linear stability of fixed point, the tangent mapping M of Eq. (2.58) is calculated, and the trace of M at the fixed point is

$$\mathrm{Tr}(M)_{\mathrm{fp}} = 2\cos(2\psi) + \frac{4\pi C y_{\mathrm{fp}}^r \cos\psi}{\epsilon_{\mathrm{fp}}\left[(1+\epsilon_{\mathrm{fp}})^{-3/2} - 1\right]^2 (1+\epsilon_{\mathrm{fp}})^{5/2}}, \tag{2.60}$$

where $\epsilon_{\mathrm{fp}} = \mathrm{sgn}(\epsilon_1)\sqrt{4(\gamma + |y_{\mathrm{fp}}^r|^2)/3}$ and $\psi = \pi f(\epsilon_{\mathrm{fp}})$. The linear stability of a fixed point is determined by this trace. If $|\mathrm{Tr}(M)_{\mathrm{fp}}| < 2$, the corresponding fixed point is an elliptic point. It is hyperbolic when $|\mathrm{Tr}(M)_{\mathrm{fp}}| > 2$, and parabolic when $|\mathrm{Tr}(M)_{\mathrm{fp}}| = 2$.

When considering the motion of MBAs, Jupiter is the main perturber. Adopting $m_p/M_\odot = 9.5479 \times 10^{-4}$, $a_p = 5.2$ AU (Jupiter mass and semi-major axis), and letting ϵ_1 vary from -0.5 to 0 while $\gamma = 3\epsilon_1^2/4$, the positions of fixed points and their linear stabilities as a function of ϵ_{fp} are plotted in Fig. 2.4. Recall the relations given in Eqs. (2.51) and (2.54), the abscissa in Fig. 2.4 is transformed to the semi-major axis a for convenience, and the ordinate is related to the eccentricity by $e = |y| = |y_{\mathrm{fp}}^r|$.

As illustrated in Fig. 2.4, when $a < 3.36$ AU, there is one stable fixed point with e initially around zero and starting to increase from $a \approx 3.15$ AU. This is basically consistent with the fact that asteroids accumulate in this region, only except some Kirkwood gaps corresponding to the 3:1, 5:2 and 7:3 MMR with Jupiter. Particularly, just before the new fixed points emerge at $a = 3.36$ AU, the fixed point leaves the origin of the complex plane, that is, the eccentricity increases to $e \sim 0.1$. It is worth noting that the accuracy of the mapping Eq. (2.58) decreases when $e = |y|$ deviates far away from 0 since the replacement of ϵ_n by ϵ_1 is valid only when e is small, provided $\gamma = 3\epsilon_1^2/4 = 3\epsilon_n^2/4 - |y_n|^2$ being a constant. In addition, orbits in this region

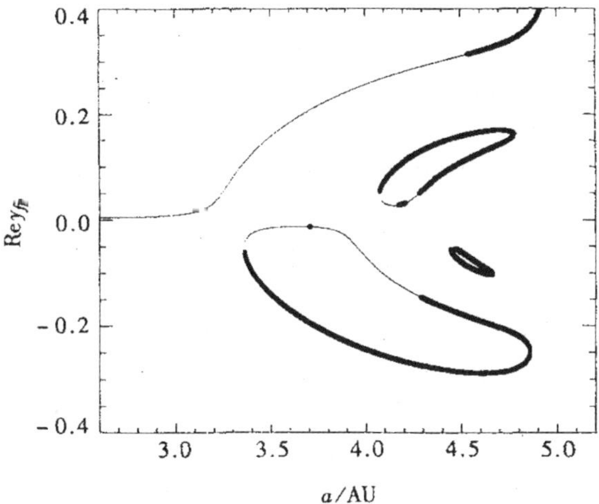

Fig. 2.4 The fixed points and their linear stabilities. The thin line stands for the stable fixed points with $|\mathrm{Tr}(M)_{\mathrm{fp}}| < 2$, while the thick line and big dots represent fixed points with $|\mathrm{Tr}(M)_{\mathrm{fp}}| \geq 2$. The abscissa and ordinate indicate the semi-major axis and eccentricity of asteroid's orbit (see text). Adapted from Zhou *et al.* (2001).

prefer to have relatively large eccentricities, indicating that these asteroids may be perturbed by other planets due to their eccentric orbits. Thus, unsurprisingly people found absence of asteroids in the 2:1 MMR around $a \approx 3.3$ AU, known as "Hecuba gap". However, it should be addressed that this simple one-planet mapping cannot explain satisfactorily the depletion of asteroids in this resonance.

A new fixed point emerges at $a = 3.36$ AU and it bifurcates immediately into two branches, one stable and the other unstable. The stable branch first stays close to $e \sim 0$ and then after a short unstable interval around $a \sim 3.69$ AU it begins to move away toward relatively large eccentricity. Accordingly, to the right of the Hecuba gap, asteroids are observed until $a \sim 3.60$ AU. Then, new stable fixed points appear again close to the y^{r} axis at $a \sim 4.00, 4.25$ AU. And these fixed points are consistent with the presence of asteroids in the 3:2 and 4:3 MMRs with Jupiter. Beyond these fixed points, no stable fixed point with small e appears, corresponding to the absence of asteroids in such region.

Two-planets mapping

The mapping can be generalized to include perturbations from more planets. Assuming an asteroid orbiting the Sun under perturbations from two nearby planets with semi-major axes a_i and a_o, and masses m_i and m_o (here the subscripts i and o indicate the inner and outer planet respectively), the mapping Eq. (2.53) can be generalized to include both perturbations from two planets. Let λ_n, t_n be the longitude and time of the nth conjunction, and z_n, a_n be values of the complex eccentricity and semi-major axis of the asteroid just before this conjunction, which is assumed to occur with planet b ($b \in \{i, o\}$). The generalized mapping reads:

$$\epsilon_{b,n} = \frac{a_n - a_b}{a_b},$$

$$z_{n+1} = z_n + \frac{ig\exp(i\lambda_n)}{\epsilon_{b,1}^2}\mathrm{sgn}(\epsilon_{b,1})\frac{m_b}{M_\odot},$$

$$\epsilon_{b,n+1} = \epsilon_{b,n}\sqrt{1 + \frac{4(|z_{n+1}|^2 - |z_n|^2)}{3\epsilon_{b,n}^2}}, \qquad (b \in \{i, o\}) \qquad (2.61)$$

$$a_{n+1} = a_b\left(1 + \epsilon_{b,n+1}\right),$$

$$\Omega_{n+1} = \sqrt{\frac{GM_\odot}{a_{n+1}^3}}.$$

The longitude and time of the $(n+1)$th conjunction are

$$\lambda_{n+1} = \lambda_n + \Omega_{n+1}\min\left(\Delta t_i, \Delta t_o\right),$$
$$t_{n+1} = t_n + \min\left(\Delta t_i, \Delta t_o\right), \qquad (b \in \{i, o\}), \qquad (2.62)$$

where

$$\Delta t_b = \begin{cases} \dfrac{Q(\Omega_b t_n - \lambda_n)}{\Omega_{n+1} - \Omega_b} & \text{for } \Omega_{n+1} - \Omega_b > 0, \\[2ex] \dfrac{Q(\lambda_n - \Omega_b t_n)}{\Omega_b - \Omega_{n+1}} & \text{for } \Omega_{n+1} - \Omega_b < 0, \end{cases} \qquad (b \in \{i, o\}), \qquad (2.63)$$

while $\Omega_b = (GM_\odot/a_b^3)^{1/2}$ and $Q(x)$ is the minimal positive value of $x + 2k\pi$ for integer k.

Among the two planets, the massive and/or nearer one is regarded as the primary perturber to the asteroid motion. For asteroids in the main belt, Jupiter plays the role of the primary perturber because both of its great mass and of the closeness of its orbit to asteroids. The change of z in once iteration of the mapping Eq. (2.53) can be estimated by calculating $|\Delta z| = |z_{n+1} - z_n|$. Taking Mars, Jupiter and Saturn as the perturber, $|\Delta z| < 5.32 \times 10^{-7}$ for Mars, $|\Delta z| > 1.60 \times 10^{-2}$ for Jupiter, and $|\Delta z| >$

1.49×10^{-3} for Saturn, if only the motion of asteroids in the outer main belt ($a > 3.3$ AU) is discussed. Therefore, it is reasonable to take Jupiter ($m_i = 9.54785 \times 10^{-4} M_\odot, a_i = 5.203$ AU) as the primary perturber and Saturn ($m_o = 2.85884 \times 10^{-4} M_\odot, a_o = 9.555$ AU) the secondary perturber.

2.3.2 *Diffusion in the mapping*

Diffusion (also known as chaotic transportation) in the phase space is a remarkable phenomenon, because in many dynamical systems the global evolution of the system can be described as a process of diffusion. The next chapter of this book devotes to this topic, nevertheless some results of the diffusion in this mapping will be briefly presented below. Please refer to next chapter for detailed discussions.

Diffusion law in different regions

The diffusion process of an orbit is determined by the structure of the phase space, thus the diffusion speed may serve as an indicator of some important properties (e.g. stability) of the dynamical system. For the mapping discussed in this section, the trajectories of mapping Eq. (2.58) can be shown in the complex y plane, in which stable islands and developed chaotic sea can be found, and the orbit diffusion happens.

The diffusion speed is generally measured by the time needed by an orbit (or a number of orbits) to diffuse through a given region in the phase space. Therefore, to study the orbit diffusion, N_0 initial points are randomly scattered in a preselected region of the complex y plane, and then their evolutions are followed. An orbit is regarded as escaped if $e = |y| > 0.25$, or stable if it has not escaped after 10^{10} years. The number N_T of orbits that have not escaped up to time T is counted, and the corresponding surviving number at time T is defined as $N(T) = N_T - N_S$, where N_S is the number of stable orbits.

For each of two different Jacobi integral values $\gamma_1 = 2.2188 \times 10^{-2}$ and $\gamma_2 = 4.0368 \times 10^{-2}$, three regions are selected in the complex y plane, and they are illustrated by squares in Fig. 2.5. These squares are of the same size, with coordinates of the lower left and upper right vertexes $(0.12, -0.02) \leftrightarrow (0.16, 0.02)$, $(0.02, -0.02) \leftrightarrow (0.06, 0.02)$ and $(-0.16, -0.02) \leftrightarrow (-0.12, 0.02)$ respectively, and they are designated with A_1, B_1, C_1 from right to left for γ_1 and A_2, B_2, C_2 for γ_2. In each square, 2,500 orbits are randomly initialized and their evolutions are followed.

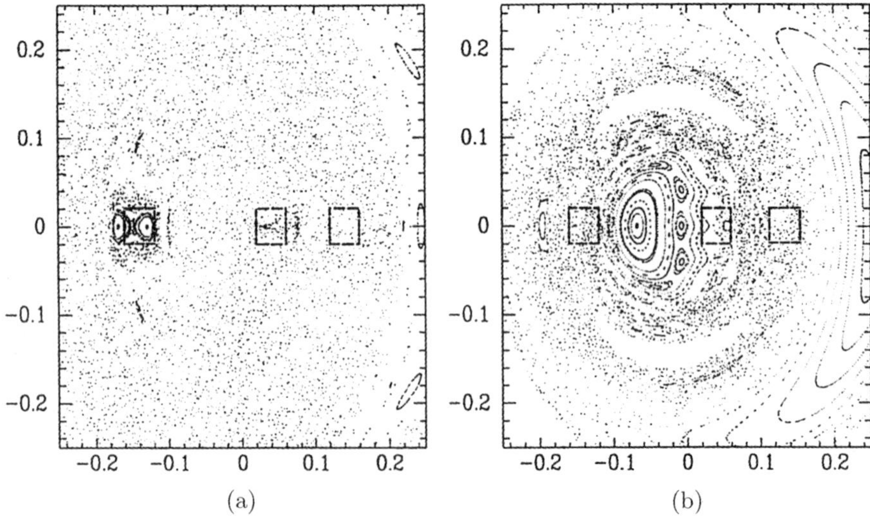

Fig. 2.5 Phase plane of mapping Eq. (2.58) for $\gamma_1 = 2.2188 \times 10^{-2}$ (left) and $\gamma_2 = 4.0368 \times 10^{-2}$ (right). Squares indicate the selected regions (see text). Adapted from Zhou *et al.* (2001).

As shown in Fig. 2.5, regions A_1, B_1 and A_2 are all in the developed chaotic sea, thus the diffusion in these regions follows the same rule. In Fig. 2.6, the variations of the number of surviving orbits $N(T)$ in regions A_1 and B_1 with respect to time T are plotted. After a short transition, the $\lg N$ versus T curves follow a straight line, indicating that the surviving number decays according to an exponential law with respect to time as

$$N(T) \propto 10^{-\alpha T}. \tag{2.64}$$

Numerical fits give the exponent α in Fig. 2.6 $\alpha = 5.51 \times 10^{-4}$ for A_1 and $\alpha = 4.41 \times 10^{-4}$ for B_1. For region A_2, similar law can be found and the exponent is $\alpha = 6.65 \times 10^{-5}$, since in region A_2 the chaos is not so developed with a bigger Jacobi integral γ_2.

On the contrary, the diffusion from regions B_2, C_2 is different because these two regions are a mixture of chaotic sea and invariant curves. In Fig. 2.7 the diffusion rates of these two cases are plotted. As in many literatures, e.g. Lai *et al.* (1992), the curves in Fig. 2.7 shall be fitted by straight lines again, but here a straight line indicates an algebraic law as

$$N(T) \propto T^{-\beta}. \tag{2.65}$$

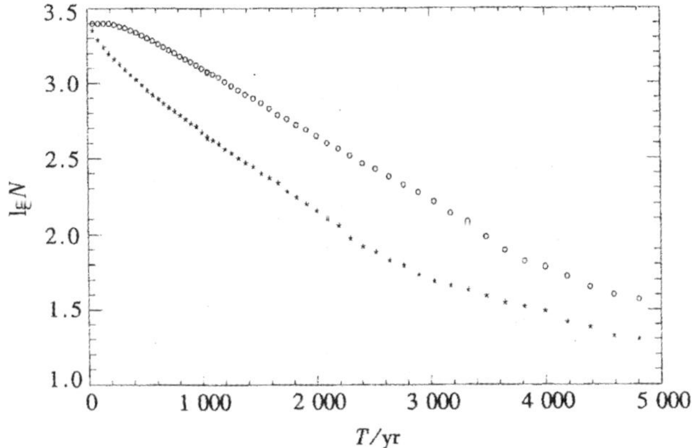

Fig. 2.6   Diffusion rate in regions A_1 (solid asterisks) and B_1 (open circles). Adapted from Zhou *et al.* (2001).

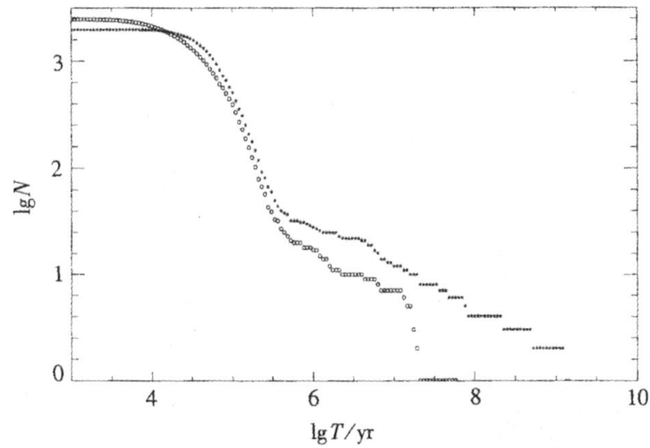

Fig. 2.7   Diffusion rate in regions B_2 (solid asterisks) and C_2 (open circles). Adapted from Zhou *et al.* (2001).

In fact, after the transition period (till $\lg T \sim 4.5$) each curve can be fitted by two segments of straight line with a "crossover" at about $\lg T \sim 5.5$. When $5.0 < \lg T < 5.5$, a least-square fit gives the algebraic decay exponent $\beta = 1.99$ for B_2 and $\beta = 2.15$ for C_2, respectively. After the crossover, the exponent becomes $\beta = 0.412$ for B_2 in the time range $5.6 < \lg T < 9.2$ and $\beta = 0.416$ for C_2 in $5.6 < \lg T < 7.1$. The crossover however, implies that in these regions orbits starting from some definite area diffuse much slower than others, and in fact such "sticky" areas are the close vicinities of invariant curves, which are common in the mixed regions B_2, C_2 and are the primary components in region C_1.

Diffusion velocity in vicinity of invariant curves

In the close vicinity of invariant curves, the diffusion may be much slower. On a straight line segment beginning at $(y^r, y^i) = (-0.122, 0)$ and ending at $(-0.070, 0)$ in the phase plane of the mapping with γ_1 as shown in Fig. 2.5, thousands of initial points are set and their evolutions are followed. Each orbit is followed for 10^9 years unless it escapes ($|y| > 0.25$) before that. On the far left of this line are the invariant curves of a stable island, and the right end is in the developed chaotic sea. So the escape time of these orbits reveals how the diffusion velocity changes with the distance to the invariant curve, although the exact position of the outermost invariant curve cannot be precisely determined.

Figure 2.8 shows the escape time of these orbits as a function of the initial position y^r (with all $y^i = 0$). In this figure, except for some interstices where the initial points locate on small invariant islands, the escape time decreases with increasing y^r. Denoting the distance of an initial point on the straight line to the outermost invariant curve of the invariant island by d, the dashed line in Fig. 2.8 indicating the linear relation between $\lg T$ and

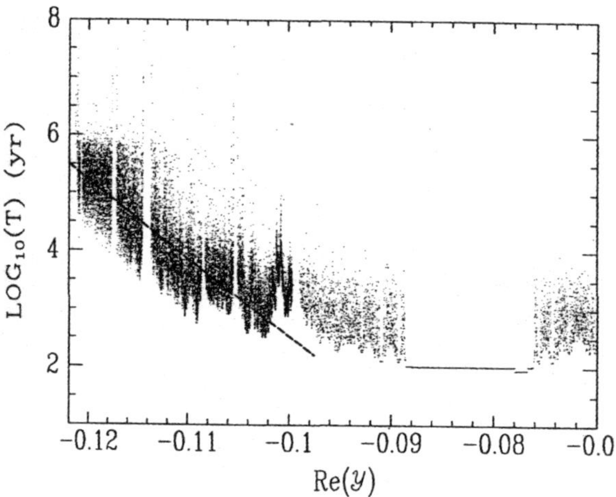

Fig. 2.8   The variation of escape time in logarithm of orbits with respect to their initial positions y^r. Adapted from Zhou *et al.* (2001).

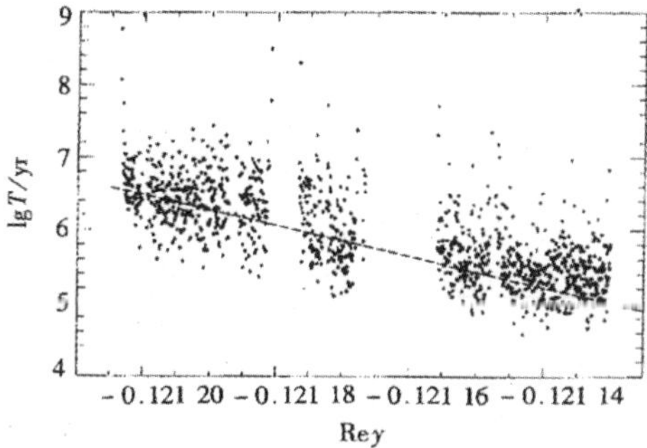

Fig. 2.9 The same as Fig. 2.8, for $y^r \in (-0.121204, -0.121130)$. Adapted from Zhou *et al.* (2001).

y^r (and thus distance d) can be written in an exponential form as

$$T \propto 10^{-kd}, \tag{2.66}$$

with k being a pending coefficient. Here the numerical fit gives $k = 135$.

When $y^r > -0.098$, the initial orbits are in a chaotic sea and they escape in a short time. Conversely, when getting closer to the invariant curve, a sharp increase of escape time can be observed in the left end of Fig. 2.8. A zooming-in on the steep edge with y^r varying from -0.121204 to -0.121130 is plotted in Fig. 2.9. A linear fit (the dashed line) is computed and the coefficient k in this case is $k = 21200$, much larger than the one in Fig. 2.8. Furthermore, in the left end of Fig. 2.9, a sharp increase of escape time can still be seen, implying that similar structures and much larger coefficient can be found when approaching closer to the invariant curve. This is consistent with the superexponential dependence of escape time on the distance to an invariant torus in its close vicinity, which is derived by successive applications of the Nekhoroshev theorem (Morbidelli & Giorgilli, 1995).

In summary, the diffusion of orbits in the developed chaotic region of phase space obeys an exponential law as Eq. (2.64), and obeys an algebraic law as Eq. (2.65) in a mixed region, while in a vicinity of invariant curve, the diffusion is much slower. In the close vicinity of an invariant curve, the diffusion time depends superexponentially on the distance to the invariant curve.

2.3.3 *Stabilities of asteroid orbits in resonances*

The fixed points of map Eq. (2.58) and their linear stabilities give a sketchy qualitative explanation of the distribution of asteroids in the main belt.

The diffusion property of the mapping may explain quantitatively different populations of asteroids in different resonances.

In one-planet case

The mapping is designed under the assumption $e \ll |\epsilon| \ll 1$, but in fact this condition can be relaxed. For $|\epsilon| \sim 0.5$, this mapping still works well (Zhou *et al.*, 2001). Below in this section, this mapping will be applied to study the stabilities of orbits in the 2:1, 3:2 and 4:3 MMRs with Jupiter, where $\epsilon = -0.370, -0.237$ and -0.175. Since a number of asteroids have been observed in the 3:2 MMR (named Hilda group after the first member 153 Hilda), the validity of this mapping can be verified by a comparison between the results of the mapping and the real orbits of Hilda asteroids.

For the simplicity of calculations, the form in Eq. (2.53) of the mapping is adopted. Thousands of initial orbits are then selected so that their initial semi-major axis a_0 and eccentricity e_0 satisfy

$$a_0 \in (a_r - 0.02, a_r + 0.02), \quad e_0 \in (0.0, 0.1), \tag{2.67}$$

where a_r is the semi-major axis in Jupiter unit in exactly the resonance center. For 3:2 MMR, it's $a_r = (2/3)^{2/3} = 0.763$. The initial angle λ_0 is randomly selected. All these orbits then evolve under the mapping. In the evolution, if an object crosses the orbit of Jupiter or Mars, it will be abandoned. The final a, e of survivors after 10^8 years are plotted in Fig. 2.10, which is interspersed by the orbital elements of the brightest members of Hilda group.

The consistence between the numerical results and the observed data in Fig. 2.10, not only verifies the validity of the mapping as a numerical tool to study the dynamics of these asteroids, but also indicates that the approximate condition about eccentricity $e \ll 1$ can be relaxed, at least to $e \sim 0.2$.

Now, for each of the 2:1, 3:2 and 4:3 resonances, N_0 initial orbits are set in a similar way according to Eq. (2.67). Each orbit is followed for 10^9 years, unless it crosses the orbit of Mars or Jupiter. An orbit satisfying

$$a \in (a_r - 0.02, a_r + 0.02), \quad e \in (0.0, 0.25), \tag{2.68}$$

is regarded as staying in the resonant zone, and the number of surviving orbits (the orbits staying in the resonant zone) at moment T, denoted by N_T, is recorded.

The numerical results are summarized in Fig. 2.11. For the 3:2 and 4:3 resonances, there is a rapid decay of N_T during the beginning hundred

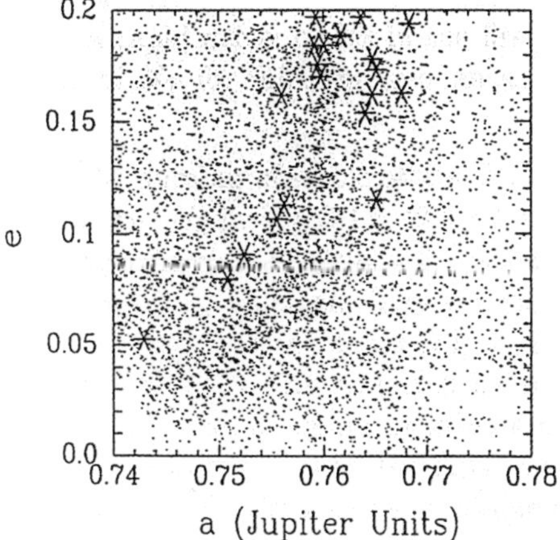

Fig. 2.10   The distribution of semi-major axis (abscissa, in Jupiter unit) and eccentricity (ordinate). Dots represent the survivors in numerical evolution of 10^8 years using the mapping Eq. (2.53), and asterisks are for the observed asteroids. Adapted from Zhou *et al.* (2001).

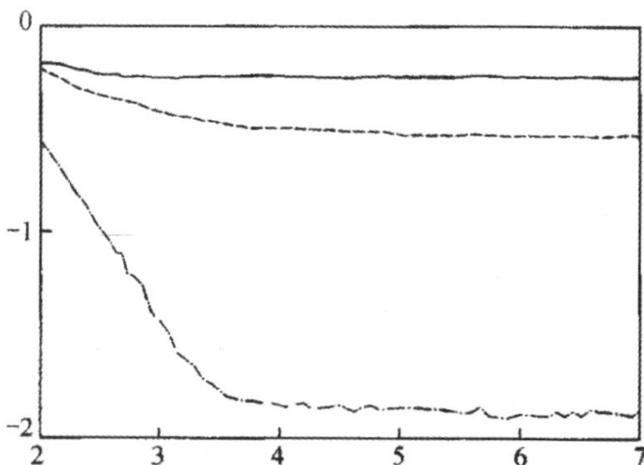

Fig. 2.11   Variation of surviving number with respect to time, both in logarithm. The solid, dashed and dot-dashed curves represent the case of the 2:1, 3:2 and 4:3 resonance respectively. Adapted from Zhou, Zhou & Sun (2002b).

years. After this short transition, the decay of N_T for different resonances shows different properties.

Define $N_S = N_T(T = 10^9\,\mathrm{yr})$ as the number of "stable" orbits, i.e. those orbits that stay in the resonance for 10^9 years. Thus N_S implies the area of stable region in resonance. In Fig. 2.11, N_S' are 40 and 864 for 4:3 and

3:2 resonance respectively, where $N_0 = 3000$. Hence the stable region in 3:2 resonance is over 20 times larger than that in the 4:3 resonance. This is consistent with the fact that there are more than 1000 asteroids (including 153 Hilda) observed in the 3:2 but only 3 (including 279 Thule) in the 4:3 resonance. Moreover, the decay of number $N(T) = N_T - N_S$ with time can be simulated well by an algebraic law as Eq. (2.65), with exponent $\beta = 0.513$ and $\beta = 1.02$ for the 3:2 and 4:3 resonance respectively. These give quantitatively the diffusion velocities of orbits in these resonant zones.

As for the 2:1 resonance where the Hecuba gap locates, unfortunately, no obvious diffusion can be found in Fig. 2.11. The number N_T for 2:1 does not decrease at all after $T = 1000\,\mathrm{yr}$. Such a simple one-planet mapping suggests that motion in 2:1 is more stable than in 3:2 and 4:3 resonances. It's simply not the truth. Therefore, the paucity of asteroids in the 2:1 resonance must have other reason than the direct perturbation from Jupiter.

It should be pointed out that the Hecuba gap at 2:1 resonance got its name because the population density of asteroids in this region is much lower than its near neighbourhood. But this "gap" is not empty. According to recent observation data (Brož *et al.*, 2005), more than 150 asteroids have been observed in this region.

In two-planet case

The diffusion property changes after Saturn is included in calculations by using the two-planet mapping Eq. (2.61). The results are shown in Fig. 2.12.

In the case of 4:3, the algebraic diffusion law, which has an exponent $\beta = 0.964$ now, is kept very well. Finally, only 1 of the 3000 initial orbits survives until $T = 10^9$ year. For 3:2, N_T decays slowly all the time and the number of surviving orbits is 29 at the moment $T = 10^9$ year. As for the 2:1 resonance, there is a fast decay of N_T before $T = 10^{4.5}$ year, during which five sixths of orbits escape. Then, the diffusion slows down, and the number N_T keeps almost the same value till $T = 10^7$ year. However, the perturbation from Saturn eventually decreases this number and only 48 orbits survive at 10^9 year. Although this number is bigger than the one in the case of 3:2, the diffusion velocity for 2:1 is larger than 3:2 around $T = 10^9$ year. So it is reasonable to argue that the Hecuba gap can only be observed at the age of the Solar System, $\sim 4.5 \times 10^9$ year.

In a similar way as being generalized to the two-planet mapping, the mapping can be further extended to include the perturbation from planet Mars. Although the perturbation from Mars is so small, calculation shows

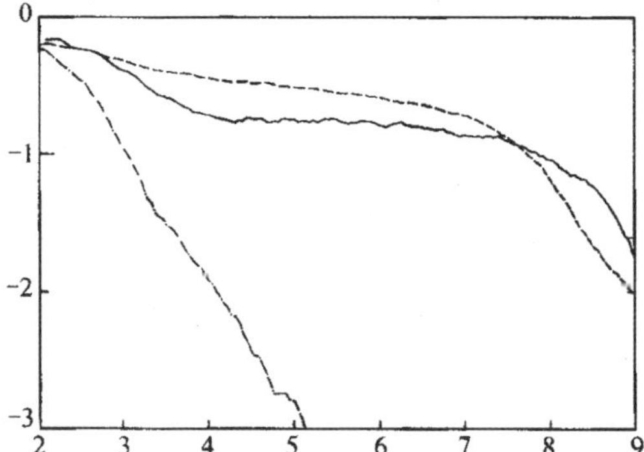

Fig. 2.12   The same as Fig. 2.11, but for the two-planet mapping. Adapted from Zhou, Zhou & Sun (2002b).

that it does accelerate a little the depletion of asteroid in the 2:1 resonance but decelerate a little the diffusion in the 3:2 and 4:3 resonances.

In summary, the one-planet mapping suggests that there is a much bigger stable region in the 3:2 resonance than in the 4:3 resonance, and the diffusion velocity in the former is smaller than the latter. The two-planet mapping tells the importance of Saturn. Saturn's perturbation brings new features to the long-term evolution of the system and gives birth to the Hecuba gap, while it does not wipe out the asteroids in the 3:2 and 4:3 resonances.

The space distribution of asteroids in the main belt could offer important clues to the formation and early evolution of the planetary orbits of the Solar System. Some secular mechanisms are found to have played important roles in sculpting the features of the belt. And the final solutions to some challenging problems, e.g. the origin of the Hecuba gap, are still being expected. The simple mapping model in this section obviously is not one of the final solutions, but it gives some helpful hints.

2.4   Shepherding of Uranian ring

Both the ground-based occultation observation and the Voyager Uranus encounter detected some narrow rings, with a typical width of several dozen kilometers, around Uranus. Typically, an unconfined narrow ring will spread within millions of years due to the particle collisions and differential rotation. The shepherding mechanism was raised to explain the presence of such narrow rings and now it is widely accepted. The Voyager discovery of two shepherding moons Cordelia and Ophelia of the epsilon (ϵ) ring of Uranus greatly supports this mechanism. However, due to the poor knowledge about some quantities, it remains an open question whether the satellites are adequate to supply sufficient torques to confine the epsilon ring against its tendency to spread. This section devotes to the shepherding mechanism of the ϵ-ring.

To understand the shepherding mechanism, the study of the dynamical evolution of particles in the ring under the perturbation of the shepherding satellites would be helpful. Meanwhile, the mapping method is an effective approach to explore the long-term evolution of a nonlinear system, because the mapping has many advantages as noted in Sec. 2.1.

Due to the presence of small drags, a mapping method for the near-conservative system is needed. Such a mapping have been constructed (Zhou, Fu & Sun, 1994), which preserves the main topological properties of the phase space both qualitatively and quantitatively. This section will present the analysis on the motion of the particles in the epsilon ring of Uranus, using the mapping method.

2.4.1 *Equations of motion and the Hamiltonian*

The evolution of particles in the edges of ϵ ring was investigated. Here the "particle" is a fluid particle, i.e. a collection of ring particles, as widely used in the streamline formalism in fluid system.

Some characteristic parameters of Uranus, the narrow ring and the shepherding satellites are listed in Table 2.1. According to the observations, the ϵ ring lies in the plane of Uranian equator. Meanwhile, neither of the shepherding satellites of the ϵ ring has a measurable inclination to the equator. Thus the investigation can be restricted in the equator plane. For this sake, a planar Cartesian coordinate system is defined in the equator of Uranus. Taking into account the oblateness of Uranus, its gravitational potential can be expressed as

$$\Phi_U = -\frac{GM_U}{r}\left[1 - \sum_{n=1}^{\infty}\left(\frac{R_U}{r}\right)^{2n} J_{2n} P_{2n}(0)\right], \qquad (2.69)$$

where $P_{2n}(0)$ is the Legendre polynomials, G the gravitational constant, M_U and R_U the mass and equatorial radius of Uranus, and J_{2n} the even zonal harmonic coefficients. Approximately, only those terms of $n = 1, 2$, i.e. J_2 and J_4, are taken into account in the gravitational potential Φ_U

Table 2.1   Dynamical parameters of planet Uranus, shepherding satellites and ring (adopted from Zhou, Sun & Hu (2000)). Subscripts U, r, s refer to Uranus, ring and satellite, respectively.

Planet (Uranus)		ϵ-Ring	
GM_U	5.79394×10^6 km$^3 \cdot$ s^{-2}	a_r	51149 km
J_2	3.3434×10^{-3}	δ_a	58.1 km
J_4	-2.885×10^{-5}	e_r	7.036×10^{-3}
R_U	26200 km	δe	0.758×10^{-3}
M_U	8.669×10^{28} g	M_r	6.1×10^{18} g
Satellites		Cordelia	Ophelia
a_s/km		49752	53764
$e_s \times 10^3$		0.47	10
$M_s \times 10^{-19}$/ g		5	9

here. For the simplicity of the equations, 5×10^4 km and 4.6448×10^8 s are defined as the units of distance and time, so that

$$GM_U = \Omega^2(r)r^3 \left[1 + \frac{3}{2} J_2 \left(\frac{R_U}{r} \right)^2 - \frac{15}{8} J_4 \left(\frac{R_U}{r} \right)^4 \right]^{-1} = 1,$$

when $r = 1$ (Ω is the angular velocity of ring particles). Meanwhile, in these units, 1 year $\approx 1081.3 \times 2\pi$ and $a \sim 1, \Omega \sim 1$ for particles moving in the ϵ ring.

For the particles of the inner (outer) edge, their evolutions are considered under the perturbation of the inner (outer) shepherding satellite Cordelia (Ophelia). For simplicity, this is called the inner case (or outer case, respectively). In each case, the satellite perturbation potential can be written as

$$\Phi_s = -\mu \left(\frac{1}{d} - \frac{\boldsymbol{r} \cdot \boldsymbol{r}'}{|\boldsymbol{r}'|^3} \right), \tag{2.70}$$

where $\mu = M'/M_U$ is the mass ratio of the satellite with respect to Uranus, $\boldsymbol{r}$ and $\boldsymbol{r}'$ are the positional vectors of the particle and the satellite respectively, and $d = |\boldsymbol{r} - \boldsymbol{r}'|$ is the distance between the particle and the satellite. Hereafter, the quantities of the satellite are always attached with a prime.

The equation of motion of a particle moving in the gravitational field of Uranus and the satellite reads

$$\frac{\mathrm{d}^2 \boldsymbol{r}}{\mathrm{d}t^2} = \frac{\partial \Phi_U}{\partial \boldsymbol{r}} + \frac{\partial \Phi_s}{\partial \boldsymbol{r}},$$

or

$$\frac{\mathrm{d}^2 x}{\mathrm{d}t^2} = \frac{\partial \Phi_U}{\partial x} + \frac{\partial \Phi_s}{\partial x},$$

$$\frac{\mathrm{d}^2 y}{\mathrm{d}t^2} = \frac{\partial \Phi_U}{\partial y} + \frac{\partial \Phi_s}{\partial y}.$$

Since

$$\frac{\partial \Phi_U}{\partial x} = \frac{\partial \Phi_U}{\partial r}\frac{\partial r}{\partial x} = \frac{GM_U}{r^3}\left[1 + \frac{3}{2}J_2\left(\frac{R_U}{r}\right)^2 - \frac{15}{8}J_4\left(\frac{R_U}{r}\right)^4\right]x = \Omega^2 x,$$

$$\frac{\partial \Phi_U}{\partial y} = \frac{\partial \Phi_U}{\partial r}\frac{\partial r}{\partial y} = \frac{GM_U}{r^3}\left[1 + \frac{3}{2}J_2\left(\frac{R_U}{r}\right)^2 - \frac{15}{8}J_4\left(\frac{R_U}{r}\right)^4\right]y = \Omega^2 y,$$

the equations of motion finally are

$$\frac{\mathrm{d}^2 x}{\mathrm{d}t^2} = -\Omega^2(r)x + \mu\left(\frac{x' - x}{d^3} - \frac{x'}{r'^3}\right),$$

$$\frac{\mathrm{d}^2 y}{\mathrm{d}t^2} = -\Omega^2(r)y + \mu\left(\frac{y' - y}{d^3} - \frac{y'}{r'^3}\right). \tag{2.71}$$

The dynamical system described by these equations are called "the original system". Below, instead of studying this original system, a mapping method will be applied to investigate the motion of particles in the ring.

To study the motion of the ring particles, the epicyclic theory in the streamline formalism (Borderies & Longaretti, 1987) is adopted. Borderies and Longaretti have developed the epicyclic theory to second order with respect to a small parameter that characterizes the deviation from circularity. In the epicyclic theory, the radius r and true longitude θ of a particle on the equatorial orbit can be expressed as

$$r = a\left[1 + \left(\frac{3\eta^2}{2\kappa^2} - 1\right)e^2 - e\cos M - \frac{\eta^2}{2\kappa^2}e^2\cos 2M\right],$$

$$\theta = \varpi + M + \frac{2\Omega}{\kappa}e\sin M + \frac{\Omega}{2\kappa}\left(\frac{3}{2} + \frac{\eta^2}{\kappa^2}\right)e^2\sin 2M.$$

In above expressions, a, e, ϖ, M are the epicyclic semi-major axis, eccentricity, periapse angle and mean anomaly, respectively; Ω and κ are the revolution and epicyclic frequencies; and η is an auxiliary quantity. And, the energy H_0 and angular momentum A of the motion are related to the epicyclic elements as follows

$$H_0 = \Phi_U(a) + \frac{1}{2}\Omega^2 a^2 + O(e^4),$$

$$A = \Omega a^2 \left(1 - \frac{\kappa^2}{2\Omega^2}e^2\right) + O(e^4). \tag{2.72}$$

Here H_0 is also the unperturbed Hamiltonian in the Hamiltonian description of the streamline formalism (Zhou & Sun, 1995).

For the ϵ ring, both two edges are located at the first-order resonances, denoted as $(m+1):m$ resonances, with the shepherding satellites. For the inner edge, which is at the 24:25 outer eccentric resonance (OER hereafter), $m = -25$; while the outer edge is at the 14:13 inner eccentric resonance (IER) thus $m = 13$. These two resonances just at the inner and outer edges of the ring, together with the shepherding satellites, maintain the ϵ ring.

After averaging, the potential at the (inner or outer) edge has only resonance terms to the first order of e, e'

$$H_s = -\frac{\mu}{a'}\left[A_3 e \cos\left(m\lambda - (m+1)\lambda' + \varpi\right)\right.$$
$$\left. + A_4 e' \cos\left(m\lambda - (m+1)\lambda' + \varpi'\right)\right], \tag{2.73}$$

where $\lambda = M + \varpi$ is the mean longitude, and A_3, A_4 are constant coefficients (see details in Shu (1984)). To simplify the system, these coefficients can be substituted by the values calculated in the specific resonance, and this approximation will be justified by the numerical results later.

In this Hamiltonian system, the canonical elements conjugating to λ and $\omega = -\varpi$ are L and ρ, defined respectively as (Zhou & Sun, 1995)

$$L = \Omega a^2 + \frac{1}{2}\left(1 - \frac{\Omega}{\kappa}\right)a^2 e^2 \kappa, \qquad \rho = \frac{1}{2}a^2 e^2 \kappa.$$

And the Hamiltonian describing the evolution of ring particles under the perturbation of the shepherding satellite is $H = H_0 + H_s$. Through a canonical transformation generated by a function

$$F = N(\omega - \omega') + S(m\lambda - (m+1)\lambda' - \omega) + L^*\lambda,$$

where $L^* = \Omega(a^*)a^{*2}$ and a^* is the semi-major axis of the involved shepherding satellite, a set of new canonical variables for the resonances are obtained:

$$N = \frac{L - L^*}{m} + \rho, \qquad \nu = \omega - \omega',$$

$$S = \frac{L - L^*}{m}, \qquad \sigma = m\lambda - (m+1)\lambda' - \omega. \tag{2.74}$$

Table 2.2 Coefficients in the Hamiltonian function Eq. (2.75). Data from Zhou, Sun & Hu (2000).

m	a^*	$\alpha_4 \times 10^4$	$\alpha_5 \times 10^5$	α_6	B_3	B_4
-25	1.02243	1.2820	-0.92636	-1793.6	19.626	-19.352
13	1.02356	-2.0284	-3.6565	-483.93	-10.289	10.544

Here σ is the resonant angle of the involved $(m + 1) : m$ mean motion resonance and $(\sigma + \nu)$ is (m times of) the resonant angle of the corotation resonance. Expressed in above canonical elements, the averaged Hamiltonian $\bar{H}$ is

$$
\begin{aligned}
\bar{H} = {} & \alpha_4 N + \alpha_5 S + \frac{1}{2}\alpha_6 S^2 \\
& - \mu \left[B_3 \sqrt{2(N - S)} \cos \sigma + B_4 e' \cos(\sigma + \nu) \right],
\end{aligned}
\tag{2.75}
$$

with the coefficients listed in Table 2.2.

2.4.2 *Viscosity in the ring and the mapping*

The inter-particle collisions in the ring transfer the energy from the orbital motion to the particles' random velocity. This brings viscosity into the ring and results in energy dissipation. The rate of viscose energy dissipation per unit mass is $\sim \bar{\nu}(ad\Omega/da)^2 \sim \bar{\nu}\Omega^2$, where $\bar{\nu}$ is the kinematic viscose coefficient, relating to the velocity dispersion c^2 and the optical depth τ by

$$
\bar{\nu} \sim \frac{c^2 \tau}{\Omega(1 + \tau^2)} \sim \frac{c^2}{\Omega},
$$

since $\tau \sim 1$ in the ring. Thus, for a particle moving in an epicyclic orbit with its energy given by Eq. (2.72), the secular change of the semi-major axis of the particle due to this viscosity is

$$
\left(\frac{\mathrm{d}a}{\mathrm{d}t}\right)_{\bar{\nu}} = \left(\frac{\mathrm{d}H_0}{\mathrm{d}t}\right)_{\bar{\nu}} \bigg/ \left(\frac{\mathrm{d}H_0}{\mathrm{d}a}\right)_{\bar{\nu}} \sim \pm \frac{\bar{\nu}\Omega^2}{a\kappa^2}.
$$

Here the sign '+' (or '−') applies to the outer (or, inner) case respectively. In resonant elements, the change caused by the viscosity is approximately

$$
\left(\frac{\mathrm{d}N}{\mathrm{d}t}\right)_{\bar{\nu}} = \frac{c^2}{|m|}, \qquad \left(\frac{\mathrm{d}S}{\mathrm{d}t}\right)_{\bar{\nu}} = \frac{c^2}{|m|}.
$$

Finally, the averaged system describing the evolution of particles in the dissipative ring reads

$$\frac{\mathrm{d}N}{\mathrm{d}t} = -\mu B_4 e' \sin(\sigma + \nu) + \frac{c^2}{|m|},$$

$$\frac{\mathrm{d}\nu}{\mathrm{d}t} = \alpha_4 - \frac{\mu B_3}{\sqrt{2(N-S)}} \cos \sigma,$$

$$\frac{\mathrm{d}S}{\mathrm{d}t} = -\mu \left[B_3 \sqrt{2(N-S)} \sin \sigma + B_4 e' \sin(\sigma + \nu) \right] + \frac{c^2}{|m|},$$

$$\frac{\mathrm{d}\sigma}{\mathrm{d}t} = \alpha_5 + \alpha_6 S + \frac{\mu B_3}{\sqrt{2(N-S)}} \cos \sigma. \tag{2.76}$$

Following a mapping construction procedure (Zhou, Fu & Sun, 1994) for near conservative system, a mapping based on the above averaged system can be constructed as:

$$N_{n+1} = N_n - \mu T B_4 e' \sin(\sigma_n + \nu_n) + \frac{c^2}{|m|},$$

$$\nu_{n+1} = \nu_n + \alpha_4 T - T \frac{\mu B_3}{\sqrt{2(N_{n+1} - S_{n+1})}} \cos \sigma_n,$$

$$S_{n+1} = S_n - \mu T \left[B_3 \sqrt{2(N_{n+1} - S_{n+1})} \sin \sigma_n + B_4 e' \sin(\sigma_n + \nu_n) \right] + \frac{c^2}{|m|},$$

$$\sigma_{n+1} = \sigma_n + (\alpha_5 + \alpha_6 S_{n+1})T + T \frac{\mu B_3}{\sqrt{2(N_{n+1} - S_{n+1})}} \cos \sigma_n. \tag{2.77}$$

In this mapping, $T = |m|2\pi/\Omega(a')$ is the commensurability period between the particle and the satellite motion, which is also the evolution time interval during one iteration of the mapping. Note that 1 year $\approx 40T$ for the inner edge and 1 year $\approx 80T$ for the outer edge of the ring.

2.4.3   *Numerical results*

In the procedure of constructing the mapping, some approximations have been adopted. For example, to simplify the system the coefficients A_3, A_4 in Eq. (2.73) have been replaced by the values in exact the specific resonances. Before applying the mapping in Eq. (2.77) to analyze the motion of particles in the ring, the plausibility of these approximations will be examined. A comparison between the original system described by equations

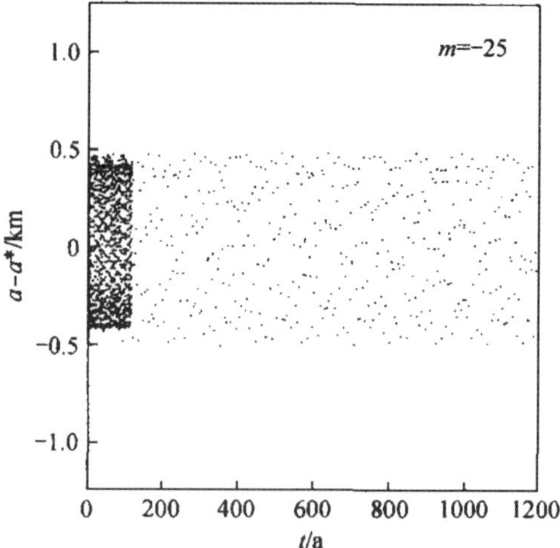

Fig. 2.13 Evolution of the same orbit, obtained from direct integration of the original equations of motion (the darker points, to 120 years) and from the mapping (light points, to 1,200 years). From Zhou, Sun & Hu (2000).

in Eq. (2.71) and the mapping in Eq. (2.77) is a direct way to verify the mapping.

Assume the dissipation is absent in the ring (thus $c = 0$), and calculate orbits from the same initial conditions but separately by integrating the equations in Eq. (2.71) and by iterating the mapping in Eq. (2.77). The results shown in Fig. 2.13 indicate that these two systems, described by differential equations of motion and by the mapping, have the same qualitative behavior. This consistence justifies the above approximation and verifies the reliability of the mapping. Using the mapping, to obtain the orbit up to 1,200 years, the computer spends less than 1/2,000 of the time it spends in obtaining the orbit up to 120 years by integrating the original differential equations, that is, the mapping method is at least 20,000 times more efficient than the full integration of original system. Therefore, it is reasonable to apply the mapping to study the ring dynamics below.

Conservative case

When the dissipative effects in the ring is ignorable, i.e. $c = 0$, the mapping Eq. (2.77) describes a conservative system. According to Table 2.1, $\mu = 1.6 \times 10^{-10}$ and 2.9×10^{-10} for the inner and outer edge of the ring respectively, and e' of Cordelia can be regarded as zero. Thus N is constant

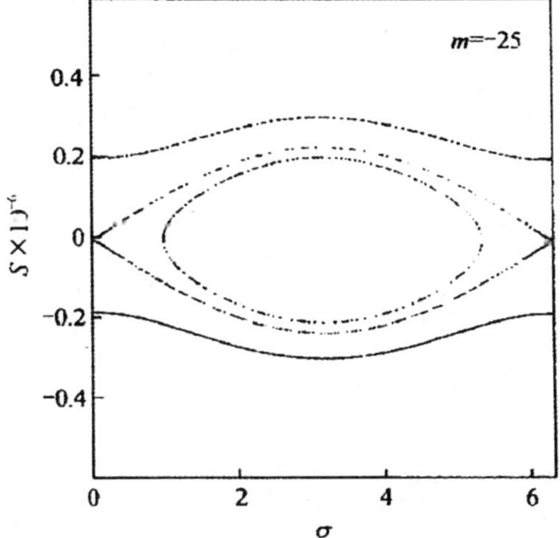

Fig. 2.14   Phase diagram of conservative mapping Eq. (2.77) with $c = 0$, for the inner case with $N = -2.92 \times 10^{-5}$. Adapted from Zhou, Sun & Hu (2000).

for the inner case and the system becomes a two-dimensional mapping. The phase portrait is like that of the simple pendulum motion, with libration and circulation regions as shown in Fig. 2.14, due to the resonance with the satellite.

For the outer case, $e' = 0.01$ (satellite Ophelia) thus N no longer remains constant, and the phase space is now four-dimensional. A surface of section is defined as $\nu = \pi, \frac{d\nu}{dt} < 0$, on which the properties of motion of particles at the outer edge of the ring can be shown, as Fig. 2.15. In this case, the resonance region looks very chaotic. The largest Lyapunov Characteristic Number (LCN) calculated for orbits initialized in this region also reveals that the motions of these particles are chaotic.

Dissipative case

When the dissipative effects are taken into account, particles in different regions of the phase space evolve differently. Adopt the parameter $c = 1\,\mathrm{cm/s}$, i.e. $c^2 \sim 10^{-12}$ in the units used in this section, and calculate orbits initially lying in libration, circulation and chaotic regions of the phase space. The results show that for these particles initially in the circulation and chaotic regions, they quickly leave the resonance region and evolve inwards (for the inner case) or outwards (for the outer case). Only those particles in the libration region remain in the libration region during the evolution.

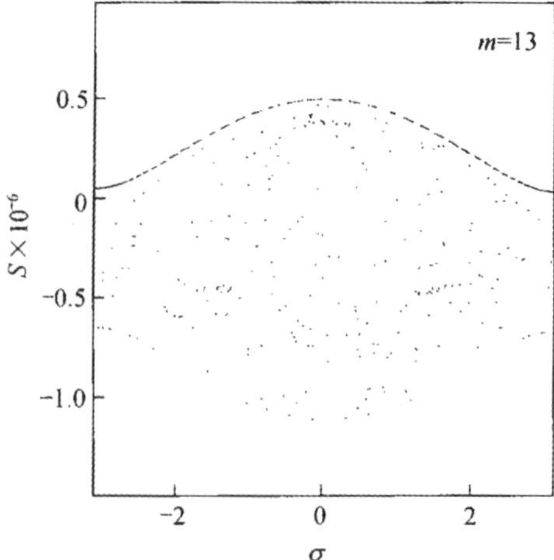

Fig. 2.15   Surface of section ($\nu = \pi, \mathrm{d}\nu/\mathrm{d}t < 0$) of conservative mapping Eq. (2.77) with $c = 0$, for the outer case. Adapted from Zhou, Sun & Hu (2000).

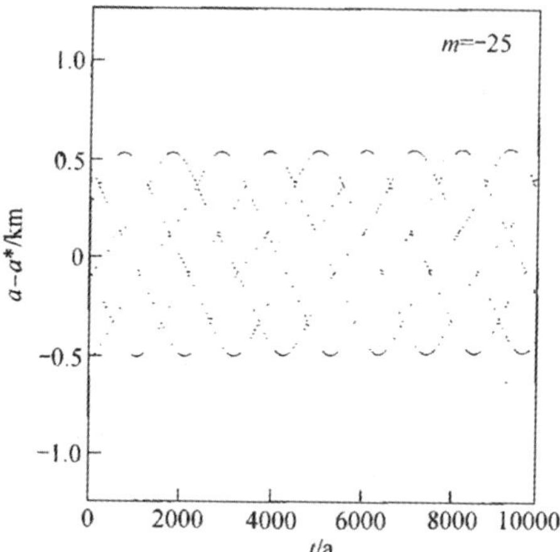

Fig. 2.16   Evolution of semi-major axis a of an orbit in the libration region of Fig. 2.14 under dissipation. Adapted from Zhou, Sun & Hu (2000).

Figure 2.16 illustrates the evolution of the semi-major axis of a particle initially in the libration region of the inner case. Meanwhile, during the evolution, the eccentricities of particles also stay low. This is because that the eccentricity excitation by the eccentric resonance must be balanced by the damping effects of viscosity in the ring.

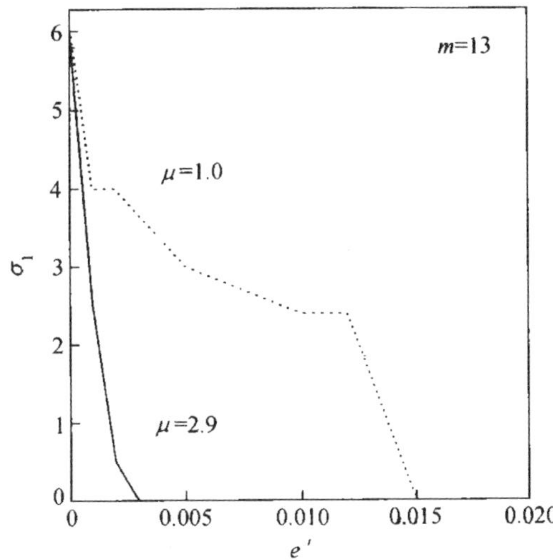

Fig. 2.17 Variation of libration range σ along with e' of the shepherding satellites of different masses (with respect to Uranus mass) $\mu(\times 10^{-10})$. Adapted from Zhou, Sun & Hu (2000).

According to these results, thanks to the small eccentricity of the inner shepherding satellite, the confinement of the inner edge of ϵ ring can be easily explained. However, for the shepherding of the outer edge, the resonance region is dominated by chaotic zone as shown in Fig. 2.15. Thus a question arises: How are the particles in the outer edge of ϵ ring shepherded?

The answer may lie in the size of the libration zone. In fact, in the conservative system without dissipation, the orbital eccentricity of the shepherding satellite plays a major role in determining the chaotic behavior of ring particles. If its eccentricity $e' = 0$, the system is integrable and the whole resonance zone is regular (libration). When e' increases, the chaotic region extends. The variation of the azimuthal range of libration region with respect to the satellite's orbital eccentricity is shown in Fig. 2.17. One can see from this figure that for $\mu = 10^{-10}$, the libration region disappears at an e' value larger than 0.015 while for $\mu = 2.9 \times 10^{-10}$ the region shrinks much quicker.

The mass of the shepherding satellite influences the size of libration region. The plot in Fig. 2.18 shows the value of the satellite eccentricity e' at which the libration region disappears as a function of the satellite's mass, ranging from 9×10^{18} g ($\mu = 10^{-10}$) to 9×10^{19} g ($\mu = 10^{-9}$) with an interval of 10^{18} g. The mass ratio $\mu = 10^{-10}$ is a third of nominal value, and

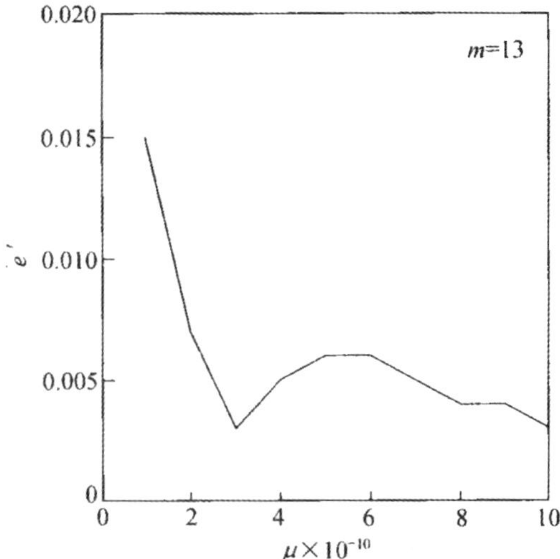

Fig. 2.18   Values of satellite eccentricity e' at which the libration region disappears as a function of the satellite-Uranus mass ratio μ. Adapted from Zhou, Sun & Hu (2000).

can correspond to the satellite Ophelia with a radius 14 km and a density of $1\,\text{g/cm}^3$, within the uncertainty of Ophelia's radius (16 ± 2 km). Numerical results display that in this value of μ, the surface of section is occupied by both the libration region and chaotic region when $e' = 0.01$. Thus, some of particles in the outer edge of ϵ ring can be shepherded by 14:13 IER. This may partly explain the confinement of outer edge of the ϵ ring, and as a consequence, the outer edge should be fuzzier than the inner one.

In the framework of the planar elliptical restricted three-body problem (Uranus, satellite and massless ring particle), together with a simple viscous model, the dynamical evolution of ring particles near the edges of the ϵ ring is investigated. Numerical results show:

1. The location of particle in the phase space of the corresponding conservative system decides the future of the particle. Only those particles initially lying in the libration region can be shepherded.
2. The mass and the orbital eccentricity of the shepherding satellite play important roles in shepherding of the ring.
3. If the satellite is in a circular orbit, all the particles in the eccentric resonance with the satellite can be shepherded. The confinement of the inner edge of ϵ ring by Cordelia is just this case.

4. If the satellite has larger orbital eccentricity, most of particle orbits in the eccentric resonance are chaotic and thus cannot be shepherded. The confinement of the outer edge of ϵ ring by Ophelia may be this case, provided it has the nominal mass of 2.5×10^{19} g. If Ophelia has a mass two thirds smaller, some of particle orbits in the resonance would be regular and be confined. This infers a smaller mass for Ophelia.

2.5 Formation of Kuiper belt

The Kuiper belt (KB), sometimes called the Edgeworth–Kuiper belt, is a region of the Solar System beyond the orbit of Neptune, occupied by numerous asteroids and extending to approximately 50 AU from the Sun. It has turned from a conjecture (Kuiper, 1951) to an observational reality since the asteroid 1992 QB1 was discovered in the early 1990s (Jewitt & Luu, 1993). At the time of writing, more than 1,200 Kuiper belt objects (KBOs) have been discovered and more than 10^5 KBOs over 100 km in diameter are believed to exist, making the KB 20 to 200 times more massive than the main asteroid belt between the orbits of Mars and Jupiter.

The main KBOs are divided into three distinct groups according to their orbital properties (see Fig. 2.19). The majority are the "Classical KBOs", locating between the 3:2 and 2:1 resonances with Neptune and with modest orbital eccentricities ($e \sim 0.1$). "Resonant KBOs" locate in mean motion resonances with Neptune (most in the 3:2 resonance with $a \approx 39.4$ AU, but also in 2:1, 7:4, 5:3 and 4:3 resonances). And the rest, which possess peculiar orbits with large eccentricities and extend as far as 100 AU, belong to the third category: "Scattered Disk Objects (SDOs)". Particularly, a few of SDOs with perihelia beyond about 40 AU, which are barely gravitationally influenced by planets, are called "Detached KBOs".

To explain the structures of KB and the orbital excitations of the KBOs, several models, such as the planet migration, the massive planet embryo scattering, and the encounter(s) between the proto-Sun and a passing star, have been proposed (see for example Gladman *et al.* (2001) for a review). All of these models have appealing facets, however the most popular model is the "planet migration and resonance sweeping" (Malhotra, 1995) in which Jovian planets experienced orbital migrations in the early epoch of the Solar System. During such a process, planetesimals in originally nearly circular orbits could be captured and "hijacked" to current zones by Neptune's mean motion resonances.

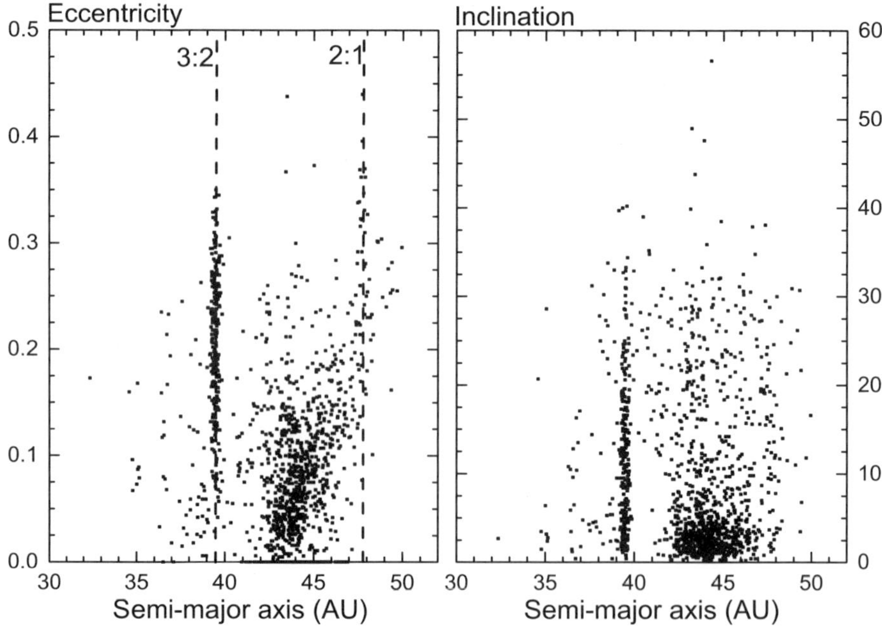

Fig. 2.19   The orbits of Kuiper belt objects. The left and right panels show their eccentricity and inclination distributions with respect to the semi-major axis. The data are taken from the IAU Minor Planet Center. The 3:2 and 2:1 resonances with Neptune are indicated by dashed lines.

2.5.1 *Planet migration and resonant capture*

The possibility is widely accepted that giant planets did not form *in situ* but migrated to current locations. In the late stages of planetary system formation, the interplanetary regions were populated by residual unaccreted planetesimal debris. The giant planets will clear these regions by gravitational perturbations and the recoil from those debris will affect planets' orbits. The numerical simulations (Fernández & Ip, 1984) show that some angular momentum and energy will be transferred from Jupiter to other Jovian planets via the scattered planetesimals and as a result, Neptune, Uranus and Saturn get significant net increases in orbital radius while Jupiter gets a net decrease.

A small body originally in a nearly circular orbit may have been captured and locked into an orbital resonance with Neptune, when these resonances sweep across a range of heliocentric distances as Neptune migrates outward. By analyzing this "resonance sweeping" mechanism and considering the eccentricity of Pluto (one of resonant KBOs in the 3:2 resonance), an estimates of the amplitudes of migrations Δa for the jovian planets are obtained (Malhotra, 1993; Malhotra, 1995). The timescale of such a migration is expected to be several to tens million year, comparable to that of accretion and much longer than the period of the resonant perturbations. The variation of the semi-major axis of Jovian planets is assumed as:

$$a(t) = a_{\mathrm{f}} - \Delta a \exp(-t/\tau), \qquad (2.78)$$

where a_{f} is the current semi-major axis and τ is the migration timescale.

Adopting this smooth migration in Eq. (2.78), some important features of the KB such as the existence of plutinos (KBOs in the 3:2 resonance are called "plutinos" because they seem to be dynamical counterparts of Pluto), the high eccentricities of them and the depletion of KBOs between 36–39 AU, are successfully explained. But some aspects are not well reproduced, in particular, the prediction of large number of objects concentrated in the 2:1 resonance simply fails. Besides, the median eccentricities and inclinations of the classical KBOs are less excited than in the observational data.

2.5.2 *Stochastic planet migration*

The early planet migration should be characterized by to-and-fro "jumps" due to the stochastic scattering of various sized planetesimals. In fact, each individual scattering event between planetesimals and planets should happen independently and stochastically. The migration must be a stochastic process although it has a mean direction.

Such a stochastic planet migration can be approximately simulated by adding an artificial force on the planet. The force is along the orbital velocity direction, $\hat{\boldsymbol{v}}$, and given by

$$\Delta \ddot{\boldsymbol{r}} = \frac{\hat{\boldsymbol{v}}}{\tau} \left\{ \sqrt{\frac{GM_{\odot}}{a_{\mathrm{f}}}} - \sqrt{\frac{GM_{\odot}}{a_{\mathrm{i}}}} \right\} (1 + \beta S_n) \exp\left(-\frac{t}{\tau}\right), \qquad (2.79)$$

where G is the gravitation constant, $M_{\odot}$ the solar mass, $a_{\mathrm{i}} = a_{\mathrm{f}} - \Delta a$ the semi-major axis at the starting point, S_n a Gaussian random variable with zero mean and standard deviation $\sigma = 1$, and β is the controllable amplitude of the variability. When $\beta = 0$, this force leads to a smooth

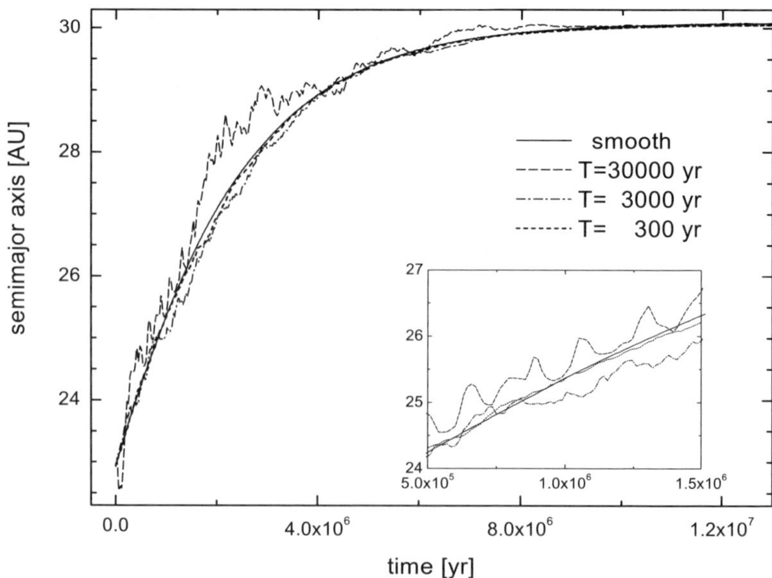

Fig. 2.20   The migration of Neptune. The solid line represents the "smooth" migration ($\beta = 0$). Other lines correspond to stochastic migrations with $\beta = 14$ and $T = 300, 3{,}000$ and $30{,}000$ yr in Eq. (2.79). Adapted from Zhou, Sun, Zhou *et al.* (2002).

migration. The subscript of S_n is determined by

$$n \equiv \mathrm{Int}[t/T], \qquad (2.80)$$

and T is a pre-selected "period". This period T means obliquely that every T-time the migrating planet suffers a "kick" from a scattering event.

Some test runs are carried out to ensure that at the end of the integration, the Jovian planets migrate to orbits which are very similar to their currently observed ones. Figure 2.20 shows the evolution of semi-major axis of Neptune.

2.5.3   *Resonance capture in stochastic migration*

The resonance capturing process in such a stochastic planet migration can be numerically simulated. In the simulations, the Solar System consists of the Sun and four Jovian planets, whose extent of migration (Δa) are $-0.2, 0.8, 3.0$, and 7.0 AU for Jupiter, Saturn, Uranus and Neptune respectively. The migration time scale is assumed to be $\tau = 2 \times 10^6$ yr (Malhotra, 1995). To illustrate the possible effects of the stochastic migration, several runs of simulation, with $T = 3,000\,$yr and different values of β in Eq. (2.79), are performed. In each run, there are 200 test particles (representing KBOs) with initial semi-major axis a distributed uniformly between 28 and 48 AU and angular orbital elements (longitude of perihelion ϖ, longitude of ascending node Ω and mean longitude λ) chosen randomly in $(0, 2\pi)$. For a definite β, there are two runs corresponding to a thin disc (in which test particles' initial eccentricities e and inclinations i equal to 0.01) and a thick disc ($e = i = 0.05$), respectively. The equations of motion of each body are integrated for $3 \times 10^7\,$yr. Simultaneously the Lyapunov characteristic indicators (LCI) are calculated, and the e-folding time $T_e = 1/\text{LCI}$ is used to judge whether an orbit is stable enough to survive and be included in the final statistics.

Figure 2.21 summarizes the a and e of the final orbits with $T_e \geq 10^5\,$yr in several cases with different values of β. And in Fig. 2.22, the distributions of KBOs in semi-major axis in these simulations are shown in histogram.

The top panels ($\beta = 0$) in Figs. 2.21 and 2.22 reproduce the features as in the smooth migration model (Malhotra, 1995): (a) The region between 28 and 36 AU is nearly empty; (b) the surviving test particles strongly concentrate in resonances with Neptune, primarily in the 3:2 and the 2:1 resonances; some other resonances such as the 4:3 and 5:3 (36.5 and 42.3 AU) also gather some objects; and (c) the orbits in the resonances may have significant eccentricities. And, no significant difference between the thin disc and the thick disc is found. Therefore, from now on, except specially mentioned, these two discs will be discussed as a whole.

When the stochastic character in the migration is enhanced as β increases, the situation changes. Some noteworthy features in Fig. 2.21 and Fig. 2.22 are summarized as follows. As β increases: (a) The survival rate decreases. As indicated in Fig. 2.22, the total number of "stable" orbits drops from 245 ($\beta = 0$) to 158 ($\beta = 24$). This number increases to 189 in the case $\beta = 32$. The reason is that the effect of "resonance sweeping" is overtaken by random effect, and the distribution of orbits is determined only by the long term stability of the system. (b) The concentration in the 3:2 resonance is well kept but fewer and fewer orbits gather in the 2:1 resonance. (c) More and more test particles enter the region between resonances, especially occupy the positions of "classical KBOs". And the case $\beta = 14$ has a semi-major axis distribution most similar to observations. (d) In the non-resonant region between 42.5 and 47.5 AU, the eccentricity is more pumped up. The average eccentricity in this region increases from $\bar{e} = 0.037$ ($\beta = 0$) to $\bar{e} = 0.054$ ($\beta = 14$), although it is still smaller than

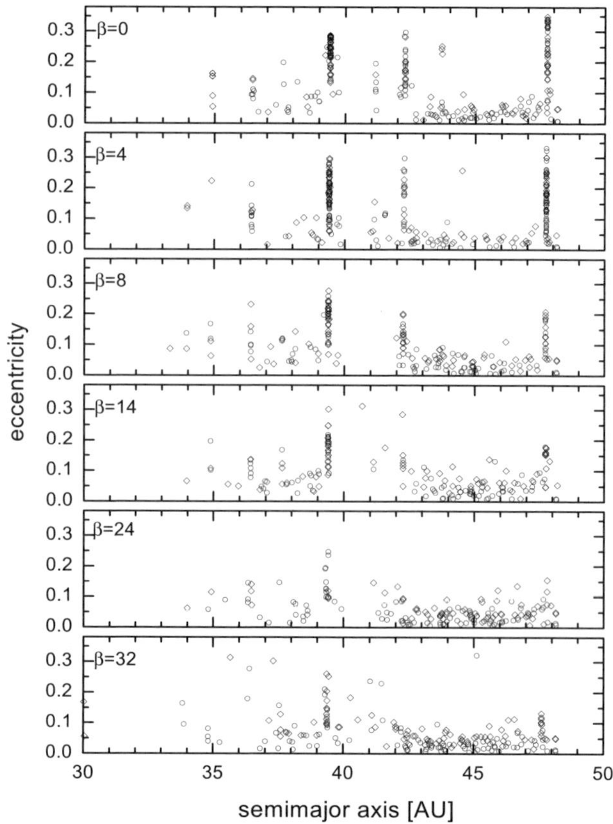

Fig. 2.21 The final orbits of the surviving test particles ($T_e \geq 10^5 \, \text{yr}$). Circles and diamonds represent particles initialized in the thin disc and the thick disc, respectively. From the top down, the panels show the cases with an increasing value of the control parameter β. Adapted from Zhou, Sun, Zhou *et al.* (2002).

the observational value. (e) Most objects remain in relatively low inclination orbits, while a correlation between larger β and higher inclinations can be found.

Figure 2.23 shows the relation between a test particle's initial and final semi-major axis, only for $\beta = 0$ and $\beta = 14$. This diagram reveals that the probability of an object being captured by a resonance is sensitive to its initial conditions. For example, the region between 37 and 39 AU has been swept by the 2:1, the 5:3 and the 3:2 resonances one by one, and test particles there have different fates due to their different initial orbits. In the case $\beta = 14$, some objects with initial a less than 33 AU get peculiar orbits

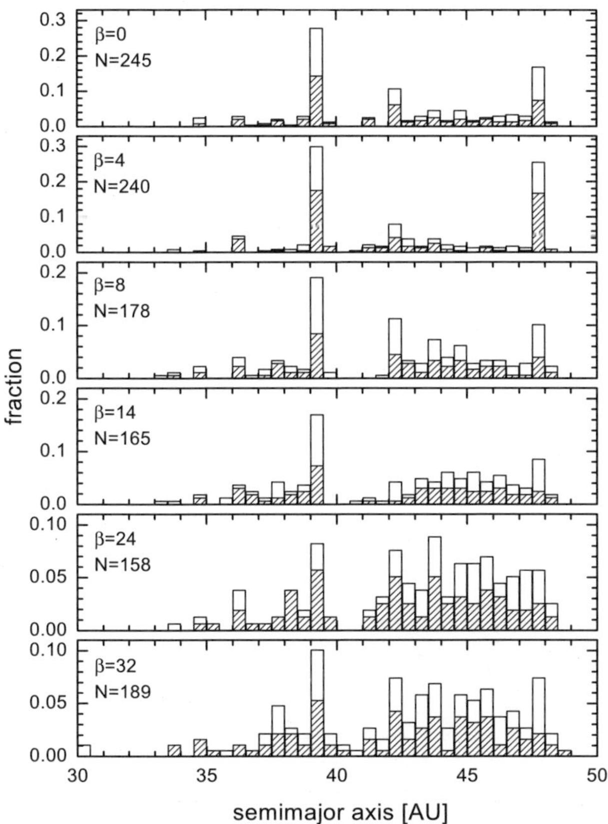

Fig. 2.22 The distribution of semi-major axis of the surviving test particles originated from the thin disc (shaded) and the thick disc (open). The number of surviving particles is indicated in each panel. Adapted from Zhou, Sun, Zhou *et al.* (2002).

with relatively high eccentricities (~ 0.3). So it seems that the stochastic effects can excite some objects to the orbits of scattered KBOs. Another conclusion inferred from this diagram is that the efficiency of resonance capture in the 3:2 resonance is higher than in the 2:1 resonance, especially when $\beta = 14$. As Neptune migrates outwards from 23.2 to 30.2 AU, the 3:2 resonance and the 2:1 resonance sweep across a range of ~ 9 and ~ 11 AU, respectively. Although the latter has swept larger range in space, the former has captured more objects.

By carefully checking hundreds of test particles' orbits in the above two cases, one finds the following details. When $\beta = 0$, the resonances begin to trap objects as soon as the migration begins. Each resonance has a certain

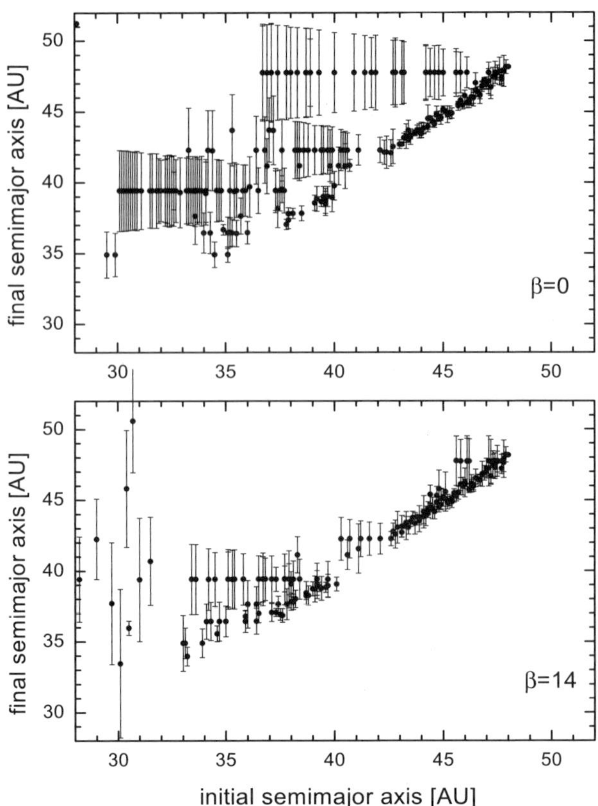

Fig. 2.23　The final semi-major axis of objects versus the initial semi-major axis, with the final eccentricities (after multiplying by 10) indicated by error bars. Points are from both the thin disc and thick disc. Adapted from Zhou, Sun, Zhou *et al.* (2002).

"most efficient" period, in which the rate of successful capturing is relatively high. Such period is from 0 to 1.5 Myr for the 3:2 resonance (1–2 Myr for 2:1) when it sweeps from about 31 to 34 AU (40–43 AU for 2:1). During such periods, a considerable fraction (~ 70 and 45 percent for the 3:2 and the 2:1 resonance, respectively) of stable test particles originally locating on the way of the sweeping resonance, are trapped into the definite resonance. Those objects leaking out of a resonance may be recaptured later by the successive resonances, and consequently the nonresonant regions interior to 43 AU are nearly cleared. The range of 43–48 AU is swept only by the (low order) 2:1 resonance, with a capture rate ~ 25 percent. Thus a fair number of objects survive there as candidates of the classical KBOs.

However, in the case $\beta = 14$, the scenario changes. In the region interior to 39 AU, more objects lose their stability and the 4:3, 3:2 and 5:3 resonances capture less objects. For the 2:1 resonance, it fails to capture any objects before 3 Myr when it had passed through the former "productive reservoir" between 40 and 43 AU. After that, in the range of 43–48 AU, the capture rate ($\sim$16 percent) is lower than when $\beta = 0$, and quite a number of objects survive near their initial locations. These survivors, now as classical KBOs, have suffered perturbation from the 2:1 resonance as it swept through, and as a result, their eccentricities and inclinations are pumped up.

Besides this relatively poorer capturing efficiency of the 2:1 resonance, objects in the 2:1 resonance are less stable than those in the 3:2 resonance after the stochastic effect was introduced. This can be shown by re-drawing the histogram as in Fig. 2.22 but with a more strict stability criterion. Figure 2.24 shows the distribution of surviving test particles with $T_e \geq 3 \times 10^5$ yr for the case $\beta = 14$. Some less stable orbits are eliminated from the statistics, and consequently the concentration of objects in the 2:1 resonance nearly disappears. And the distribution of objects are more similar to the reality (observation data).

Another parameter, the "period" T in Eq. (2.80), is also controlling the property of the migration. A smaller T induces more frequent fluctuations during the migration, but the deviation from the smooth route is relatively small. While a larger T means that the planet suffers a migration composed of relatively larger but less frequent fluctuations.

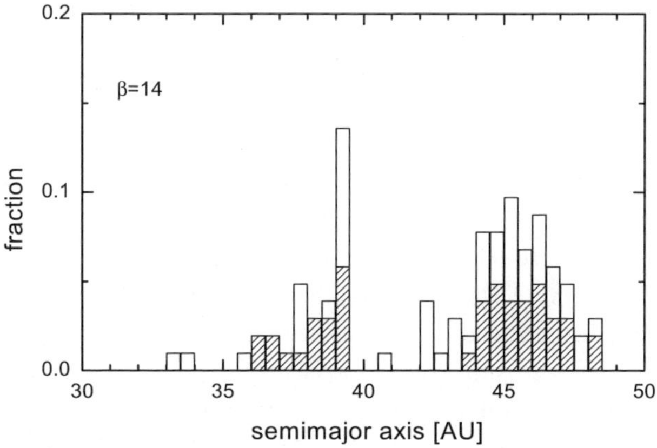

Fig. 2.24   The same as Fig. 2.22, but the survival criterion is now $T_e \geq 3 \times 10^5$ yr. Adapted from Zhou, Sun, Zhou *et al.* (2002).

In addition to $T = 3{,}000\,\mathrm{yr}$ applied above, another two runs with $T = 300$ and $T = 30{,}000\,\mathrm{yr}$, both with $\beta = 14$, are performed. The time variations of Neptune's a have been shown in Fig. 2.20. The noteworthy features in the final distributions of the orbital elements are as follows. The survival rate in the former case ($T = 300\,\mathrm{yr}$) is by one thirds lower than the latter case ($T = 30{,}000\,\mathrm{yr}$). Thus the frequent fluctuations in the orbital expand are more efficient in depleting the KB. And the latter case favors more the concentration of objects in the 3:2 resonance and depletion in the 2:1 resonance.

Thus far, all runs have a migration timescale $\tau = 2 \times 10^6\,\mathrm{yr}$. When a longer timescale is adopted, the 2:1 resonance in a smooth migration is more efficient in trapping objects, retaining over 40 percent of survivors, and leaving relative few of them in the 3:2 resonance and the classical belt. However, the stochastic effects in this slow migration reduce the global survival rate, increase the population in the 3:2 resonance relatively to the 2:1 resonance, and leave more objects in the nonresonant classical belt. The stochastic effects also do good to the excitation of orbital eccentricities and inclinations although the inclination is much less excited compared with the real KBOs' orbits, as in the cases with a shorter timescale τ. In summary, a slower migration makes the 2:1 resonance much more "sticky", but the stochastic effects surely still play an important role.

The dynamical structure of the KB and the radial distribution of the KB objects can be well explained by the planet migration and resonance sweeping mechanism. In a smooth migration model, the 2:1 mean motion resonance with Neptune capture a quite large number of objects. As a result, the concentration of KBOs in the 2:1 resonance is prominent. This is not consistent with the reality: Hundreds of KBOs have found concentrated in the 3:2 resonance but much less objects have been observed in the 2:1 resonance.

The real process of migration should have some stochastic characters, and the stochastic character plays an important role in determining the orbital distribution of the KB. The numerical simulations show that the distribution of KB objects changes considerably when the stochastic effects were introduced. The concentration of objects in the 3:2 resonance is still distinct, but the efficiency of resonance capture in the 2:1 resonance decreases dramatically. Numerous objects flee from being captured by the 2:1 resonance and enter the region of classical KBOs. The final radial distribution of objects match the observed "reality" quite well when the proper stochastic character was included. The eccentricities and inclinations of orbits are also excited by the stochastic effects, although the inclinations were not excited enough in comparison with the observations. There should be some other mechanisms, probably the low order secular resonances, being responsible for the inclination excitation. The stochastic migration also scattered some objects to peculiar orbits with high eccentricities, contributing to the source of the scattered KB objects.

2.6 Dynamics of Neptune Trojans

The orbital stability of asteroids trapped around the triangular equilibrium points L_4, L_5 has been investigated in Sec. 2.2. Asteroids in the vicinities of L_4 and L_5 of a parent planet are called Trojans after the group of asteroids found around Jupiter's Lagrange points. Objects on Trojan-like orbits around Mars and (temporarily) around Earth have also been observed.

As for Neptune, the possibility of stable orbits around the triangular Lagrange points have been verified before the discovery of the first Neptune Trojan, 2001 QR322 (Pittichova *et al.*, 2003). At the time of this writing, 9 Neptune Trojans have been discovered, and their orbital properties are listed in Table 2.3. Taking into account of the discovering efficiency and the surveyed sky area, people believe that the population of Neptune Trojans may be much larger (by at least one order of magnitude) than Jupiter Trojans or even than the main belt asteroids (Sheppard & Trujillo, 2006). Both the observing and the theoretical analysis on their orbital dynamics could give important clues on how these objects were trapped into their current orbits, where and when the planet Neptune formed and how the

Table 2.3   Orbits of 9 observed Neptune Trojans, given at epoch Dec. 9, 2014, with respect to the mean ecliptic and equinox at J2000.0. The mean anomaly M, perihelion argument ω, ascending node Ω and inclination i are in degrees.

Designation	M	ω	Ω	i	e	a (AU)	L_4/L_5
2004 UP10	346.09	7.2	34.8	1.4	0.027	30.130	L_4
2005 TO74	278.72	305.3	169.4	5.3	0.055	30.120	L_4
2001 QR322	73.90	158.1	151.7	1.3	0.027	30.208	L_4
2004 KV18	65.13	294.9	235.6	13.6	0.187	30.202	L_5
2005 TN53	297.98	88.3	9.3	25.0	0.068	30.127	L_4
2006 RJ103	254.38	24.9	121.0	8.2	0.032	30.045	L_4
2007 VL305	3.08	218.6	188.7	28.1	0.065	30.079	L_4
2008 LC18	176.12	9.4	88.5	27.6	0.085	29.922	L_5
2011 HM102	28.43	150.6	101.0	29.4	0.080	30.059	L_5

orbits of the outer planets evolved in the early stage of the formation of the Solar System.

In this section, a global investigation on the dynamical properties of Neptune Trojans will be presented.

2.6.1 *Model and method*

To study the dynamics of Neptune Trojans, the orbital evolutions of thousands of test particles around the Lagrange points L_4 and L_5 of Neptune are numerically simulated using a Lie-series integrator (Hanslmeier & Dvorak, 1984), and their orbital stability are investigated using a method based on the spectral analysis on the outcome of numerical integrations.

Dynamical model

The dynamical model includes the Sun, four Jovian planets and the massless Trojans, i.e. the motion of massless Trojans is a restricted many-body problem. A Neptune Trojan asteroid shares the same orbit with Neptune and they are in fact in a 1:1 mean motion resonance. The critical argument of this resonance is

$$\sigma = \lambda - \lambda_8, \tag{2.81}$$

where $\lambda = \omega + \Omega + M$ is the mean longitude and the subscript '8' denotes Neptune (hereafter the subscripts '5, 6, 7' and '8' refer to Jupiter, Saturn, Uranus and Neptune as usual).

The initial σ_0 of fictitious Neptune Trojans for the numerical simulations is set as $\sigma_0 = \sigma_c$, where σ_c is the libration center. This center is near $\pm 60°$ for planar near-circular orbit and varies with the eccentricity and inclination of the Trojan. The value $\sigma_c(e_0, i_0)$ can be determined using a numerical method (Zhou, Dvorak & Sun, 2011). To ensure a coverage of the representative region around the Lagrange points, the choice of the initial angular variables for the fictitious Trojans is made as: $\omega_0 = \omega_8 + 60°$ for L_4 ($-60°$ for L_5), $\Omega_0 = \Omega_8$, $M_0 = M_8 + \sigma_c - 60°$ for L_4 ($M_0 = M_8 + \sigma_c + 60°$ for L_5). After some test simulations, the initial semi-major axes a_0 of the test particles are determined to be evenly sampled in the range from 29.72 to 30.32 AU for L_4 (and 29.90 to 30.50AU for L_5). Note these two ranges of a are different from each other (see a discussion below).

Apparently in the above arrangement, the mean longitude of the test particle is $\lambda = \lambda_8 + \sigma_c$ and the initial value of the resonant argument σ is always at the libration center. For a Trojan, the argument σ generally librates around σ_c with a nonzero amplitude D, and simultaneously the semi-major axis a oscillates with an amplitude d around a mean value given by the semi-major axis of Neptune. In the simplest approximation the values of d and D are related to each other by:

$$\frac{d}{D} = \sqrt{3\mu_8}\, a_8 + \cdots \approx 0.373, \tag{2.82}$$

where μ_8 is the mass of Neptune measured in the solar mass, d is measured in AU and D in radians. For this reason, having different a_0 values, it is not necessary to test different initial value of σ.

Spectral analysis

During the integration of the system, the orbits of the fictitious Trojans are recorded for further analyzing. The system must be integrated to a time long enough to reveal the secular behavior, but, for such a long time span, the data to be analyzed would be too large. The amount of the output data may be reduced by increasing the time interval of outputs, but the risk of losing valuable information and/or introducing artificial features arises from the large sampling period. As a compromise, a low-pass digital filter is introduced to remove the short-period terms, which generally are not crucial and can be cut off by the average procedure.

For the purpose of understanding the dynamics of Neptune Trojans, an output interval of $4\,\mathrm{yr}$ is chosen, and after the filtering process a sampling interval of $\Delta = 512\,\mathrm{yr}$ is adopted. Finally, the systems are integrated to $3.4 \times 10^7\,\mathrm{yr}$ and thus totally $N = 65,536\,(= 2^{16})$ lines of the filtered orbital variables are recorded for further analyzing. The frequency resolution through an FFT is $f_{\mathrm{res}} = \frac{1}{N\Delta} = 2.980 \times 10^{-8}\,\mathrm{yr}^{-1}$.

Generally, a regular orbit is characterized by quasi-periodic behaviors, so that the power spectra of the orbital elements of this orbit are dominated by a countable (small) number of frequency components. On the contrary, a typically chaotic orbit is not quasi-periodic and the independent frequencies of the motion change with time, such that the power spectrum consists of broadband components and is characterized by strong noise. To tell the difference between the spectra of regular and chaotic motion, one may simply count the spectral peaks above a given noise level, and the number obtained is defined as the "spectral number" (SN). Obviously, a small SN indicates a regular motion while a large one implies a chaotic motion (Michtchenko & Ferraz-Mello, 1995).

Symmetry between L_4 and L_5

Since long ago, the asymmetrical distribution of stable regions around Neptune's L_4 and L_5 points puzzles. For example, a shift in the semi-major axis of stable regions around L_4 and L_5 was found when test particles in the outer solar system were integrated to 20 Myr (Holman & Wisdom, 1993). In the numerical simulations, the initial conditions of fictitious Neptune Trojans are set referring to the osculating orbit of Neptune, symmetrically distributed around the L_4 and L_5 points. This symmetry is only valid in the frame consisting of the Sun, Neptune and asteroid. But the real system evolves symmetrically with respect to the barycenter, which is not exactly in the Sun. As a consequence, a shift in the initial semi-major axis of the libration center around L_4 and L_5 appears. This can be demonstrated numerically using the spectral analysis method (Zhou, Dvorak & Sun, 2009).

Therefore, under current planetary configuration of the Solar System, there is no asymmetry between L_4 and L_5, and it is reasonable to focus on only one of these two points in the dynamical analyses that will be presented below. However, this does not expel the possibility that a specific formation history of Trojan clouds may affect this symmetry.

2.6.2　*Dynamical maps*

By constructing dynamical maps in the space of orbital elements, a global understanding of the dynamics of Neptune Trojans can be obtained.

Dynamical map in (a, i)-plane

The inclinations of observed Neptune Trojans spread in a wide range up to $\sim 30°$ (see Table 2.3). To check the dependence of Neptune Trojans' orbital stability on the orbital inclination, a dynamical map on the (a, i)-plane is constructed as below. The initial eccentricity of test particles are set as $e_0 = e_8 = 0.00625 \approx 0$ (nearly circular), and other initial orbital elements ω, Ω, M are chosen to be the representative values as mentioned before.

Adopting the SN as an indicator of the regularity of orbits, the dynamical map is constructed using 5757 orbits starting from a 101×57 grid on the (a_0, i_0) plane. The power spectrum of $\cos \sigma$ for each orbit is calculated and the number of peaks, which are over one percent of the highest peak, is defined as the SN. To limit the number into a manageable range, an SN is forced to be 100 if it was originally larger than that. Those orbits that

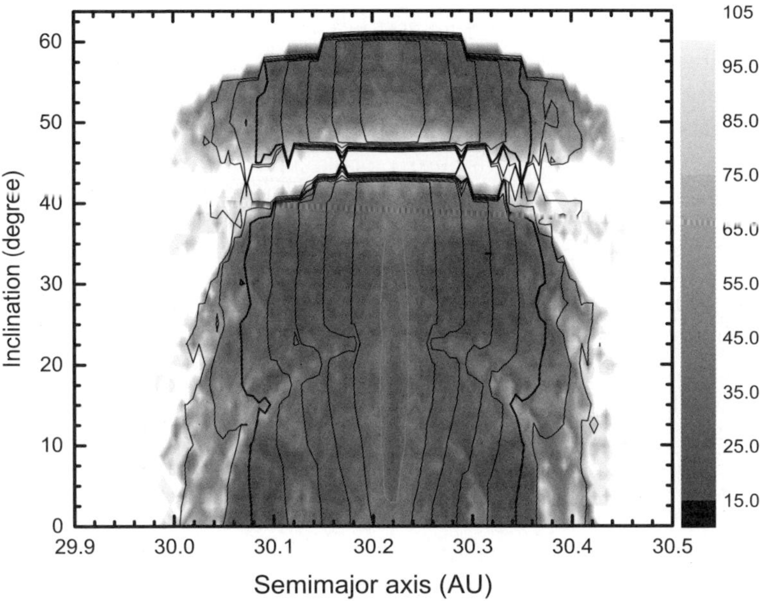

Fig. 2.25 The dynamical map around L_5 point. The grey scale indicates the SN. Dark indicates regular motion, while light indicates chaotic motion. The libration amplitude $\Delta\sigma$ of the resonant angle is also mapped by contour curves. Adapted from Zhou, Dvorak & Sun (2009).

escape from the 1:1 resonance region are excluded by assigning an SN of 110 to them. Such a dynamical map for L_5 is shown in Fig. 2.25.

The grey scale in the dynamical map indicates the SN. In the dark region where SN is relatively small, the motion is dominated by a few dominating frequencies and thus it is more regular; but in the light grey region, the spectrum of $\cos\sigma$ is characterized by strong noise and thus the motion is chaotic; the white color with $SN = 110$ indicates escaped orbits.

Judging from the dynamical map, the most regular orbits mainly exist in three regions in inclination: A: $0°-12°$, B: $22°-36°$, and C: $51°-59°$, approximately. The regions A and B are connected by a less regular region, but region C is apparently separated from others by an unstable gap at $i_0 \sim 44°$. Region C at high inclination is isolated from regions A and B, therefore any Trojans found in region C must be primordial.

The coplanar orbits are not necessarily more regular than the inclined ones. In fact, there are two less regular "holes" at low inclination $i_0 < 5°$ and around $a_0 \sim 30.13, 30.31\,\mathrm{AU}$ in Fig. 2.25. The central area, in which the libration amplitude is small, is likewise not necessarily more safe for Trojans. Different secular resonances are crowded in this area and their overlappings may introduce chaos.

The libration amplitude of the resonant argument σ, i.e. the difference between the maximum and minimum values of σ during the integration time, $\Delta\sigma = \sigma_{max} - \sigma_{min}$, is also plotted in Fig. 2.25 as contours. The regular region in Fig. 2.25 is approximately shaped by the contour curve of $\Delta\sigma = 60°$. But for small inclination $i_0 < 10°$ there are some regular orbits having $\Delta \sim 70°$. Submit $\Delta\sigma = 60°$ ($D = \pi/3$) into Eq. (2.82), one gets the oscillating amplitudes of semi-major axis $d \approx 0.39\,\mathrm{AU}$. It is a little larger than the value obtained from Fig. 2.25 ($\sim 0.31\,\mathrm{AU}$), which is shrunk by perturbations from other planets besides Neptune.

Some other fine structures can be easily recognized in the dynamical map in Fig. 2.25. These notable features are also accompanied by the deformations of the $\Delta\sigma$ contour curves.

Dynamical maps in (a, e)-plane

The dependence of orbital stability on the eccentricity can also be portrayed by constructing dynamical maps on the (a, e)-plane. For this sake, the initial inclinations i_0 is fixed at given values and the fictitious trojans are initialized on the grid on the (a_0, e_0)-plane while other orbital elements are given the representative values as described previously.

In Fig. 2.26, the dynamical maps for $i_0 = 0°, 10°, 20°, 30°, 40°$ and $i_0 = 50°$ are presented. Surely it is impossible and also unnecessary to calculate each slice at every possible initial inclination. These slices in Fig. 2.26 are enough to show the variation of dynamical map with respect to the initial inclination. Note that those cases in which most of the Trojan orbits are unstable and therefore less interested, e.g. $i_0 \sim 45°$, are abandoned. However, for the sake of providing more information about the Trojan motion at other inclinations, other two dynamical maps for $i_0 = 5°, 35°$ will be used in Fig. 2.29 and Fig 2.30 as examples to present the frequency analysis results below.

The first conclusion drawn from these dynamical maps may be that the most stable motion does not happen at any inclination, but only at $i_0 = 0°, 5°, 10°, 30°, 35°$ and $55°$. This is consistent with the finding in Fig. 2.25 of three most stable regions in inclination.

A second conclusion is that the initial eccentricity e_0 for stable orbits must be small. Although for some initial inclinations (e.g. $i_0 = 30°, 35°$) some orbits with e_0 as large as 0.25 survive in the integrations and may

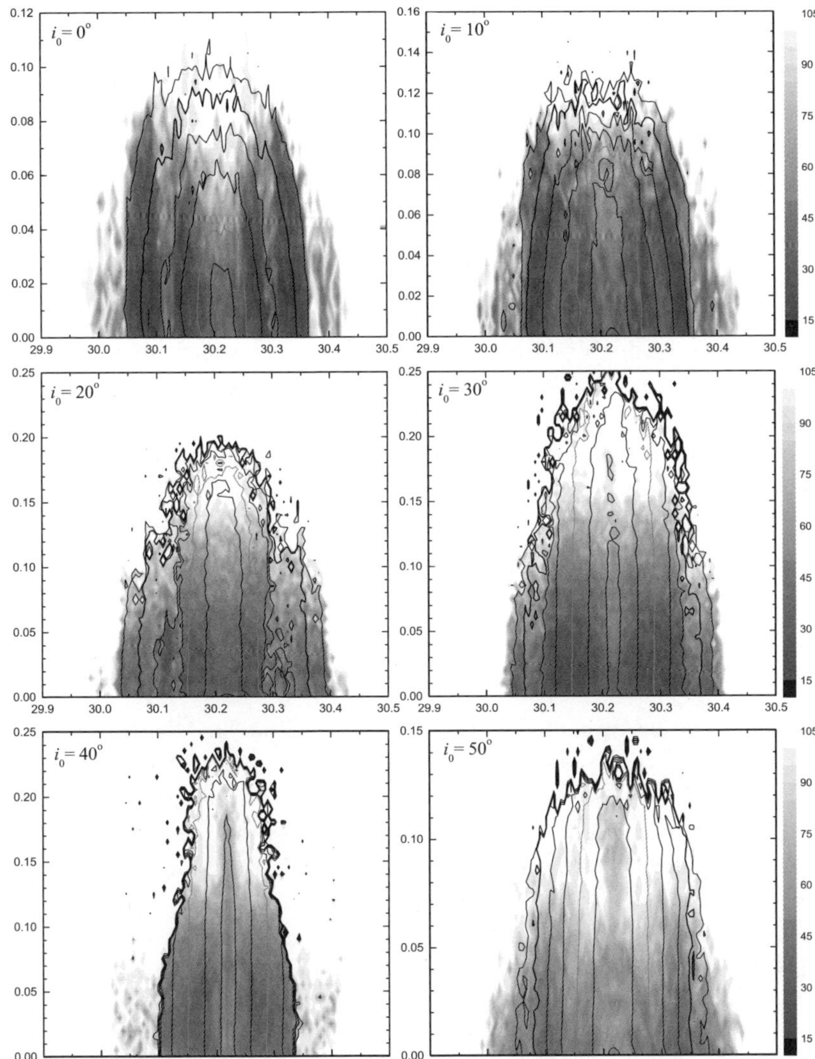

Fig. 2.26 Dynamical maps on (a_0, e_0)-plane, for initial inclinations $0°, 10°, 20°, 30°, 40°$ and $50°$ (left to right, top to bottom). The abscissae and ordinates are initial semi-major axes and eccentricities. Note the ordinates may have different scales. The contour curves represent the libration amplitude of resonant angle $\sigma = \lambda - \lambda_8$. From inside out are contours for $\Delta\sigma = 10°$ to $\Delta\sigma = 70°$ with an increment of $10°$. Adapted from Zhou, Dvorak & Sun (2011).

survive even for much longer time, their SN's indicate that most probably they will not remain on the Trojan-like orbits in the Solar System age. These orbits with high e_0 may suffer perturbations through resonances whose effects appear significantly only in a long time span. For the most stable orbits, the largest initial eccentricities are around 0.08 in Region A, 0.07 in Region B and 0.02 in Region C.

The libration amplitude of the resonant angle $\Delta\sigma$ increases monotonically with the distance of initial semi-major axis a_0 from its value at libration center ($a_c \sim 30.22\,\mathrm{AU}$). The maximum $\Delta\sigma$ is around $\Delta\sigma \sim 70°$ for stable orbits with low (Region A) and medium inclination (Region B), but shrinks a little for high inclination (Region C) to $\Delta \sim 50°$. At given i_0 and a_0, the libration amplitude increases only very little as the initial eccentricity increases.

Contrary to the profiles of the libration amplitude contours that are more or less plain, the dynamical maps bear plenty of fine structures. Most distinctly, there are vertical gaps of irregular motion on slices of small inclination ($i_0 = 0°, 5°, 10°$), and a "spike" of relatively regular orbits with small $\Delta\sigma$ can be clearly seen on slices of large inclination ($i_0 = 40°, 50°$).

2.6.3   *Frequency analysis and resonances map*

To understand the mechanisms that are responsible for the forming of the fine structures in the dynamical maps, one needs to find different resonances associated with these structures. One natural and simple way to determine a resonance is to check the behavior of the corresponding critical resonant angle, whose librating character is an evident sign of being inside a resonance. But this method is valid only for low order mean motion resonances or secular resonances of second order. It does not work when the resonance involves more than two degrees of freedom or when it is of high order. In fact, a typical power spectrum of a complex motion generally is a composition of an amount of terms with comparable amplitudes and different frequencies, this means, the regular variation of a critical angle of a high-order resonance may be enshrouded by other high-amplitude terms even when the motion is deep inside the resonance. Therefore, most of the time, to check whether the resonant angle is librating or circulating does not work for detecting the resonances. Instead, the power spectra of some variables of motion are calculated and used as a tool for detecting the involved resonances.

Dynamical spectrum

Typically, a power spectrum is very informative but very complicated as well. It is a complicated composition of peaks at all the forced frequencies, free frequencies, their harmonics and their combinations. Therefore a direct look at a specific power spectrum may not be very helpful. However, when examining the continuous change of the spectrum with a parameter, some important features emerge. This continuous change of a spectrum, defined as the "dynamical spectrum", is constructed as follows.

For each Neptune Trojan orbit in the simulation, the power spectra of three basic variables in the motion, namely $a\cos\sigma$, $e\cos\varpi$ and $i\cos\Omega$ are calculated. They give information about the most important rates of the resonant argument, the perihelion longitude and the nodal longitude, respectively. These three proper frequencies are denoted by f_σ, g and s hereafter. From each power spectrum, the leading terms in the spectrum are picked out. Here "leading terms" refers to those frequencies at which the peaks in the power spectrum are among the highest ones.

Vary the orbital parameter, record these leading frequencies at each parameter value, plot all these frequencies together on one figure, and finally one may find how the dominating frequencies of the motion change with respect to the parameter. Meanwhile, the fundamental secular frequencies of the outer Solar System, i.e. the apsidal and nodal precession frequencies of Jupiter, Saturn, Uranus and Neptune are known. These frequencies, denoted as usual by $g_5, g_6,\ g_7, g_8$ for apsidal processions and $s_5, s_6,\ s_7, s_8$ for nodal processions, can be found in literatures or directly from the frequency analysis on the numerical results. Thus, the forced frequencies in the spectrum of Neptune Trojans' motion can be easily recognized. Besides these secular rates, the quasi-2:1 mean motion resonance between Neptune and Uranus, which has a frequency of $f_{2:1} = 2.3606 \times 10^{-4}\, 2\pi/\mathrm{yr}$, also plays an important role in the motion of Neptune Trojans. Since the proper frequencies of the motion generally change continuously with the orbital parameter, they can also be recognized. In a dynamical spectrum, along with the varying parameter, a continuously varying frequency may meet and cross another frequency, where a resonance associated with these frequencies happens. Thus, a dynamical spectrum can be used to detect and locate the resonances in the motion, and more importantly, it helps us determine the proper libration and precession frequencies of the motion, f_σ, g and s.

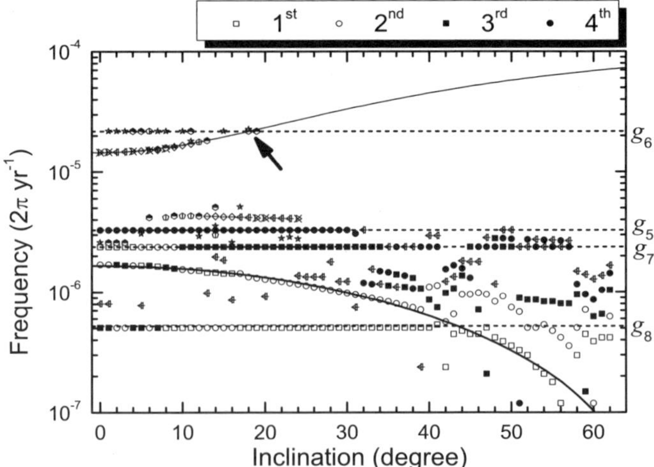

Fig. 2.27 Dynamical spectra of the apsidal variable $e \cos \varpi$. The initial orbits are taken from a vertical line $a_0 = 30.098\,\mathrm{AU}$ on the (a_0, e_0) plane (Fig. 2.25). The frequency of the highest peak is denoted by open squares, and the frequencies of the 2nd, 3rd and 4th highest peaks are open circles, open triangles and solid circles respectively. Other less prominent frequencies are given in different type of symbols. The dashed lines give the values of the fundamental frequencies in the outer Solar Systems g_5, g_6, g_7, g_8. The thick solid curves are the numerical fit of the proper frequency of $e \cos \varpi$ and the thin solid curve stands for $f_{2:1} = 2f_\sigma$ from the numerical fit. Adopted from Zhou, Dvorak & Sun (2009).

Figure 2.27 shows an example of dynamical spectrum, in which the variation of the frequencies of $e \cos \varpi$ with respect to the initial inclination is plotted. Note the strength of the peak is denoted by its order in the sequence of peaks' heights. The peaks at the fundamental frequencies are the forced terms, and their frequencies do not change with the parameters (i_0). The proper frequency can be easily recognized in Fig. 2.27 because it changes continuously with the parameter. Note the proper frequency of the apsidal precession g is always below g_5, g_6 and g_7, but it meets g_8 at $i_0 \sim 43.5°$ where the ν_8 secular resonance ($g = g_8$) is located. Also, a frequency at low inclination can be seen approaching g_6 with increasing i_0 and meeting g_6 finally at $i_0 \sim 18°$, as indicated by an arrow in Fig. 2.27, indicating another resonance happens here.

Semi-analytical model

With the data of the proper frequencies (f_σ, g, s) at different parameter values in hand, one can obtain the empirical formulations for the proper frequencies with respect to orbital parameters (e.g. the semi-major axis, inclination and eccentricity) on the whole parameter space. With these empirical formulations and those already-known frequencies, one may calculate all kinds of equations of frequencies that define the resonances in the parameter space.

After carefully checking the dynamical maps, the dynamical spectra, and the empirical expressions of the proper frequencies, two types of resonances with their frequencies characterized by two equations as below are found tightly related to the motion of Neptune Trojans:

$$S: \quad pg + qs + \sum_{j=5}^{8}(p_j g_j + q_j s_j) = 0,$$

$$C: \quad h f_\sigma + k f_{2:1} + pg + qs + \sum_{j=5}^{8}(p_j g_j + q_j s_j) = 0,$$

$$(2.83)$$

where h, k, p, q, p_j, q_j are integers. The "S" stands for "Secular", indicating they are secular resonances in the usual sense. While the "C" for "Combined" indicates that these resonances are combinations of mean motions and secular frequencies. The d'Alembert rule for S-type resonance is $p + q + \sum_{j=5}^{8}(p_j + q_j) = 0$, and it's $k + p + q + \sum_{j=5}^{8}(p_j + q_j) = 0$ for C-type. In both equations, $(q + \sum_{j=5}^{8} q_j)$ must be even so that the symmetry of inclination with respect to the reference plane can be guaranteed.

Resonances map

In Fig. 2.28, the major secular resonances on the initial (a, i)-plane is presented. The dynamical map is plotted as the background simultaneously, providing a convenient comparison between the secular resonances and the dynamical stability.

The very good agreement, between the fine structures in dynamical map and the locations of different resonances, is obvious. As shown in Fig. 2.28, it is clear that the unstable gap around $i_0 \sim 44°$ is determined by the ν_8 resonance, the Kozai mechanism is responsible for the depletion of Neptune Trojans at high inclination, and the nodal resonance ν_{18} takes place at low inclination. Around the exact location, every resonance has a "width" in which the resonance may happen. But, whether a resonance actually happens depends on whether the motion is strongly influenced by mechanisms other than the resonance itself. For example, along the V-shape ν_{18} secular resonance curves in Fig. 2.28, only at low inclination with $i_0 < 1.5°$ this

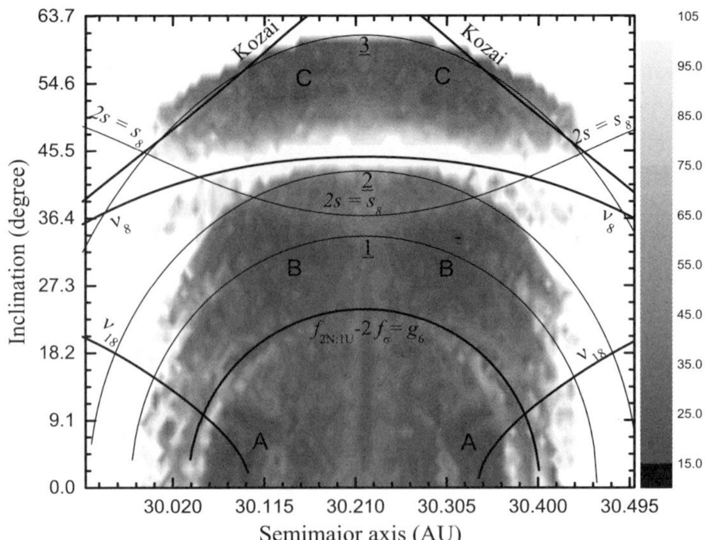

Fig. 2.28 The main resonances for Trojans around the L_5 point on the (a_0, i_0) plane. The resonances are labeled along the curves. The S-type resonances includes the apsidal resonance ν_8 $(g - g_8 = 0)$, the nodal resonance ν_{18} $(s - s_8 = 0)$ and the Kozai mechanism where $g - s = 0$. Three thin curves labeled $\underline{1}$, $\underline{2}$ and $\underline{3}$ represent the locations of C-type resonances where $f_{2:1} - 2f_\sigma = \frac{3}{2}g_6$, $2g_6$ and $3g_6$, respectively. Adapted from Zhou, Dvorak & Sun (2009).

resonance happens, and at higher inclination it is hidden by other mechanisms.

The curve $f_{2:1} - 2f_\sigma = g_6$ in Fig. 2.28 coincides very well with the stable arc on the dynamical map. Further calculations show that different commensurabilities between $f_{2:1} - 2f_\sigma$ and g_6 are most probably responsible for the multiple arc structures, as other three curves labelled $\underline{1}$, $\underline{2}$ and $\underline{3}$ show in Fig. 2.28. Note that these commensurabilities are not exactly the resonances since the d'Alembert rule is not fulfilled. This is because the contributions of all the longitude precessions, especially those with lowest frequencies, is difficult to identify. However, the plot of these commensurabilities gives good estimate of the position of the real resonances.

On the (a, e)-plane with different inclinations, such resonance maps can also be constructed. Two examples, for $i_0 = 5°$ and $i_0 = 35°$ respectively will be presented below.

The most distinguishable structure in the dynamical map of $i_0 = 5°$ (and also other slices at low inclination, e.g. $i_0 = 0°, 10°$ in Fig. 2.26) is

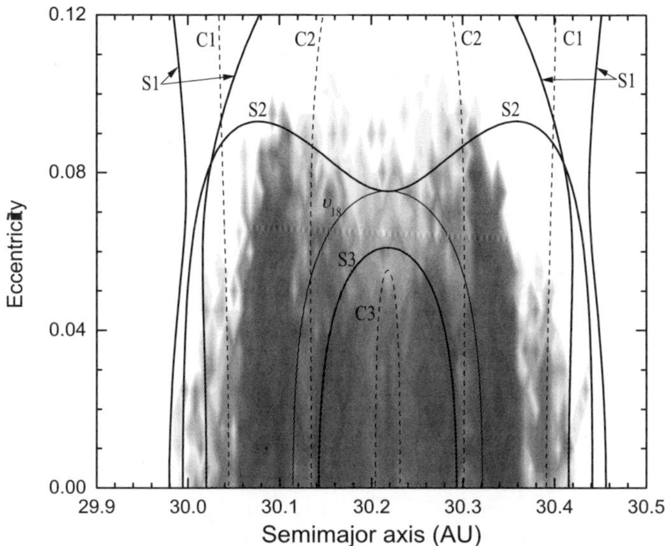

Fig. 2.29 The locations of resonances plotted over the dynamical map for the case with initial inclination $i_0 = 5°$. The C-type resonances are indicated by dashed curves while the solid curves stand for the S-type resonances, particularly, the secular resonance ν_{18} is plotted by the thin solid curve. See text to find the meaning of labels C1, C2, S1, S2 and S3. Adapted from Zhou, Dvorak & Sun (2011).

the less stable vertical "stripes". They arise from the effects of some C-type resonances, among which three major ones are shown in Fig. 2.29. The dashed curves labelled by C1, C2 and C3 are locations of the following resonances:

$$C1 : 2f_\sigma - f_{2:1} + g_6 = 0,$$
$$C2 : 4f_\sigma - 2f_{2:1} + g_6 + g_7 = 0, \tag{2.84}$$
$$C3 : 6f_\sigma - 3f_{2:1} + g_5 + g_6 + g_7 = 0.$$

Apparently, their locations match the positions of vertical structures very well. It is worth stressing that each resonance listed above (and below too) is just a representative of a bunch of resonances, characterized by higher order combinations of integers p, q, p_j, q_j in Eq. (2.83). For example, beside the resonance C1 that reaches the abscissa at $a_0 = 30.044$ and $30.392\,\mathrm{AU}$, one can easily find another two resonances as C1$'$: $2f_\sigma - f_{2:1} - g + g_8 + g_6 = 0$ and C1$''$: $2f_\sigma - f_{2:1} + g - g_8 + g_6 = 0$. The resonance C1 itself is favorable for the stability of Trojan orbit. In fact, the asteroid 2001 QR322 listed in Table 2.3 is deep inside such a resonance and its orbit is stable over the Solar System age. On the contrary, the resonances C1$'$ and C1$''$ are responsible for the unstable gaps in both sides of resonance C1, especially the apparent ones around $a_0 = 30.050\,\mathrm{AU}$ and $30.375\,\mathrm{AU}$ approximately.

Another mechanism that may contribute to the formation of the vertical structures is the secular nodal resonance ν_{18}. Its location is also plotted in Fig. 2.29, as well as other three S-type resonances:

$$\begin{aligned}
&\text{S1}: \ 2g - g_7 - g_8 = 0, \\
&\text{S2}: \ g - 2s + g_8 - 2s_8 + 2s_5 = 0, \\
&\text{S3}: \ 3s - 2g_8 - s_8 = 0.
\end{aligned} \tag{2.85}$$

However, no apparent details in the dynamical map can be found around the location of ν_{18} resonance. This resonance does not show any prominent effects in protecting or destroying the orbital stability of Neptune Trojans. Other secular resonances, e.g. the S1, S2 and S3 in Eq. (2.85), however, may have more evident dynamical effects. The resonances S1 and S2 define the edge of the survival region, and the curvature of S2 in the eccentricity range of 0.07–0.09 matches the outline of the stable region. The border between the stable region and chaotic region makes a V-shape area from $e_0 = 0.05$ to $e_0 = 0.09$, in which several crossovers or close encounters between resonances S2, S3, C2, C3 and ν_{18} happen. Thus it seems reasonable to argue that the border is made of a net of these resonances.

Another observing from Fig. 2.29 and other dynamical maps at different initial inclinations in Fig. 2.26 is that the C1-type resonance with $h = 2, k = -1$ in Eq. (2.84) has quite promising dynamical effects in all initial inclination values, but the influences from C2 and C3 decreases as i_0 increases. It will be seen clearly in next example of $i_0 = 35°$.

As in the previous case, some C-type resonances involve in the motion of orbits with $i_0 = 35°$, of which several typical ones are plotted in Fig. 2.30 and two are listed below.

$$\begin{aligned}
&\text{C1}: \ 2f_\sigma - f_{2:1} - g + 2g_6 = 0, \\
&\text{C2}: \ 2f_\sigma - f_{2:1} - s + 2s_6 - s_8 + g_8 = 0.
\end{aligned} \tag{2.86}$$

The low-order C-type resonance like the C1 defined in Eq. (2.84) does not appear in the region around $i_0 = 35°$. Among the resonances in Eq. (2.86), only the one with the lowest order (C1) has the distinguishable dynamical effects, shaping the sharp edge of stable region at low eccentricity

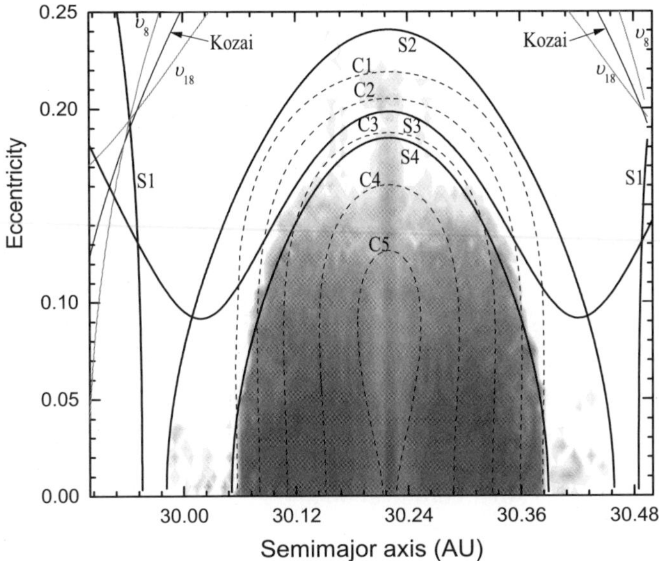

Fig. 2.30   The same as Fig. 2.29 but for $i_0 = 35°$. Except the resonances labelled by C1–C5 (dashed curves) and S1–S4 (solid curves), the locations of ν_8, ν_{18} resonances (solid curves) and Kozai mechanism (thin solid curve) have also been plotted. Adapted from Zhou, Dvorak & Sun (2011).

($e_0 \leq 0.05$). Another interesting resonance is C2, in which the nodal precessions (s, s_8 and s_6) are involved. In fact, as the inclination of Trojan orbit increases, it is natural to expect that the nodal precession is becoming more important. The influences from the C-type resonances are less important in slice $i_0 = 35°$ than in slice $i_0 = 5°$. Relatively, the effects from the S-type resonances are remarkable. Beside the ν_8, ν_{18} and Kozai mechanism that locate outside the survival region as shown in Fig. 2.30, several other S-type resonances as below are determined and illustrated.

$$\begin{aligned} \text{S1}: \quad & g + s - g_8 - s_8 = 0, \\ \text{S2}: \quad & g - 2s + g_8 - s_8 + s_5 = 0, \\ \text{S3}: \quad & g + 2s - g_6 + s_6 - 3s_5 = 0, \\ \text{S4}: \quad & g - 2s + g_8 + s_8 + g_7 - g_5 - s_5 = 0. \end{aligned} \tag{2.87}$$

Clearly the S2 resonance combined with S3 defines the boundary of the survival region, and the resonance S4 establishes the edge of the stable region up to $e_0 = 0.12$. Again, on the top part of the dynamical map, the

C-type and S-type resonances gather, resulting in overlaps among them and leading to chaotic motion.

At different inclinations, different resonances participate in featuring the dynamical maps. Basically, the secondary resonances (C-type) as well as the secular resonances (S-type) are the most important dynamical mechanisms that form the fine details in the dynamical maps and determine the complicated orbital behavior of Neptune Trojans. Compared to the cases in low inclination, those highly inclined orbits are more deeply affected by the nodal-type resonances. In all cases, different resonances with different orders make a resonance net in the parameter space of initial conditions. Thus an orbit may diffuse slowly along the net and migrate on the parameter space from one configuration to another. Of particular interest is the slow diffusion from low inclination orbits to high inclination orbits, which may contribute to the high inclinations observed in some Neptune Trojans' orbits such as 2005 TN53, 2007 VL305 and 2008 LC18 as listed in Table 2.3.

2.6.4 *Origin of Neptune Trojans*

The high inclination of some Neptune Trojans may also be inherited from its "predecessor", or obtained during the process of being captured into current orbits. Since there must be a large number of Neptune Trojans to be discovered in future, it would be worth to discuss the origin of them, especially those Neptune Trojans with high inclination. In fact, the origin of asteroids on Trojan-like orbits is always a challenge.

In a sense, Neptune Trojans are a kind of "resonant KBOs" trapped in the 1:1 mean motion resonance with Neptune (Sec. 2.5). But their origin cannot be simply explained by the planet migration and resonance sweeping model. In this scenario, the planetesimals have either been captured by preceding resonances, such as the 2:1 and 3:2 resonances, or been scattered away, leaving nothing for the subsequent 1:1 resonance. However, the capturing of Trojans is possible if Neptune was "suddenly thrown" into a highly eccentric orbit in the planetesimals disc and it captured planetesimals into the Trojan orbits quickly before they are scattered away.

In fact, this could happen as a result of encounters of proto-Neptune with Jupiter and/or Saturn (Thommes, Duncan & Levison, 1999; Tsiganis *et al.*, 2005), or of the fact that Neptune was temporarily trapped in a mean motion resonance with a hypothetical massive planetary embryo or one of the other planets (Morbidelli & Levison, 2004). By the dynamical friction due to the planetesimal disc, Neptune's eccentricity was damped quickly in less than 10^6 years. A byproduct of this eccentricity damping process is a quick decline in the semi-major axis of several AU due to energy dissipation. Although this mechanism is delicate and would need very careful simulating, the capturing efficiency can be tested through simple simulations.

Simply assume a time variation of Neptune's eccentricity as:

$$e(t) = e_i \exp(-t/\tau_{\rm e}), \tag{2.88}$$

where e_i is the initial eccentricity, $e(t)$ the eccentricity at epoch t, and $\tau_{\rm e}$ the eccentricity damping time-scale. This eccentricity damping and the accompanied rapid inward migration of Neptune can be mimicked by adding an artificial force on the planet along the radius vector $\hat{\boldsymbol{r}}$:

$$\Delta\ddot{\boldsymbol{r}} = -\frac{2\boldsymbol{r}\cdot\boldsymbol{v}}{\tau_{\rm e}} \cdot \frac{\hat{\boldsymbol{r}}}{r}, \tag{2.89}$$

where $\boldsymbol{r}$ and $\boldsymbol{v}$ are Neptune's heliocentric position and velocity. The initial eccentricity of Neptune is set to 0.3, within the range proposed in the literature (Thommes, Duncan & Levison, 1999). The linear perturbation theory gives the estimate of the magnitude of orbital migration related to the decrease in the eccentricity of a planet:

$$\ln\left(\frac{a_f}{a_i}\right) = e_f^2 - e_i^2, \tag{2.90}$$

where a_i and a_f indicate the initial and final semi-major axes of the planet. To simplify the model, only the Sun, Jupiter, Saturn and Neptune are included in the model. Their initial semi-major axes are assumed to be 5.4, 8.7, 23.2 AU for Jupiter, Saturn and Neptune respectively, consistent with the estimation calculated from the planet migration model introduced in Sec. 2.5.

To reproduce a similar orbital distribution as the observations, 1000 test particles with random $e \in (0, 0.1)$ and $i \in (0, 30°)$, is set from 20.5 to 23.5 AU to simulate a dynamically hot planetesimal disc. The initial (e, i) space of test particles approximately covers the same volume as the observed Trojans. All other angular elements are chosen randomly in the range of $[0, 2\pi)$.

Set $\tau_{\rm e} = 2 \times 10^5$ and 3×10^5 years, the system is integrated for 2×10^6 years. In the numerical simulations, the test particles that entered a Hill sphere radius of any planets is discarded, and a test particle is regarded as a captured Trojan if its semi-major axis $a \approx a_N$ and the critical argument $\sigma = \lambda - \lambda_N$ librates. Figure 2.31 summarizes the eccentricities and inclinations of the captured Trojans at the end of numerical integration for the case $\tau_{\rm damp} = 2 \times 10^5$ years.

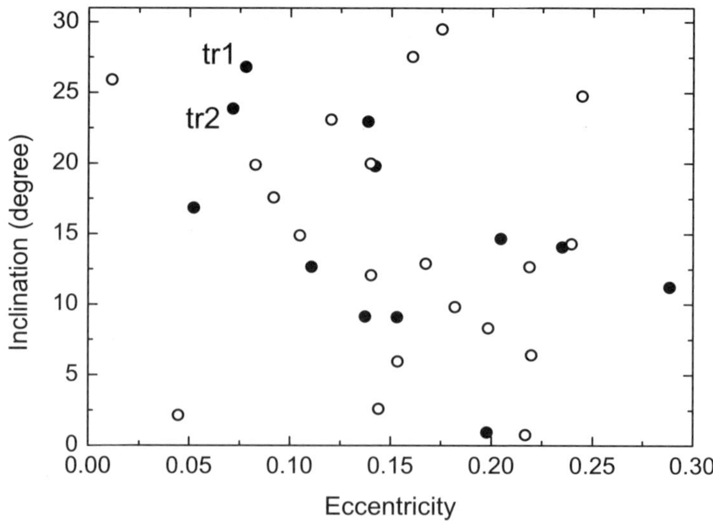

Fig. 2.31   The eccentricities and inclinations of the captured Trojans for the case $\tau_e = 2 \times 10^5$ years. Only those with librating amplitudes $\Delta\sigma < 75°$ are shown. The filled circles stand for Trojans librating around the L4 point, and the open circles around the L5 point. Adapted from Li, Zhou & Sun (2007).

There are two Trojans (labeled as "tr1" and "tr2") have orbits analogue to 2005 TN53 and 2007 VL305, that is $i > 20°$ and $e < 0.1$, librating around Neptune's L4 point. The total number of the test particles trapped in the Trojan region is 32, indicating a capture rate $\sim 3\%$. A careful checking of the orbits of trapped Trojans reveals that (1) the capture of Trojans starts as soon as the inward migration begins, despite Neptune's high eccentricity; (2) Trojans' inclinations have no significant change throughout the integration time; (3) the high-inclination ($i > 5°$) test particles are more likely to survive, then be captured as Trojans when Neptune sweeps through.

When a longer damping timescale is adopted, $\tau_e = 3 \times 10^5$ years, the capture efficiency reduces a little to $\sim 2\%$. In comparison with the short τ_e case, a distinguishable difference is the deficit of Trojans in the region $e < 0.1$. This might be due to the relatively longer time gravitational perturbation of the eccentric Neptune. So a longer damping timescale does not favor such a Trojan capturing scenario.

Thus far, the initial eccentricity of Neptune has been assumed to be 0.3. Some other starting eccentricities of Neptune $e_i^N = 0.1$, 0.2, 0.4 (an even higher value is unlikely to happen) are also examined. It is found that

Neptune's ability to capture test particles into the Trojan-like orbits, in the process of the rapid inward migration, is not sensitive to its starting eccentricity. However, this migration is closely related to the eccentricity damping (the orbital shift is a function of e_i as described in Eq. (2.90)), so that a smaller e_i^N leads to less capturing because a much narrower region would be swept through by Neptune in this case.

After the orbit being circularizing by the dynamical friction of the planetesimal disc, Neptune joined in the global migration with other planets as described in Sec. 2.5, in which its orbit expanded from $\sim 23\,\mathrm{AU}$ to $\sim 30\,\mathrm{AU}$. During such migration, about 70% of the captured Trojans leak out the 1:1 resonance, and the rest 30% migrate along with the planet current location. Particularly, those Trojans with relatively large eccentricities are more susceptible to lose, but no visible correlation between the surviving probability and the inclination can be found. Consequently, the final orbital distribution at the end of the system's evolution is similar to the observational data, i.e. Trojans on orbits of low eccentricity and wide range of inclination.

The orbits of Neptune Trojans around two triangular Lagrange points L_4 and L_5 have the same dynamical behaviors. The only difference between them is in the value of the osculating semi-major axis of orbits at the libration center in the Trojan clouds around the L_4 and L_5 points. This difference was found due to the asymmetrical selection of initial conditions.

There are three most stable areas in the (a, i)-plane with $i_0 \in (0°, 12°)$, $(22°, 36°)$ and $(51°, 59°)$. The dynamical map on the (a, e)-plane is also enriched by fine structures, indicating the diversity of orbital behavior of Neptune Trojans. The eccentricity for stable orbits is small.

Using the dynamical spectrum method based on the frequency analysis, the mechanisms creating the fine structures in the dynamical maps is revealed. Two types of resonance are found may involve deeply in the dynamics of Neptune Trojans. One is the secondary resonance (C-type) concerning mainly two frequencies: The libration frequency of the resonant angle $\sigma = \lambda - \lambda_8$ and the frequency of the quasi 2:1 mean motion resonance between Neptune and Uranus $2\lambda_8 - \lambda_7$. The other is the secular resonance in general sense (S-type), characterized by the commeasurability between different combinations of secular frequencies, such as the frequencies of apsidal and/or nodal precessions of Neptune Trojans and/or planets. Among the well-known secular resonances, the ν_8 resonance is very effective in driving a Trojan to highly eccentric orbit thus make a deep unstable gap in inclination around $44°$. On the contrary, the ν_{18} resonance, found nearly everywhere at low inclination, is so weak that it has hardly any influence on the dynamics of Neptune Trojans.

All the resonances may comprise a "resonant net" that connect different parts of the whole orbital parameter space. Correspondingly, there must be such a net in the phase space. With the help of this net, an orbit may diffuse in the phase space, resulting in significant change in the orbital behavior. Long-term diffusion may be expectable along the net. The most attractive diffusion, probably is the slow transferring of an orbit from low inclination to high inclination. Although under current configuration of the outer Solar System, such inclination-increasing phenomenon seemingly does not happen. However, this is still a very important issue. The importance arises not only from the interesting dynamics itself, but also from the fact that the orbital configuration of planets in the outer Solar System must have changed in the early stage. Particularly, the C-type resonances are very powerful in featuring the dynamics of Neptune Trojans, meanwhile C-type resonances are sensitively determined by the frequency $f_{2:1}$. If the planetary orbits changed, even only very little, $f_{2:1}$ would change and this would lead to significant varying in the position and strength of the C-type resonances.

2.7 Apsidal and nodal resonances in multiple planetary systems

In recent two decades, hundreds of planets have been discovered around the extra-solar stars. Particularly, the Kepler mission increases dramatically the number of extra-solar planets (also known as "exoplanet") to more than one thousand. Many of these exoplanets are found in multiple-planet systems. For a multiple-planet system, like the Solar System, the dynamical stability of the system under planetary interaction is an important issue concerning the dynamical evolution as well as the possible existence of a habitable zone in the system.

2.7.1 *Apsidal resonance*

There are many dynamical effects that can affect the stability of a multiple-planet system. For the orbits of planets with small or modest eccentricities and inclinations, mean motion resonances (MMR) between planets can sometimes lead to stable configurations. Another effect is the secular resonance. An apsidal resonance (also known as "apsidal corotation") occurs when the relative apsidal longitudes of two planets orbits $\Delta\varpi$ librates about 0 (aligned resonance) or π (antialigned resonance) during the evolution. Because of the aligned apsidal resonance, the two planets on elliptic orbits may greatly reduce the possibility of close encounters, thus it is believed that aligned apsidal resonance can stabilize the orbits of interactive planets.

In this section, a two-planet system will be discussed. Hereafter, subscripts "0", "1" and "2" will be used to indicate quantities of the host star, the inner planet and the outer planet, respectively. Thus in such a system, three objects have masses m_0, m_1 and m_2, where $m_1, m_2 \ll m_0$. According to the theory of planet formation, all planets in a planetary system are formed from the same circumstellar disk thus have small relative inclinations. Therefore, a planar model is adopted here, in which two planets with osculating orbital elements $(a_1, e_1, \varpi_1, M_1)$ and $(a_2, e_2, \varpi_2, M_2)$, where a, e, ϖ and M are the semi-major axis, eccentricity, longitude of periastron and mean anomaly, as usual. Note that the inclination and ascending node of orbits do not appear since the orbits of two planets are in the same reference plane. The commonly used unit system is adopted, in which the mass unit is the Solar mass, the length unit is $1\,\mathrm{AU}$ and the time unit is $1\,\mathrm{yr}/(2\pi)$. In such a unit system, the gravitational constant $G = 1$.

According to the linear secular theory (e.g. Murray & Dermott (1999)), for coplanar planets m_1 and m_2, the perturbation function for their motions are given by

$$R_1 = n_1 a_1^2 \left[\frac{1}{2} A_{11} e_1^2 + A_{12} e_1 e_2 \cos(\varpi_1 - \varpi_2) \right],$$

$$R_2 = n_2 a_2^2 \left[\frac{1}{2} A_{22} e_2^2 + A_{21} e_1 e_2 \cos(\varpi_1 - \varpi_2) \right], \tag{2.91}$$

where $n_{1,2}$ are the mean motions of planets $m_{1,2}$ respectively and A_{ij} are elements of the matrix A defined as

$$A = \begin{pmatrix} A_{11} & A_{12} \\ A_{21} & A_{22} \end{pmatrix} = \begin{pmatrix} c_1 & -c_0 c_1 \\ -c_0 c_2 & c_2 \end{pmatrix}. \tag{2.92}$$

In this equation, $c_k > 0$ $(k = 0, 1, 2)$ are functions of a_1, a_2, m_0, m_1 and m_2:

$$c_0 = \frac{b_{3/2}^{(2)}(\alpha)}{b_{3/2}^{(1)}(\alpha)} \approx \frac{5}{4}\alpha \left(1 - \frac{1}{8}\alpha^2\right),$$

$$c_1 = \frac{1}{4} n_1 \frac{m_2}{m_0 + m_1} \alpha^2 b_{3/2}^{(1)}(\alpha),$$

$$c_2 = \frac{1}{4} n_2 \frac{m_1}{m_0 + m_2} \alpha b_{3/2}^{(1)}(\alpha), \tag{2.93}$$

with $b_{3/2}^{(i)}(\alpha)$ $(i = 1, 2)$ being the Laplace coefficients, and α is the semi-major axis ratio $\alpha = a_1/a_2 < 1$. Define

$$\xi = \frac{c_2}{c_1} = \frac{1}{\alpha} \frac{n_2 m_1 (m_0 + m_1)}{n_1 m_2 (m_0 + m_2)} \approx q \alpha^{1/2}, \tag{2.94}$$

where $q = m_1/m_2$. In the above approximation, the terms with orders of $O(m_{1,2}/m_0)$ or higher are neglected, since typically $m_{1,2}/m_0 \sim 10^{-3}$ in a planetary system. Denote the eigenvalues of matrix A in Eq. (2.92) by g_1, g_2, and the corresponding eigenvectors are

$$S_i \begin{pmatrix} \cos\theta_i \\ \sin\theta_i \end{pmatrix},$$

where $\theta_i \in (-\pi/2, \pi/2)$ and

$$\cos\theta_i = \frac{c_0 c_1}{\sqrt{(c_1 - g_i)^2 + c_0^2 c_1^2}},$$

$$\sin\theta_i = \frac{c_1 - g_i}{\sqrt{(c_1 - g_i)^2 + c_0^2 c_1^2}}, \qquad (i = 1, 2), \tag{2.95}$$

with eigenvalues g_1 and g_2 being

$$g_1 = \frac{1}{2}\left[(c_1 + c_2) + \sqrt{(c_1 - c_2)^2 + 4c_0^2 c_1 c_2} \right],$$

$$g_2 = \frac{1}{2}\left[(c_1 + c_2) - \sqrt{(c_1 - c_2)^2 + 4c_0^2 c_1 c_2} \right]. \tag{2.96}$$

Further define

$$\rho_1 \equiv \tan\theta_1 = \frac{1}{2c_0}\left[1 - \xi - \sqrt{(1 - \xi)^2 + 4c_0^2 \xi} \right],$$

$$\rho_2 \equiv \tan\theta_2 = \frac{1}{2c_0}\left[1 - \xi + \sqrt{(1 - \xi)^2 + 4c_0^2 \xi} \right], \tag{2.97}$$

therefore $\rho_1 < 0$ and $\rho_2 > 0$, or $-\pi/2 < \theta_1 < 0 < \theta_2 < \pi/2$. The scaling factor S_i $(i = 1, 2)$ are expressed in terms of initial eccentricities e_{10}, e_{20}

and $\Delta\varpi_0 = \varpi_{20} - \varpi_{10}$ as

$$S_1 = \frac{(\rho_2^2 e_{10}^2 - 2\rho_2 e_{10} e_{20}\cos\Delta\varpi_0 + e_{20}^2)^{1/2}}{|\rho_1 - \rho_2|\cos\theta_1} \equiv \frac{F}{|\rho_1 - \rho_2|\cos\theta_1},$$

$$S_2 = \frac{(\rho_1^2 e_{10}^2 - 2\rho_1 e_{10} e_{20}\cos\Delta\varpi_0 + e_{20}^2)^{1/2}}{|\rho_1 - \rho_2|\cos\theta_2} \equiv \frac{G}{|\rho_1 - \rho_2|\cos\theta_2}. \tag{2.98}$$

The secular system with perturbation functions in Eq. (2.91) is integrable and the solutions can be written as

$$e_1 = \frac{\left(F^2 + 2FG\cos\Delta\psi + G^2\right)^{1/2}}{|\rho_1 - \rho_2|},$$

$$e_2 = \frac{\left(\rho_1^2 F^2 + 2\rho_1\rho_2 FG\cos\Delta\psi + \rho_2^2 G^2\right)^{1/2}}{|\rho_1 - \rho_2|},$$

$$e_1 e_2 \sin\Delta\varpi = -\frac{FG\sin\Delta\psi}{\rho_1 - \rho_2}, \tag{2.99}$$

$$e_1 e_2 \cos\Delta\varpi = \frac{\rho_1 F^2 + (\rho_1 + \rho_2)FG\cos\Delta\psi + \rho_2 G^2}{(\rho_1 - \rho_2)^2},$$

where $\Delta\psi = \psi_2 - \psi_1 = (g_2 t + \beta_2) - (g_1 t + \beta_1)$ with t being the time and $\beta_{1,2}$ given by

$$\sin\beta_1 = \frac{1}{F}(h_{10}\rho_2 - h_{20}), \quad \sin\beta_2 = -\frac{1}{G}(h_{10}\rho_1 - h_{20}),$$

$$\cos\beta_1 = \frac{1}{F}(k_{10}\rho_2 - k_{20}), \quad \cos\beta_2 = -\frac{1}{G}(k_{10}\rho_1 - k_{20}), \tag{2.100}$$

with $h_{i0} = e_{i0}\sin\varpi_{i0}$, $k_{i0} = e_{i0}\cos\varpi_{i0}$ $(i = 1, 2)$. From Eq. (2.99) it is easy to verify that the evolution of e_1, e_2 is confined by an integral

$$\frac{e_1^2}{A_{12}} + \frac{e_2^2}{A_{21}} = D, \tag{2.101}$$

where D is a constant that depends only on the initial parameters. Moreover, from Eqs. (2.98) and (2.99), the maximum of e_1 and the minimum of e_2 occur when $\cos\Delta\psi = 1$ (as $\rho_1 < 0$), and

$$e_{1\,\max} = \frac{F + G}{|\rho_1 - \rho_2|}, \quad e_{2\,\min} = \frac{|\rho_1 F + \rho_2 G|}{|\rho_1 - \rho_2|}. \tag{2.102}$$

The minimum of e_1 and maximum of e_2 happen at $\cos\Delta\psi = -1$, and

$$e_{1\,\min} = \frac{|F - G|}{|\rho_1 - \rho_2|}, \quad e_{2\,\max} = \frac{\rho_2 G - \rho_1 F}{|\rho_1 - \rho_2|}. \tag{2.103}$$

Thus the maximum excursions of e_1 and e_2 for any given $e_{10}, e_{20}, \Delta\varpi_0$ are

$$
\begin{aligned}
\Delta e_1 &= \frac{(F+G) - |F-G|}{|\rho_1 - \rho_2|}, \\
\Delta e_2 &= \frac{(\rho_2 G - \rho_1 F) - |\rho_1 F + \rho_2 G|}{|\rho_1 - \rho_2|}.
\end{aligned}
\tag{2.104}
$$

With the help of the last equation in Eq. (2.99), the criterion for the apsidal resonance in literature Laughlin *et al.* (2002) can be expressed as

$$
S = \left| \frac{(\rho_1 + \rho_2) F G}{\rho_1 F^2 + \rho_2 G^2} \right| < 1.
\tag{2.105}
$$

When $S < 1$, $\Delta\varpi$ cannot reach $\pi/2$ or $3\pi/2$ ($\cos\Delta\varpi \neq 0$), thus it must librate around 0 or π. On the contrary, when $S > 1$, $\Delta\varpi$ may reach $\pi/2$ or $3\pi/2$ and it may circulate in $[0, 2\pi]$.

After some algebra manipulations, Eq. (2.105) is equivalent to

$$
\frac{F}{G} > \max\left(1, -\frac{\rho_1}{\rho_2}\right), \quad \text{or} \quad 0 < \frac{F}{G} < \min\left(1, -\frac{\rho_1}{\rho_2}\right).
\tag{2.106}
$$

In view of the definitions of F, G in Eq. (2.98), the above relations are equivalent to

$$
\frac{e_{20}}{e_{10}} < \frac{2\rho_1\rho_2}{\rho_1 + \rho_2} \cos\Delta\varpi_0,
\tag{2.107}
$$

or

$$
\frac{e_{20}}{e_{10}} > \frac{\rho_1 + \rho_2}{2\cos\Delta\varpi_0} > 0.
\tag{2.108}
$$

Substitute Eqs. (2.93), (2.94) and (2.97) into the above expressions, and one finally obtains

$$
\frac{e_{20}}{e_{10}} < -\frac{5q\alpha^{3/2}\left(1 - \alpha^2/8\right)\cos\Delta\varpi_0}{2(1 - q\alpha^{1/2})},
\tag{2.109}
$$

or

$$
\frac{e_{20}}{e_{10}} > -\frac{2(1 - q\alpha^{1/2})}{5\alpha\left(1 - \alpha^2/8\right)\cos\Delta\varpi_0}.
\tag{2.110}
$$

These are the explicit criteria for the occurence of the apsidal secular resonance. Equations (2.107) and (2.108) are obtained with the linear secular perturbation theory, while to get Eqs. (2.109) and (2.110), the approximations of c_0 and ξ in Eqs. (2.93) and (2.94) have been used.

Here the libration region defined by the inequality in Eq. (2.109) is called the "down-libration region" and the one defined by Eq. (2.110) the

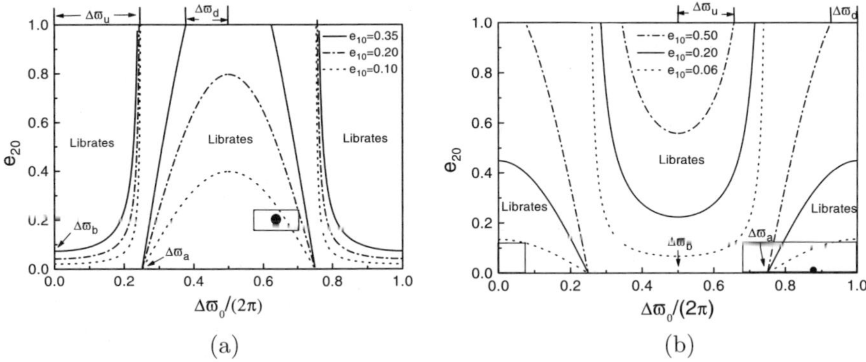

Fig. 2.32 Libration region in the $e_{20} - \Delta\varpi_0$ plane defined by Eqs. (2.109) and (2.110) with different e_{10} for α and q equal to (a) the planetary system HD12661 (left panel) and (b) the planetary system 47 Uma (right panel). The black dots show the present configuration of the two planets in both systems, around which the boxes show the uncertainties of the orbital elements. The values $\Delta\varpi_a, \Delta\varpi_b$ are the lower integration limits in Eq. (2.113) and Eq. (2.116), respectively. $\Delta\varpi_d, \Delta\varpi_u$ are defined in Eqs. (2.112) and (2.115). Figures taken from Zhou & Sun (2003).

"up-libration region". Whether the down-libration or up-libration is the aligned or antialigned resonance depends on the sign of $(1 - q\alpha^{1/2})$. For $q\alpha^{1/2} < 1$, the down-libration occurs only when $\pi/2 < \Delta\varpi_0 < 3\pi/2$ (thus it is antialigned resonance) and the up-libration occurs when $\Delta\varpi_0 < \pi/2$, $\Delta\varpi_0 > 3\pi/2$ (the aligned resonance). This is the case in the HD12661 system. If $q\alpha^{1/2} > 1$, the conclusions are reversed, and this is the case in the 47 Uma system. For the critical case $q\alpha^{1/2} = 1$, according to the inequality in Eq. (2.110), all the orbits are in libration except those with $\Delta\varpi_0 = \pi/2, 3\pi/2$. Figure 2.32 displays in the $e_{20} - \Delta\varpi_0$ plane the libration regions defined by Eqs. (2.109) and (2.110) with some different values of e_{10}.

There are two limiting cases for criteria in Eqs. (2.109) and (2.110).

1. $\alpha \to 0$. The minimum e_{20} for the up-libration and maximum e_{20} for down-libration can be obtained by setting $\cos\Delta\varpi_0 = 1$ in Eqs. (2.109) and (2.110). When $\alpha \to 0$, the minimum e_{20} for up-libration tends to very large and the maximum e_{20} for down-libration tends to zero. Because of $0 \leq e_{20} < 1$, when the two planets are far away, the libration regions in the $e_{20} - \Delta\varpi_0$ plane can be negligibly small.

2. $e_{i0} \to 0$. This happens when one of the planets is in a nearly circular orbit, and it is just the case discussed in the literature (Malhotra, 2002),

namely, for two planets initially in nearly circular orbit, an impulse perturbation may impart a finite eccentricity to one planet's orbit.

When $e_{10} \to 0$, criterion Eq. (2.110) is always fulfilled if $-\pi/2 < \Delta\varpi_0 < \pi/2$ for $q\alpha^{1/2} < 1$ or $\pi/2 < \Delta\varpi_0 < 3\pi/2$ for $q\alpha^{1/2} > 1$.

From the above criteria, one can calculate the probability that two planets fall in apsidal resonance in the space of initial orbital elements. The probability can be defined as the area of the libration region in the $e_{20} - \Delta\varpi_0$ plane for a given e_{10}. For the down-libration case, according to the criterion Eq. (2.109), it is possible that the peak value of the $e_{20} - \Delta\varpi_0$ curve can be above unit for larger e_{10} (as the case of $e_{10} = 0.35$ in Fig. 2.32(a) and the case of $e_{10} = 0.50$ in Fig. 2.32(b)). Let $\Delta\varpi_d$ be the half-width of the down-libration region where the boundary curve reaches $e_{20} = 1$. Such half-width $\Delta\varpi_d$ are shown in Fig. 2.32 for $e_{10} = 0.35$ and $e_{10} = 0.50$, respectively. Define

$$Q_d = \frac{5e_{10}q\alpha^{3/2}\left(1 - \alpha^2/8\right)}{2|1 - q\alpha^{1/2}|}, \tag{2.111}$$

then

$$\Delta\varpi_d = \begin{cases} \arccos\left(\dfrac{1}{Q_d}\right), & \text{if } Q_d > 1 \\[2mm] 0, & \text{if } Q_d \leq 1 \end{cases} \tag{2.112}$$

and the area ratio of the down-libration area to the total area of the studied region in $e_{20} - \Delta\varpi_0$ plane is

$$\begin{aligned} P_d &= \frac{1}{\pi}\left(\Delta\varpi_d + \int_{\Delta\varpi_a}^{\Delta\varpi_a + \pi/2 - \Delta\varpi_d} Q_d \cos\Delta\varpi \, d\Delta\varpi\right) \\ &= \frac{1}{\pi}\left[\Delta\varpi_d + Q_d\left(1 - \sin\Delta\varpi_d\right)\right], \end{aligned} \tag{2.113}$$

where the lower integration limit is the beginning point of the down-libration region in the $\Delta\varpi_0$-axis ($\Delta\varpi_a = \pi/2$ for $q\alpha^{1/2} < 1$ and $\Delta\varpi_a = 3\pi/2$ for $q\alpha^{1/2} > 1$). The $\Delta\varpi_a$ for the cases of $e_{10} = 0.35$ and $e_{10} = 0.50$ have been shown in Fig. 2.32(a) and Fig. 2.32(b), respectively. As one can see, the area ratio for the down-libration increases linearly with e_{10} when e_{10} is small, since in the interval one has $\Delta\varpi_d = 0$ in Eq. (2.113). However, for larger e_{10}, the increase of area ratio is no longer linear since $\Delta\varpi_d \neq 0$ and its a function of e_{10}.

Similarly, for the up-libration resonance, let $\Delta\varpi_u$ be the half-width of the up-libration region when the boundary curve meets $e_{20} = 1$

(Fig. 2.32(b)) and define

$$Q_u = \frac{2e_{10}|1 - q\alpha^{1/2}|}{5\alpha\left(1 - \alpha^2/8\right)}.$$ (2.114)

Then one has

$$\Delta\varpi_u = \arccos(Q_u),$$ (2.115)

and the area ratio of the up-libration area to the total area in $e_{20} - \Delta\varpi_0$ plane is

$$\begin{aligned}
P_u &= \frac{1}{\pi}\left(\Delta\varpi_u - \int_{\Delta\varpi_b}^{\Delta\varpi_b + \Delta\varpi_u} \frac{Q_d}{\cos\Delta\varpi}\mathrm{d}\Delta\varpi\right) \\
&= \frac{1}{\pi}\left[\Delta\varpi_u - Q_u \ln\left(\frac{1 + \sin\Delta\varpi_u}{\cos\Delta\varpi_u}\right)\right],
\end{aligned}$$ (2.116)

where the lower integration limit $\Delta\varpi_b$ is the center of the up-libration region ($\Delta\varpi_b = 0$ for $q\alpha^{1/2} < 1$ and $\Delta\varpi_b = \pi$ for $q\alpha^{1/2} > 1$, see Fig. 2.32(b)).

Since the linear secular perturbation theory is an approximation to the real three-body system, the above criteria obtained from the linear approximation has its limitation. To verify the criteria of the apsidal secular resonance, the orbits in a general coplanar three-body system are numerically integrated. The Runge-Kutta-Fehlberg integrator RKF7(8) with adaptive step size is employed to integrate the orbits. The results show that most of the orbits in the libration region predicted by the criteria Eqs. (2.109) and (2.110) are in the real libration region when e_{20} is small. The discrepancies between the linear system and the three-body system mainly occur for larger values of e_{20} and around the boundary between the libration and circulation region.

2.7.2 *Nodal secular resonance*

Similar to the case of apsidal resonance, here the nodal secular resonance is discussed. The perturbing function for the planets m_1 and m_2 in the $I - \Omega$ degree of freedom are given as

$$\begin{aligned}
R_1' &= n_1 a_1^2\left[\frac{1}{2}B_{11}I_1^2 + B_{12}I_1 I_2 \cos(\Omega_1 - \Omega_2)\right], \\
R_2' &= n_2 a_2^2\left[\frac{1}{2}B_{22}I_2^2 + B_{21}I_1 I_2 \cos(\Omega_1 - \Omega_2)\right],
\end{aligned}$$ (2.117)

where I_i, Ω_i are the inclinations and longitudes of ascending nodes of the orbits of planets m_i $(i = 1, 2)$, B_{ij} are the elements of matrix given by

$$\begin{pmatrix} B_{11} & B_{12} \\ B_{21} & B_{22} \end{pmatrix} = \begin{pmatrix} -c_1 & c_1 \\ c_2 & -c_2 \end{pmatrix},$$

with $c_i > 0$ $(i = 1, 2)$ as defined in Eq. (2.93). The two eigenvalues of this matrix are

$$f_1 = 0, \quad f_2 = -(c_1 + c_2),$$

and the corresponding eigenvectors are

$$T_i \begin{pmatrix} \cos \theta_i \\ \sin \theta_i \end{pmatrix} \quad (i = 1, 2),$$

where θ_i, T_i are defined by

$$\tan \theta_1 \equiv \rho_1 = 1 \quad (\theta_1 = \frac{\pi}{4}),$$
$$\tan \theta_2 \equiv \rho_2 = -\frac{c_2}{c_1} \approx -q\alpha^{\frac{1}{2}}, \tag{2.118}$$

and

$$T_1 = \frac{\left(\rho_2^2 I_{10}^2 - 2\rho_2 I_{10} I_{20} \cos \Delta\Omega_0 + I_{20}^2\right)^{1/2}}{(1 - \rho_2) \cos \theta_1} \equiv \frac{F'}{(1 - \rho_2) \cos \theta_1},$$
$$T_2 = \frac{\left(I_{10}^2 - 2 I_{10} I_{20} \cos \Delta\Omega_0 + I_{20}^2\right)^{1/2}}{(1 - \rho_2) \cos \theta_2} \equiv \frac{G'}{(1 - \rho_2) \cos \theta_2}, \tag{2.119}$$

with $\Delta\Omega_0 = \Omega_{20} - \Omega_{10}$. The secular system with the perturbing functions as Eq. (2.117) has a solution:

$$I_1 I_2 \cos \Delta\Omega = \frac{F'^2 + (1 + \rho_2) F' G' \cos \Delta\psi + \rho_2 G'^2}{(1 - \rho_2)^2}, \tag{2.120}$$

where $\Delta\psi = f_2 t + \gamma_2 - \gamma_1$ with t being the time and γ_1, γ_2 constants.

Due to the similarity of Eq. (2.120) and the last formula in Eq. (2.99), the criteria for nodal resonance can be obtained by replacing $e_{i0}, \Delta\varpi_{i0}$ in Eqs. (2.107, 2.108) by $I_{i0}, \Delta\Omega_{i0}$ $(i = 1, 2)$ respectively, so that one has

$$\frac{I_{20}}{I_{10}} < \frac{2\rho_2 \cos \Delta\Omega_0}{1 + \rho_2}, \quad \text{or} \quad \frac{I_{20}}{I_{10}} > \frac{1 + \rho_2}{2 \cos \Delta\Omega_0} > 0. \tag{2.121}$$

By substituting Eqs. (2.118) and (2.119) into the above expressions, one obtains the criteria:

$$\frac{I_{20}}{I_{10}} < \frac{2q\alpha^{1/2} \cos \Delta\Omega_0}{q\alpha^{1/2} - 1}, \quad \text{or} \quad \frac{I_{20}}{I_{10}} > \frac{1 - q\alpha^{1/2}}{2 \cos \Delta\Omega_0} > 0. \tag{2.122}$$

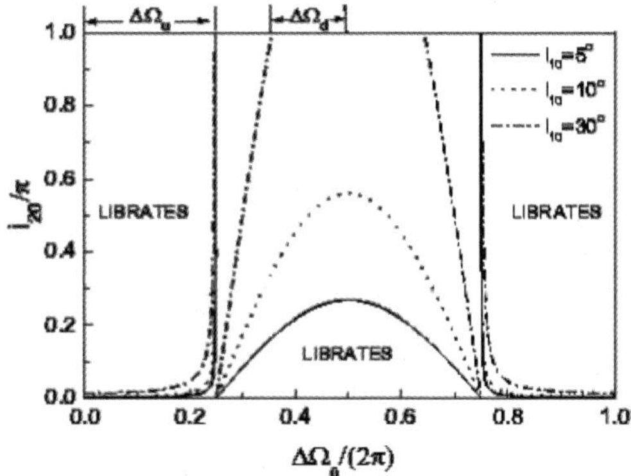

Fig. 2.33 Libration region in the $I_{20} - \Delta\Omega_0$ plane defined by Eq. (2.122) with different I_{10}. Adapted from Zhou & Sun (2004).

Inequalities in Eq. (2.122) are the explicit criteria for the occurrence of nodal resonance. As in the case of apsidal resonance, the nodal resonance defined by the first and the second part of Eq. (2.122) are called the down-libration and the up-libration, respectively. And in Fig. 2.33 these libration regions are illustrated.

Similarly, the area occupied by the libration region defined by the above criteria can be measured in the $I_{20} - \Delta\Omega_0$ plane for a given I_{10}. This can be achieved by integrating the libration defined by Eq. (2.122) (or simply by replacing the $e_{i0}, \Delta\varpi_0$ with $I_{i0}/\pi, \Delta\Omega_0$ in the corresponding formulas in Eqs. (2.113), (2.116), etc.). In this regard, define

$$Q_d = \frac{2I_{10}q\alpha^{1/2}}{\pi|q\alpha^{1/2} - 1|}, \quad Q_u = \frac{I_{10}(1 - q\alpha^{1/2})}{2\pi}, \tag{2.123}$$

for the down-libration resonance (denoted by subscript "d") and the up-libration resonance (by subscript "u"), then

$$\Delta\Omega_d = \begin{cases} \arccos\left(\dfrac{1}{Q_d}\right) & \text{if } Q_d > 1, \\[2mm] 0 & \text{if } Q_d < 1, \end{cases} \tag{2.124}$$

$$\Delta\Omega_u = \arccos(Q_u)$$

are the values of $\Delta\Omega_0$ when Eq. (2.122) can reach $I_{20} = 1$. Finally, the area ratio of the down-libration or up-libration region to the total area in the

$I_{20} - \Delta\Omega_0$ plane is respectively

$$P_d = \frac{1}{\pi}[\Delta\Omega_d + Q_d(1 - \sin\Delta\Omega_d)],$$
$$P_u = \frac{1}{\pi}\left[\Delta\Omega_u - Q_u \ln\left(\frac{1 + \sin\Delta\Omega_u}{\cos\Delta\Omega_u}\right)\right]. \tag{2.125}$$

Numerical experiments have been performed in a three-body system to verify the criteria of nodal secular resonance. And the above analytical criteria obtained by linear approximations are in good agreement with the numerical results when I_{20} is small.

2.8 Apsidal corotation in 3:1 mean motion resonance

Generally the motion of planets in a multiple planetary system may be complicated. Besides the gravitation from the central star that is dominating the system, the gravitational interactions between planets is also critical. Particularly, when two planets are trapped in an MMR, they always demonstrate interesting orbital dynamics. This is one of the reasons why the "multiple planet systems" among the extra-solar systems attract much attention. In Sec. 2.7, the apsidal secular resonance of planets, in which the periastron longitudes (ϖ, secular variable) of two planets are locked with each other, have been investigated. Compared with the MMR, such secular resonance takes significant effects in long term. In relatively short term, the MMR is the most important mechanism that affects the configuration and stability of planetary orbits. However, for two planets in MMR, their longitudes of periastron (ϖ_1, ϖ_2) may also be locked to each other, i.e. the $\Delta\varpi = \varpi_2 - \varpi_1$ librates around a certain value. This phenomenon, generally called "apsidal corotation", though different from the case discussed in Sec. 2.7, is also an important mechanism. In this section, through an example, the apsidal corotation in the 3:1 MMR will be analyzed.

The planetary system around the 55 Cancri (also known as ρ Cancri) is quite famous, not only because it is one of the earliest known multiple planetary systems harboring at least 5 planets, but also because two planets in this system were once regarded as in a 3:1 MMR. Two planets in this system, designated as Companion B and Companion C (also often denoted by "55 Cancri b" and "55 Cancri c"), were found to have orbital periods of 14.65 and 44.27 days, and most probably were locked in a 3:1 MMR. Although follow-up observations reduce the likelihood of this resonance, it is still interesting to check the orbital dynamics of two planets trapped in

Table 2.4 The orbital elements of 55 Cancri b and 55 Cancri c. The mass of the central star is $1.03 M_\odot$. By assuming $i = 90°$ as usual, where i is the angle between the planetary orbital plane and the line of sight, the planet masses (in Jupiter mass) are just the values listed here. Adapted from Zhou *et al.* (2004).

Parameter	$M \sin i$ (M_J)	P (days)	a (AU)	e
Companion B	0.83	14.65	0.115	0.03
Companion C	0.20	44.27	0.241	0.41

a 3:1 MMR and to understand when the apsidal corotation may happen and how it may influence the orbital stability in an MMR. For this is a topic of general interests, only those two planets in question are included in the model in this section and suppose they are locked in the 3:1 resonance, while other planets are abandoned. The masses, semi-major axes and eccentricities of planets are listed in Table 2.4.

The motion of a planet is dominated by the central star and its orbit is a conic section with small deviations due to the gravitational perturbation from other planet(s). This perturbation can be described by the perturbing function, which can be expanded in terms of the orbital elements. Use $a, e, i, \varpi, \Omega, \lambda$ to denote the semi-major axis, eccentricity, inclination, longitude of periastron, longitude of ascending node, and mean longitude, respectively. The perturbing function for a planet with mass m_1 and orbital elements $(a_1, e_1, i_1, \varpi_1, \Omega_1, \lambda_1)$, perturbed by another planet (indicated by subscripts '2'), can be written as

$$R = \mu_2 \sum F(a_1, a_2, e_1, e_2, i_1, i_2) \cos \varphi.$$

Here $\mu_2 = Gm_2$ and φ is a linear combination of angles

$$\varphi = j_1 \lambda_1 + j_2 \lambda_2 + j_3 \varpi_1 + j_4 \varpi_2 + j_5 \Omega_1 + j_6 \Omega_2$$

with $j_1, j_2, \ldots, j_6$ being integers satisfying $j_1 + j_2 + \cdots + j_6 = 0$.

Particularly, when the orbital periods satisfy $P_2/P_1 \approx 3$, that is, $\dot{\lambda}_1 - 3\dot{\lambda}_2 \approx 0$ (dots denote the time derivation), one has $\dot{\varphi}_{(j_1=1, j_2=-3)} \approx 0$ as generally $\dot{\varpi}_1, \dot{\varpi}_2, \dot{\Omega}_1, \dot{\Omega}_2 \ll \dot{\lambda}_1, \dot{\lambda}_2$. Then in the perturbing function R, those terms containing $\lambda_1 - 3\lambda_2$ can not be eliminated by the averaging technique and become the leading terms. And if the inclinations i_1, i_2 are not too large, the leading terms in R are of $O(e_1^2), O(e_2^2), O(e_1 e_2)$ and the corresponding angles φ, now called "resonant arguments" (or resonant angles), are

$$\theta_1 = \lambda_1 - 3\lambda_2 + 2\varpi_1, \quad \theta_2 = \lambda_1 - 3\lambda_2 + 2\varpi_2, \quad \theta_3 = \lambda_1 - 3\lambda_2 + \varpi_1 + \varpi_2.$$

Therefore, the libration of any one of $\theta_{1,2,3}$ indicates the two planets are in an (eccentricity-type) 3:1 MMR. Note that no more than two of them are linear independent and $\theta_3 = (\theta_1 + \theta_2)/2$.

Beside these resonant arguments, the relative apsidal longitude $\Delta\varpi = \varpi_1 - \varpi_2$ is also a critical argument often to be discussed. In a 3:1 MMR, if both θ_1 and θ_2 librate with small amplitudes, $\Delta\varpi = (\theta_1 - \theta_2)/2$ will also necessarily librate. In this sense, the apsidal corotation is not independent and should be different from the term of "apsidal resonance", which is in the context of a secular perturbation (Sec. 2.7).

Hereafter for convenience, the subscripts $1, 2$ refer to the orbital elements of companion B and C respectively.

2.8.1 *Numerical simulations*

Using a symplectic integrator (Mikkola & Palmer, 2000), this system is numerically simulated. The orbital evolution of each body is followed and simultaneously the stability of this orbit indicated by the LCI is calculated. In practice, the largest component from the Lyapunov exponent spectrum (see Sec. 4.2 for details) is regarded as the LCI. The masses, semi-major axes and eccentricities of planets are taken from Table 2.4. The inclinations of companion B and C with respect to the same reference plane are set to be a very small value of 10^{-5} degrees, that is, their orbits are nearly coplanar. Other angular orbital elements $(\varpi_{1,2}, \Omega_{1,2}, \lambda_{1,2})$ are randomly generated from $[0, 2\pi)$. Then, four hundred simulations, with different initial conditions, are integrated up to 10^6 years. During the integrations, if the distance between two planets became smaller than half of the criterion for the "Hill stability" (Gladman, 1993), the system was considered collapsed and the simulation was terminated.

About one third (133 out of 400) of the simulations collapse, and all the remainders survive for the 10^6 yr integration. Most of the survivors have very short ($< 10^2$ yr) e-folding time T_e ($= 1/\text{LCI}$, the reciprocal of LCI), while 38 of them have $T_e \geq 10^3$ yr, and they are regarded as stable. In these stable systems, both planets have final inclinations smaller than $0°.1$, that is, the stable system prefers to retain to be coplanar.

All the stable systems are associated with the 3:1 MMR. According to different configurations of $\theta_{1,2,3}$ and $\Delta\varpi$, the stable systems can be divided into three groups, with representative cases illustrated in Fig. 2.34. The e-folding times of these systems are also summarized in Table 2.5.

Case **a** (see Fig. 2.34) shows a strong resonance with θ_1, θ_2 and θ_3 librating around $215°$, $75°$ and $325°$, respectively. Simultaneously, the $\Delta\varpi$

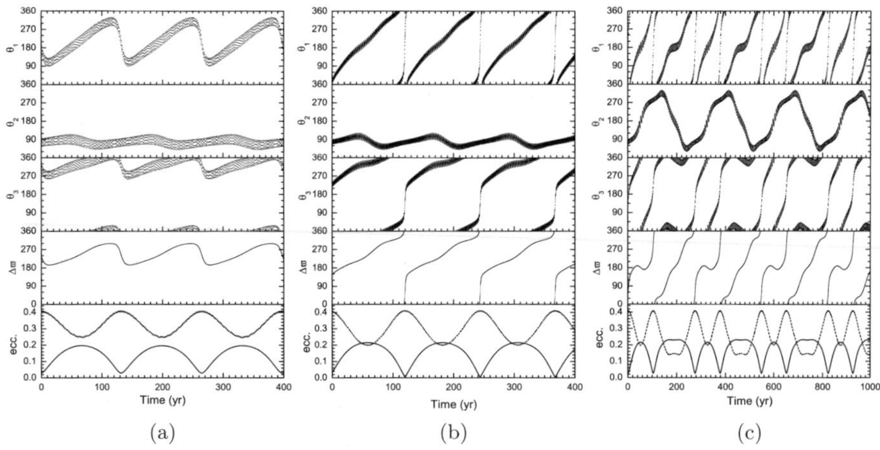

Fig. 2.34   The temporal evolution of the critical angles $\theta_1, \theta_2, \theta_3, \Delta\varpi$, and the eccentricities of planets e_1, e_2 (in the bottom panel, the dashed curves indicate e_2). Case **a**, **b**, and **c** are from different initial conditions. From Zhou *et al.* (2004).

Table 2.5   The distribution of T_e (see text) of the stable systems. The numbers with subscripts **a**, **b** or **c** are the sums of systems with different configurations **a**, **b** or **c** as shown in Fig. 2.34.

T_e (yr)	$\sim 10^3$	$\sim 10^4$	$\sim 10^5$	$\sim 10^6$
Number	$12_\mathbf{a} + 2_\mathbf{b} + 3_\mathbf{c}$	$8_\mathbf{a} + 1_\mathbf{b}$	$7_\mathbf{a} + 2_\mathbf{b}$	$1_\mathbf{a} + 2_\mathbf{b}$

also librates around 250°. The well known symmetric configurations of $\Delta\varpi$ librating around 0° (alignment) or 180° (anti-alignment) implies the conjunctions of the two planets always occur at a certain position and this attributes to the stability of the system (Lee & Peale, 2002). However the asymmetric configuration with $\Delta\varpi$ librating around angles other than 0° or 180°, has also been predicted (Beaugé *et al.*, 2003).

Case **b** differs from case **a** in the behavior of θ_1, θ_3 and $\Delta\varpi$, which circulate now. The θ_2 still librates around 80° with an amplitude of 40°. In fact, when θ_2 librates with a small amplitude, one may approximately assume θ_2 as a constant, that is, $\dot{\theta}_2 \approx 0$. From $\dot{\theta}_2 = \dot\lambda_1 - 3\dot\lambda_2 + 2\dot\varpi_2 \approx 0$ one can derive

$$\frac{n_2 - \dot\varpi_2}{n_1 - \dot\varpi_2} \approx \frac{1}{3},$$

where $n_{1,2} = \dot\lambda_{1,2}$ are the mean motions of planets. In a reference frame co-rotating with the periastron of the second planet, the relative mean motions are $n'_k = n_k - \dot{\varpi}_2$ ($k = 1, 2$), and from $n'_2/n'_1 \approx 1/3$ one knows that the conjunctions of planets always happen at a certain direction. Meanwhile, given $n_k \gg \dot{\varpi}_2$, one has $n_2/n_1 \approx n'_2/n'_1$ and there is still a 3:1 relation between the two orbital periods ($n_{1,2}$ are generally larger than $\dot{\varpi}_2$ by at least two orders of magnitude). The numerical simulations give more examples with configuration as case **a** than case **b** (28 versus 7, see Table 2.5), but among the most stable sets with $T_e \sim 10^6$ yr (the same order as the integration time), there are two sets having configurations as case **b** but only one as case **a**. This prevents a conclusion of which type, case **a** or **b**, is more stable.

In case **c** however, θ_1, θ_3 and $\Delta\varpi$ circulate, while θ_2 librates around $180°$ with a large amplitude of $160°$. It is less stable than the above two cases. The only three examples with this configuration all have $T_e < 5 \times 10^3$ yr.

Besides the angles $\theta_{1,2,3}$ and ϖ, the orbital eccentricities e_1, e_2 also have remarkably different behaviors in these three cases. Another noteworthy feature of numerical results is that the system has a symmetry over angles θ_k and $\Delta\varpi$, that is, when ϕ is a libration center of θ_k or $\Delta\varpi$, the argument $(360° - \phi)$ is also.

Some extended calculations

Three examples shown in Fig. 2.34 and other 5 examples randomly selected from the 17 systems with $T_e \sim 10^3$ yr, relatively short T_e in stable systems (see Table 2.5), are integrated to 1×10^7 yr. All of them, except two examples having motion configuration as case **c**, survive and keep the configurations all the time. Because systems with larger T_e are more stable than the ones with smaller T_e, one may argue without more calculations that those systems with $T_e \geq 10^4$ yr should have longer surviving lives.

Some other simulations for higher inclinations of planets show that as long as the relative inclination of these two planets is smaller than $10°$ the stability of the system does not decrease. And if smaller than 10 degrees, the inclinations have nearly no influence on the stability of the system. The fact that all the stable examples found here are associated with the 3:1 MMR reveals again the importance of this resonance in stabilizing this planetary system, even when the orbits are not coplanar.

2.8.2 *Analytical model*

The usual analytical perturbation models, such as the Laplace expansions, are not convergent for high eccentricities, and expansions with low-order truncations may yield imprecise results. Thus, a new expansion of the Hamiltonian suitable for high-eccentricities developed by Beaugé and Michtchenko (Beaugé & Michtchenko, 2003) is applied to this system.

The Hamiltonian

Suppose there are three bodies with mass M_0, m_1, m_2 orbiting around their common center of mass. $M_0 \gg m_{1,2}$ is the central star and $m_{1,2}$ are the planets. Near the 3:1 MMR, a set of canonical variables are defined as:

$$
\begin{aligned}
\lambda_1; & \qquad J_1 = L_1 + \frac{1}{2}(I_1 + I_2), \\
\lambda_2; & \qquad J_2 = L_2 - \frac{3}{2}(I_1 + I_2), \\
\sigma_1 = \frac{3}{2}\lambda_2 - \frac{1}{2}\lambda_1 - \varpi_1; & \quad I_1 = L_1 - G_1, \\
\sigma_2 = \frac{3}{2}\lambda_2 - \frac{1}{2}\lambda_1 - \varpi_2; & \quad I_2 = L_2 - G_2,
\end{aligned}
\tag{2.126}
$$

where $L_k = m'_k \sqrt{\mu_k a_k}$ and $G_k = L_k \sqrt{1 - e_k^2}$ are Delaunay variables, $a_k, e_k, \lambda_k, \varpi_k$ are the orbital elements of the kth planet ($k = 1, 2$), $\mu_k = G(m_k + M_0)$, and $m'_k = \frac{m_k M_0}{m_k + M_0}$ is the reduced mass. These canonical variables define a four degree-of-freedom system. And the critical angles $\theta_{1,2,3}$ and $\Delta \varpi$ are related to these variables through $\theta_k = -2\sigma_k$ ($k = 1, 2$), $\theta_3 = -\sigma_1 - \sigma_2$ and $\Delta \varpi = \sigma_2 - \sigma_1$.

After considering the conservation of angular momentum and averaging over the fast-angle, two integrals of motion $J_{\mathrm{sum}} = J_1 + J_2$ and $J_{\mathrm{dif}} = J_1 - J_2$ are obtained. And the problem is reduced to a two degree-of-freedom system in canonical variables (I_k, σ_k). The resulting Hamiltonian reads

$$
H = -\sum_{k=1}^{2} \frac{\mu_k^2 m'^3_k}{2L_k^2} + H_1,
\tag{2.127}
$$

where H_1 is the truncated perturbing function,

$$
H_1 = \frac{Gm_1 m_2}{a_2} \sum_{n=0}^{3} \sum_{j=0}^{j_{\max}} \sum_{k=0}^{k_{\max}} \sum_{u=0}^{u_{\max}} \sum_{l=-l_{\max}}^{l_{\max}} R_{n,j,k,u,l}(\alpha - \alpha_0)^n
\tag{2.128}
$$

$$
\times\, e_1^j e_2^k \cos[2u\sigma_1 + l(\sigma_2 - \sigma_1)].
$$

Here $\alpha = a_1/a_2$ is the ratio between the two semi-major axes and α_0 is the nominal value. The coefficients $R_{n,j,k,u,l}$ can be determined beforehand (Beaugé & Michtchenko, 2003). Considering the high value of eccentricities, the upper limits of sums $u_{\max} = l_{\max} = 12$ and $j_{\max} = k_{\max} = 15$ are adopted. For convenience, the Hamiltonian has been expressed explicitly in (a_k, e_k) instead of (I_k, J_k), therefore the integrals J_{sum} and J_{dif} don't appear explicitly but exist as constrains. Note the total energy H is an integral too, and the Hamiltonian is symmetric:

$$H(\sigma_1, \sigma_2) \equiv H(-\sigma_1, -\sigma_2) \equiv H(180^\circ + \sigma_1, 180^\circ + \sigma_2).$$

The Hamiltonian can be applied to calculate the planets' orbits. A comparison between this Hamiltonian expansion and the direct numerical integration is shown in Fig. 2.35, in which the most unstable case of Fig. 2.34c is chosen as an example. The agreements in Fig. 2.35 evidently imply the reliability and suitability of this Hamiltonian.

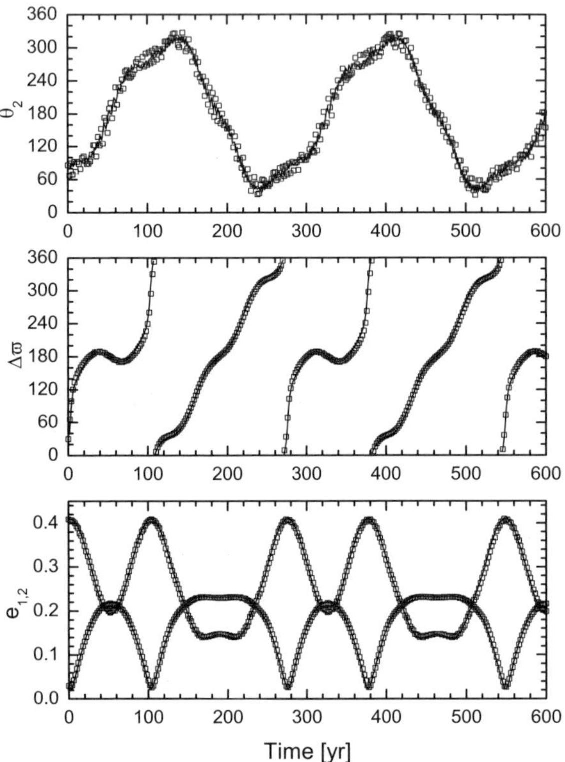

Fig. 2.35   The temporal evolution of θ_2, $\Delta\varpi$ and eccentricities $e_{1,2}$ of the orbits as Fig. 2.34(c). Squares show the results of a numerical integration of the exact dynamical equations, while the curves are from the Hamiltonian equations for a 3-body model. From Zhou *et al.* (2004).

2.8.3 *Apsidal corotation*

The Hamiltonian introduced above can be used to analyze the dynamical properties of the system.

The eccentricities and the resonant angles

Since J_{sum} and J_{dif} are integrals, $J_1 = \frac{1}{2}(J_{\mathrm{sum}}+J_{\mathrm{dif}})$ and $J_2 = \frac{1}{2}(J_{\mathrm{sum}}-J_{\mathrm{dif}})$ are invariables too. They constrain the semi-major axes by

$$3J_1 + J_2 = 3m_1'\sqrt{\mu_1 a_1} + m_2'\sqrt{\mu_2 a_2} = \text{constant}.$$

When discussing the dynamics of planets trapped in an MMR, one may simply assume the semi-major axes are constants, otherwise the resonance would be destroyed. Simple calculations also show that

$$J_{\mathrm{sum}} = m_1'\sqrt{\mu_1 a_1(1 - e_1^2)} + m_2'\sqrt{\mu_2 a_2(1 - e_2^2)}. \tag{2.129}$$

This confines the variation of the eccentricities when the semi-major axes a_1 and a_2 are fixed. In addition, the variation range of the eccentricities is also bounded by other constrains, e.g. the total energy of the system. Assuming $a_1 = 0.115\,\mathrm{AU}$ and $a_2 = 0.241\,\mathrm{AU}$, then Eq. (2.129) defines a family of curves on the (e_1, e_2)-plane (Fig. 2.36). Each curve in Fig. 2.36

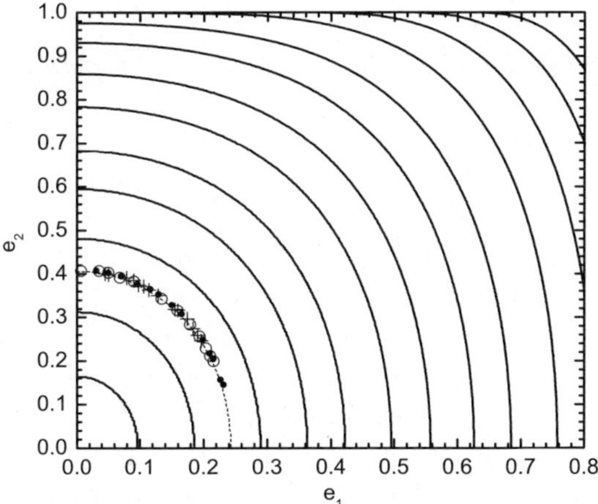

Fig. 2.36 The possible variation of eccentricities of the planets. The dotted curve corresponds to $J_{\mathrm{sum}} = 3.951733\times 10^{-4}$. Results from the numerical simulations of Fig. 2.34(a) (plus), (b) (open circle) and (c) (black dot) are plotted to compare. From Zhou *et al.* (2004).

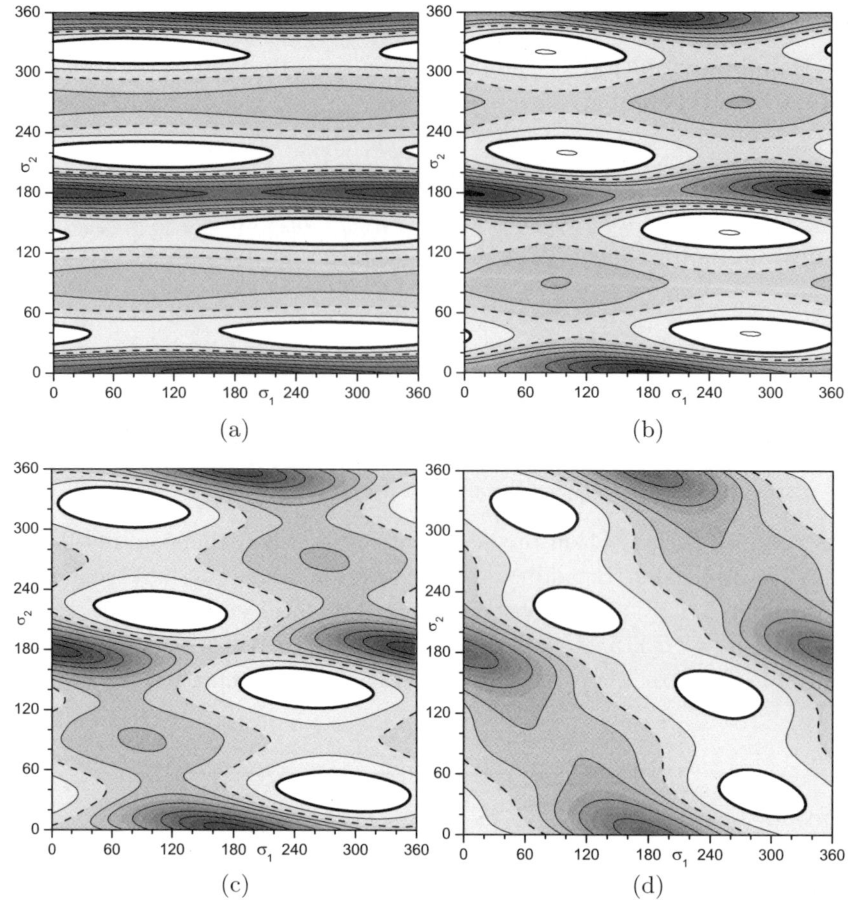

Fig. 2.37 The energy level curves of the Hamiltonian on the (σ_1, σ_2)-plane. A gray level is used to measure energy values, the darker regions indicate smaller energy value, and vice versa. **a**, **b**, **c** and **d** are the case for $e_1 = 0.03, 0.10, 0.15$ and 0.20 respectively, with e_2 obtained from Eq. (2.129). From Zhou *et al.* (2004).

corresponds to a definite value of the integral J_{sum}. The orbital elements in Table 2.4 lead to $J_{\text{sum}} = 3.591733 \times 10^{-4}$, and the eccentricities e_1, e_2 beginning from these initial conditions are restricted on the corresponding curve (dotted curve in Fig. 2.36) during the dynamical evolution.

Assuming $a_{1,2}$ fixed, for certain $e_{1,2}$ values, one can calculate the energy level curves of the Hamiltonian on the (σ_1, σ_2)-plane. Figure 2.37 shows such contours. Since the total energy H is an integral, a system starting from a point on any curve in Fig. 2.37 can never jump to other curves with different energy values. Meanwhile, the variation of eccentricities is constrained by the integral J_{sum}. Therefore, as the eccentricities vary following Eq. (2.129), an energy curve on the (σ_1, σ_2)-plane will extend to an energy surface (e.g. a cylinder) and the system will evolve on this surface. To illustrate this, fix $a_1 = 0.115\,\mathrm{AU}$ and $a_2 = 0.241\,\mathrm{AU}$, set $e_1 = 0.03$ (and sequently $0.10, 0.15, 0.20$), evaluate the corresponding e_2 from Eq. (2.129) with $J_{\mathrm{sum}} = 3.591733 \times 10^{-4}$, then calculate contours on the (σ_1, σ_2)-plane for different pairs of (e_1, e_2). Finally one gets a series of section of the energy surface in Fig. 2.37. In this way, one can follow the dynamical evolution of a system starting from definite initial conditions.

For example, the thick curves in Fig. 2.37, representing an energy $H = -3.957679 \times 10^{-3}$, keep the closed oblate shape when e_1 evolves from 0.03 to 0.20. The oblate profile of the curve implies that the angle σ_1 (θ_1) has a larger reachable range than σ_2 (θ_2) during the temporal evolution of a system with initial conditions located on this curve. The angle σ_1 could vary in a range of $\sim (160°, 400°)$, while σ_2 is bounded between $20°$ and $60°$. Correspondingly, the resonant angle θ_1 and θ_2 could vary in ranges with widths of $480°$ (circulation) and $80°$ (libration), respectively. This is just the situation of case **b** in Fig. 2.34. Another example indicated by dashed curves in Fig. 2.37 with $H = -3.957722 \times 10^{-3}$ exhibits the possibility of circulating of both θ_1 and θ_2.

Denote the possible variation of σ_k by $\Delta\sigma_k$. According to the foregoing analysis, $\Delta\sigma_2 < \Delta\sigma_1$ (Fig. 2.37). If $\Delta\sigma_1 \leq 180°$, both θ_1 and θ_2 would librate ($\Delta\theta_2 < \Delta\theta_1 \leq 360°$) and the system could run as Fig. 2.34(a). It could run as Fig. 2.34(b), if $\Delta\sigma_1 \geq 180°$ while σ_2 is still bounded in a narrow range, say $\Delta\sigma_2 \leq 90°$. And if $90° \leq \Delta\sigma_2 \leq 180°$, Fig. 2.34(c) could be followed. So the measures of such areas in the whole (σ_1, σ_2)-plane tell the probabilities of different types of motion. Calculations show the proportions of these regions in the whole plane, i.e. the probabilities of motion type **a**, **b** and **c**, approximatively are $10\%, 16\%$ and 4.5%, respectively. These are coarse estimates. Taking into account the stability, these estimates must decrease in different extents.

In a word, given a_k, e_k, the dynamical evolution of a system is determined by the initial values of σ_k ($k = 1, 2$), which consequently correspond to different energy H.

The occurrence of apsidal corotation

When two planets are in a 3:1 MMR, at least one of the three resonant angles librates around a definite value. Since σ_2 generally has a smaller

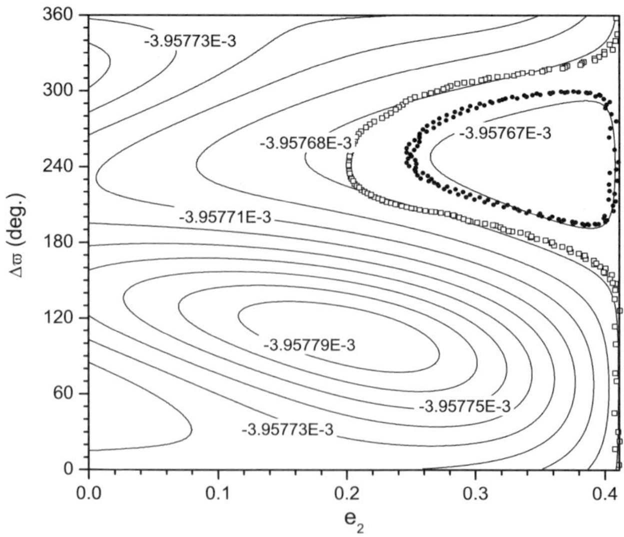

Fig. 2.38   The contour of Hamiltonian on the $(e_2, \Delta\varpi)$-plane. Some Hamiltonian values are labelled. Dots and squares represent the case **a** and **b** in Fig. 2.34 from numerical integrations. From Zhou *et al.* (2004).

libration amplitude than σ_1, one can further simplify the Hamiltonian by safely setting $\sigma_2 \equiv \sigma_2^0$. The value of σ_2^0 can be determined by finding the stable stationary solutions of the Hamiltonian equations, or by estimating the position of the Hamiltonian extremum in Fig. 2.37, or simply by adopting $\sigma_2^0 = 320°$ ($\theta_2 = 80°$) from the numerical results. Recalling the constant semi-major axes and the constrain on eccentricities, the Hamiltonian can be explicitly expressed as $H = H(a_2^0, a_1^0, e_2, e_1(J_{\mathrm{sum}}, e_2), \sigma_2^0, \Delta\varpi)$, with only two "free" variables e_2 and $\Delta\varpi$. The contour of such Hamiltonian on the $(e_2, \Delta\varpi)$-plane can be calculated, and it is shown in Fig. 2.38. From the Hamiltonian equations, $\mathrm{d}\Delta\varpi/\mathrm{d}t$ can also be calculated and the contour is shown in Fig. 2.39.

A system approximately satisfying the assumptions mentioned above will evolve along one of the curves in Fig. 2.38. For two planets with initial orbital elements listed in Table 2.4 ($e_1^0 = 0.03$, $e_2^0 = 0.41$), it corresponds to the right boundary of the box in Fig. 2.38. Two different fates of a system starting from these initial conditions are distinguished: One is characterized by a libration of $\Delta\varpi$ around $\sim 250°$ and the other a circulation of $\Delta\varpi$. In the latter case, the eccentricity e_2 can reach a smaller value (correspondingly e_1 a higher value). These agree pretty well with the directly numerical

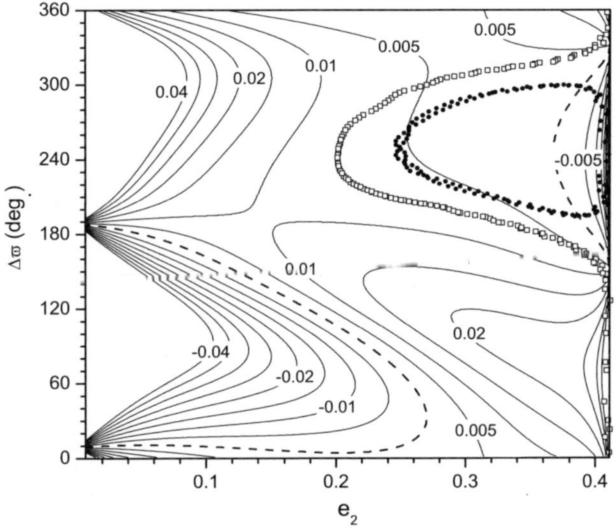

Fig. 2.39   The contour of $\mathrm{d}\Delta\varpi/\mathrm{d}t$. Some velocity values are labelled and the scale is $360°\,\mathrm{yr}^{-1}$. Dashed curves indicate the zero velocity curve. Dots and squares are the same as in Fig. 2.38. From Zhou *et al.* (2004).

integrations. Such agreements are shown in Fig. 2.38 by a comparing plots of the numerical results.

Figure 2.39 shows the velocity of $\Delta\varpi$. Along the right boundary of the box, the absolute value of velocity could be very large, so that $\Delta\varpi$ increases or decreases quickly. Write the perturbing part of the Hamiltonian as

$$H_1 = \sum_{j=0}^{j_{\max}} \sum_{k=0}^{k_{\max}} R_{j,k} e_1^j e_2^k,$$

where $R_{j,k}$ contains the coefficients $R_{n,j,k,u,l}$ and sums over n, u, l in Eq. (2.128). Then a calculation shows

$$\frac{\mathrm{d}\Delta\varpi}{\mathrm{d}t} = \frac{\sqrt{1-e_1^2}}{e_1^2 L_1} \sum_{j=1}^{j_{\max}} \sum_{k=0}^{k_{\max}} j R_{j,k} e_1^j e_2^k - \frac{1-e_2^2}{e_2^2 L_2} \sum_{j=0}^{j_{\max}} \sum_{k=1}^{k_{\max}} k R_{j,k} e_1^j e_2^k.$$

$$(2.130)$$

Obviously, when $e_k \ll 1$, the dominating term on the right hand side has the order $\sim 1/e_k$. Near the right boundary of the box in Fig. 2.39, $e_1 \sim 0.03 \ll 1$. That's the reason why $\Delta\varpi$ has a quick change when e_1 approaches its minimum value as shown in Fig. 2.34b,c.

Numerical simulations reveal that the quick varying of $\Delta\varpi$ in Fig. 2.34b,c is caused by a swift varying of ϖ_1 when $e_1 \to 0$. In fact, when an orbit is nearly circular ($e \sim 0$), it would be relatively easy to change its direction (ϖ), thus the two planets could loose the locking of periastrons if they were not in a strong coupling. Conversely, whether they can strongly couple with each other depends also on how close they can approach to each other, that is, how large their eccentricities (especially e_2) are. Bearing this in mind one can understand why in case **a** e_2 has a higher low-limit than in cases **b** and **c**. Of course, a small e_2 can also cause a quick variation of $\Delta\varpi$, but this situation does not happen because the energy integral prevents e_2 from approaching a very small value.

Different behaviors of $\Delta\varpi$ (circulation or libration) have different varying directions along the right edge of the box in Figs. 2.38 and 2.39. By numerically solving the equation $\mathrm{d}\Delta\varpi/\mathrm{d}t = 0$, two zero velocity points can be spotted on the boundary. They define the boundaries of the initial conditions leading to a librating or circulating $\Delta\varpi$. The calculations show that the apsidal corotation happens if $\Delta\varpi^0 \in (160°, 330°)$, while if $\Delta\varpi^0 \in (-30°, 160°)$, $\Delta\varpi$ circulates.

The case **c** has a Hamiltonian value a little further from the extremum in Fig. 2.37, so that σ_2 (θ_2) has a large amplitude of libration and can not be assumed as a constant. As a result, there is no corresponding curve for it in Fig. 2.38. Of course, if a series of Hamiltonian contour with different value of σ_2 was plotted (adding another dimension to Fig. 2.38), the dynamical evolution of this case can also be depicted.

Figure 2.40 summarizes the initial conditions of the 38 stable examples found in the numerical simulations. For clarity, the symmetric configurations have been mapped to one quadrant according to the symmetry of the Hamiltonian. Except the three examples as case **c**, all the points gather along the line of $\sigma_2^0 = 320°$, and points for librating and circulating $\Delta\varpi$ (case **a** and **b**) are separated by the dotted lines of $\Delta\varpi^0 = 160°, 330°$, in good agreements with the analytical results via the Hamiltonian method.

An estimate of the probability of a system following different evolving ways as case **a**, **b** and **c** has been given previously. The analyses here tell that all the initial conditions leading to stable configurations gather in a narrow strip along the line in Fig. 2.40. That is, for systems having given $a_{1,2}, e_{1,2}$ as in Table 2.4, only those with initial conditions satisfying $\sigma_2^0 \approx 320°$ and $\sigma_1^0 + \Delta\varpi^0 \approx 320°$ can survive for long time.

The stability of these three different orbital configurations is investigated using the technique of surface of section in the Hamiltonian model.

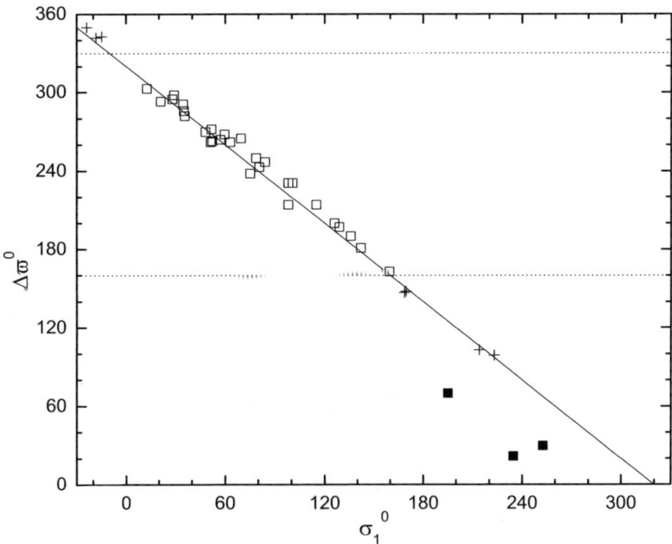

Fig. 2.40 The initial conditions of the 38 stable systems. Open squares, pluses and solid squares indicate the systems evolving as case **a**, **b** and **c**. The dotted lines give the positions of $\Delta\varpi^0 = 160°, 330°$ and the solid line is $\sigma_2^0 = \Delta\varpi^0 + \sigma_1^0 = 320°$. From Zhou *et al.* (2004).

The results show that the motion as case **c** is chaotic but both case **a** and case **b** are regular. That is, no matter whether the apsidal corotation happens (as case **a**) or not (as case **b**), the system trapped in the 3:1 MMR could be stable. This is consistent with the results revealed by the LCI (Table 2.5). From this point of view, the apsidal corotation only has a limited contribution to the stability of the system. It should be pointed out that, the structure of the phase space and thus the stability of the system depend sensitively on the total energy H and other two integrals (J_{sum} and J_{dif}), the application of the surface-of-section technique is limited by the high dimension of the system.

Systems with different motion configurations have different energy (H) levels. The apsidal corotation happens when the Hamiltonian approaches the extreme value. So, if these two planets were captured into current orbital configuration through orbital migration caused by the action of non-conservative forces, the system should have an extremum of energy and the apsidal corotation should happen. This is also a reason why the analyses presented here is valuable.

Chapter 3

Chaotic motion of orbits

The chaos theory begun to be one of the frontier sciences only after the middle of the 20th century despite the initial insights had been made by Poincaré in 1880s while studying the three-body problem. As the excellent examples of nonlinear dynamical system, the N-body problem in celestial mechanics is one of the most studied objects in nonlinear science. In this chapter, we will discuss the chaotic motion of celestial bodies, particularly of comets as an example. Some useful tools in the study of nonlinear system, like the symplectic integrator, are also discussed.

3.1 Conservative dynamical system

Usually, a dynamical system is described by the differential equation as follows.

$$\frac{\mathrm{d}\boldsymbol{x}}{\mathrm{d}t} = \boldsymbol{f}(\boldsymbol{x}), \tag{3.1}$$

where $\boldsymbol{x} = (x_1, x_2, \ldots, x_m)^T$, $\boldsymbol{f}(\boldsymbol{x}) = (f_1(\boldsymbol{x}), f_2(\boldsymbol{x}), \ldots, f_m(\boldsymbol{x}))^T$, with T denoting the transpose of the vector. Suppose $\boldsymbol{f}(\boldsymbol{x})$ is an analytical function in $\boldsymbol{x}$ space, the independent variable t is the time, then Eq. (3.1) is called the continuous dynamical system, and $\boldsymbol{x}$ space the phase space. $\boldsymbol{f}(\boldsymbol{x})$ determines a vector field in the phase space $\boldsymbol{x}$. For an initial value $\boldsymbol{x}_0$, its evolution with time produces a set called an orbit.

Generally, a dynamical system is called a conservative dynamical system if the system possesses a conservative force field. This definition is not suitable for the discrete dynamical system (a mapping), because there is no "force field" in this case. According to the fact that the flows of conservative force field are incompressible, one may generalize the definition of conservative system to discrete ones by defining a discrete system that possesses incompressible flow as a conservative system. Otherwise, if a discrete does not have incompressible flow, it is a dissipative system.

The rate of volume-changing in phase space is given by Liouville's formula. Let $\boldsymbol{x} = \boldsymbol{x}(t, \boldsymbol{\xi}^*)$ be the solution to Eq. (3.1), with its initial value $\boldsymbol{x}(0, \boldsymbol{\xi}) = \boldsymbol{\xi}^*$, and let $X = (x_{k\xi_l})$ be an $m \times m$ matrix, with the elements being the derivatives of the components of $\boldsymbol{x} = \boldsymbol{x}(t, \boldsymbol{\xi}^*)$ with respect to the components of the initial vectors. Make the derivatives of Eq. (3.1) to ξ_l, one gets

$$\frac{\mathrm{d}x_{k\xi_l}}{\mathrm{d}t} = \sum_{i=1}^{m} f_{kx_i}\left(\boldsymbol{x}(t, \boldsymbol{\xi}^*)\right) x_{i\xi_l} \quad (k = 1, 2, \ldots, m). \tag{3.2}$$

Denote matrix $F = (f_{kx_i}\left(\boldsymbol{x}(t, \boldsymbol{\xi}^*)\right)) \quad (i = 1, 2, \ldots, m)$. The evolution of X satisfies the variation equation

$$\frac{\mathrm{d}X}{\mathrm{d}t} = FX. \tag{3.3}$$

Denote the determinant of X by Δ, i.e. $\Delta = |X|$. According to the rule for the derivation calculus of determinant, one has

$$\frac{\mathrm{d}\Delta}{\mathrm{d}t} = \sum_{i=1}^{m} \begin{vmatrix} \cdots \cdots \\ \dfrac{\mathrm{d}x_{i\xi_k}}{\mathrm{d}t} \\ \cdots \cdots \end{vmatrix} = \sum_{i=1}^{m} \begin{vmatrix} \cdots \cdots \\ \displaystyle\sum_{j=1}^{m} f_{ix_j} x_{j\xi_k} \\ \cdots \cdots \end{vmatrix} = \sum_{i=1}^{m} \begin{vmatrix} \cdots \cdots \\ f_{ix_i} x_{i\xi_k} \\ \cdots \cdots \end{vmatrix} = (\mathrm{Tr}F)\,\Delta.$$

$$\tag{3.4}$$

Moreover, $X(0) = E$ when $t = 0$, hence the initial value of $\Delta(t)$ equals to 1, i.e. $\Delta(0) = 1$. Thus, one has

$$\Delta = \exp\left(\int_0^t (\mathrm{Tr}F)\,\mathrm{d}t\right). \tag{3.5}$$

This is just the Liouville formula.

Equation (3.1) can be regarded as the evolution equations of the flow, in which $\boldsymbol{x}$ is the coordinate of fluid element. As the time t does not appear explicitly in the right-hand side of Eq. (3.1), it describes a steady flow. Let $\boldsymbol{\xi}$ be the coordinate of fluid element at $t = 0$ and at time t the fluid element moves to $\boldsymbol{x}(t, \boldsymbol{\xi})$. Thus a mapping from $\boldsymbol{\xi}$ to $\boldsymbol{x}$ is defined, and the Jacobian of this mapping is $(x_{k\xi_l}) = X(t, \boldsymbol{\xi})$, with a determinant Δ. If $\Delta \equiv 1$, the mapping is volume-preserving (measure-preserving) and the corresponding flow is incompressible. From Eq. (3.5), if

$$\mathrm{Tr}F = \sum_{i=1}^{m} f_{ix_i} = 0, \tag{3.6}$$

Eq. (3.1) is a conservative system.

The Hamiltonian system is an important sort of continuous conservative system. It is widely adopted to study the motion of celestial bodies in celestial mechanics. As well-known, the Hamiltonian equations read

$$\frac{\mathrm{d}q_i}{\mathrm{d}t} = \frac{\partial H}{\partial p_i}, \quad \frac{\mathrm{d}p_i}{\mathrm{d}t} = -\frac{\partial H}{\partial q_i}, \quad (i = 1, 2, \ldots, m), \tag{3.7}$$

where $H(\boldsymbol{p}, \boldsymbol{q}, t)$ is the Hamiltonian function, $\boldsymbol{p} = (p_1, p_2, \ldots, p_m)^T, \boldsymbol{q} = (q_1, q_2, \ldots, q_m)^T$ are n-dimensional vectors. Usually, $\boldsymbol{p}$ is the momentum while $\boldsymbol{q}$ is the coordinate. From Eq (3.7), one has immediately

$$\mathrm{Tr}F = \sum_{i=1}^{m} \left(\frac{\partial^2 H}{\partial p_i \partial q_i} - \frac{\partial^2 H}{\partial q_i \partial p_i} \right) = 0. \tag{3.8}$$

Denote

$$\boldsymbol{z} = (z_1, z_2, \ldots, z_{2m})^T = (p_1, p_2, \ldots, p_m, q_1, q_2, \ldots, q_m)^T. \tag{3.9}$$

Now the Hamiltonian equations become

$$\frac{\mathrm{d}\boldsymbol{z}}{\mathrm{d}t} = J\nabla_{\boldsymbol{z}} H, \tag{3.10}$$

where $\nabla_{\boldsymbol{z}}$ is the gradient with respect to $\boldsymbol{z}$, and J is the Poisson matrix

$$J = \begin{pmatrix} 0 & -E_m \\ E_m & 0 \end{pmatrix}_{2m \times 2m}, \tag{3.11}$$

with E_m being an $m \times m$ unit matrix and 0 a zero matrix. Apparently, J is orthogonal and antisymmetric, i.e.

$$J^{-1} = J^T = -J, \tag{3.12}$$

with J^{-1}, J^T denoting the inverse matrix and transpose matrix respectively of J. The determinant is $|J| = 1$ and $J^2 = -E_{2m \times 2m}$.

Let $\boldsymbol{z} = \boldsymbol{z}_0$ is the equilibrium solution (fixed point) of the Hamiltonian equation. Then the linearized equation of Eq. (3.10) at $\boldsymbol{z} = \boldsymbol{z}_0$ is

$$\frac{\mathrm{d}\delta z}{\mathrm{d}t} = J\frac{\partial^2 H}{\partial z^2}(t, \boldsymbol{z}_0)\delta z = JS\delta z \equiv A\delta z. \tag{3.13}$$

In this equation, $A = JS$ and S is a symmetric matrix. In fact, such a matrix like A, which is a product of J and a symmetric matrix S, is called a "Hamiltonian matrix". A Hamiltonian matrix A satisfies $A = JA^T J$ and its characteristic polynomial satisfies $p(\lambda) = |A - \lambda E| = |JA^T J - \lambda E| = |JA^T J - \lambda JJ| = |J||(A + \lambda E)^T||J| = |A + \lambda E| = p(-\lambda)$. Therefore, if λ is an eigenvalue of a Hamiltonian matrix, $-\lambda$ must be too. Meanwhile, because A is a real matrix, the conjugate number of λ is also an eigenvalue. That is, if λ is an eigenvalue, $-\lambda, \bar{\lambda}$ and $-\bar{\lambda}$ are all eigenvalues of A. Due to this property of the eigenvalues and the incompressibility of the Hamiltonian flow, there does not exist the asymptotic stability of equilibrium points for a Hamiltonian system.

3.2 Ordered and chaotic motion

The Hamiltonian systems can be divided into two categories: the integrable systems and the nonintegrable systems. For the former, the motions in the whole phase space are always ordered. Among the nonintegrable systems, there is a relatively simpler class of Hamiltonian system: near-integrable Hamiltonian system. A near-integrable system can be regarded as an originally integrable system perturbed by small conservative forces. Usually such a system possesses both the ordered and the chaotic motions in phase space.

Consider a set of Hamiltonian equations

$$\frac{\mathrm{d}q_i}{\mathrm{d}t} = \frac{\partial H}{\partial p_i}, \quad \frac{\mathrm{d}p_i}{\mathrm{d}t} = -\frac{\partial H}{\partial q_i}, \quad (i = 1, 2, \ldots, m), \tag{3.14}$$

where $H(\boldsymbol{p}, \boldsymbol{q})$ is the Hamiltonian function. A function $F(\boldsymbol{p}, \boldsymbol{q})$ is a "first integral" of Eq. (3.14), if and only if the corresponding Poisson bracket is zero, i.e.

$$\{F, H\}(\boldsymbol{p}, \boldsymbol{q}) \equiv \sum_{i=1}^{m} \left[\frac{\partial F}{\partial q_i} \frac{\partial H}{\partial p_i} - \frac{\partial F}{\partial p_i} \frac{\partial H}{\partial q_i} \right] = 0. \tag{3.15}$$

In fact, along the solution of Eq. (3.14),

$$\frac{\mathrm{d}F}{\mathrm{d}t} = \sum_{i=1}^{m} \left[\frac{\partial F}{\partial q_i} \frac{\mathrm{d}q_i}{\mathrm{d}t} + \frac{\partial F}{\partial p_i} \frac{\mathrm{d}p_i}{\mathrm{d}t} \right] = \{F, H\}. \tag{3.16}$$

This proves that, if and only if $\{F, H\} = 0$, one has $\frac{\mathrm{d}F}{\mathrm{d}t} = 0$, i.e. F is a first integral of Eq. (3.14). Geometrically, $\{F, H\} = 0$ means that the normal direction of the surface defined by $F(\boldsymbol{p}, \boldsymbol{q}) = c$ (c is a constant) is orthogonal to the vector field defined by Eq. (3.14) everywhere. If the Poisson bracket of $F_1(\boldsymbol{p}, \boldsymbol{q})$ and $F_2(\boldsymbol{p}, \boldsymbol{q})$ at the motion surface of Eq. (3.14) equals to zero, i.e. $\{F_1, F_2\} = 0$, F_1 and F_2 are said to be involutive.

Liouville theorem proves that if m independent involutive integrals exist in a Hamiltonian system of m degrees of freedom, this system is integrable. And for an integrable system, one can find a set of canonical conjugate coordinates $(\boldsymbol{J}, \boldsymbol{\phi})$ such that all the first integrals contain only the action

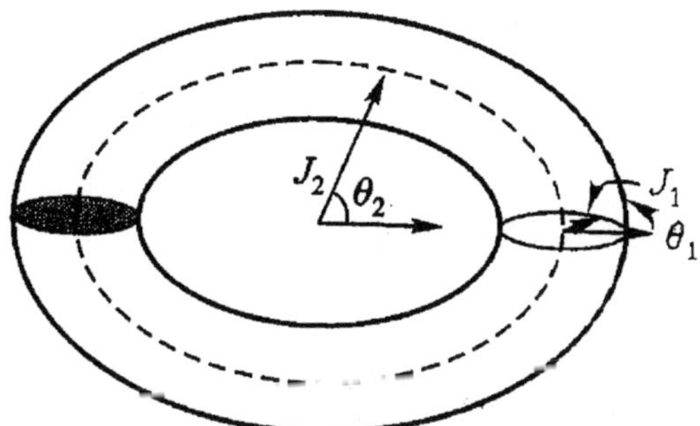

Fig. 3.1   A sketch of a two-dimensional invariant torus.

variable $\boldsymbol{J}$. Particularly, the Hamiltonian is a function of only $\boldsymbol{J}$ but not $\boldsymbol{\phi}$, i.e. $H = H(\boldsymbol{J})$. In such a situation, the Hamiltonian equations are

$$\frac{\mathrm{d}\boldsymbol{J}}{\mathrm{d}t} = -\frac{\partial H(\boldsymbol{J})}{\partial \boldsymbol{\phi}} = 0,$$
$$\frac{\mathrm{d}\boldsymbol{\phi}}{\mathrm{d}t} = \frac{\partial H(\boldsymbol{J})}{\partial \boldsymbol{J}} = \boldsymbol{\omega}(\boldsymbol{J}),$$

(3.17)

where $\boldsymbol{J} = (J_1, J_2, \ldots, J_m)^T$, $\boldsymbol{\phi} = (\phi_1, \phi_2, \ldots, \phi_m)^T$ and $\boldsymbol{\omega} = (\omega_1, \omega_2, \ldots, \omega_m)^T$. Correspondingly, the solutions are

$$\phi_i = \omega_i(J_1, J_2, \ldots, J_m)t + \beta_i, \quad (i = 1, 2, \ldots, m)$$

(3.18)

with ω_i, β_i being constants. Therefore, $\boldsymbol{\phi} = (\phi_1, \phi_2, \ldots, \phi_m)^T$ is the angular variables on an m-dimensional torus. Because this torus is invariant for the flow of Eq. (3.17), it is called the invariant torus.

Figure 3.1 illustrates a sketch map of a two-dimensional invariant torus, on which two characteristic frequencies ω_1, ω_2 dominate the motion. If ω_1 and ω_2 are commensurable, the orbit is periodic, otherwise it is a quasi-periodic orbit that is dense on the torus.

Since the integrable systems are quite rare while the near-integrable systems are much more common and thus more important in practice, a two-dimensional near-integrable Hamiltonian system will be discussed below in this section.

Consider a Hamiltonian system with a Hamiltonian function as

$$H = H_0(J_1, J_2) + \epsilon H_1(J_1, J_2, \theta_1, \theta_2),$$

(3.19)

where ϵ is a perturbation parameter. A canonical transformation will be searched for to obtain a set of new canonical variables $(\boldsymbol{I}, \boldsymbol{\phi})$, so that the

     From Ordered to Chaotic Motion

new Hamiltonian function contains only the action variable $\boldsymbol{I}$. Define a generating function

$$S = I_1\theta_1 + I_2\theta_2 + \epsilon S_1\left(I_1, I_2, \theta_1, \theta_2\right). \tag{3.20}$$

It gives the following transformation

$$
\begin{aligned}
J_1 &= \frac{\partial S}{\partial \theta_1} = I_1 + \epsilon \frac{\partial S_1}{\partial \theta_1}, \\
J_2 &= \frac{\partial S}{\partial \theta_2} = I_2 + \epsilon \frac{\partial S_1}{\partial \theta_2}, \\
\phi_1 &= \frac{\partial S}{\partial I_1} = \theta_1 + \epsilon \frac{\partial S_1}{\partial I_1}, \\
\phi_2 &= \frac{\partial S}{\partial I_2} = \theta_2 + \epsilon \frac{\partial S_1}{\partial I_2}.
\end{aligned}
\tag{3.21}
$$

And the new Hamiltonian function reads

$$
\begin{aligned}
H'(\boldsymbol{I}, \boldsymbol{\phi}) &= H_0\left(I_1 + \epsilon\frac{\partial S_1}{\partial \theta_1}, I_2 + \epsilon\frac{\partial S_1}{\partial \theta_2}\right) + \epsilon H_1(I_1, I_2, \phi_1, \phi_2) + O(\epsilon^2) \\
&= H_0(I_1, I_2) + \epsilon\frac{\partial H_0}{\partial J_1}\frac{\partial S_1}{\partial \theta_1} + \epsilon\frac{\partial H_0}{\partial J_2}\frac{\partial S_1}{\partial \theta_2} + \epsilon H_1(I_1, I_2, \phi_1, \phi_2) + O(\epsilon^2) \\
&= H_0(I_1, I_2) + \epsilon\left[\omega_1\frac{\partial S_1}{\partial \theta_1} + \omega_2\frac{\partial S_1}{\partial \theta_2} + H_1(I_1, I_2, \phi_1, \phi_2)\right] + O(\epsilon^2),
\end{aligned}
\tag{3.22}
$$

where $\omega_1 = \frac{\partial H_0}{\partial J_1}$, $\omega_2 = \frac{\partial H_0}{\partial J_2}$ are the unperturbed frequencies. A proper selection of the generating function S makes it possible that no angle variable appears in the new Hamiltonian H' up to $O(\epsilon)$. For this sake, the S_1 in Eq. (3.20) must be carefully chosen so that the terms in the square brackets of Eq. (3.22) should be eliminated. Expand S_1 and H_1 in Fourier series:

$$
\begin{aligned}
S_1 &= \sum S_{1m,n}(I_1, I_2)\exp\left[i(m\theta_1 + n\theta_2)\right], \\
H_1 &= \sum H_{1m,n}(I_1, I_2)\exp\left[i(m\theta_1 + n\theta_2)\right].
\end{aligned}
\tag{3.23}
$$

If the coefficients of all the terms containing any angle variables $(m, n \neq 0)$ in the square brackets of Eq. (3.22) are all zero, one has

$$
\begin{aligned}
&i(m\omega_1 + n\omega_2)\sum S_{1m,n}(I_1, I_2)\exp\left[i(m\theta_1 + n\theta_2)\right] \\
&= -\sum H_{1m,n}(I_1, I_2)\exp\left[i(m\theta_1 + n\theta_2)\right].
\end{aligned}
\tag{3.24}
$$

Thus,

$$S_{1m,n} = -\mathrm{i}\frac{H_{1m,n}}{m\omega_1 + n\omega_2}. \tag{3.25}$$

When ω_1/ω_2 is a rational number, there must exist such m, n that $m\omega_1 + n\omega_2 = 0$. Or, if ω_1/ω_2 is an irrational number but close enough to a rational number, one still has $m\omega_1 + n\omega_2 \approx 0$. Therefore, the resonance will lead to a divergent series for S_1. This is the problem of small denominator in mathematics. Consequently, in this situation, the system cannot be expressed in an integrable form. Geometrically, the invariant torus in the unperturbed system does not persist when the perturbation was introduced. When the torus is broken, almost all the orbits on it become chaotic orbits. However, according to the KAM theorem, if ω_1/ω_2 is an irrational number that satisfies the Diophantine condition, most of the invariant tori will survive with the perturbation. The Diophantine condition reads

$$|\boldsymbol{k} \cdot \boldsymbol{\omega}| \geq \gamma|\boldsymbol{k}|^{-\mu}, \quad \forall \boldsymbol{k} \in \boldsymbol{Z}^2\backslash\{0\}, \tag{3.26}$$

where $\boldsymbol{\omega} = (\omega_1, \omega_2)$ is a frequency vector, $\boldsymbol{k} = (k_1, k_2)$ is an integer vector, $|\cdot|$ is the norm of vector, and $\gamma > 0, \mu > 1$ are constants. Meanwhile, chaotic motion appear after the tori were broken. Usually, the ordered and chaotic motion will exist together. For the ordered motion, the distance between any nearby orbits increases linearly, while for the chaotic orbits, the distance between any initially nearby orbits will increase exponentially. Due to the definition of Lyapunov Characteristic Numbers (LCNs), one may use the LCNs as indicators to determine the kind of motion. If the LCNs are zero, the motions are ordered, while nonzero LCNs correspond to chaotic motions.

3.3 Poincaré surface of section

All dynamical systems fall into two categories: continuous dynamical systems (generally described by differential equations) and discrete dynamical systems (described by mapping). The mapping method has advantages both in the sense of numerical analysis and in theoretical analysis. For this reason, in the study of long-term evolution of a dynamical system, one often adopts the mapping method. Poincaré surface of section is just such a technique to reduce a differential equation to a mapping.

Consider a differential equation

$$\frac{\mathrm{d}\boldsymbol{x}}{\mathrm{d}t} = \boldsymbol{f}(\boldsymbol{x}), \tag{3.27}$$

where $\boldsymbol{x} = (x_1, x_2, \ldots, x_m)^T$, $\boldsymbol{f} = (f_1, f_2, \ldots, f_m)^T$, $f_k(\boldsymbol{x})$ $(k = 1, 2, \ldots, m)$ are analytical functions in a domain G of $\boldsymbol{x}$ space, and they satisfy

$$\mathrm{Tr}\left(\frac{\partial \boldsymbol{f}}{\partial \boldsymbol{x}}\right) = \sum_{k=1}^{m} \frac{\partial f_k}{\partial x_k} = 0, \tag{3.28}$$

implying that the system described by Eq. (3.27) is a conservative dynamical system. Suppose $\boldsymbol{x} = \boldsymbol{x}(t, \boldsymbol{\xi})$ is a solution of Eq. (3.27) with an initial value $\boldsymbol{x}(0, \boldsymbol{\xi}) = \boldsymbol{\xi}$. The conservativity means that the mapping $\boldsymbol{\xi} \to \boldsymbol{x}(t, \boldsymbol{\xi})$ is volume-preserving for any given time t. Consider a solution $\boldsymbol{x} = \boldsymbol{x}(t, \boldsymbol{\xi}^*)$ of Eq. (3.27). Assume $\boldsymbol{x}(t, \boldsymbol{\xi}^*) \in G$ and $\boldsymbol{x} = \boldsymbol{\xi}^*$ is not an equilibrium solution, i.e. $\boldsymbol{f}(\boldsymbol{\xi}^*) \neq \boldsymbol{0}$. Without loss of generality, one may assume that $f_m(\boldsymbol{\xi}^*) \neq 0$ and this solution $\boldsymbol{x}(t, \boldsymbol{\xi}^*)$ intersects the surface $x_m = \xi_m^*$ again at $t = \tau^* > 0$. This surface $x_m = \xi_m^*$ is called Poincaré surface of section.

In a sufficiently small neighborhood U of $\boldsymbol{\xi}^*$, one may trace a solution curve emanating from U at $t = 0$. Due to the continuity of function $\boldsymbol{f}(\boldsymbol{x})$ and the continuous dependence of a solution on the initial value, each of these solution curves will cross the Poincaré surface of section $(x_m = \xi_m^*)$ after a time approximately equal to τ^*. All these sequential intersection points then form another neighborhood U_1 of $\boldsymbol{x}(\tau^*, \boldsymbol{\xi}^*)$, which is just the image of U under the mapping described above. For a sufficiently small $t_0 > 0$, denote by B and B_1 two regions in G, respectively defined by $\boldsymbol{x} = \boldsymbol{x}(t, \boldsymbol{\xi}), 0 \leq t \leq t_0, \boldsymbol{\xi} \in U$ and $\boldsymbol{\xi}^* \in U_1$. Obviously, B and B_1 can be regarded as two cylinders with U and U_1 as their bottoms. Consider a tube R that is formed by the solution curves connecting U and U_1. After a time t_0, the tube R with its interior move to $R + B_1 - B$. Because of the preservation of phase volume, R and $R + B_1 - B$ must have the same m-dimensional volume, hence B_1 and B have the same volume. Denoting the volume element $\mathrm{d}x_1 \mathrm{d}x_2 \cdots \mathrm{d}x_m$ by $\mathrm{d}\boldsymbol{x}$, this volume preserving property reads

$$\int_B \mathrm{d}\boldsymbol{x} = \int_{B_1} \mathrm{d}\boldsymbol{x}. \tag{3.29}$$

New variables $\xi_1, \xi_2, \ldots, \xi_m$ can be introduced through the substitution $x_k = x_k(t, \boldsymbol{\xi})$, and the corresponding transformation matrix has the

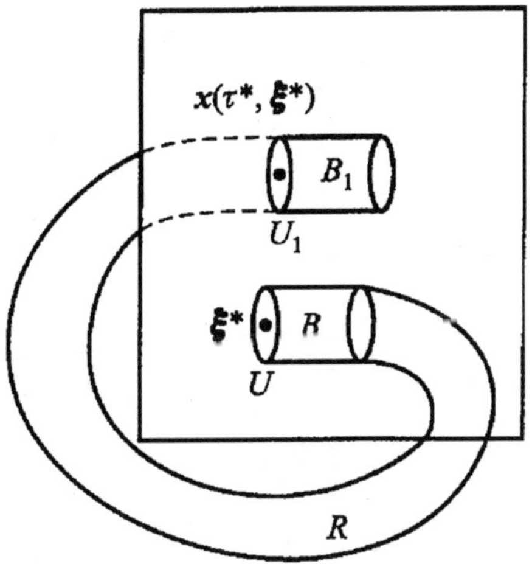

Fig. 3.2   A sketch of Poincaré surface of section.

following form:

$$
\Gamma = \begin{pmatrix}
x_{1\xi_1} & x_{1\xi_2} & \cdots & x_{1\xi_{m-1}} & f_1 \\
x_{2\xi_1} & x_{2\xi_2} & \cdots & x_{2\xi_{m-1}} & f_2 \\
\vdots & \vdots & \vdots & \vdots & \vdots \\
x_{m\xi_1} & x_{m\xi_2} & \cdots & x_{m\xi_{m-1}} & f_m
\end{pmatrix}.
\tag{3.30}
$$

Since the $m \times m$ matrix $(\frac{\partial \boldsymbol{x}}{\partial \boldsymbol{\xi}})|_{t=0} = E$ (E being $m \times m$ unit matrix), when $t = 0$ the determinant $|\Gamma| = f_m(\boldsymbol{\xi}) \neq 0$. Through the above transformation, Eq. (3.29) becomes

$$
\int_0^{t_0} \left[\int_U |\Gamma| \mathrm{d}\xi_1 \mathrm{d}\xi_2 \cdots \mathrm{d}\xi_{m-1} \right] \mathrm{d}t = \int_0^{t_0} \left[\int_{U_1} |\Gamma| \mathrm{d}\xi_1 \mathrm{d}\xi_2 \cdots \mathrm{d}\xi_{m-1} \right] \mathrm{d}t.
\tag{3.31}
$$

Apply the mean value theorem for integrals, divide both sides of Eq. (3.31) by t_0 and let $t_0 \to 0$, one obtains the limit

$$
\int_U f_m(\boldsymbol{\xi}) \mathrm{d}\boldsymbol{\xi} = \int_{U_1} f_m(\boldsymbol{\xi}) \mathrm{d}\boldsymbol{\xi}, \quad \mathrm{d}\boldsymbol{\xi} = \mathrm{d}\xi_1 \mathrm{d}\xi_2 \cdots \mathrm{d}\xi_{m-1}.
\tag{3.32}
$$

One may further assume that $\psi(\boldsymbol{x})$ is a time-independent integral of the system described by Eq. (3.27) with nonzero derivative $\phi_{x_{m-1}} \neq 0$ both in U and U_1. Then $\psi(\boldsymbol{x}) = \gamma$ is constant along each solution curve. Introduce γ via the substitution $\psi(\boldsymbol{\xi}) = \gamma$ as a new variable instead of ξ_{m-1} and keep all the rest variables $\xi_1, \ldots, \xi_{m-2}$ unchanged, i.e. define a transformation as

$$\gamma = \psi(\xi_1, \xi_2, \ldots, \xi_{m-1}, \xi_m^*),$$
$$\xi_1 = \xi_1,$$
$$\vdots$$
$$\xi_{m-2} = \xi_{m-2},$$

(3.33)

and one has now

$$\mathrm{d}\xi_1 \mathrm{d}\xi_2 \cdots \mathrm{d}\xi_{m-2}\mathrm{d}\gamma = \left| \frac{\partial(\xi_1, \ldots, \xi_{m-2}, \psi)}{\partial(\xi_1, \ldots, \xi_{m-2}, \xi_{m-1})} \right| \mathrm{d}\xi_1 \mathrm{d}\xi_2 \cdots \mathrm{d}\xi_{m-2}\mathrm{d}\xi_{m-1}$$
$$= \psi_{\xi_{m-1}}(\boldsymbol{\xi})\mathrm{d}\xi_1 \mathrm{d}\xi_2 \cdots \mathrm{d}\xi_{m-2}\mathrm{d}\xi_{m-1}.$$

(3.34)

Formally, $\psi_{\xi_{m-1}}(\boldsymbol{\xi})\mathrm{d}\xi_{m-1} = \mathrm{d}\gamma$. In particular, choose U to be the product of an $(m-2)$-dimensional neighborhood F of $\xi_k = \xi_k^*$ $\ (k = 1, 2, \ldots, m-2)$ with an interval containing the point $\psi(\boldsymbol{\xi}) = \gamma^*$. As $\psi(\boldsymbol{x}) = \gamma$ is invariant along each solution, the interval of γ remains pointwise fixed in the mapping from U to U_1. Corresponding to $\psi(\boldsymbol{\xi}) = \gamma$, the image of F is a certain set $F_1 = F_1(\gamma)$. From Eq. (3.32), one has

$$\int_{F \times \gamma} \frac{f_m(\boldsymbol{\xi})}{\psi_{x_{m-1}}(\boldsymbol{\xi})} \mathrm{d}\xi_1 \mathrm{d}\xi_2 \cdots \mathrm{d}\xi_{m-2}\mathrm{d}\gamma$$
$$= \int_{F_1 \times \gamma} \frac{f_m(\boldsymbol{\xi})}{\psi_{x_{m-1}}(\boldsymbol{\xi})} \mathrm{d}\xi_1 \mathrm{d}\xi_2 \cdots \mathrm{d}\xi_{m-2}\mathrm{d}\gamma.$$

(3.35)

Introduce in addition

$$g = g(\xi_1, \xi_2, \ldots, \xi_{m-2}, \gamma) = \frac{f_m(\boldsymbol{\xi})}{\psi_{x_{m-1}}(\boldsymbol{\xi})} \quad (\xi_m = \xi_m^*), \qquad (3.36)$$

and let $\mathrm{d}v = \mathrm{d}\xi_1 \mathrm{d}\xi_2 \cdots \mathrm{d}\xi_{m-2}$, a simpler form of Eq. (3.35) reads

$$\int_{\gamma_1}^{\gamma_2} \left[\int_F g\mathrm{d}v \right] \mathrm{d}\gamma = \int_{\gamma_1}^{\gamma_2} \left[\int_{F_1} g\mathrm{d}v \right] \mathrm{d}\gamma. \qquad (3.37)$$

Equation (3.37) is valid for any interval $[\gamma_1, \gamma_2]$ and the integrand is a continuous function of γ. Therefore, one has

$$\int_F g\mathrm{d}v = \int_{F_1} g\mathrm{d}v. \qquad (3.38)$$

For a conservative Hamiltonian system

$$\frac{\mathrm{d}q_k}{\mathrm{d}t} = \frac{\partial H}{\partial p_k}, \quad \frac{\mathrm{d}p_k}{\mathrm{d}t} = -\frac{\partial H}{\partial q_k}, \quad (k = 1, 2, \ldots, n) \qquad (3.39)$$

one has $m = 2n$, and the integral is $\psi(\boldsymbol{q}, \boldsymbol{p}) = H(\boldsymbol{q}, \boldsymbol{p}) = \gamma$. A suitable ordering of the coordinates may make $f_m = -H_{q_n}, \psi_{q_n} = H_{q_n}$, and thus $g = -1$, implying that the mapping from F to F_1 is also volume-preserving. This indicates that a conservative Hamiltonian system of n degrees of freedom can be reduced to a volume-preserving mapping on a $(2n - 2)$-dimensional Poincaré surface of section. The ordered and chaotic regions in phase space of the Hamiltonian system correspond to ordered and chaotic regions on the surface of section. Specifically, the periodic orbits correspond to fixed points and the n-dimensional invariant tori correspond to $(n - 1)$-dimensional invariant tori on the surface of section.

In fact, a famous example of the conversion and correspondence between a Hamiltonian system and a mapping is the "standard mapping" (also known as Chirikov mapping). The standard mapping (Chirikov, 1979) is a very well-known model in nonlinear dynamics. It is deduced from pendulum equation by discretization of the equation of motion. The Hamiltonian function of the pendulum motion is

$$H = \frac{I^2}{2} + K \cos \theta. \tag{3.40}$$

According to the averaging principle, the basic property of this Hamiltonian system will not change even if some high-frequency terms were added to this function

$$\begin{aligned} H' &= \frac{I^2}{2} + K \cos \theta + 2K \cos \theta \cos(2\pi t) + 2K \cos \theta \cos(4\pi t) + \cdots \\ &= \frac{I^2}{2} + K \cos \theta \sum_{m=-\infty}^{+\infty} \cos(2\pi m t). \end{aligned} \tag{3.41}$$

On the other hand, one notes that the Dirac delta function has a Fourier expansion as

$$\delta_1(t) = \sum_{m=-\infty}^{+\infty} \cos(2\pi m t) = 1 + 2 \sum_{m=1}^{+\infty} \cos(2\pi m t). \tag{3.42}$$

Hence, H' can be now expressed as

$$H' = \frac{I^2}{2} + K \cos \theta \, \delta_1(t). \tag{3.43}$$

The corresponding Hamiltonian equations are

$$\begin{aligned} \frac{\mathrm{d}I}{\mathrm{d}t} &= -\frac{\partial H'}{\partial \theta} = K \sin \theta \, \delta_1(t), \\ \frac{\mathrm{d}\theta}{\mathrm{d}t} &= \frac{\partial H'}{\partial I} = I. \end{aligned} \tag{3.44}$$

A mapping can be obtained by integrating the above equations in a period:

$$I_{n+1} = I_n + \int_0^1 K \sin \theta_n \, \delta_1(t) \mathrm{d}t = I_n + K \sin \theta_n,$$

$$\theta_{n+1} = \theta_n + \int_0^1 I_{n+1} \mathrm{d}t = \theta_n + I_{n+1}. \tag{3.45}$$

Here in the second formula I_n is replaced by I_{n+1} for the sake of keeping the volume(area)-preserving property. This mapping can be used to study the qualitative behavior of the pendulum motion.

This is an example of conversion from a Hamiltonian system to a mapping. Conversely, a mapping can also be converted (extended) to a Hamiltonian system too.

3.4 Ordered and chaotic motion of stars

The motion of stars in galaxy is an important topic in celestial mechanics and also is a concrete example of dynamical system. Because the gravitational potential in a galaxy can be well approximated by its average, the motion of a star in a steady averaged potential field is a Hamiltonian system with three degrees of freedom, typically expressed in the cylindrical coordinates (R, ϕ, z). It is a dynamical system of six-dimensional phase space $(R, \phi, z, \dot{R}, \dot{\phi}, \dot{z})$. In this system, there are two integrals of motion: the energy integral and the angular momentum integral. The motion of star will be numerically investigated, especially the third integral will be searched.

If there exists an integral for such a system

$$I(R, \phi, z, \dot{R}, \dot{\phi}, \dot{z}) = c, \quad \text{where } c \text{ is a constant}, \tag{3.46}$$

the motion will be confined on a five-dimensional surface defined by this integral. More integrals will set more confinements for the motion. According to Liouville theorem of Hamiltonian system, if a Hamiltonian system of $2n$ dimensions (n degrees of freedom) possesses n independent and involutive integrals, it is integrable. The Hamiltonian system describing the motion of stars in a galaxy with averaged potential is of three degrees of freedom, thus it is integrable if there exist three independent and involutive integrals.

In particular, for the motion of stars in a steady and axisymmetric gravitational potential, the energy integral and the angular momentum integral around z-axis can be expressed by

$$I_1 = \frac{1}{2}\left(\dot{R}^2 + R^2\dot{\phi}^2 + \dot{z}^2\right) + U_g(R,z) \equiv T + U_g,$$

$$I_2 = R^2\dot{\phi},$$

(3.47)

where $U_g(R,z)$ is the axisymmetric gravitational potential. It is easy to show that the above two integrals are independent and involutive. In fact, in a cylindrical coordinate system the generalized momentum is defined by

$$p_1 = \frac{\partial T}{\partial \dot{R}} = \dot{R},$$

$$p_2 = \frac{\partial T}{\partial \dot{\phi}} = R^2\dot{\phi},$$

$$p_3 = \frac{\partial T}{\partial \dot{z}} = \dot{z}.$$

(3.48)

And the above two integrals can be expressed as

$$I_1 = T + U_g = \frac{1}{2}\left(p_1^2 + \frac{p_2^2}{R^2} + p_3^2\right) + U_g(R,z),$$

$$I_2 = p_2.$$

(3.49)

Their Poisson bracket is

$$\{I_1, I_2\} = \sum_{i=1}^{3}\left[\frac{\partial I_1}{\partial q_i}\frac{\partial I_2}{\partial p_i} - \frac{\partial I_1}{\partial p_i}\frac{\partial I_2}{\partial q_i}\right] = 0,$$

(3.50)

where $(q_1, q_2, q_3) = (R, \phi, z)$. This shows that I_1, I_2 are involutive.

Therefore, if a third integral that is involutive with I_1 and I_2 respectively, the corresponding Hamiltonian system will be integrable. This new integral is called the "third integral".

After introducing the effective potential

$$U(R,z) = U_g(R,z) + \frac{p_2^2}{2R^2},$$

(3.51)

the original system of three degrees of freedom may be reduced to a system of two degrees of freedom. In fact this reduction is a reflection of the angular momentum integral. Define new variables as $(x, y, \dot{x}, \dot{y}) = (R, z, \dot{R}, \dot{z})$ in the reduced system, the first integral (the energy integral) is

$$I_1 = \frac{1}{2}\left(\dot{x}^2 + \dot{y}^2\right) + U(x,y).$$

(3.52)

Hence, the motion in the four-dimensional phase space will be confined onto a three-dimensional manifold. If the third integral I_3 exists (the second integral has been used to reduce one degree of freedom), the motion will be further confined to the two-dimensional manifold. By choosing a suitable surface of section, the orbits may intersect it on straight lines. Conversely, if I_3 does not exist, an orbit can occupy a two-dimensional region on the surface of section. Thus, in terms of the behavior of intersection of orbits with the surface of section, one may determine whether the third integral exists or not.

Choose an effective potential function as

$$U(x, y) = \frac{1}{2}\left(x^2 + y^2 + 2x^2 y - \frac{2}{3}y^3\right). \tag{3.53}$$

Then the corresponding equations of motion are

$$\frac{\mathrm{d}^2 x}{\mathrm{d}t^2} = -\frac{\partial U}{\partial x} = -x - 2xy,$$

$$\frac{\mathrm{d}^2 y}{\mathrm{d}t^2} = -\frac{\partial U}{\partial y} = -y - x^2 + y^2. \tag{3.54}$$

Take the Poincaré surface of section as

$$E = c, x = 0, \dot{x} > 0, \tag{3.55}$$

where $E = c$ is the energy integral. Then, on the Poincaré surface of section, the motion is confined in the following region

$$U(0, y) + \frac{1}{2}\dot{y}^2 \leq E.$$

Figure 3.3 displays the phase diagram on the Poincaré section $(y, \dot{y})$ for $E = 0.05$. Each set of points linked by a curve corresponds to one computed orbit. It seems that all the orbits lie on curves around four elliptical fixed points inside. Moreover, there exist three hyperbolic fixed points. In this case, the third integral I_3 exists.

When the energy increases, the structure of phase space changes. Figure 3.4 is the phase diagram for $E = 0.14$. Now the curves do not fill the whole Poincaré section near the hyperbolic fixed points. On the contrast, the chaotic regions appear from where the curves disappear. Still, some "islands" of closed curves exist. If the energy increases further, for example to $E = 0.16$ as shown in Fig. 3.5, the islands deform and shrink, leaving more regions to the chaotic sea.

In a similar numerical experiments, Hénon & Heiles proposed a method to decide whether a point P_1 belongs to a curve or to a chaotic orbit. A second initial point P_1' is taken very close to P_1 (usually at a distance of

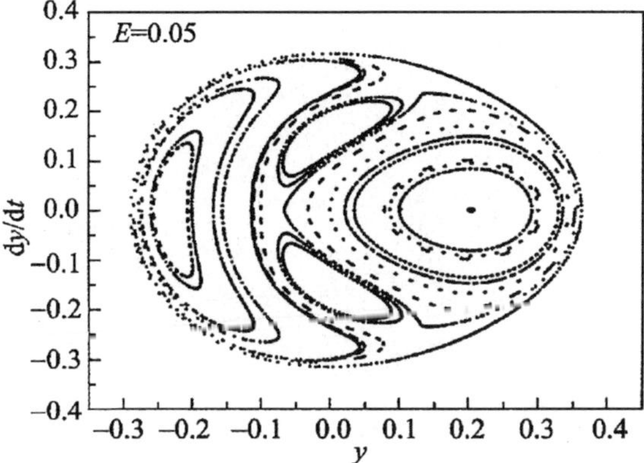

Fig. 3.3 The phase diagram for $E = 0.05$. From Sun & Zhou (2008).

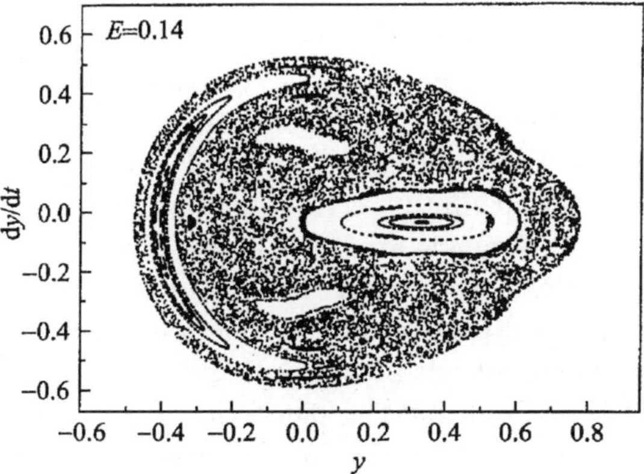

Fig. 3.4 The phase diagram for $E = 0.14$. From Sun & Zhou (2008).

10^{-7}), then a number (usually 25) of successive transforms of both P_1 and P_1' are computed. If the distance between the images of these two points $\overline{P_i P_i'}$ increases only slowly, about linearly with the transform time i, P_1 and P_1' are in a region occupied by curves, that is, ordered region. On the contrary, if P_1 and P_1' are in chaotic region, the distance $\overline{P_i P_i'}$ increases rapidly, roughly exponentially. The quantity

$$\mu = \sum_{i=1}^{25} \left(\overline{P_i P_i'}\right)^2 \tag{3.56}$$

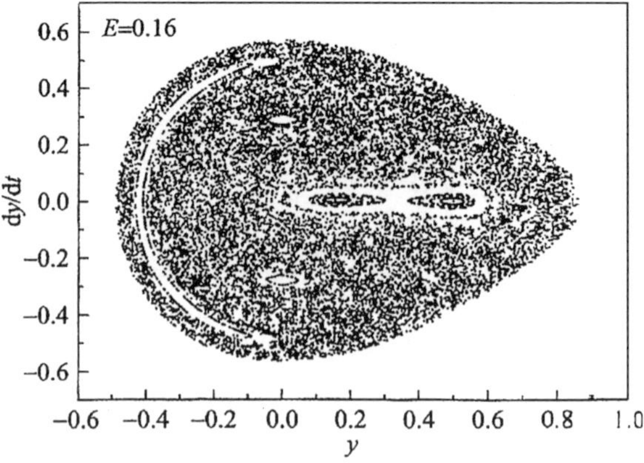

Fig. 3.5   The phase diagram for $E = 0.16$. From Sun & Zhou (2008).

is computed and regarded as an indicator. The point P_1 as well as its successive points are considered as belonging to the chaotic region if $\mu > \mu_c$, or the ordered region if $\mu < \mu_c$, where μ_c is a chosen constant. The value of μ may cover a very wide range from about 10^{-12} to 10^{+1}, and the criterion seems very sensitive. Here, in this case, $\mu_c \approx 10^{-4}$ is an appropriate choice.

Figure 3.6 shows that up to a critical energy (about $E = 0.11$) there is no chaotic region in the whole area. For higher energy values, the ordered region shrinks rapidly. Thus the situation could be described very roughly by saying that the third integral I_3 exists only for orbits below a "critical energy", but not for orbits with higher energy.

It is worthy to point out that this pioneering method proposed by Hénon & Heiles had adopted exactly the same concept as the Lyapunov character numbers, which is now widely used in the study of nonlinear dynamics.

3.5   Application of mapping method to comet motion

Comets are a special kind of bodies in solar system, with a variety of orbit types. Although all the elliptic, parabolic and hyperbolic orbits may be recognized in the comet motion, most of comets move on the near-parabolic orbits, i.e. very oblate elliptic orbits with very long orbital periods. In numerically simulating the orbit of a comet, the long orbital period and high eccentricity make the computation very expensive. However the mapping method is advantageous to long-term numerical computations, hence it is an ideal method for the study of the motion of long periodic comets.

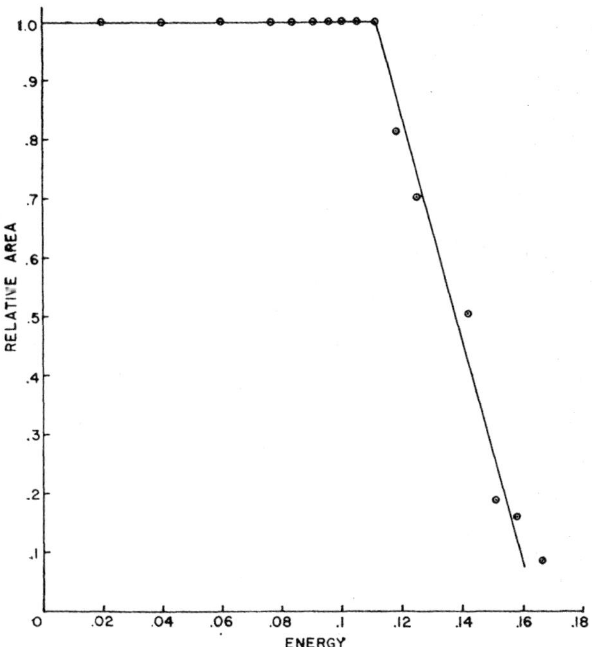

Fig. 3.6   Relative area covered by the curves as a function of energy, as computed by the method described in the text. From Hénon & Heiles (1964).

3.5.1   *Basic model and mapping*

In an inertial coordinate system with origin at the center of mass of the Sun and Jupiter. Let m, m_s, m_j be the masses of a comet, the Sun and Jupiter, respectively. In the framework of planar restricted three-body problem, the differential equations of motion of the comet are

$$m\frac{\mathrm{d}^2x}{\mathrm{d}t^2} = Gmm_s\frac{x_s - x}{R_{13}^3} + Gmm_j\frac{x_j - x}{R_{23}^3},$$

$$m\frac{\mathrm{d}^2y}{\mathrm{d}t^2} = Gmm_s\frac{y_s - y}{R_{13}^3} + Gmm_j\frac{y_j - y}{R_{23}^3}, \tag{3.57}$$

where G is the gravitational constant as usual, and $R_{13}^2 = (x_s - x)^2 + (y_s - y)^2$, $R_{23}^2 = (x_j - x)^2 + (y_j - y)^2$.

Take the distance between the Sun and Jupiter a_s and $M = m_s + m_j$ as the units of distance and mass, select the appropriate unit of time such that $G = 1$, and define $\mu = m_j$ ($\mu \ll 1$). The equations of motion now read

$$\frac{\mathrm{d}^2 x}{\mathrm{d}t^2} = \frac{(1-\mu)(x_s - x)}{R_{13}^3} + \frac{\mu(x_j - x)}{R_{23}^3},$$

$$\frac{\mathrm{d}^2 y}{\mathrm{d}t^2} = \frac{(1-\mu)(y_s - y)}{R_{13}^3} + \frac{\mu(y_j - y)}{R_{23}^3}.$$

$$(3.58)$$

Under such units, Jupiter revolves around the Sun with unit angular velocity, and the motion of Jupiter and the Sun are described by

$$x_j = (1-\mu)\cos(t - t_0), \quad y_j = (1-\mu)\sin(t - t_0); \tag{3.59}$$

$$x_s = -\mu\cos(t - t_0), \quad y_s = -\mu\sin(t - t_0). \tag{3.60}$$

Here, t_0 is the initial time. Substituting Eqs. (3.59) and (3.60) into Eq. (3.58) and expanding the right-hand side of the equation to $O(\mu)$, one has

$$\frac{\mathrm{d}^2 x}{\mathrm{d}t^2} = \frac{-x}{(x^2 + y^2)^{3/2}} + \mu F(x, y, t, t_0),$$

$$\frac{\mathrm{d}^2 y}{\mathrm{d}t^2} = \frac{-y}{(x^2 + y^2)^{3/2}} + \mu G(x, y, t, t_0),$$

$$(3.61)$$

where

$$F(x, y, t, t_0) = \frac{x}{(x^2 + y^2)^{3/2}} - \frac{\cos(t - t_0)}{(x^2 + y^2)^{3/2}}$$

$$+ \frac{\cos(t - t_0) - x}{\{[x - \cos(t - t_0)]^2 + [y - \sin(t - t_0)]^2\}^{3/2}}$$

$$+ \frac{3x[x\cos(t - t_0) + y\sin(t - t_0)]}{(x^2 + y^2)^{5/2}},$$

$$(3.62)$$

$$G(x, y, t, t_0) = \frac{y}{(x^2 + y^2)^{3/2}} - \frac{\sin(t - t_0)}{(x^2 + y^2)^{3/2}}$$

$$+ \frac{\sin(t - t_0) - y}{\{[x - \cos(t - t_0)]^2 + [y - \sin(t - t_0)]^2\}^{3/2}}$$

$$+ \frac{3y[x\cos(t - t_0) + y\sin(t - t_0)]}{(x^2 + y^2)^{5/2}}.$$

When $\mu = 0$, the energy of the comet is

$$\frac{K}{2} = \frac{1}{2}(\dot{x}^2 + \dot{y}^2) - \frac{1}{(x^2 + y^2)^{1/2}}, \tag{3.63}$$

and $K = $ constant is an integral. However, the perturbation from Jupiter will make the energy vary with time.

$$\frac{\mathrm{d}K}{\mathrm{d}t} = 2\mu \left[\dot{x} F(x, y, t, t_0) + \dot{y} G(x, y, t, t_0) \right].$$ (3.64)

Suppose that the orbit of the comet is near-parabolic, thus a parabolic orbit can be taken as an appropriate approximate orbit and one has

$$r = q \sec^2 \frac{f}{2},$$ (3.65)

where q and f are the perihelion distance and the true anomaly of the orbit. The relation between the true anomaly and time is given by

$$2 \tan \frac{f}{2} + \frac{2}{3} \tan^3 \frac{f}{2} = \sqrt{2} q^{-\frac{3}{2}} (t - \tau).$$ (3.66)

Here, τ denotes the time when the comet passes through the perihelion, which can be simply assumed $\tau = 0$ for convenience.

Let $E(q, t) = \tan \frac{f}{2}$. From Eq. (3.66), one has

$$E(q, t) = \left(\frac{3}{2\sqrt{2}} q^{-\frac{3}{2}} t + \sqrt{\frac{9}{8} q^{-3} t^2 + 1} \right)^{\frac{1}{3}}$$
$$+ \left(\frac{3}{2\sqrt{2}} q^{-\frac{3}{2}} t - \sqrt{\frac{9}{8} q^{-3} t^2 + 1} \right)^{\frac{1}{3}}$$ (3.67)

and

$$x = r \cos f = q \left[1 - E^2(q, t) \right],$$
$$y = r \sin f = 2q E(q, t);$$ (3.68)

$$\dot{x} = - 2q E(q, t) \dot{E}(q, t),$$
$$\dot{y} = 2q \dot{E}(q, t).$$ (3.69)

When $t = \tau = 0$, the comet is passing through the perihelion. Suppose Jupiter has a phase angle g at this moment, thus $g = -t_0$ (see Fig. 3.7 for a sketch). Substituting Eqs. (3.68) and (3.69) into Eq. (3.64) and integrating it from $t = -\infty$ to $t = +\infty$, one obtains the change of energy for each comet's return:

$$\Delta K = 2\mu \psi(q, g) = 2\mu \int_{-\infty}^{+\infty} \left[\dot{x} F(x, y, t, g) + \dot{y} G(x, y, t, g) \right] \mathrm{d}t.$$ (3.70)

Note that the above discussion is restricted to the direct motion of the comet (the direction of motion is consistent with that of Jupiter). Concerning the retrograde orbit (with the opposite direction of motion to

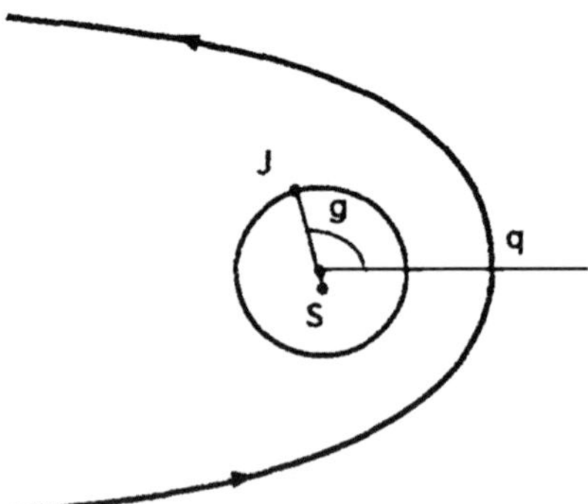

Fig. 3.7   The configuration when the comet is passing through the perihelion From Liu & Sun (1994).

Table 3.1   Coefficients of the Fourier expansion of $\psi(q = 1.5, g)$.

Term No.	Direct orbit	Retrograde orbit
1	$b_1 = 0.261$	$b_1' = 0.0428$
2	$b_2 = 1.132$	$b_2' = -0.162$
3	$b_3 = 0.779$	$b_3' = -0.589$
4	$b_4 = 0.525$	$b_4' = -0.461$
5	$b_5 = 0.350$	$b_5' = -0.0531$
6	$b_6 = 0.230$	$b_6' = -0.00131$
7	$b_7 = 0.148$	$b_7' = -0.177$
8	$b_8 = 0.0875$	$b_8' = -0.0879$
9	$b_9 = 0.0409$	$b_9' = -0.0320$

that of Jupiter), the discussion is similar, except for a change of sign in Eq. (3.67).

The function $\psi(q, g)$ in Eq. (3.70) is a 2π-periodic function of g, so it can be expanded into Fourier series

$$\psi(q, g) = \frac{a_0}{2} + \sum_{n=1}^{\infty} \left[a_n(q) \cos ng + b_n(q) \sin ng \right]. \tag{3.71}$$

Numerical computations show that the function $\psi(q, g)$ is symmetric about $g = 0, \pi$. Hence, the terms of cosine function in the Fourier expansion of Eq. (3.71) are zero. As for the coefficients $b_n(q)$, although the first one $b_1(q)$ is not the largest, the general trend of its quickly decreasing as the order n is confident. As listed in Table 3.1, those coefficients $b_k(q)$ $(k > 7)$ are much smaller than those of $k \le 7$. Therefore, the function $\psi(q, g)$ with $q = 1.5$ for comets on the direct and retrograde orbit can be well approximated respectively by

$$\psi(q,g) \approx \sum_{k=1}^{7} b_k \sin kg, \quad \text{(direct)} \tag{3.72}$$

$$\psi(q,g) \approx b_2' \sin 2g + b_3' \sin 3g + b_4' \sin 4g + b_7' \sin 7g. \quad \text{(retrograde)} \tag{3.73}$$

It can be shown that the change of q during each perihelion passage is of order μ and its effect on ΔK is of order μ^2. Therefore, q can be treated as a constant during the evolution of the energy of a comet.

According to Kepler's third law, the period of time between two successive perihelion passages is $\Delta t = 2\pi(-K)^{-3/2}$, where $K = -1/a$ and a is the semimajor axis of an orbit. The change of the phase angle of Jupiter during this period is then $\Delta g = \Delta t$, thus

$$\Delta g = 2\pi(-K)^{-\frac{3}{2}} \quad (\text{mod } 2\pi). \tag{3.74}$$

Combining with Eq. (3.70), the two-dimensional mappings, describing the motion of a comet on the direct orbit (M_1) and the retrograde orbit (M_2) respectively, are obtained

$$M_1 : \begin{cases} K_{n+1} = K_n + \mu \displaystyle\sum_{n=1}^{7} b_i \sin ig_n, \\[2em] g_{n+1} = g_n + \dfrac{2\pi}{(-K_{n+1})^{\frac{3}{2}}} \quad (\text{mod } 2\pi), \end{cases} \tag{3.75}$$

$$M_2 : \begin{cases} K_{n+1} = K_n + \mu\,(b_2' \sin 2g + b_3' \sin 3g \\[0.5em] \qquad\qquad\qquad +b_4' \sin 4g + b_7' \sin 7g), \\[1em] g_{n+1} = g_n + \dfrac{2\pi}{(-K_{n+1})^{\frac{3}{2}}} \quad (\text{mod } 2\pi). \end{cases} \tag{3.76}$$

It will be shown below that the mappings M_1 and M_2 are both area-preserving.

3.5.2 *Properties of the mapping*

Both mapping M_1 and M_2 are area-preserving, since both of them satisfy

$$\det A = 1 \quad \text{with} \quad A = \begin{pmatrix} \dfrac{\partial K_{n+1}}{\partial K_n} & \dfrac{\partial K_{n+1}}{\partial g_n} \\[1.5em] \dfrac{\partial g_{n+1}}{\partial K_n} & \dfrac{\partial g_{n+1}}{\partial g_n} \end{pmatrix}. \tag{3.77}$$

The fixed points of mapping M_1 can be easily found:

$$g^* = g_1, g_2, g_3, g_4, g_5,$$
$$K^* = -\frac{1}{m} \qquad (m : \text{positive integers}), \tag{3.78}$$

where $g_1 = 0, g_2 = 1.387, g_3 = 3.142, g_4 = 4.883, g_5 = 6.284$.

The stability of these fixed points depends on the absolute value of the trace of A. When $|\text{Tr}A|$ is less than, equal to or greater than 2, the corresponding fixed points are elliptic, parabolic or hyperbolic, respectively. Thus, the fixed points $(K^*, g_1), (K^*, g_3)$ and (K^*, g_5) are stable elliptic fixed points for small m, and they become unstable hyperbolic fixed points when $m > 0.270\mu^{-3/2}$.

Similar calculations show that the fixed points of the mapping M_2 are (K^*, g_i') $(i = 1, 2, \ldots, 9)$, with $g_1' = 0, g_2' = 0.969, g_3' = 1.891, g_4' = 2.568$, $g_5' = 3.142, g_6' = 3.725, g_7' = 4.379, g_8' = 5.316$ and $g_9' = 6.284$. The points (K^*, g_i') are unstable hyperbolic fixed points for $i = 2, 4, 6, 8$; (K^*, g_5') is a parabolic fixed point, and (K^*, g_i') of $i = 1, 3, 7, 9$ are stable fixed points if m is small. When $m > 0.140\mu^{-3/5}$, two of them, (K^*, g_1') and (K^*, g_9') become hyperbolic points. And later, when m becomes larger than $0.210\mu^{-3/5}$, (K^*, g_3') and (K^*, g_7') turn to hyperbolic fixed points.

To analyze the structure of phase space in the vicinity of the fixed points, the mapping now can be expanded with respect to the energy. Below only the case of direct motion will be analyzed, and for the retrograde motion, the discussions are similar. Let $K = K^* + U$ and $g = \theta$, Eq. (3.75) can be transferred to

$$\begin{cases} K_{n+1} = K^* + U_{n+1} = K^* + U_n + \mu f(\theta_n), \\[2ex] g_{n+1} = \theta_{n+1} = \theta_n + \dfrac{2\pi}{(-K^*)^{\frac{3}{2}}} + \dfrac{3\pi U_{n+1}}{(-K^*)^{\frac{5}{2}}} \quad (\text{mod } 2\pi), \end{cases} \tag{3.79}$$

where

$$f(\theta_n) = \sum_{i=1}^{7} b_i \sin i\theta_n. \tag{3.80}$$

Let $J = 3\pi U(-K^*)^{-5/2}$, the mapping becomes

$$T : \begin{cases} J_{n+1} = J_n + Q f(\theta_n), \\ \theta_{n+1} = \theta_n + J_{n+1} \quad (\text{mod } 2\pi), \end{cases} \tag{3.81}$$

with $Q = 3\pi\mu(-K^*)^{-5/2}$. Just as the well-known standard mapping, the mapping T is defined on a two-dimensional torus. For this mapping, five 1-periodic fixed points are located on the line $J = 0$. Since $f(\theta)$ is an odd function, it is easy to get 2-periodic and 3-periodic fixed points as follows.

$$4\theta^* + Qf(\theta^*) = 2k\pi, \quad J^* = 2\theta^* - 2m\pi, \tag{3.82}$$

$$3\theta^* + Qf(\theta^*) = k\pi, \quad J^* = 2\theta^* - 2m\pi, \tag{3.83}$$

where m and k are both positive integers. The stability of these fixed points depends on the absolute value of the trace of the following matrix

$$B = \prod_{i=1}^{n} \begin{pmatrix} 1 & Qf'(\theta_i) \\ 1 & 1 + Qf'(\theta_i) \end{pmatrix}, \tag{3.84}$$

where n is the period number and (J_i, θ_i) are the coordinates of the corresponding fixed points.

The solutions to Eqs. (3.82) and (3.83) can be found numerically. Denote four 2-periodic solutions by O_1, O_2, O_3, O_4 and six 3-periodic solutions by $S_1, S_2, S_3, S_4, S_5, S_6$ respectively. Among them, the fixed points O_1, O_3, S_1, S_3 and S_5 are unstable for any value of parameter Q, because $|\mathrm{Tr}B|$ is larger than 2. However, O_2 and O_4 are stable if Q is smaller than 2.0, and S_2, S_4, S_6 become unstable a little earlier when Q is larger than 0.4. It seems that the higher-order period fixed points can be changed from stable to unstable by perturbations more easily.

3.5.3 *Phase plane of mapping and orbital stability of comets*

In a two-dimensional near-integrable system, the chaos occurs near the separatrix, i.e. the boundary between the resonant and ordered region. When the perturbation parameter Q is not equal to zero, there exist chaotic regions for the mapping T. But invariant tori (KAM tori) that exist between the primary resonant regions, such as the 1/1 and 2/1 resonant regions prevent the orbital diffusion along the direction of J axis. So the phase plane, unconnected chaotic regions and periodic islands appear. As Q increases, these KAM tori will be destroyed successively. After a critical value Q_c, the last KAM torus breaks and the chaotic regions become connected when $Q \geq Q_c$, thus the orbits may diffuse in the whole phase plane and the motion is unstable.

The critical value Q_c can be found (for example) by means of resonance overlapping method. First, the periodic delta function and its Fourier series are introduced

$$\delta_1(n) = \sum_{m=-\infty}^{+\infty} \delta_1(n-m) = 1 + 2\sum_{s=1}^{\infty} \cos 2\pi s n. \tag{3.85}$$

Then the mapping T is equivalent to the following equations

$$\frac{\mathrm{d}J}{\mathrm{d}n} = Qf(\theta)\delta_1(n),$$
$$\frac{\mathrm{d}\theta}{\mathrm{d}n} = J, \tag{3.86}$$

and the corresponding Hamiltonian function is

$$H = \frac{1}{2}J^2 + QV(\theta)\delta_1(t) = \frac{1}{2}J^2 + QV(\theta) + 2QV(\theta)\sum_{s=1}^{+\infty} \cos 2\pi s t, \tag{3.87}$$

with $V(\theta)$ defined as

$$V(\theta) = -\int f(\theta)\mathrm{d}\theta = \sum_{i=1}^{7} \frac{b_i \cos i\theta}{i}. \tag{3.88}$$

For the $1/1$ resonance, $J = J^* = 2\pi$, the resonant terms are

$$b_1 \cos\theta \cos 2\pi t = \frac{b_1}{2}\left[\cos(\theta - 2\pi t) + \cos(\theta + 2\pi t)\right],$$
$$\frac{b_2}{2} \cos 2\theta \cos 4\pi t = \frac{b_2}{4}\left[\cos(2\theta - 4\pi t) + \cos(2\theta + 4\pi t)\right],$$
$$\cdots\cdots = \cdots\cdots \tag{3.89}$$
$$\frac{b_7}{7} \cos 7\theta \cos 14\pi t = \frac{b_7}{14}\left[\cos(7\theta - 14\pi t) + \cos(7\theta + 14\pi t)\right].$$

Through a generating function $L = (\theta - 2\pi t)I$, a canonical transformation of variables (J, θ) to (I, ψ) as

$$J = I, \quad \psi = \theta - 2\pi t \tag{3.90}$$

can be performed and the new Hamiltonian function is $H_1 = H - 2\pi I$. H_1 can be simplified by averaging over the fast variable, and it becomes $H_1 = \frac{1}{2}I^2 - 2\pi I + Qh_1(\psi)$. Let $I = 2\pi + \Delta I$ and expand the Hamiltonian function at $I = 2\pi$, one obtains eventually

$$H_1 = \frac{\Delta I^2}{2} + Qh_1(\psi). \tag{3.91}$$

Similar calculations can be made concerning other $1/m$ $(m = 2, 3, \ldots, 7)$ resonances and the corresponding Hamiltonian functions may be obtained:

$$H_1 = \frac{1}{2}\Delta I^2 + Qh_m(\psi), \tag{3.92}$$

where ψ represents different resonant angles and $h_m(\psi)$ are formulated respectively as

$$h_1(\psi) = b_1 \cos\psi + \frac{b_2}{2}\cos 2\psi + \cdots + \frac{b_7}{7}\cos 7\psi,$$

$$h_2(\psi) = \frac{b_2}{2}\cos 2\psi + \frac{b_4}{4}\cos 4\psi + \frac{b_6}{6}\cos 6\psi,$$

$$h_3(\psi) = \frac{b_3}{3}\cos 3\psi + \frac{b_6}{6}\cos 6\psi, \tag{3.93}$$

$$h_k(\psi) = \frac{b_k}{k}\cos k\psi \quad (k = 4, 5, 6, 7).$$

The half widths $\Delta I^m_{\max}$ of the resonant regions can be calculated from the expression of $h_m(\psi)$ following the procedure introduced by Chirikov (1979), and the results are listed in Table 3.2.

According to the resonance overlapping method, the condition for destruction of the last KAM tori is

$$\sum_{m=1}^{7} \Delta I^m_{\max} \geq \pi. \tag{3.94}$$

The above inequality gives the critical value Q_c, over which the last KAM tori disappears

$$Q \geq Q_c = 0.113. \tag{3.95}$$

Numerical exploration may verify the theoretical analysis. The phase plane for $Q = 0.100$, which is slightly below Q_c, is numerically computed from the mapping and illustrated in Fig. 3.8. There are 1-period islands, 2-period islands, $\ldots$, and 7-period islands. An undestroyed KAM torus that prevents the diffusion of chaotic orbits can be found between the 6-period islands and the 7-period islands. When $Q = Q_c$, all the KAM tori are destroyed and the chaotic regions become connected as shown in Fig. 3.9.

Table 3.2   The half widths ($\Delta I_{\max}$) of different resonant regions.

Winding number	1/1	1/2	1/3	1/4	1/5	1/6	1/7
	1/1	1/2	2/3	3/4	4/5	5/6	6/7
$\Delta I_{\max}(\times\sqrt{Q})$	1.913	1.549	1.000	0.725	0.529	0.392	0.291

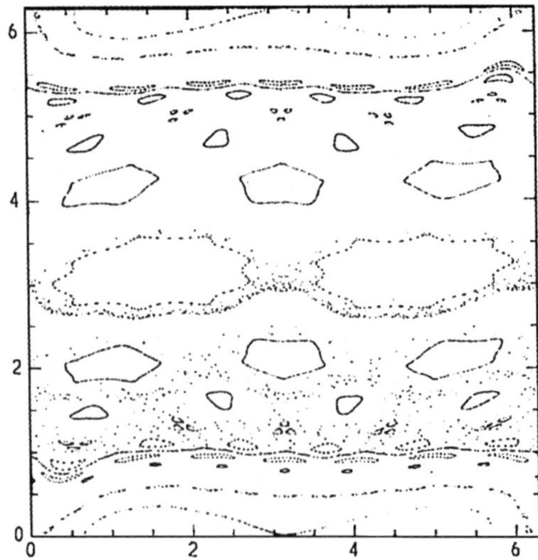

Fig. 3.8   The phase plane of mapping T for $Q = 0.100$. KAM tori can be seen between the 6-period islands and 7-period islands. From Liu & Sun (1994).

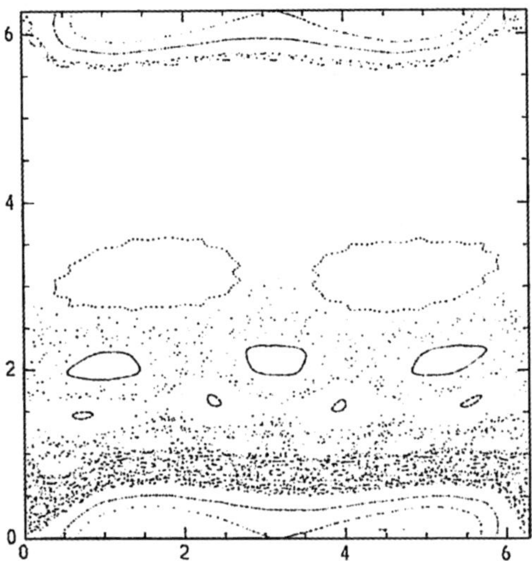

Fig. 3.9   The phase plane of mapping T for $Q = 0.113$. The last KAM torus has been destroyed and the chaotic regions become connected. From Liu & Sun (1994).

A careful numerical exploration reveals that the winding number corresponding to the last KAM torus should be between 5/6 and 6/7 (or 1/6 and 1/7). This result differs from that of the standard mapping, in which the last KAM tours has a golden mean winding number.

Substitute the critical values Q_c corresponding to the direct and retrograde orbits into $Q_c = 3\pi\mu(-K_c)^{-5/2}$, one gets respectively for the direct motion

$$K_c = -\left(\frac{3\pi\mu}{Q_c}\right)^{\frac{2}{5}} = -5.87\mu^{\frac{2}{5}}, \tag{3.96}$$

and for the retrograde motion

$$K'_c = -\left(\frac{3\pi\mu}{Q'_c}\right)^{\frac{2}{5}} = -4.58\mu^{\frac{2}{5}}. \tag{3.97}$$

The K_c and K'_c define the boundaries that restrict connected chaotic region for the mappings M_1 and M_2. According to the condition for destabilization of 1-periodic solution, there is another critical value about the energy K

$$K_\mu = -2.39\mu^{\frac{2}{5}}, \quad \text{for direct motion,} \tag{3.96}'$$

$$K'_\mu = -2.83\mu^{\frac{2}{5}}, \quad \text{for retrograde motion.} \tag{3.97}'$$

When the comet's energy K is less than K_c (or K'_c for retrograde motion), usually its orbit is regular or bounded and it will stay in the Solar System forever. When $K > K_\mu$ or $K > K'_\mu$ respectively for direct or retrograde orbit, the stochastic regions are connected, implying that the comets' orbits may diffuse and the comets may probably escape from the Solar System. Between K_μ and K_c (or, between K'_μ and K'_c for retrograde orbit), the chaotic regions and regular islands coexist simultaneously. If a comet's orbit belongs to the regular islands, the resonant effect will guarantee its stable staying in the Solar System. Otherwise, it will escape eventually.

3.5.4 *Elliptical case*

All the above discussions are based on the framework of the circular restricted three-body problem. Since Jupiter's orbit is an ellipse, the elliptic restricted three-body problem would be a better model in reality.

For a comet on a direct orbit (similar discussions can be made for retrograde orbit), the perihelion orientation of the comet's orbit and thus the mean longitude of perihelion ω changes a little at each return (a sketching

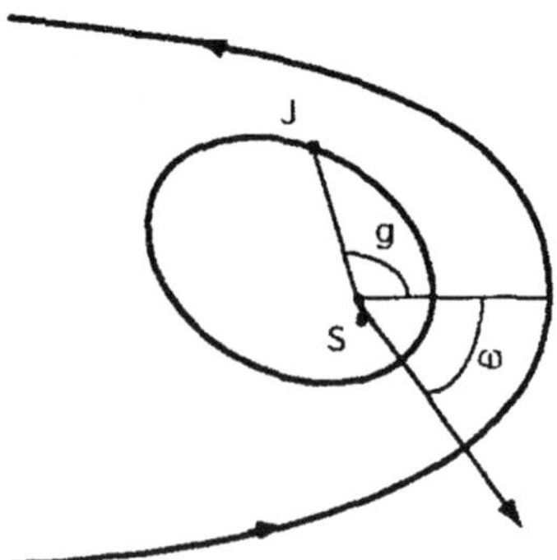

Fig. 3.10   The configuration for the elliptic model when the comet is at perihelion. From Liu & Sun (1994).

configuration can be seen in Fig. 3.10). The energy change during such a perihelion fly is

$$\Delta K \cong \mu f(g - \omega) \cong \mu \sum_{i=1}^{7} b_i \sin i(g - \omega). \qquad (3.98)$$

Considering the effect of eccentricity of the perturbing planet, even in the case of $g = \omega$, the integral in Eq. (3.70) is not zero because the orbit is not isotropic now. It means that the Fourier expansions of ΔK should include cosine terms. For the sake of simplification, only the first cosine term will be kept and a qualitative discussion will be made below. Based on Eq. (3.98), ΔK can be written as

$$\Delta K \cong \mu \left[\sum_{i=1}^{7} b_i \sin i(g - \omega) + V \cos(g - \omega) \right], \qquad (3.99)$$

where V is related to the perturbing body's orbit eccentricity. On the other hand, in the case of $g = \omega = 0$, the integral in Eq. (3.70) vanishes, so V can be taken as

$$V \sim \rho \sin \omega. \qquad (3.100)$$

Thus ΔK becomes

$$\Delta K \cong \mu \left[\sum_{i=1}^{7} b_i \sin i(g - \omega) + \rho \sin \omega \cos(g - \omega) \right], \qquad (3.101)$$

where ρ is a parameter related to the orbital eccentricity of the perturbing planet.

The change of the phase angle of the perturbing planet can be also described by the same formula like in the circular restricted three-body model in Eq. (3.74). But in this case, the mean longitude of comet at each comet's return changes.

Using Laplace integral and the relation with orbital elements, for the near-parabolic orbits of a comet the eccentricity can be approximated to be 1. Under this assumption, one has

$$F_1 = -\cos\omega, \quad F_2 = -\sin\omega, \tag{3.102}$$

where

$$F_1 = -Dy + \frac{x}{(x^2 + y^2)^{1/2}}, \quad F_2 = Dx + \frac{y}{(x^2 + y^2)^{1/2}}, \tag{3.103}$$

with D being the constant of angular momentum integral.

When $\mu = 0$, apparently F_1 and F_2 are the first integrals of equations of motion [Eq. (3.61)]. When $\mu \neq 0$, they vary with time

$$\frac{dF_1}{dt} = \mu D G(x, y, t, g),$$

$$\frac{dF_2}{dt} = \mu D F(x, y, t, g). \tag{3.104}$$

From Eqs. (3.102, 3.104), one has

$$\frac{d\omega}{dt} = \frac{dF_2}{dt} F_1 - \frac{dF_1}{dt} F_2. \tag{3.105}$$

Integrating this equation from $-\infty$ to $+\infty$ leads to

$$\Delta\omega = \int_{-\infty}^{+\infty} \left(\frac{dF_2}{dt} F_1 - \frac{dF_1}{dt} F_2 \right) dt. \tag{3.106}$$

Substituting Eqs. (3.68, 3.69, 3.104) into Eq. (3.106) and making the Fourier expansion, one obtains

$$\Delta\omega = \mu p(g - \omega) = \mu \left[\frac{a_0}{2} + \sum_{i=1}^{7} a_i \cos i(g - \omega) \right], \tag{3.107}$$

where $a_0/2 = 1.676, a_1 = 1.077, a_2 = 2.141, a_3 = 1.500, a_4 = 0.934, a_5 = 0.582, a_6 = 0.363, a_7 = 0.227$.

Finally, from Eqs. (3.101), (3.74), and (3.107), a new mapping is obtained for the elliptic restricted three-body model:

$$M_3 : \begin{cases} K_{n+1} = K_n + \mu[f(g_n) + \rho \sin\omega_n \cos g_n], \\ g_{n+1} = g_n + 2\pi(-K_{n+1})^{-\frac{3}{2}} \quad (\text{mod } 2\pi), \\ \omega_{n+1} = \omega_n + \mu p(g_n) \quad (\text{mod } 2\pi), \end{cases} \tag{3.108}$$

where the function $f(g)$ and $p(g)$ are defined by Eqs. (3.98, 3.107). In the above deduction, $g - \omega$ has been replaced by g' and the prime has then been omitted for convenience.

The mapping in Eq. (3.108) is a three-dimensional mapping which gives the change in comet's energy at each return in the elliptic model. When $\rho = 0$, the mapping M_3 becomes the product of the area-preserving mapping M_1 multiplying the one-dimensional mapping on the ω-direction. In this sense, M_3 is the three-dimensional extension of M_1. The above discussions are for the direct motion of comet, and for the retrograde motion similar discussions can be made.

It is easy to verify that the mapping M_3 is volume-preserving. A detailed discussion concerning the three-dimensional volume-preserving mapping will be made in later sections. At this moment, we would point out that if there exist the two-dimensional invariant tori, the motion of comet will be confined in the bounded regions, implying that the motion of comets is stable and will stay in the Solar System. Otherwise, the comets will probably escape from the Solar System.

First of all, set the parameter $\mu = 0.01$, and the invariant curves of the mapping M_3 will be searched. In the phase space of a three-dimensional volume-preserving mapping, there usually exist two-dimensional invariant tori in the neighborhood of the invariant curves. Let $g = g(\omega), K = K(\omega)$ be the invariant curve of the mapping M_3 so that from Eq. (3.108),

$$\begin{cases} K(\omega_{n+1}) = K(\omega_n) + \mu R\left(g(\omega_n), \omega_n\right) \\ g(\omega_{n+1}) = g(\omega_n) + 2\pi\left[-K(\omega_{n+1})\right]^{-\frac{3}{2}} \quad (\mathrm{mod}\ 2\pi), \end{cases} \tag{3.109}$$

where $R(\theta, \omega) = f(\theta) + \rho \sin\omega \cos\theta$. Approximately, $K(\omega) = -(1/m)^{2/3}$, $R(g(\omega), \omega) = 0$. Therefore, the one-dimensional manifolds in the 2/1 resonant zone are described by $K(\omega) = -0.630$ and $R(g(\omega), \omega) = 0$.

The invariant curves corresponding to the 3/2 resonance can be calculated as follows. Using Eq. (3.79) the mapping M_3 can be expanded at $K_{3/2} = -0.763$ and one gets:

$$\begin{cases} J_{n+1} = J_n + QR(\theta_n), \\ \theta_{n+1} = \theta_n + J_{n+1} + 3\pi \quad (\mathrm{mod}\ 2\pi), \end{cases} \tag{3.110}$$

with $Q = 0.185$. The 2-periodic fixed points of this mapping in Eq. (3.110) satisfy the following equations

$$\begin{aligned} &R(\theta) + R\left[\theta + J + QR(\theta) + 3\pi\right] = 0, \\ &2J + 2QR(\theta) + QR\left[\theta + J + QR(\theta) + 3\pi\right] = 2k\pi, \quad (k \in \mathbb{Z}^+). \end{aligned} \tag{3.111}$$

When $\rho = 0$ the above equations are equivalent to Eq. (3.82). Let (J^*, θ^*) be the solutions of the equations [Eq. (3.111)], thus

$$\theta^* + \frac{1}{2}Qf(\theta^*) + k\pi = -\theta^*,$$

$$2J^* + Qf(\theta^*) = 2k\pi. \tag{3.112}$$

Further, let $J = J^* + J(\omega)$ and $\theta = \theta^* + \theta(\omega)$ be the solutions of Eq. (3.111), one has

$$\theta(\omega) = \frac{A}{B}, \tag{3.113}$$

where

$$A = -R(\theta^*) - R\left[\theta^* + \frac{1}{2}QR(\theta^*) + k\pi\right],$$

$$B = R'(\theta^*) + R'\left[\theta^* + \frac{1}{2}QR(\theta^*) + k\pi\right][1 + QR'(\theta^*)], \tag{3.114}$$

with the prime denoting the time derivative $R' = \frac{dR}{dt}$. Substituting Eq. (3.113) into Eq. (3.111), one obtains the expression of $J(\omega)$, and the invariant curve corresponding to the 3/2 resonant zone is

$$K(\omega) = -0.763 + \frac{\mu}{Q}J(\omega),$$

$$g(\omega) = \theta^* + \theta(\omega). \tag{3.115}$$

Concerning the 4/3 resonant case, the mapping M_3 can be expanded near $K_{4/3} = -0.825$ and one gets

$$\begin{cases} J_{n+1} = J_n + QR(\theta_n), \\ \theta_{n+1} = \theta_n + J_{n+1} + \frac{8}{3}\pi \quad (\text{mod } 2\pi). \end{cases} \tag{3.116}$$

with $Q = 0.152$. While the 3-periodic fixed points should satisfy

$$R(\theta) + R\left[\theta + J + QR(\theta) + \frac{8}{3}\pi\right]$$

$$+ R\left\{\theta + 2J + 2QR(\theta) + \frac{16}{3}\pi + R\left[\theta + J + QR(\theta) + \frac{8}{3}\pi\right]\right\} = 0,$$

$$3J + 2QR(\theta) + QR\left[\theta + J + QR(\theta) + \frac{8}{3}\pi\right] = 2k\pi, \quad (k \in \mathbb{Z}^+).$$
$$\tag{3.117}$$

When $\rho = 0$ the above equations become

$$J^* + \frac{2}{3}\pi = 2\theta^*,$$

$$3\theta^* + Qf(\theta^*) = k\pi. \tag{3.118}$$

Assume the solution of Eq. (3.117) is

$$J = J^* + J(\omega),$$
$$\theta = \theta^* + \theta(\omega). \tag{3.119}$$

Substitute this solution into Eq. (3.117), one obtains

$$\theta(\omega) = \frac{A}{B},$$
$$J(\omega) = \frac{[QR'(-\theta^*) - QR'(\theta^*)]\,\theta(\omega)}{3 + QR'(\theta^*)}, \tag{3.120}$$

where

$$A = R(\theta^*) + R(-\theta^*) + R\left[3\theta^* + QR(\theta^*)\right],$$

$$\begin{aligned} B = & -\{R'(\theta^*) + R'[3\theta^* + QR(\theta^*)][1 + QR'(\theta^*)] + R'(-\theta^*)\} \\ & -\frac{\{R'[3\theta^* + QR(\theta^*)] - R'(-\theta^*)\}\,[QR'(-\theta^*) - QR'(\theta^*)]}{3 + QR'(-\theta^*)}. \end{aligned} \tag{3.121}$$

Finally the approximate expression for the invariant curves corresponding to the 4/3 resonance is obtained

$$K(\omega) = -0.825 + \frac{\mu}{Q}J(\omega),$$
$$g(\omega) = \theta^* + \theta(\omega). \tag{3.122}$$

These invariant curves can be calculated numerically, and an example of the 2/1 resonance case is plotted in Fig. 3.11. In the neighborhood of these invariant curves, there could exist invariant tori.

The two-dimensional invariant tori in the three-dimensional phase space can be illustrated by "slice-cutting method". Nine "slices" in the phase space are defined as

$$|\omega - s| \leq 0.01, \quad s = \frac{(k - 5)\pi}{4}, \quad k = 1, 2, \ldots, 9.$$

When a trajectory in the phase space of the mapping M_3 falls in one of these slices, its coordinates (K, g) are recorded. In the neighborhood of the invariant curve defined by Eq. (3.122), such nine slices of a trajectory are plotted in Fig. 3.12. The closed curves on each slice indicate that there does exist an invariant torus. The LCNs are calculated and their values are $6.17 \times 10^{-3}, 8.93 \times 10^{-5}$ and -6.26×10^{-3}. This confirms that the trajectory belongs to an invariant torus. More numerical experiments show that the invariant tori will be destroyed as the parameter ρ increases, and the global resonant zone is unstable at last.

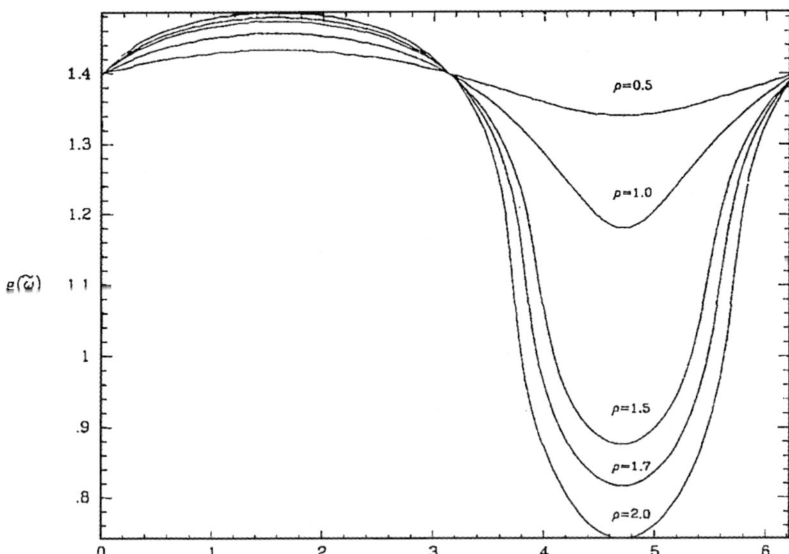

Fig. 3.11   The invariant curves corresponding to the 2/1 resonant zone for several ρ values as indicated by the labels near the curves. The abscissa and ordinate are ω and $g(\omega)$. From Liu & Sun (1994).

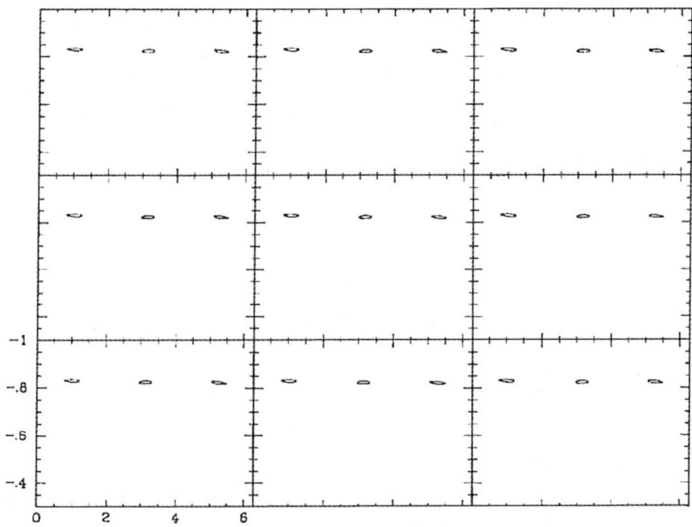

Fig. 3.12   The slices for the trajectory originated from $(K, g, \omega) = (-0.83, 0.8, 0)$ when $\mu = 0.01, \rho = 0.1$. From Liu & Sun (1994).

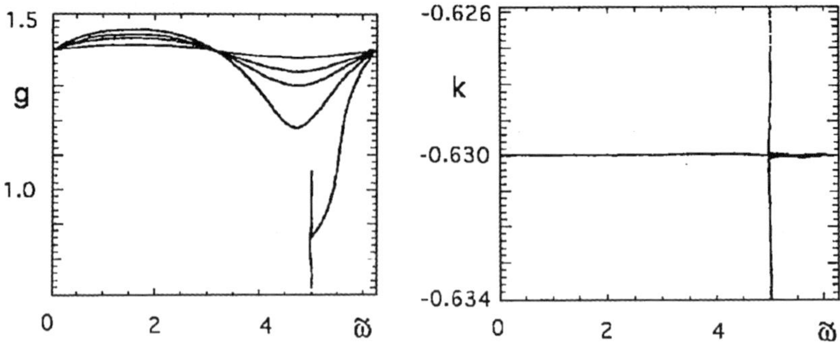

Fig. 3.13   The invariant curves corresponding to the 2/1 resonant zone are projected on the plane $(\omega, g(\omega))$ and the plane $(\omega, K(\omega))$. Different curves are for different parameters $\rho = 0.2, 0.5, 0.7, 1.0, 1.5$ and when $\rho = 1.5$, a diffusion phenomenon happens. From Liu & Sun (1994).

Besides, the existence of invariant tori depends on ω too. In fact, from Eq. (3.108) one has

$$\Delta\omega \approx \mu p(g). \tag{3.123}$$

When $g_1 = 0.861, g_2 = 1.637, g_3 = 4.627, g_4 = 5.422$, the function $p(g)$ equals to zero. According to the results Sun (1984), if $\Delta\omega$ does not keep a definite sign along an invariant curve $(K(\omega), g(\omega))$, there would be a surface of section $\omega = \omega^*$, along which the orbits will diffuse, due to the conservation of volume for the mapping M_3. This leads to the breaking of the invariant curve, and also the breaking of the invariant tori around it. So, the necessary condition for the existence of an invariant curve can be concluded as

$$0.861 \leq g(\omega) \leq 1.637. \tag{3.124}$$

Figure 3.13 is a diagram of invariant curves for different ρ values in the 2/1 resonant zone. It shows that when $\rho = 1.5$ the diffusion occurs at $\omega^* = 5.0$.

With the above analysis, the conditions for the destruction of the 2/1, 4/3, 3/2 resonant zones are revealed and they are listed in Table 3.3.

Finally, some conclusions can be summarized as follows.

(1) In the framework of planar circular restricted three-body problem, there are critical values of comet's energy K_c, K_μ (for the direct motion) and K'_c, K'_μ (for retrograde motion). When $K > K_c$ or $K > K'_c$, the chaotic regions of motion are connected and when $K < K_c$ or $K < K'_c$ the orbits

Table 3.3   The critical parameter ρ for the destruction of several resonant zones.

Resonant zone	Critical parameter	
	Theoretical value	Numerical value
4/3	3.0	2.0
3/2	≥ 7.0	6.0
2/1	1.5	1.0

usually are regular or bounded, thus will stay in the Solar System forever. When K is greater than K_μ or K'_μ, the comets will eventually escape from the Solar System.

(2) Within the model of planar elliptic restricted problem, the stability of comet motion depends not only on the energy but also on the eccentricity of the perturbing body. A critical parameter related to the eccentricity of the perturbing body is defined, by which the stability of a comet's orbit can be decided.

Last but not least, it should be noticed that all the above discussions are for a concrete problem with the background of the comet's motion in the Solar System, but for convenience the parameters are not completely the same as of the real system. Consequently, the results obtained in this section are reliable only qualitatively, but not quantitatively.

3.6   Global applicability of symplectic integrator

The symplectic integration method is effective for long-time evolution of dynamical systems, because symplectic integration algorithms preserve the symplectic structure, and consequently the phase-space volume is preserved. Since a numerical integration method is a discretization of continuous dynamical systems, it can be regarded as a mapping approach. A method to construct a mapping model for a Hamiltonian system has been proposed in Hadjidemetriou (1991), in which the main idea is to preserve the basic feature of an actual system in the mapping, i.e. to preserve the local topology of phase space, not only qualitatively but also quantitatively. This can be achieved by preserving local fixed points and their stability indices.

However it is found that the mapping constructed in Hadjidemetriou (1991) is exactly the same as the first-order scheme of the symplectic integrator. So, the first-order scheme can preserve the local topology of phase space. Though higher-order schemes of integrator are generally more accurate than the first-order one, the fixed points they possess may be different from those of the first-order one. If the differences are only of higher order of step size, and they have the same stability indices, high-order schemes should be better than the first-order one, otherwise the situation is complicated. Moreover, for the mapping system there may exist n-periodic fixed points $(n \geq 2)$, but for the actual Hamiltonian system, only equilibrium solutions can be found. So when the mapping system has n-periodic fixed points $(n \geq 2)$, one cannot conclude that it has the same topology of phase space with the actual system. Thus the validity of the scheme becomes a question. This relates to the problem of global validity for the symplectic integration method and mapping approach. This problem will be studied using a simple dynamical model in this section.

3.6.1 *Model*

The simple pendulum is adopted as the dynamical model. The Hamiltonian and the corresponding Hamiltonian equations are

$$H = \frac{1}{2}p^2 + \cos q,$$

$$\frac{\mathrm{d}q}{\mathrm{d}t} = \frac{\partial H}{\partial p}, \quad \frac{\mathrm{d}p}{\mathrm{d}t} = -\frac{\partial H}{\partial q}. \tag{3.125}$$

Obviously $p = p_0 = 0$, $q = q_0 = k\pi$ $(k = 0, \pm 1, \ldots)$ are equilibria of the Hamiltonian system. Later on, q will be restricted $q \in [0, 2\pi]$. They are the periodic solutions with any period, and the corresponding monodromy matrix is $M = \mathrm{e}^{A\tau}$, where

$$A = \begin{pmatrix} 0 & \cos q_0 \\ 1 & 0 \end{pmatrix}$$

is the matrix of the variational equations. The stability index is defined as

$$K = \mathrm{Tr}M = \sum_{i=1}^{2} \mathrm{e}^{\lambda_i \tau},$$

where λ_i $(i = 1, 2)$ are the eigenvalues of A. Thus one has $K = 2 + \tau^2 \cos q_0 + O(\tau^4) = 2 + (-1)^k \tau^2 + O(\tau^4)$, $k = 0, 1$. The equilibrium is stable when $K < 2$, and unstable when $K > 2$. While $K = 2$ is the critical case.

3.6.2 *Integration schemes*

The first-, second- and third-order schemes of symplectic integration are respectively,

$$\text{SI}_1 : \begin{cases} p' = p + \tau \sin q, \\ q' = q + \tau p', \end{cases} \tag{3.126}$$

$$\text{SI}_2 : \begin{cases} p' = p + \tau \sin q + \dfrac{\tau^2}{2} p' \cos q, \\[2mm] q' = q + \tau p' - \dfrac{\tau^2}{2} \sin q, \end{cases} \tag{3.127}$$

and

$$\text{SI}_3 : \begin{cases} p' = p + \tau \sin q + \dfrac{\tau^2}{2} p' \cos q - \dfrac{\tau^3}{6} \left(p'^2 \sin q + \sin 2q \right), \\[2mm] q' = q + \tau p' - \dfrac{\tau^2}{2} \sin q - \dfrac{\tau^3}{3} p' \cos q, \end{cases} \tag{3.128}$$

where τ is the step size. For the above schemes, generally n-periodic fixed points exist.

3.6.2.1 1-periodic fixed points

The 1-periodic fixed points of SI_1, SI_2 and SI_3 have the following general expressions, respectively

$$\begin{cases} q_{01} = k\pi, \ k = 0, 1, \\[2mm] p_{01} = \dfrac{2m\pi}{\tau}, \ m \in \mathbb{Z}, \end{cases} \tag{3.129}$$

$$\begin{cases} q_{02} = k\pi - \arctan(m\pi) + (-1)^k \tau^2 \dfrac{m\pi}{4(1 + m^2\pi^2)^{3/2}} + O(\tau^4), \ k = 0, 1, \\[2mm] p_{02} = \dfrac{2m\pi}{\tau} - (-1)^k \dfrac{\tau}{2} \dfrac{m\pi}{\sqrt{1 + m^2\pi^2}} + O(\tau^3), \ m \in \mathbb{Z}, \end{cases} \tag{3.130}$$

and

$$\begin{cases} q_{03} = q^* + \tau^2 q^{**} + O(\tau^4), \\[2mm] p_{03} = \dfrac{2m\pi}{\tau} + \left(\dfrac{1}{2} \sin q^* + \dfrac{2}{3} m\pi \cos q^* \right) \tau + O(\tau^3), \ m \in \mathbb{Z}, \end{cases} \tag{3.131}$$

where

$$
\begin{cases}
q^* = k\pi + \arctan \dfrac{m\pi}{\frac{2}{3}m^2\pi^2 - 1}, \ k = 0, 1, \\[4mm]
q^{**} = (-1)^k \dfrac{m\pi \left[\left(\frac{1}{12} + \frac{4}{9}m^2\pi^2\right) \left(\frac{2}{3}m^2\pi^2 - 1\right) + \frac{2}{3}\left(\frac{2}{3}m^2\pi^2 - 1\right)^2 + \frac{m^2\pi^2}{3} \right]}{\left[\left(\frac{2}{3}m^2\pi^2 - 1\right)^2 + m^2\pi^2 \right]^{3/2}}, \\[4mm]
m \in \mathbb{Z}.
\end{cases}
$$

$$
\tag{3.132}
$$

When $m = 0$, $p_{01} = p_{02} = p_{03} = 0$, $q_{01} = q_{02} = q_{03} = k\pi$ ($k = 0, 1$), coinciding with the equilibrium solution of Hamiltonian system [Eq. (3.125)]. For $m \geq 1$, there are two classes of fixed points, corresponding to $k = 0$ and $k = 1$ respectively. Their stabilities can be judged from the stability index defined as the trace of matrix A, denoted by $K = \text{Tr}A$, where A is the matrix of the linearized mapping near the fixed points. After some manipulations, one obtains

$$
\begin{aligned}
K_1 &= 2 + \tau^2 \cos q_{01}, \\
K_2 &= 2 + \tau^2 (\cos q_{02} - m\pi \sin q_{02}) + O(\tau^4), \\
K_3 &= 2 + \tau^2 \left[\left(1 - \frac{2m^2\pi^2}{3}\right) \cos q_{03} - m\pi \sin q_{03} \right] + O(\tau^4),
\end{aligned}
\tag{3.133}
$$

for SI_1, SI_2 and SI_3, respectively. Thus, for $m \geq 1$, (q_{01}, p_{01}) and (q_{02}, p_{02}) are stable when $k = 1$ and unstable when $k = 0$, while (q_{03}, p_{03}) is stable when $k = 0$ and unstable when $k = 1$. For $m = 0$, one has $K_1 = 2 + (-1)^k \tau^2$, $K_2 = 2 + (-1)^k \tau^2 + O(\tau^4)$, $K_3 = 2 + (-1)^k \tau^2 + O(\tau^4)$. In this case, for all the schemes SI_1, SI_2 and SI_3, the fixed point $(0, \pi)$ is stable, while $(0, 0)$ is unstable. In addition, one can see that the stability indices of $(0, \pi)$ and $(0, 0)$ for SI_1, SI_2 and SI_3 are, up to $O(\tau^4)$, the same as those of the Hamiltonian system [Eq. (3.125)]. Thus in the neighborhood of the fixed points $(0, \pi)$ and $(0, 0)$ the phase space of the Hamiltonian system and of the symplectic integration schemes possess the same topology. But for $m \geq 1$, the topologies of them are different, since no such equilibrium solution exists for the Hamiltonian system [Eq. (3.125)].

3.6.2.2   2-periodic fixed points

For SI_1, there are two sets of 2-periodic fixed points:

$$
\begin{cases}
q_{01}^{(2a)} = k\pi, \ k = 0, 1, \\[2mm]
p_{01}^{(2a)} = \dfrac{m\pi}{\tau}, \ m = 1, 3, 5, \ldots,
\end{cases}
\tag{3.134}
$$

and

$$\begin{cases} q_{01}^{(2b)} = \left(k - \dfrac{m}{2}\right)\pi - \dfrac{\tau^2}{4}\sin\left(k - \dfrac{m}{2}\right)\pi + O(\tau^4), \ \ k = 0,1, \\[2mm] p_{01}^{(2b)} = \dfrac{m\pi}{\tau} - \dfrac{\tau}{2}\sin\left(k - \dfrac{m}{2}\right)\pi + O(\tau^3). \ m = 1,3,5,\dots \end{cases} \tag{3.135}$$

The stability index can be computed easily

$$\begin{aligned} K = 2 + \{2\tau^2\left[\cos q + \cos(q + \tau p - \tau^2\sin q)\right] \\ + \tau^4\cos q\cos(q + \tau p + \tau^2\sin q)\}\big|_{(q_0,p_0)}, \end{aligned} \tag{3.136}$$

where (q_0, p_0) stands for $(q_{01}^{(2a)}, p_{01}^{(2a)})$ or $(q_{01}^{(2b)}, p_{01}^{(2b)})$. Thus $K = 2 - \tau^4 + o(\tau^4)$ for $(q_{01}^{(2a)}, p_{01}^{(2a)})$ and $K = 2 + \tau^4 + o(\tau^4)$ for $(q_{01}^{(2b)}, p_{01}^{(2b)})$, indicating that the fixed point $(q_{01}^{(2a)}, p_{01}^{(2a)})$ is always stable while $(q_{01}^{(2b)}, p_{01}^{(2b)})$ is always unstable.

For the 2-periodic fixed points of SI_2 and SI_3, they are rather laborious to find, therefore they will be searched by numerical calculations.

3.6.3 *Numerical results*

The structure of the phase space near fixed points for the Hamiltonian system Eq. (3.125) and the symplectic integration schemes listed in Eqs. (3.126)–(3.128) are investigated numerically. First, $\tau = 0.2$ is chosen. The diagrams of the Hamiltonian system in the neighborhood of $p = 0, p = \pi/\tau = 15.7$ and $p = 2\pi/\tau = 31.4$ are plotted in Fig. 3.14. This is a well-known picture, while in Figs. 3.15, 3.16, and 3.17 are the corresponding diagrams of SI_1, SI_2 and SI_3 respectively. Apparently, in the vicinity of $p = 0$, the Hamiltonian system and the SI_1, SI_2, SI_3 have the same topology, coinciding with the theoretical analysis. But in the strip near $p = \pi/\tau$ and $p = 2\pi/\tau$, far from the q-axis, there exist 1-periodic and 2-periodic fixed points for symplectic integrations as indicated theoretically, while according to Fig. 3.14 no equilibrium solution exists for the Hamiltonian system, and the topologies of the Hamiltonian system and the symplectic integration schemes are different.

Some $n > 2$ periodic fixed points are also present (see for example in Fig. 3.18, 3-periodic islands of SI_1). However, when a smaller step size, e.g. $\tau = 0.05$, is adopted, the topologies of the phase space near $p = 15.7$ of SI_1, SI_2 and SI_3 become the same as that of the Hamiltonian system, because in this case there is no n-periodic fixed point of SI_1, SI_2 and SI_3 in the region (see Fig. 3.19). But the n-periodic fixed points still exist, with larger p, as shown by an example in Fig. 3.20. It implies that a proper selection of the

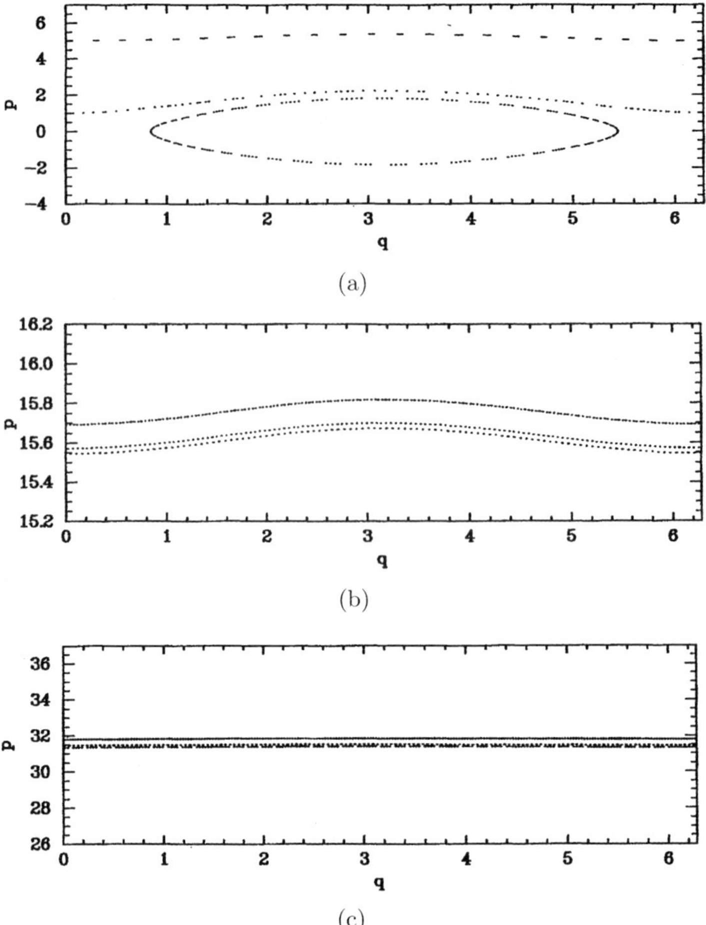

Fig. 3.14   Diagrams of the Hamiltonian system Eq. (3.125) in the vicinity of (a) $p = 0$, (b) $p = 15.7$, and (c) $p = 31.4$ with $\tau = 0.2$. From Sun & Zhou (1996).

step size may remove the spurious fixed points artificially created by the symplectic integration schemes out from the studied region.

The numerical results reveal that a higher-order integration scheme is not necessarily better than a lower-order one if the mapping (integration scheme) possesses some other fixed points than the real system, but in the region where the symplectic integration schemes and the Hamiltonian system have the same topology, the higher-order schemes will be better than the lower-order ones. This phenomenon is illustrated in

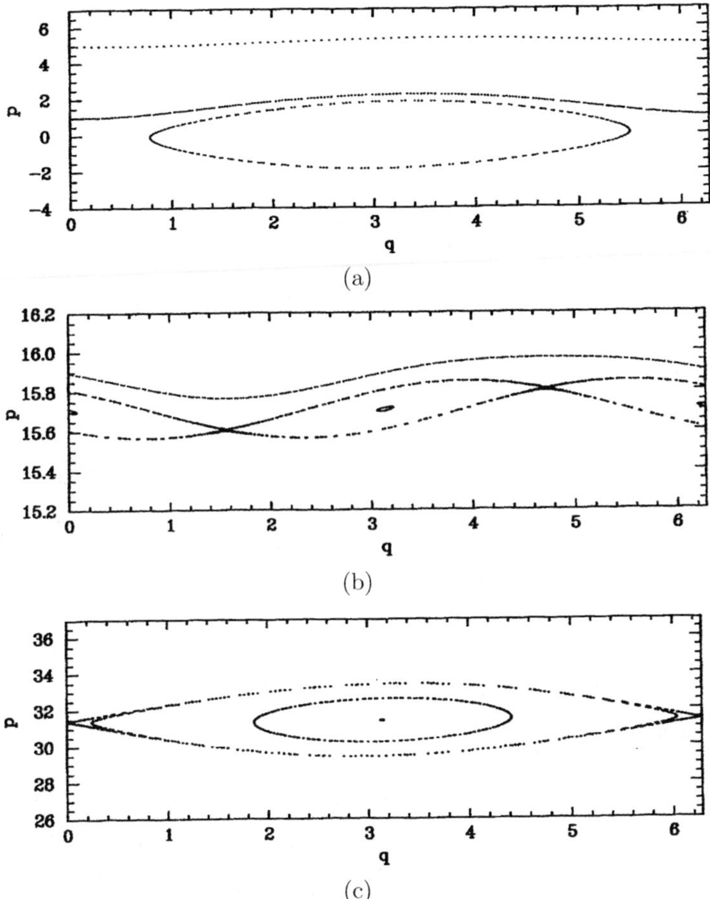

Fig. 3.15 Diagrams of the first-order scheme SI_1 in the vicinity of (a) $p = 0$, (b) $p = 15.7$, (c) $p = 31.4$ with $\tau = 0.2$. From Sun & Zhou (1996).

Figs. 3.14(a)$-$3.17(a), in which by carefully checking one can find that the invariant curves of schemes near the fixed point $(0, \pi)$ are closer and closer to those of the Hamiltonian system with increasing order of schemes.

The investigations presented above show:

(1) When the symplectic integration schemes and the actual system possess the same topology of phase space, the symplectic integration method is effective and the higher-order schemes are better than the lower ones.

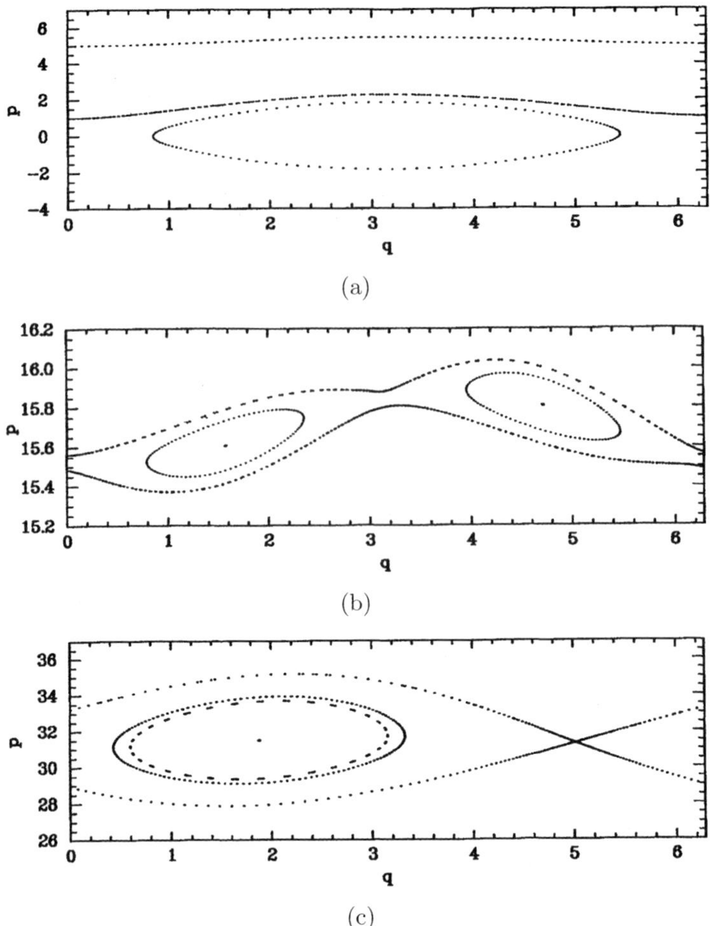

Fig. 3.16 Diagrams of the second-order scheme SI_2 in the vicinity of (a) $p = 0$, (b) $p = 15.7$, and (c) $p = 31.4$ with $\tau = 0.2$. From Sun & Zhou (1996).

(2) When the symplectic schemes possess some other fixed points than the ones of the actual system, the topologies of phase space are different from the actual system, and in this case the symplectic integration method is generally not effective globally and one cannot conclude that the higher-order schemes are better than the lower ones.

According to the above statements, when the symplectic integration method (or, mapping approach) is applied, one has to examine whether or not the symplectic integration schemes have the same topology of phase

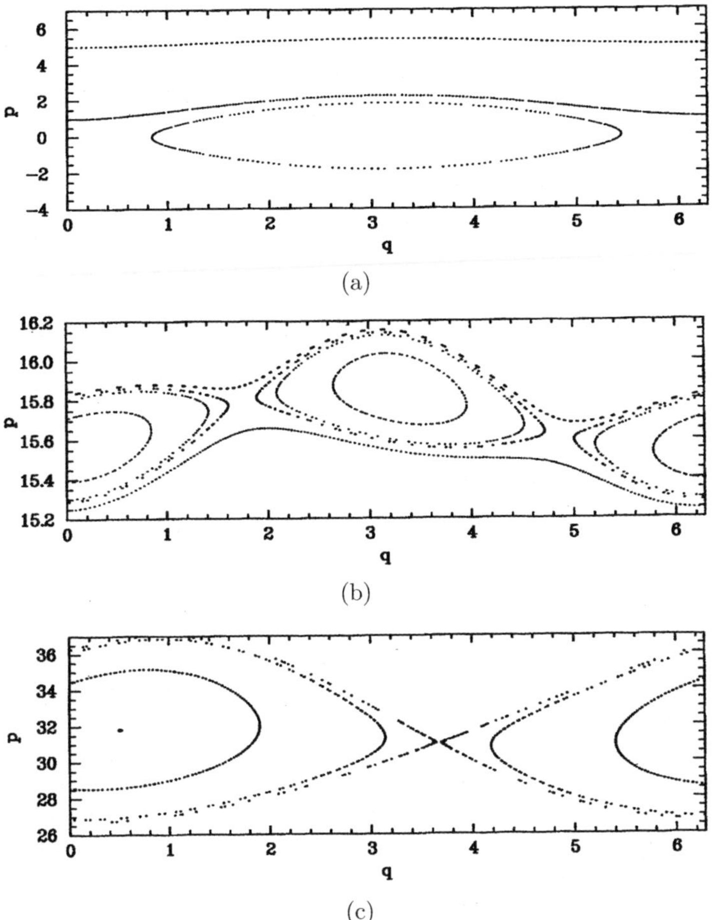

Fig. 3.17 Diagrams of the third-order scheme SI_3 in the vicinity of (a) $p = 0$, (b) $p = 15.7$, and (c) $p = 31.4$ with $\tau = 0.2$. From Sun & Zhou (1996).

space as the actual system. If the schemes permit some additional fixed points, in the vicinity of these fixed points the topology of phase space is distorted, and the symplectic integration method or mapping approach cannot be applied. Because the fixed point depends strongly on the step size, one should choose it carefully so that there are no other fixed points than those of the actual system in the interested region. How these spurious fixed points are generated is explained, and how to avoid them in applications has been proposed in this section.

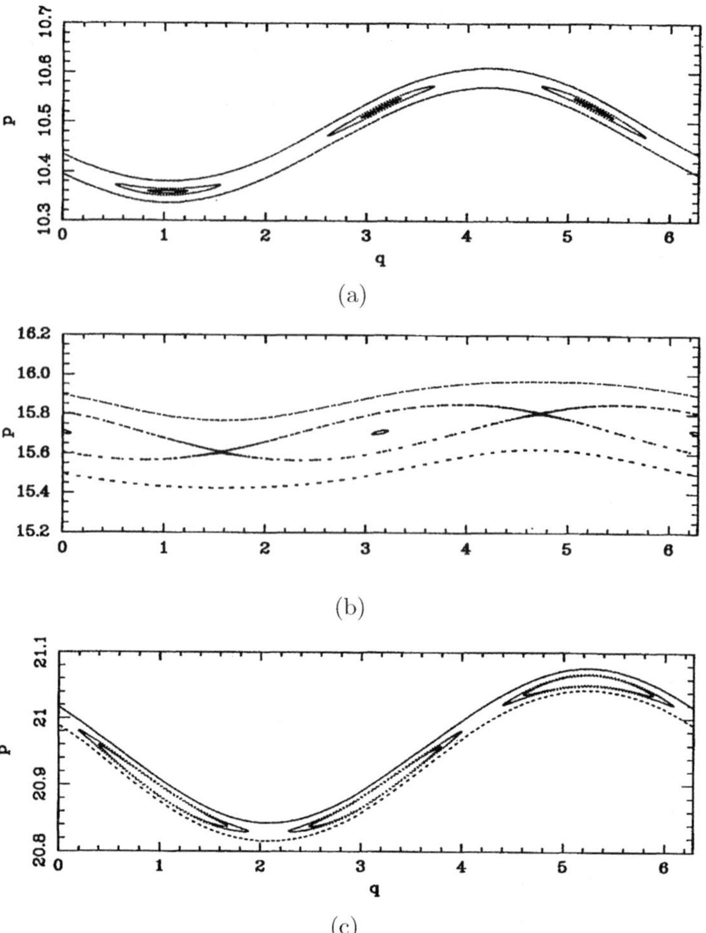

Fig. 3.18   3-periodic islands of SI_1 with $\tau = 0.2$, with comparison of size to those of the 2-periodic islands. From Sun & Zhou (1996).

3.7   Transfer of comet orbit

The capture of comets from the Oort cloud is a subject with a long history and fruitful results. At the time of writing, there are nearly 5,000 comets with computed orbits recorded, of which more than 80% are long-period comets (period $P > 200\,\mathrm{yr}$). Most of the discovered long-period comets (LPCs) dive deeply into the inner part of the Solar System with perihelion distance q smaller than a few Astronomy Units (AU). Only a very small

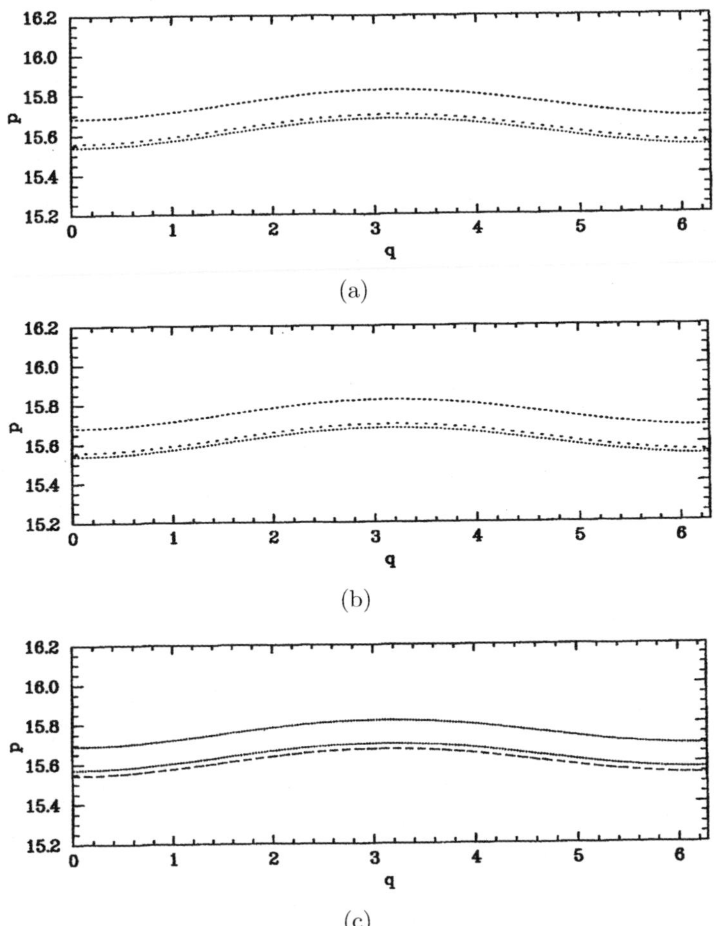

Fig. 3.19   Diagrams of schemes SI_1, SI_2, SI_3 in the vicinity of $p = 15.7$ with $\tau = 0.05$. (a) SI_1, (b) SI_2, (c) SI_3. From Sun & Zhou (1996).

fraction of them have perihelia beyond Jupiter's orbit ($q > 5.2\,\text{AU}$), e.g. the comet C/1991 R1 has a $q = 7\,\text{AU}$. The study of the orbital evolution of LPCs subjected to the perturbation of outer planets leads to the idea that the short-period comets (SPCs) could be the end products of the dynamical evolution of LPCs. The SPCs are divided into two families: Halley-family comets (HFCs, $20\,\text{yr} < P < 200\,\text{yr}$ or $T < 2$, where T is Tisserand parameter with respect to Jupiter) and Jupiter-family comets (JFCs, $P < 20\,\text{yr}$ or $T > 2$). Due to the flat near-ecliptic distribution, the JFCs are believed

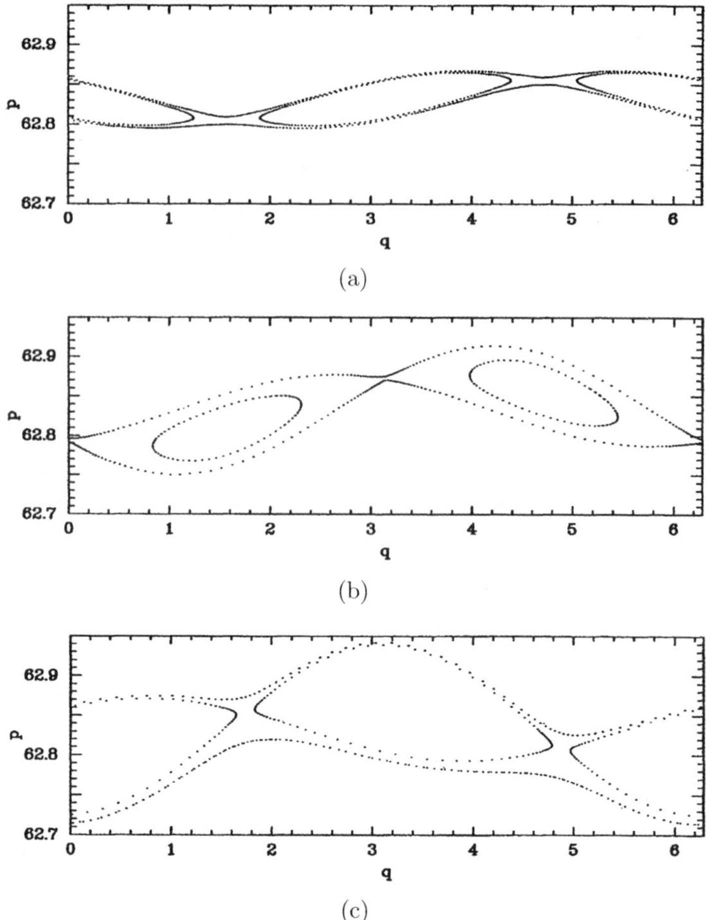

Fig. 3.20   Diagrams of schemes SI_1, SI_2, SI_3 in the vicinity of $p = 62.8$ with $\tau = 0.05$. (a) SI_1, (b) SI_2, (c) SI_3. From Sun & Zhou (1996).

to originate from a flat trans-Neptunian source. As the HFCs may come from the Oort cloud, a question then arises as to whether this mechanism is sufficient to maintain a steady-state population of HFCs.

One of the major difficulties in the statistical study of the comet dynamics is that one needs to integrate the orbits of millions of comets over the lifetime of the Solar System, which is beyond the capability of existing computers. To circumvent this difficulty, Monte Carlo simulations have been often used. Another alternative strategy is the mapping approach, which is very efficient in simulating the long-term evolution and thus more popular. A differential equation system can be approximated by a mapping system, with the main properties of the original system being conserved. By using the mapping, instead of integrating the original differential equations directly, not only a lot of computing time can be saved, but also great conveniences are brought both in analytical and in numerical investigations, as pointed out in Sec. 3.3.

In this section, the motion of comets on near-parabolic orbits will be studied by using the mapping method, particularly, the transfer of comets from the near-parabolic orbits to short-period orbits will be investigated.

3.7.1 *Mapping model*

According to Eqs. (3.70) and (3.74) in Sec. 3.5, a two-dimensional area-preserving mapping can be applied to describe the motion of comets on both direct and retrograde orbits:

$$\begin{cases} K_{n+1} = K_n + 2\mu\psi(g_n, q), \\[2mm] g_{n+1} = g_n + \dfrac{2\pi}{(-K_{n+1})^{\frac{3}{2}}}, \end{cases} \qquad (3.137)$$

where the subscript n numbers the perihelion passages.

Unlike in Sec. 3.5, where the function ψ was expanded into Fourier series in g, in this section, the integral

$$\psi(g, q) = \int_{-\infty}^{+\infty} [\dot{x}(t)F(t) + \dot{y}(t)G(t)]\, \mathrm{d}t, \qquad (3.138)$$

will be calculated by numerical quadrature. In this integral, $F(t), G(t)$ are defined by Eq. (3.62), and $x, y, \dot{x}, \dot{y}$ by Eqs. (3.68) and (3.69) in Sec. 3.5. This method has two advantages as follows. (1) It avoids the problem of series truncation. The convergence of series is very slow when q is close to one. (2) It can be generalized to $q < 1$, which is important in the transfer of comets. While the Fourier series diverges in this situation, the numerical integration of Eq. (3.138) is still valid.

In practice, the integral $\int_{-\infty}^{+\infty}(\dot{x}F + \dot{y}G)\mathrm{d}t$ in Eq. (3.138) can be approximated by $\int_{-20}^{+20}(\dot{x}F + \dot{y}G)\mathrm{d}t$. The discarded part $|t| > 20$ corresponds to the situation where the comet is more than $100\,\mathrm{AU}$ away from Jupiter, and thus the perturbations by Jupiter can be safely neglected. Taking into account the fact that $\psi(g)$ is a periodic function with period of 2π, the

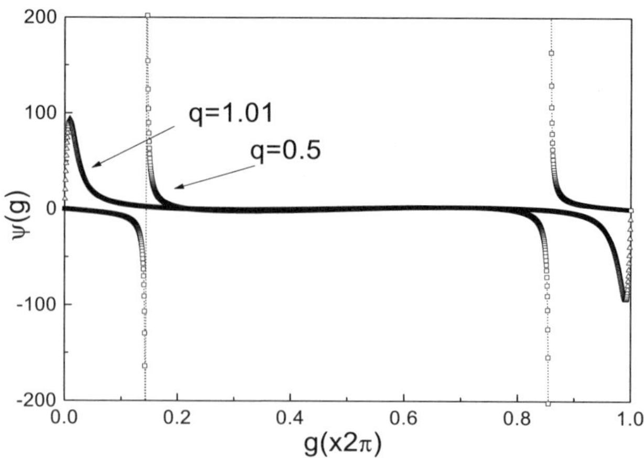

Fig. 3.21   The function $\psi(g)$ for $q = 1.01$ and $q = 0.5$. Each curve is plotted at 2000 discrete values of g uniformly distributed in $[0, 2\pi]$. From Zhou *et al.* (2000).

integral $\psi(g, q)$ for a given q value is numerically calculated on 2×10^4 g values uniformly spaced in the interval $[0, 2\pi]$. The ψ values with g other than these points can be computed by linear interpolation.

Two examples of function $\psi(g)$ for $q = 1.01$ and $q = 0.5$ are illustrated in Fig. 3.21 (the semi-major axis of Jupiter has been set as 1). As one can see, ψ is continuous for $q > 1$, while for $q < 1$, there are two values of g where ψ is discontinuous due to the comet–Jupiter collision. The width of these interpolation intervals around the collision is $10^{-4}\pi$, which corresponds to the length of the same order of magnitude as the equator radius of Jupiter. If a comet falls in one of these collision intervals, it is assumed to collide with Jupiter and its orbital evolution stops.

To test the validity and the computing efficiency of the mapping in Eq. (3.137), the orbit of a comet is calculated both by integrating the original system [Eq. (3.61)] and by the mapping [Eq. (3.137)]. The results are plotted in Fig. 3.22. It shows that the final qualitative results (escape, remain, or transfer) of the evolution of individual orbits in the original system is preserved by the mapping. However, for the regular orbit (presented in the lower panel of Fig. 3.22) evolving up to 10^7 comet passages of Jupiter, the mapping takes only one 30th of the computing time needed by the direct integration to finish 10^5 passages, that is, considering the computing time, the mapping approach is about 3,000 times quicker than the direct integration.

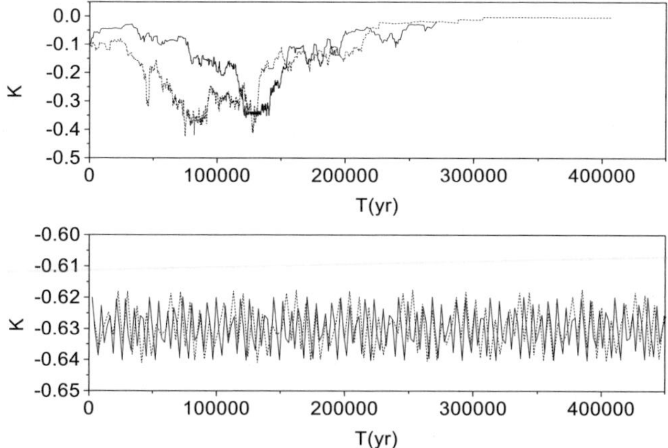

Fig. 3.22   Evolution of the energy of a comet by the integration of the original system (dashed curves) and by the mapping (solid curves) with initial value $K = -0.64, g = 0.4\pi$ (lower panel) for a regular orbit, and $K = -0.10, g = 0$ (upper panel) for an escape orbit. From Zhou *et al.* (2000).

The statistical validity of the mapping can also be verified. Orbits of 1,000 comets with the same initial energy are simulated to 10^7 years both by using the mapping and by direct numerical integration, and the transfer probability (the fraction of comets whose final energy $K < -0.152$) and the transfer time (number of passages before the transfer) are recorded and compared in Fig. 3.23. Apparently, the results from the mapping coincide quite well with the results from the direct integration.

Since the orbits of comets are taken approximately as parabolic ones, the mapping is not valid when the orbital eccentricities of comets are small. According to the numerical experiments, the minimum eccentricity that guarantees the validity of the mapping is ~ 0.3. But, thanks to the large semi-major axes and small perihelion distances, the eccentricities of LPCs are always large (at least 0.7). Thus, the limitation of large eccentricity does not affect the analysis presented in further sections.

3.7.2   *Orbit transfer*

The mapping Eq. (3.137) is now used to study the transfer of comets from long-period orbits to short-period orbits. The phase diagram of the mapping is shown in Fig. 3.24, which obviously has both regular and chaotic

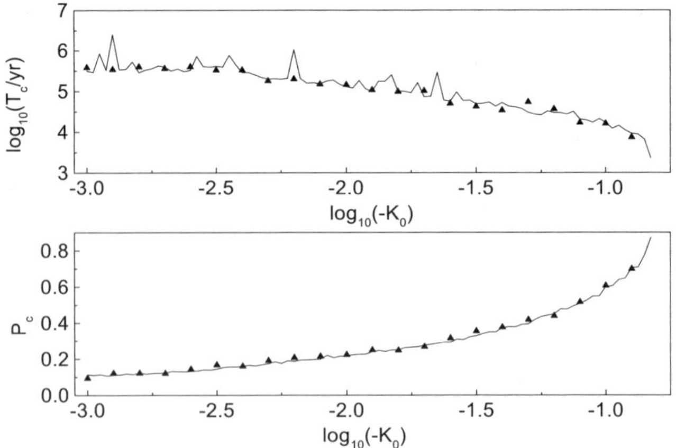

Fig. 3.23   The variations of the transfer probability (lower window) and the average transfer time (upper window) with the initial energy, obtained by direct integration of the original system (triangles) or by the mapping method (solid curves). All the orbits have $q = 0.8$. From Zhou *et al.* (2000).

orbits. The islands correspond to the locations where a comet is in the exterior mean motion resonance with Jupiter. For example, the first resonance visible at $K = -0.32$ corresponds to a $n : n_{\rm J} = 2 : 11$ resonance. The chaoticity of orbits in the phase space with $-0.3 < K < 0$ comes from the fact that, according to the mapping Eq. (3.137), for small $|K|$, the phase angles g of two consecutive perihelion passages are almost independent.

To study the possibility of orbital transfer of comets with near zero energy, it is reasonable to concentrate on cases when a comet obtains enough (negative) energy from the planet to convert itself from an LPC to an SPC. The relation between the energy K and the period P in the adopted units is $K = -(11.86\,{\rm yr}/P)^{2/3}$. The boundary of period between LPCs and SPCs is 200 year, corresponding to the energy of about $K_f = -0.152$. Since the Oort Cloud is supposed to locate at 10^4–10^5 AU, the initial energy of comets is assumed to be -0.001 below.

In all of the following experiments, several sets of 10^4 orbits each with initial energy $K = -0.001$ are selected uniformly in $g \in [0, 2\pi]$, and the evolution of each orbit is computed with the mapping Eq. (3.137), until it either escapes ($K > 0$) or is transferred to $K < -0.152$. Then statistics are made for these 10^4 orbits. Note that for comet orbits with $q < 1$, collision with Jupiter is also possible, though the possibility is very small (0.07–0.2% according to calculations). Three quantities related the transfer dynamics

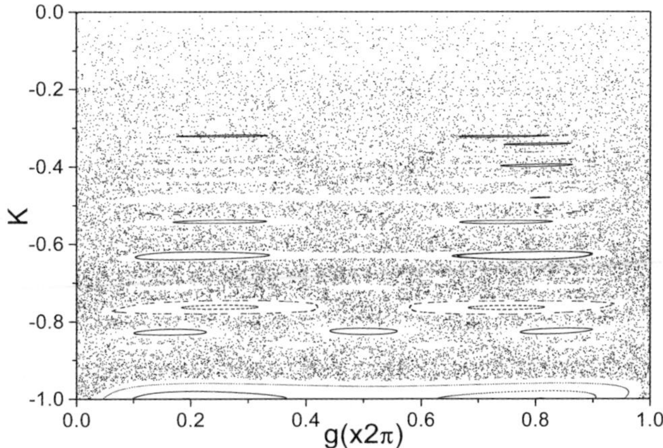

Fig. 3.24   Phase diagram of the mapping Eq. (3.137) with $\mu = \mu_J$ and $q = 0.8$. From Zhou *et al.* (2000).

of the comets are determined: the transfer probability (P_c), the average transfer time (n_f) and the average energy change of transferred comets per passage (σ). P_c is defined as the ratio of transferred orbits (with final energy $K < K_f$) to the total number of the initial orbits (10^4). For the transferred comets, n_f is defined as the average number of passages needed for accomplishing the transfer, and σ is defined as the root-mean-square average of the energy change per perihelion passage. The dependence of these quantities on the perihelion distance of the comet orbit (which is assumed constant during the energy evolution) and the planet mass will be analyzed in the following sections.

3.7.2.1   Dependence on perihelion distance

Fix $\mu = \mu_J = 1/1047.355$ (Jupiter mass) and vary the perihelion distance q from 0.2 to 1.6. For each q value, evolutions of 10^4 orbits are followed and the transfer probability P_c is computed. The variation of P_c with q is summarized in Fig. 3.25. For $q < 1$, P_c has values ranging from 0.09 to 0.13 for comets on direct orbits and it is about 0.07 for retrograde comets. In both cases the variation of P_c is small. In contrast, for comets with perihelion distances $q > 1$, the variation of P_c is large. The transfer probability P_c decreases from 0.11 to 0.01 as q increases from 1 to 1.6 for direct motion, and it decreases drastically for retrograde motions ($P_c \sim 10^{-3}$ at $q = 1.05$). Thus the transfer of comets with $q > 1$ on retrograde orbits is negligible.

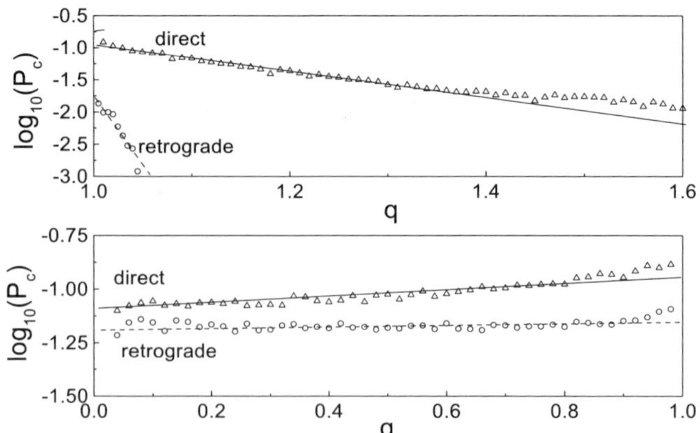

Fig. 3.25   Dependence of the transfer probability P_c on q with $\mu = \mu_{\rm J}$. The unit of q is 5.2 AU. The lines are linear fits to the numerical results. From Zhou *et al.* (2000).

For comets on direct motion with $q > 1$, only those with $q \sim 1$ can be transferred efficiently.

According to the fitting lines in Fig. 3.25, the variations of P_c obey exponential laws with the forms of

$$
\begin{aligned}
P_d^{<1} &\simeq 0.11 \times 10^{0.2(q-1)} & (q < 1), \\
P_d^{>1} &\simeq 0.11 \times 10^{-2(q-1)} & (q > 1), \\
P_r^{<1} &\simeq 0.069 \times 10^{0.025(q-1)} & (q < 1), \\
P_r^{>1} &\simeq 0.016 \times 10^{-22(q-1)} & (q > 1),
\end{aligned}
\tag{3.139}
$$

where the subscripts "d" and "r" denote comets on direct and retrograde orbits, respectively. Adopting these exponential laws, the total transfer probability can be integrated as

$$
\begin{aligned}
I_d^{<1} &= \int_0^1 P_d^{<1}(q)\mathrm{d}q, & I_d^{>1} &= \int_1^\infty P_d^{>1}(q)\mathrm{d}q, \\
I_r^{<1} &= \int_0^1 P_r^{<1}(q)\mathrm{d}q, & I_r^{>1} &= \int_1^\infty P_r^{>1}(q)\mathrm{d}q.
\end{aligned}
\tag{3.140}
$$

And calculations give

$$
\begin{aligned}
I_d^{<1} &= 0.092, & I_d^{>1} &= 0.024 & \Longrightarrow\ & I_d^{<1} + I_d^{>1} \approx 0.12, \\
I_r^{<1} &= 0.067, & I_r^{>1} &= 0.0003 & \Longrightarrow\ & I_r^{<1} + I_r^{>1} \approx 0.067.
\end{aligned}
\tag{3.141}
$$

Thus the flux of transferred comets from direct orbits is ~ 1.8 times as large as that from retrograde orbits. This may be the reason why most of the

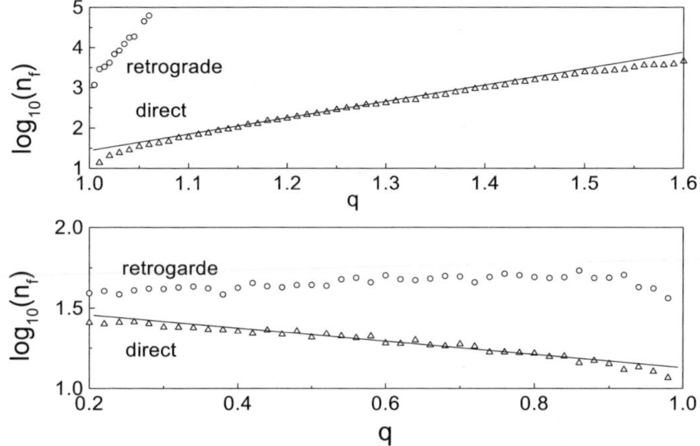

Fig. 3.26 Dependence of the average transfer time n_f on q with $\mu = \mu_J$. The unit of q is 5.2 AU. From Zhou *et al.* (2000).

observed HFCs are on direct orbits. Another interesting conclusion is that the flux of transferred comets from Jupiter-crossing orbits ($q < 1$) is about $(I_d^{<1} + I_r^{<1})/(I_d^{>1} + I_r^{>1}) \approx 6.7$ times as large as that from non-crossing orbits ($q > 1$).

The variation of the average transfer time n_f for the transferred orbits with q is shown in Fig. 3.26. The variation of n_f is smaller for $q < 1$ than that for $q > 1$. For $q < 1$, n_f varies with q very slowly and equals to 10–25 passages for direct orbits and to $\sim$60 passages for retrograde orbits. For $q > 1$, the number of passages needed for transfer from direct motion increases from 10 to 3,000 as q increases from 1 to 1.6. For comets in direct motion, the dependence of n_f on q can be roughly fitted with the exponential laws as:

$$
\begin{aligned}
n_f^{<1} &\simeq 13 \times 10^{-0.4(q-1)} \quad (q < 1), \\
n_f^{>1} &\simeq 35 \times 10^{4(q-1)} \qquad (q > 1).
\end{aligned}
\tag{3.142}
$$

Moreover, for the transferred comets, the dependence of the average energy change per passage σ on q is shown in Fig. 3.27. The values of σ for $q < 1$ are large, sometimes even close to 1. This is because the encounters between Jupiter and comets with $q < 1$ can be much closer than those with $q > 1$. From the graph of ψ (Fig. 3.21), one can also see that the energy exchange of the comet with $q < 1$ is much larger than those with $q > 1$.

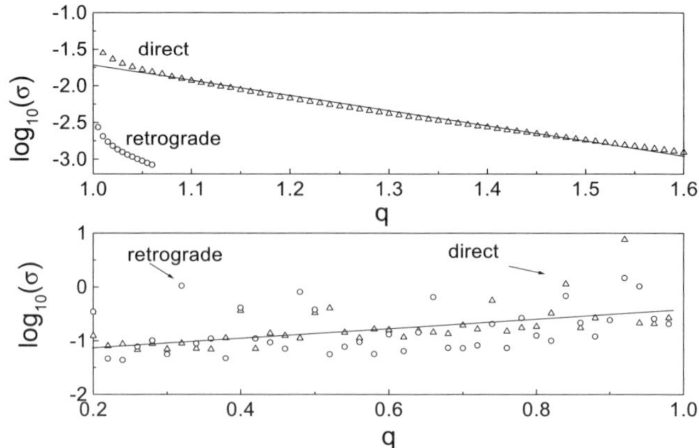

Fig. 3.27   Dependence of the average energy change σ on q with $\mu = \mu_{\rm J}$. From Zhou *et al.* (2000).

For the direct motions σ varies with q exponentially following

$$
\begin{aligned}
\sigma^{<1} &\simeq 0.28 \times 10^{0.8(q-1)}, & (q < 1), \\
\sigma^{>1} &\simeq 0.023 \times 10^{-2(q-1)}, & (q > 1).
\end{aligned}
\tag{3.143}
$$

The exponential dependence of σ on q gives an explanation of the exponential forms of Eqs. (3.139) and (3.142). In fact, according to Fig. 3.27, the typical values of σ is about 10^{-3}–10^{-2} for $q > 1$ in the direct motions, which is much less than the energy decrement required for a comet with near zero initial energy to reach $K_f = -0.152$. Thus the evolution of energy for $q > 1$ obeys the diffusion approximation. According to the diffusion approximation, the transfer probability and average transfer time for a comet with initial energy near zero to reach energy K_f obey [Fernandez & Gallardo (1994)]

$$
P_c \simeq -\frac{\sigma}{2K_f}, \qquad n_f \simeq \left(\frac{K_f}{\sigma}\right)^2.
\tag{3.144}
$$

Thus the exponential laws in Eqs. (3.139, 3.142) for $q > 1$ can be deduced from the Eq. (3.143). However, for $q < 1$, due to the large average energy change per passage, which is of the same order as $|K_f|$, the condition of the diffusion approximation is no more fulfilled.

3.7.2.2   Dependence on planet mass

In the above study, the planet mass is fixed $\mu = \mu_{\rm J}$ indicating the perturbing planet is Jupiter. In the following, the study is generalized to general case of

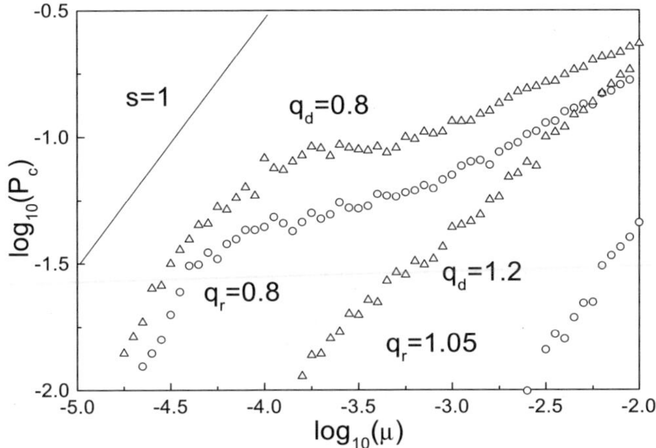

Fig. 3.28 Dependence of the transfer probability P_c on μ with different q. The straight line in the up-left corner has a slope $s = 1$. q_d stands for the perihelion distance of comets in the direct motion, q_r stands for that in the retrograde motion. From Zhou *et al.* (2000).

planet of any mass. To see the influence of the planet mass on the transfer of comet, the evolution of above 10^4 orbits with μ varying from 10^{-2} to 10^{-5}.

The variations of the transfer probability P_c with μ are shown in Fig. 3.28. One can see that the probabilities for $q < 1$ are larger than those for $q > 1$ as long as $\mu < 10^{-2}$. This is the case for our Solar System, where the largest perturbation for comet motions comes from Jupiter $(\mu_J \sim 10^{-3})$.

As μ increases, approximately $P_c \sim \mu$ in the cases of (a) $\mu < 10^{-4}$ for $q < 1$ and (b) $\mu < 10^{-2}$ for $q > 1$. However, in the case of $q < 1$ with $\mu > 10^{-4}$, this relationship does not hold (Fig. 3.28). This is caused by the differences in the average energy change σ. To demonstrate this, the dependence of σ on μ is plotted in Fig. 3.29, where the relation $\sigma \sim \mu$ can be found in all the studied cases. This is just as expected since the change of energy is proportional to μ (recall Eq. (3.70) in Sec. 3.5).

With the relation of $\sigma \sim \mu$, the variation of P_c on μ can be explained. For the situations (a) and (b) listed above, the average energy changes are small so that the diffusion approximation holds, with $P_c \sim \mu \sim \sigma$, which obeys the relation Eq. (3.144). For other situations, the diffusion approximation is not fulfilled due to the large energy change, thus the relation Eq. (3.144) does not hold since $P_c \sim \mu$ does not hold.

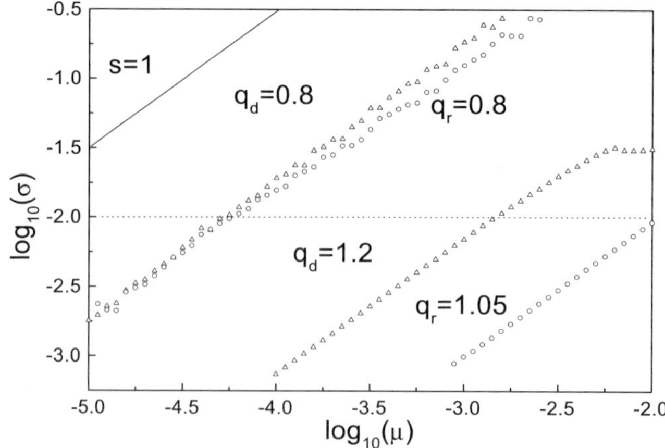

Fig. 3.29 Dependence of the average energy change σ on μ with different q. The straight line in the up-left corner has a slope $s = 1$. The dotted line shows the critical value σ_c below which the diffusion approximation holds. From Zhou *et al.* (2000).

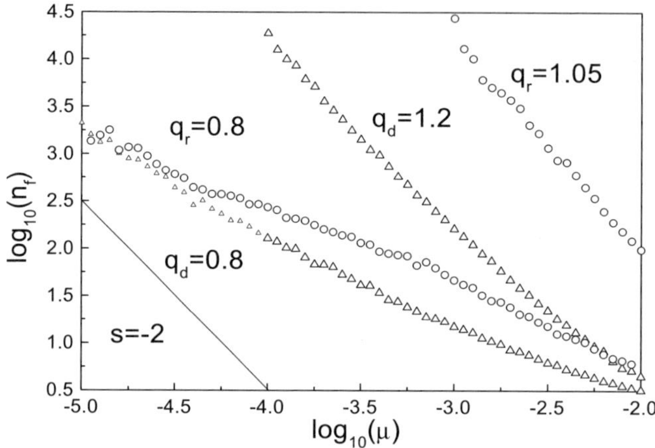

Fig. 3.30 Dependence of the average transfer time n_f on μ with different q. The straight line in the lower-left corner has a slope $s = -2$. From Zhou *et al.* (2000).

Finally, the dependence of the average transfer time n_f on μ is presented in Fig. 3.30. For the situations (a) and (b) listed above but with $\mu < 10^{-4.5}$ for $q < 1$, one can identify roughly a power-law with the form of $n_f \sim \mu^{-2}$, which is consistent with the relation Eq. (3.144). In other cases, when conditions of the diffusion approximation are not fulfilled, the relation $n_f \sim \mu^{-2}$ is no more obeyed.

Table 3.4 Integral transfer probability for different planet perturbations.

Planet	K_f	$I_d^{<1}$	$I_r^{<1}$	$I_d^{>1}$	I_{sum}
Jupiter	-0.152	0.0920	0.0672	0.0244	0.184
Saturn	-0.279	0.0457	0.0296	0.0045	0.080
Uranus	-0.560	0.0005	0.0029	0	0.003
Neptune	-0.875	0.0010	0.0021	0	0.003

Note: K_f in the second column is defined by $K_f = (\frac{P\,\mathrm{yr}}{200\,\mathrm{yr}})^{2/3}$; $I_d^{<1}, I_r^{<1}, I_d^{>1}$ are defined in a similar way as in Eq. (3.140) and $I_{\mathrm{sum}} = I_d^{<1} + I_r^{<1} + I_d^{>1}$. Data from Zhou *et al.* (2000).

In the above qualitative discussions, when the mass of planet μ is changed, the energy boundary between LPCs and HFCs is fixed as $K_f = -0.152$. However, when the only perturbing planet is changed (e.g. Saturn, Uranus or Neptune), K_f should be also modified due to the different orbital periods of planets. The transfer probabilities in the cases of different planet perturbations are calculated, and the results are summarized in Table 3.4. Only Saturn contributes by a non-negligible amount to the transferred comets while the contributions from Uranus and Neptune are negligible.

Under the perturbation of a large planet (mainly Jupiter in our Solar System), long-period comets may transfer from near-parabolic orbits to short period orbits ($P < 200\,\mathrm{yr}$) after some perihelion passages. Using a mapping model, such dynamical transfer of comets has been analyzed. Numerical results showed that due to the difference of average energy change per perihelion passage, the transfer of comets on planet-crossing orbits with perihelion distance $q < 1$ (the semi-major axis of the disturbing planet is defined as the length unit) is more efficient than the transfer of comets in non-crossing orbits ($q > 1$). The comets with $q < 1$ are easier to be transferred in both direct and retrograde motions. For comets with $q > 1$, only those on direct orbits with $q \sim 1$ have a relatively higher probability of transfer. The flux of transferred comets with $q < 1$ is about 4 times as large as that of $q > 1$ on direct orbits.

The transfer of comets is a subject of orbital energy evolution under the perturbations of planets. In the case of Jupiter as the perturber, the evolution of the orbital energy for a comet with $q > 1$ obeys the diffusion approximation, i.e. the relation in Eq. (3.144), due to the small average exchange per passage as compared with the total transferred energy -0.152. However for the comets with $q < 1$, the diffusion approximation may not be followed by the energy evolution, due to the larger average energy change.

In fact, for a comet on a near-parabolic orbit, the evolution of its energy under a planet's perturbation can be treated as a "random walk" in the one-dimensional space, which is the topic of the following section.

3.8 Random walk in comet motion

For a comet on a near-parabolic orbit, the evolution of its energy under Jupiter's perturbation can be treated as a random walk in the one-dimensional energy space. This is the random assumption first introduced by Oort and it is widely adopted in calculating the orbital evolution of comets on near-parabolic orbits.

A well-known example of random walk is the Brownian motion and it has a Gaussian probability distribution. Consider an N-step random walk in one-dimension, with each step of random walk governed by the same probability distribution $p(\lambda)$ of zero mean. Lévy posed a question: when does the probability $P_N(x)$ for the sum of N steps $x = x_1 + x_2 + \cdots + x_N$ have the same distribution $p(x)$ (up to a scale factor) as the individual steps? Apparently, the sum of N Gaussian is again a Gaussian, but Lévy proved that there exist other solutions to this question besides the Gaussian distribution. Additionally, the Cauchy distribution shows the connection between a one-step and an N-step distribution. Below in this section, using a mapping model, the random walk in comet walk will be investigated.

The mapping adopted in this section is the two-dimensional area-preserving mapping [Eq. (3.137)], which has been used in several sections. This time $\mu = 1/1047.355$, corresponding to the mass ratio of Jupiter to the Sun. q is the perihelion distance of the comet orbit, and its unit is chosen as the average distance between the Sun and Jupiter (5.2 AU) so that when $q > 1$ the orbit of comet does not cross Jupiter's orbit while $q < 1$ implies the comet penetrates Jupiter's orbit. The function $\psi(g_n, q)$ is a 2π-periodic function of g_n, and its profile has been shown in Fig. 3.21. The function $\psi(g_n, q)$ describes the increment of the comet energy after each passage through the perihelion.

Judging from the mapping, when the magnitude of $-K$ is small, the consecutive phase angles g are almost "lost of correlation", thus the change of comet energy will be chaotic and the comet energy K evolves as random walks.

First of all, some statistics on the evolution of a group of comets under the mapping are made: 10^5 orbits, with the same initial energy $K_0 = -0.2$

and different initial angles g_0 uniformly distributed in $[0, 2\pi]$, are calculated up to $n = 50$. The fate of these 10^5 orbits under the evolution of the mapping are listed in Table 3.5, where "escape" or "collision" means the numbers of orbits escaping to infinity or colliding with Jupiter before or at iteration number $n = 50$, respectively. The probability distribution of energy for the surviving comets (except those escaped or collided with Jupiter) at $n = 50$ are plotted in Fig. 3.31 for $q = 1.2$ and Fig. 3.32 for $q = 0.9$, respectively.

For $q > 1$, the probability distributions of energies of the surviving comets can be fitted with the Gaussian distribution

$$P(K) \sim \exp\left(-\frac{(K - K_0)^2}{2\sigma^2}\right). \tag{3.145}$$

For example, the distribution in Fig. 3.31 for $q = 1.2$ can be fitted by Eq. (3.145) with $K_0 = -0.2$ and $\sigma = 0.05$. However, the distribution for

Table 3.5   Number of comets with different evolution fate.

q	Collision	$K < -3$	$-3 < K < 0$	Escape
1.2	0	0	99,989	11
0.9	395	3,119	51,843	44,643

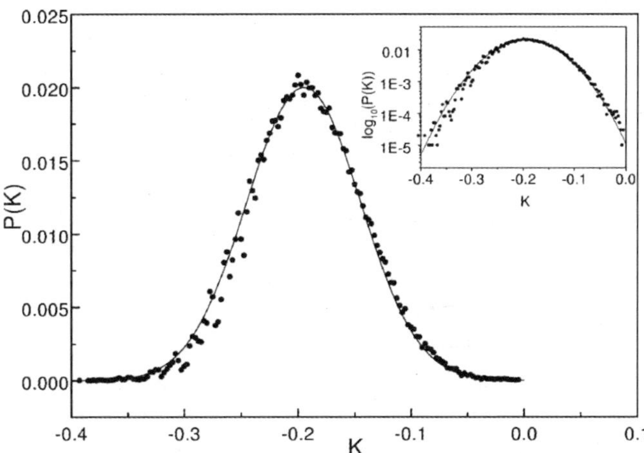

Fig. 3.31   Probability distribution of energy for the surviving comets at iteration number $n = 50$. Initial energy $K = -0.2$, parameter $q = 1.2$. The fitted curve is Gaussian (see text). From Zhou & Sun (2001).

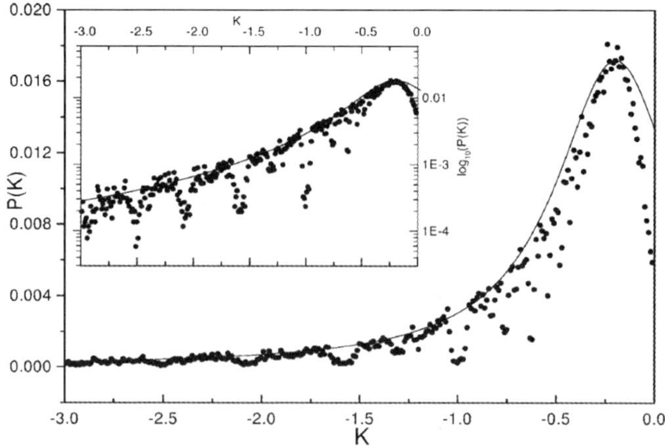

Fig. 3.32 Probability distribution of energy for the surviving comets at iteration number $n = 50$. Initial energy $K = -0.2$, parameter $q = 0.9$. The fitted curve is Cauchy distribution (see text). From Zhou & Sun (2001).

$q < 1$ is no more Gaussian. In fact, they can be well fitted with the Cauchy distribution as

$$P(K) \sim \frac{\alpha}{(K - K_0)^2 + \alpha^2}. \tag{3.146}$$

Such a distribution can be found in Fig. 3.32 where the energy distribution at $n = 50$ for $q = 0.9$ is shown, and the distribution can be well fitted by Eq. (3.146) with $K_0 = -0.2$ and $\alpha = 0.37$.

Since the Cauchy distribution is a kind of Lévy distribution, the evolution of a comet's energy follows a Lévy random walk when $q < 1$. Moreover, the oscillations in $P(K)$ in Fig. 3.32 superposed to the Cauchy power-law decay are in fact caused by the presence of those rarely-happened but very-large-amplitude individual walking steps (increments of energy), so that many small clusters are formed. In these clusters, there are mainly small walking steps and thus each cluster obeys more like Gaussian distributions (see embedded panel in the figure). This phenomenon of oscillations appears in many dynamical systems with Lévy flight walks too.

The Gaussian distribution of energy evolution is the character of usual random walks, e.g. the Brownian motion, while the Cauchy distribution is the character of Lévy flight motion. Due to the difference on the average energy exchanges between $q > 1$ and $q < 1$, the average numbers of passages for comets to reach a given energy K_f are also different. Below, the evolutions of another 10^4 orbits uniformly distributed in $g \in [0, 2\pi]$ but

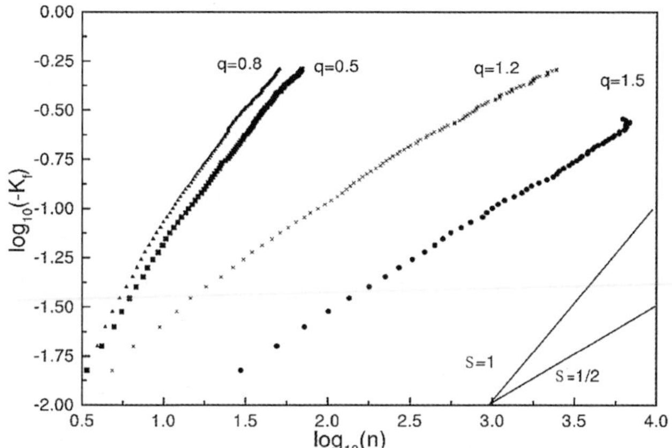

Fig. 3.33   Average number of passages n for 10^4 comets with initial energy $K_0 = -0.001$ to reach a final energy K_f. Two lines in the graph have the slopes $s = 1$ and $s = 1/2$, respectively. From Zhou & Sun (2001).

with initial energy of $K_0 = -0.001$ are studied using the mapping. The average number of passages for those orbits that can reach a final energy K_f is shown in Fig. 3.33. From this figure, one may see that for a comet in a non-crossing orbit ($q > 1$), the final energy that the comet reaches is proportional to the square root of the passage numbers, i.e.

$$K_f \sim n^{1/2}. \tag{3.147}$$

This is the usual diffusion of random walk and is called the diffusion approximation in comet dynamics.

However, for comets in Jupiter-crossing orbits ($q < 1$), since their energy evolutions obey the Cauchy distribution, the increment of energies mainly come from some large jumps due to close encounters with Jupiter, thus the speeds of diffusion in the energy space are enhanced. In this situation, the final energies are proportional to the numbers of passages

$$K_f \sim n. \tag{3.148}$$

This phenomenon of diffusion enhancement on comet energy due to the Lévy flights is observed for the first time in this work. Due to the Lévy random walk behavior, the evolutions of such comets in the energy space are faster, and the capture of these comets by Jupiter is much easier, that is, with larger capture probability and shorter capture time.

The main difference between the motions of comets with $q > 1$ and $q < 1$ are the presence of singularities in the energy changing function ψ (see

Fig. 3.21) of the mapping model. Due to the singularities, the increment of energy during each passage can be very large, which is called "jump" here. However, the larger the jump is, the smaller the probability that it happens would be, or in other words, the time waiting for the occurrence of large jumps would be longer. For this reason, though the exact variance of Lévy distribution may be divergent, the finite variances and power-law increments of mean square displacements can always been observed in many experiments. This may be one of the reasons why Lévy flight dynamics is so widely used. Despite of the small probability of large jumps, the increments of energy comes mainly from those large jumps, which is another important character of Lévy flight dynamics.

In the above discussions, two regimes of random walk in comet motion have been found. One regime is the diffusion approximation with Gaussian distribution for the noncrossing orbits of comets, while the other regime is the Lévy flight model with Cauchy distribution for those Jupiter-crossing comets.

The evolution of the energy of comets on non-crossing orbits, but with q close to 1, lies inbetween the above two regimes, that is, the energy evolution follows a Lévy random walk for small final energy increments, but obeys a diffusion law for large energy increments.

Since the main cause of the differences between a Gaussian and a Lévy random walk is the existence of large steps in the latter, and while in the mapping model the presence of large steps for $q < 1$ is obviously a result of the singularities in the step-function $\psi(g, q)$, the occurrence of Lévy random walks in this model attributes reasonably to the close encounters between the comets and Jupiter. Therefore, if a spatial restricted three-body model is employed, the differences between these two kinds of random walk motion may become much smaller.

3.9　Chaotic region of encounter-type orbit

When a comet approaches to Jupiter closely enough, a collision could happen between them, and then for dynamical studies, a collision singularity occurs. As shown in Fig. 3.21 in Sec. 3.7, the function describing the energy change due to this encounter $\psi(g)$ in the mapping Eq. (3.137) is discontinuous. As a result, the energy K varies drastically and it induces also the drastic change of g. Because the phase angles g of two consecutive perihelion passages are almost independent, this drastic change of g then means chaotic motion. In this section, the chaotic region of encounter-type orbit will be investigated.

3.9.1 *Equations of motion*

Let m_i $(i = 1, 2, 3)$ be the mass of the ith body and suppose $m_1 \gg m_2 \gg m_3$. This is a model of restricted three-body problem. Let us start with the planar problem. In the planar inertia coordinate system $(\boldsymbol{X}, \boldsymbol{Y})$, the equations of motion for body m_1 read

$$\frac{\mathrm{d}^2 X_1}{\mathrm{d}t^2} = \frac{Gm_2(X_2 - X_1)}{R_{12}^3} + \frac{Gm_3(X_3 - X_1)}{R_{13}^3},$$

$$\frac{\mathrm{d}^2 Y_1}{\mathrm{d}t^2} = \frac{Gm_2(Y_2 - Y_1)}{R_{12}^3} + \frac{Gm_3(Y_3 - Y_1)}{R_{13}^3}.$$

(3.149)

The equations of motion $\frac{\mathrm{d}^2 X_2}{\mathrm{d}t^2}, \frac{\mathrm{d}^2 Y_2}{\mathrm{d}t^2}$ and $\frac{\mathrm{d}^2 X_3}{\mathrm{d}t^2}, \frac{\mathrm{d}^2 Y_3}{\mathrm{d}t^2}$ for bodies m_2 and m_3 are analogous. In these equations, G is the gravitation constant as usual, and R_{ij} $(i, j = 1, 2, 3)$ is the distance between the ith body and the jth body,

$$R_{ij} = \sqrt{(X_i - X_j)^2 + (Y_i - Y_j)^2}.$$

(3.150)

Let a_p be the orbital semi-major axis of the second body, $m = m_1 + m_2 + m_3$ be the total mass of three bodies, $n = \sqrt{Gm/a_p^3}$, and introduce dimensionless variables as

$$X_i' = \frac{X_i}{a_p}, \quad Y_i' = \frac{Y_i}{a_p}, \quad m_i' = \frac{m_i}{m}, \quad t' = nt.$$

(3.151)

In these new variables, the gravitation constant, the total mass of the three bodies, the orbital semi-major axis of the second body, and the mean angular velocity of it, are all equal to 1.

Since $m_1 \gg m_2 \gg m_3$ and $m = m_1 + m_2 + m_3 = 1$, a small number μ can be introduced so that

$$m_1 = 1 - \mu, \quad m_2 = \mu, \quad m_3 = 0.$$

(3.152)

Choose a new coordinate system that rotates with an unit angular velocity around the origin at m_1. In this coordinate system, the positions of m_2 and m_3 are (x_2, y_2) and (x_3, y_3) as:

$$x_2 = (X_2' - X_1') \cos(t - t_0) + (Y_2' - Y_1') \sin(t - t_0),$$

$$y_2 = -(X_2' - X_1') \sin(t - t_0) + (Y_2' - Y_1') \cos(t - t_0),$$

(3.153)

and

$$x_3 = (X_3' - X_1') \cos(t - t_0) + (Y_3' - Y_1') \sin(t - t_0),$$

$$y_3 = -(X_3' - X_1') \sin(t - t_0) + (Y_3' - Y_1') \cos(t - t_0).$$

(3.154)

Thus the equations of motion for the third body m_3 are

$$\frac{\mathrm{d}^2 x_3}{\mathrm{d}t^2} = 2\frac{\mathrm{d}y_3}{\mathrm{d}t} + x_3 - \frac{(1-\mu)x_3}{R_{13}^3} + \frac{\mu(x_2 - x_3)}{R_{23}^2} - \frac{\mu x_2}{R_{12}^3},$$

$$\frac{\mathrm{d}^2 y_3}{\mathrm{d}t^2} = -2\frac{\mathrm{d}x_3}{\mathrm{d}t} + y_3 - \frac{(1-\mu)y_3}{R_{13}^3} + \frac{\mu(y_2 - y_3)}{R_{23}^2} - \frac{\mu y_2}{R_{12}^3}, \tag{3.155}$$

where

$$R_{12}^2 = x_2^2 + y_2^2, \quad R_{13}^2 = x_3^2 + y_3^2, \quad R_{23}^2 = (x_3 - x_2)^2 + (y_3 - y_2)^2. \tag{3.156}$$

Define new coordinates (ξ, η) as

$$\begin{aligned} x_2 &= 1 + \xi_2, & x_3 &= 1 + \xi_3, \\ y_2 &= \eta_2, & y_3 &= \eta_3. \end{aligned} \tag{3.157}$$

In fact, this is the Hill coordinate system. The relation between this new coordinate system and the original one is displayed in Fig. 3.34. In this system, the equations of motion for m_3 read

$$\frac{\mathrm{d}^2 \xi_3}{\mathrm{d}t^2} = 2\frac{\mathrm{d}\eta_3}{\mathrm{d}t} + 3\xi_3 + \frac{\mu(\xi_2 - \xi_3)}{\rho^3} + O(\mu^{4/3}),$$

$$\frac{\mathrm{d}^2 \eta_3}{\mathrm{d}t^2} = -2\frac{\mathrm{d}\xi_3}{\mathrm{d}t} + \frac{\mu(\eta_2 - \eta_3)}{\rho^3} + O(\mu^{4/3}), \tag{3.158}$$

where

$$\rho^2 = (\xi_2 - \xi_3)^2 + (\eta_2 - \eta_3)^2. \tag{3.159}$$

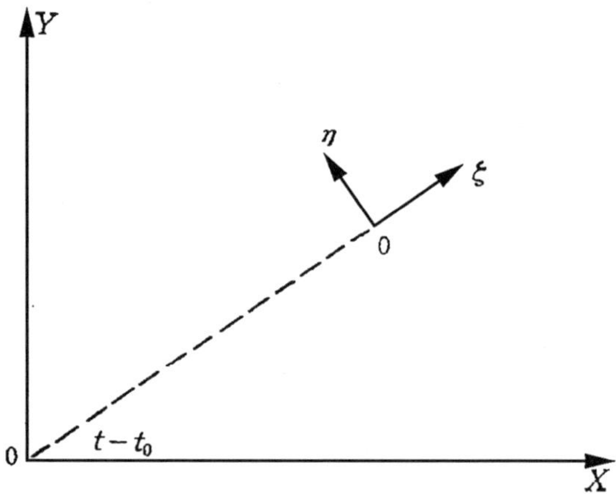

Fig. 3.34 Transformations between coordinates (X, Y) and (ξ, η). From Liu & Sun (1991).

Neglecting $O(\mu^{4/3})$, one finally obtains the equations of motion for m_3 in the Hill coordinate system

$$\frac{\mathrm{d}^2\xi_3}{\mathrm{d}t^2} = 2\frac{\mathrm{d}\eta_3}{\mathrm{d}t} + 3\xi_3 + \frac{\mu(\xi_2 - \xi_3)}{\rho^3},$$

$$\frac{\mathrm{d}^2\eta_3}{\mathrm{d}t^2} = -2\frac{\mathrm{d}\xi_3}{\mathrm{d}t} + \frac{\mu(\eta_2 - \eta_3)}{\rho^3}. \tag{3.160}$$

As usual, define $\mathrm{d}\xi_3/\mathrm{d}t = W, \mathrm{d}\eta_3/\mathrm{d}t = V$, the equations of motion Eq. (3.160) can be changed into

$$\frac{\mathrm{d}\xi_3}{\mathrm{d}t} = W,$$

$$\frac{\mathrm{d}\eta_3}{\mathrm{d}t} = V,$$

$$\frac{\mathrm{d}W}{\mathrm{d}t} = 2V + 3\xi_3 + \frac{\mu(\xi_2 - \xi_3)}{\rho^3},$$

$$\frac{\mathrm{d}V}{\mathrm{d}t} = -2W + \frac{\mu(\eta_2 - \eta_3)}{\rho^3}. \tag{3.161}$$

When $\mu = 0$ (or $\rho \to \infty$), these equations are reduced to linear equations and they have the following solutions

$$\xi_3 = D_1 \cos t + D_2 \sin t + D_3,$$

$$\eta_3 = -2D_1 \sin t + 2D_2 \cos t - \frac{3}{2}D_3 t + D_4,$$

$$W = -D_1 \sin t + D_2 \cos t, \tag{3.162}$$

$$V = -2D_1 \cos t - 2D_2 \sin t - \frac{3}{2}D_3,$$

with D_1, D_2, D_3, D_4 being integration constants.

Taking into account the nonlinear terms in Eq. (3.161), $D_i\ (i = 1, 2, 3, 4)$ will be no longer constants but functions of time. Using the variation of constants, one obtains

$$\frac{\mathrm{d}D_1}{\mathrm{d}t} = -\frac{\mu(\xi_2 - \xi_3)}{\rho^3}\sin t - \frac{2\mu(\eta_2 - \eta_3)}{\rho^3}\cos t,$$

$$\frac{\mathrm{d}D_2}{\mathrm{d}t} = \frac{\mu(\xi_2 - \xi_3)}{\rho^3}\cos t - \frac{2\mu(\eta_2 - \eta_3)}{\rho^3}\sin t,$$

$$\frac{\mathrm{d}D_3}{\mathrm{d}t} = \frac{2\mu(\eta_2 - \eta_3)}{\rho^3},$$

$$\frac{\mathrm{d}D_4}{\mathrm{d}t} = -\frac{2\mu(\xi_2 - \xi_3)}{\rho^3} + \frac{2\mu(\eta_2 - \eta_3)}{\rho^3}t. \tag{3.163}$$

3.9.2 *Orbital variation in close encounter*

Since D_i $(i = 1, 2, 3, 4)$ in Eqs. (3.162) and (3.163) determine the orbit of the third body, the variations of D_i during encounters of bodies reflect the changes of the orbit.

3.9.2.1 Planar case

In this case, m_2 is moving on an elliptic orbit. The approximated motions of m_2 and m_3 in the $(\boldsymbol{\xi}, \boldsymbol{\eta})$ coordinate system will be given first.

In the inertia coordinate system $(\boldsymbol{X}, \boldsymbol{Y})$, the orbit of m_2 is an ellipse with small eccentricity e_p. In the dimensionless coordinate system, the coordinates of m_2 are (neglecting $o(e_p)$)

$$\begin{aligned}
X_2 &= \cos E - e_p, \\
Y_2 &= \sin E,
\end{aligned} \tag{3.164}$$

where E is the eccentric anomaly. From Eqs. (3.153, 3.154, 3.157), one obtains

$$\begin{pmatrix} \xi_2 + 1 \\ \eta_2 \end{pmatrix} = \begin{pmatrix} \cos(t - t_0) & \sin(t - t_0) \\ -\sin(t - t_0) & \cos(t - t_0) \end{pmatrix} = \begin{pmatrix} X_2 \\ Y_2 \end{pmatrix}. \tag{3.165}$$

And the eccentric anomaly can be expanded to a series as

$$\begin{aligned}
\sin E &= \sin M + \frac{1}{2}e_p \sin 2M + o(e_p), \\
\cos E &= \cos M + \frac{1}{2}e_p \cos 2M - \frac{1}{2}e_p + o(e_p).
\end{aligned} \tag{3.166}$$

Neglecting $o(e_p)$ terms in Eq. (3.166) and substituting Eqs. (3.164) and (3.166) into Eq. (3.165), one obtains the approximated motion of m_2 in the Hill coordinate system as

$$\begin{aligned}
\xi_2 &= -e_p \cos(t - t_0), \\
\eta_2 &= 2e_p \sin(t - t_0).
\end{aligned} \tag{3.167}$$

Denote the semi-major axis of the near circular orbit of m_3 by a and the eccentricity by e, let $h = a - 1$ and suppose $e \ll h$. When $\mu \ll 1$, m_3 moves on a near circular orbit. According to Eq. (3.162), near the encounter with m_2 the motion of m_3 can be approximated by

$$\xi_3 = h,$$
$$\eta_3 = -\frac{3}{2}h(t - \tau),$$
$$W = 0,$$
$$V = -\frac{3}{2}h,$$

(3.168)

where τ is the time of encounter. For simplicity and without loss of generality, let $\tau = 0$. Institute Eqs. (3.167) and (3.168) into Eq. (3.159), one obtains

$$\rho^2 = \left(h^2 + \frac{9}{4}h^2 t^2\right) + [2h\cos(t - t_0) + 6ht\sin(t - t_0)]\,e_p + o(e_p).$$

(3.169)

Now substitute Eqs. (3.167), (3.168) and (3.169) into Eq. (3.163) and integrate the equations from $t = -\infty$ to $t = +\infty$, the variations of D_i are obtained as follows.

$$\Delta D_1 = \mu h^{-3} e_p \sin t_0 \,\mathrm{sgn}(h) \left[\int_{-\infty}^{+\infty} \frac{\sin^2 t + 4\cos^2 t}{(1 + 9t^2/4)^{3/2}} dt \right.$$
$$\left. + \int_{-\infty}^{+\infty} \frac{18t \sin t \cos t - 3\sin^2 t - 27t^2 \cos^2 t}{(1 + 9t^2/4)^{5/2}} dt \right],$$

$$\Delta D_2 = -\mu h^{-2}\,\mathrm{sgn}(h) \int_{-\infty}^{+\infty} \frac{3t \sin t + \cos t}{(1 + 9t^2/4)^{3/2}} dt$$
$$+ \mu h^{-3} e_p \cos t_0 \,\mathrm{sgn}(h) \left[-\int_{-\infty}^{+\infty} \frac{\cos^2 t + 4\sin^2 t}{(1 + 9t^2/4)^{3/2}} dt \right.$$
$$\left. + \int_{-\infty}^{+\infty} \frac{3\cos^2 t + 18t \sin t \cos t + 27t^2 \sin^2 t}{(1 + 9t^2/4)^{5/2}} dt \right],$$

(3.170)

$$\Delta D_3 = -\mu h^{-3} e_p \sin t_0 \,\mathrm{sgn}(h) \left[\int_{-\infty}^{+\infty} \frac{4\cos t}{(1 + 9t^2/4)^{3/2}} dt \right.$$
$$\left. + \int_{-\infty}^{+\infty} \frac{9t \sin t - 27t^2 \cos t}{(1 + 9t^2/4)^{5/2}} dt \right],$$

where $\mathrm{sgn}(x)$ indicates the sign of x. Because τ has been set $\tau = 0$, from Eqs. (3.162) and (3.168) one has $\Delta D_4 = 0$.

According to literature (Hénon & Petit, 1986), a higher order correction term $3.34\mu^2 h^{-5}\mathrm{sgn}(h)$ may be added to ΔD_3 to polish the approximation. With this correction, integrate Eq. (3.170) one gets

$$\Delta D_1 = \mu h^{-3} e_p \sin t_0 \, \mathrm{sgn}(h) \left[-\frac{10}{3} + \frac{128}{27} k_0 \left(\frac{4}{3}\right) \right.$$

$$\left. + \frac{68}{27} k_1 \left(\frac{4}{3}\right) + \frac{32}{27} k_2 \left(\frac{4}{3}\right) \right],$$

$$\Delta D_2 = -\mu h^{-2} \, \mathrm{sgn}(h) \left[\frac{16}{9} k_0 \left(\frac{2}{3}\right) + \frac{8}{9} k_1 \left(\frac{2}{3}\right) \right]$$

$$+ \mu h^{-3} e_p \cos t_0 \, \mathrm{sgn}(h) \left[-2 + \frac{104}{27} k_1 \left(\frac{4}{3}\right) + \frac{32}{27} k_2 \left(\frac{4}{3}\right) \right],$$

$$\Delta D_3 = -\mu h^{-3} e_p \sin t_0 \, \mathrm{sgn}(h) \left[\frac{64}{27} k_0 \left(\frac{2}{3}\right) + \frac{32}{27} k_1 \left(\frac{2}{3}\right) \right]$$

$$+ 3.34\mu^2 h^{-5}\mathrm{sgn}(h),$$

$$(3.171)$$

where $k_i(x)$ $(i = 0, 1, 2)$ are all Bessel functions (see Liu & Sun (1991) for details).

It should be remarked: If m_3 is moving on a circular orbit, Eq. (3.168) is a good approximated solution. But, if the orbit is an elliptic one, there exist two deviations. (i) The component of velocity in $\boldsymbol{\xi}$ direction is not zero ($W \neq 0$). (ii) The position at encounter should be corrected to $\xi_3 = h + O(e)$. A further analysis shows that the correction on W is of the order $O(e)$ too. Thus, under the condition $e \ll h$, the approximation solution Eq. (3.168) is reasonable.

3.9.2.2 Spatial case

The above analyses on the planar problem can be generalized to the spatial one, provided the inclination of m_3's orbit is small.

With the same method as in the planar case, the equations of motion of m_3 in the spatial coordinate system $(\boldsymbol{\xi}, \boldsymbol{\eta}, \boldsymbol{\zeta})$ read

$$\frac{d^2\xi_3}{dt^2} = 2\frac{d\eta_3}{dt} + 3\xi_3 + \frac{\mu(\xi_2 - \xi_3)}{\rho^3},$$

$$\frac{d^2\eta_3}{dt^2} = -2\frac{d\xi_3}{dt} + \frac{\mu(\eta_2 - \eta_3)}{\rho^3}, \qquad (3.172)$$

$$\frac{d^2\zeta_3}{dt^2} = -\zeta - \frac{\mu\zeta_3}{\rho^3},$$

with

$$\rho^2 = (\xi_2 - \xi_3)^2 + (\eta_2 - \eta_3)^2 + \zeta_3^2. \tag{3.173}$$

As in the planar case, define $\mathrm{d}\xi_3/\mathrm{d}t = W$, $\mathrm{d}\eta_3/\mathrm{d}t = V$, $\mathrm{d}\zeta_3/\mathrm{d}t = U$, then one has

$$\frac{\mathrm{d}\xi_3}{\mathrm{d}t} = W,$$

$$\frac{\mathrm{d}\eta_3}{\mathrm{d}t} = V,$$

$$\frac{\mathrm{d}\zeta_3}{\mathrm{d}t} = U,$$

$$\frac{\mathrm{d}W}{\mathrm{d}t} = 2V + 3\xi_3 + \frac{\mu(\xi_2 - \xi_3)}{\rho^3},$$

$$\frac{\mathrm{d}V}{\mathrm{d}t} = -2W + \frac{\mu(\eta_2 - \eta_3)}{\rho^3},$$

$$\frac{\mathrm{d}U}{\mathrm{d}t} = -\zeta_3 - \frac{\mu\zeta_3}{\rho^3}. \tag{3.174}$$

When $\mu = 0$ (or $\rho \to \infty$), Eq. (3.174) is a linear and integrable system, with a solution as

$$\xi_3 = D_1 \cos t + D_2 \sin t + D_3,$$

$$\eta_3 = -2D_1 \sin t + 2D_2 \cos t - \frac{3}{2}D_3 t + D_4,$$

$$\zeta_3 = D_5 \cos t + D_6 \sin t,$$

$$W = -D_1 \sin t + D_2 \cos t,$$

$$V = -2D_1 \cos t - 2D_2 \sin t - \frac{3}{2}D_3,$$

$$U = -D_5 \sin t + D_6 \cos t, \tag{3.175}$$

with D_i $(i = 1, 2, \ldots, 6)$ being constants.

When the nonlinear terms in Eq. (3.174) is taken into account, the variations of D_i can be obtained in the same way as in the planar case.

$$\frac{\mathrm{d}D_1}{\mathrm{d}t} = -\frac{\mu(\xi_2 - \xi_3)}{\rho^3} \sin t - \frac{2\mu(\eta_2 - \eta_3)}{\rho^3} \cos t,$$

$$\frac{\mathrm{d}D_2}{\mathrm{d}t} = \frac{\mu(\xi_2 - \xi_3)}{\rho^3} \cos t - \frac{2\mu(\eta_2 - \eta_3)}{\rho^3} \sin t,$$

$$\frac{\mathrm{d}D_3}{\mathrm{d}t} = \frac{2\mu(\eta_2 - \eta_3)}{\rho^3},$$

$$\frac{\mathrm{d}D_4}{\mathrm{d}t} = -\frac{2\mu(\xi_2 - \xi_3)}{\rho^3} + \frac{2\mu(\eta_2 - \eta_3)}{\rho^3}t,$$

$$\frac{\mathrm{d}D_5}{\mathrm{d}t} = \frac{\mu\zeta_3}{\rho^3}\sin t, \qquad\qquad (3.176)$$

$$\frac{\mathrm{d}D_6}{\mathrm{d}t} = -\frac{\mu\zeta_3}{\rho^3}\cos t.$$

Also in a similar way as for the planar case, the motion of m_3 near the encounter position can be described approximately as follows.

$$\xi_3 = h, \qquad \eta_3 = -\frac{3}{2}h(t - \tau), \qquad \zeta_3 = l\sin(t - t_1),$$

$$W = 0, \qquad V = -\frac{3}{2}h, \qquad U = l\cos(t - t_1), \qquad (3.177)$$

where $h = a - 1$, $l = a\sin i$, $t_1 = -\Omega$ and i, Ω are the inclination and ascending node. On the other hand, the motion of m_2 is still the same as Eq. (3.167) in the planar case and $\zeta_2 = 0$. Substitute Eqs. (3.167) and (3.177) into Eq. (3.173), one obtains

$$\rho^2 = h^2 + \frac{9}{4}h^2 t^2 + [2h\cos(t - t_0) + 6ht\sin(t - t_0)]\,e_p + o(e_p) + o(l).$$

$$(3.178)$$

Adopting the linear approximation and neglecting the higher orders term, ΔD_i ($i = 1, 2, 3$) are the same as in the planar case, i.e. Eq. (3.171), while ΔD_5 and ΔD_6 can be obtained by substituting the approximation solutions into Eq. (3.176) and then integrating the equations from $-\infty$ to $+\infty$. They are

$$\Delta D_5 = \mu l h^{-3}\mathrm{sgn}(h)\cos t_1 \left[\frac{2}{3} - \frac{8}{9}k_1\left(\frac{4}{3}\right)\right],$$

$$(3.179)$$

$$\Delta D_6 = \mu l h^{-3}\mathrm{sgn}(h)\sin t_1 \left[\frac{2}{3} + \frac{8}{9}k_1\left(\frac{4}{3}\right)\right].$$

Finally, the configuration of orbits during the encounter is plotted in Fig. 3.35.

3.9.3 D_i and orbital elements

The relation between variables D_i and orbital elements can be revealed by some calculations. Through Eqs. (3.153) and (3.157), the coordinates can

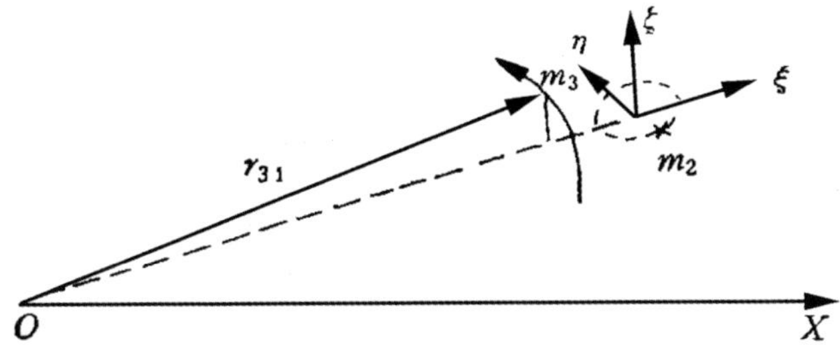

Fig. 3.35   Configuration of encounter-type orbits. From Liu & Sun (1991).

be converted from $(\boldsymbol{\xi}, \boldsymbol{\eta}, \boldsymbol{\zeta})$ back to inertia coordinate system $(\boldsymbol{X}, \boldsymbol{Y}, \boldsymbol{Z})$ as follows.

$$\begin{pmatrix} X \\ Y \\ Z \end{pmatrix} = \begin{pmatrix} \cos(t - t_0) & -\sin(t - t_0) & 0 \\ \sin(t - t_0) & \cos(t - t_0) & 0 \\ 0 & 0 & 1 \end{pmatrix} = \begin{pmatrix} s \\ m \\ n \end{pmatrix}, \qquad (3.180)$$

where

$$s = \xi_3 + 1 = D_1 \cos t + D_2 \sin t + D_3 + 1,$$

$$m = \eta_3 = -2D_1 \sin t + 2D_2 \cos t - \frac{3}{2} D_3 t + D_4, \qquad (3.181)$$

$$n = \zeta_3 = D_5 \cos t + D_6 \sin t.$$

As well-known, in the inertia coordinate system, the position vector $\boldsymbol{r}$ of an object on a spatial orbit can be expressed as a function of the orbital elements as

$$\boldsymbol{r} = r \cos f \boldsymbol{P} + r \sin f \boldsymbol{Q} = a(\cos E - e)\boldsymbol{P} + a\sqrt{1 - e^2} \sin E \boldsymbol{Q}, \qquad (3.182)$$

where $\boldsymbol{P}, \boldsymbol{Q}$ are vectors

$$\boldsymbol{P} = \begin{pmatrix} \cos \omega \cos \Omega - \sin \omega \sin \Omega \cos i \\ \cos \omega \sin \Omega + \sin \omega \cos \Omega \cos i \\ \sin \omega \sin i \end{pmatrix}, \qquad (3.183)$$

$$\boldsymbol{Q} = \begin{pmatrix} -\sin \omega \cos \Omega - \cos \omega \sin \Omega \cos i \\ -\sin \omega \sin \Omega + \cos \omega \cos \Omega \cos i \\ \cos \omega \sin i \end{pmatrix}, \qquad (3.184)$$

and a, e, ω, Ω, i are the orbital elements (semi-major axis, eccentricity, argument of pericenter, ascending node and inclination, respectively), f the true anomaly, E the eccentric anomaly. Denote

$$A = a(\cos E - e), \quad B = a\sqrt{1 - e^2}\,\sin E. \tag{3.185}$$

From Eqs. (3.180–3.185), one derives

$$\begin{pmatrix} \cos t_0 & -\sin t_0 & 0 \\ \sin t_0 & \cos t_0 & 0 \\ 0 & 0 & 1 \end{pmatrix} \begin{pmatrix} AP_x + BQ_x \\ AP_y + BQ_y \\ AP_z + BQ_z \end{pmatrix} = \begin{pmatrix} \cos t & -\sin t & 0 \\ \sin t & \cos t & 0 \\ 0 & 0 & 1 \end{pmatrix} \begin{pmatrix} s \\ m \\ n \end{pmatrix}, \tag{3.186}$$

where P_x, P_y, P_z and Q_x, Q_y, Q_z are the three components of vectors $\boldsymbol{P}$ and $\boldsymbol{Q}$, respectively. And then D_i can be calculated from this equation as

$$\begin{aligned} D_1 &= -ae\cos(\omega + \Omega + t_0), \\ D_2 &= -ae\sin(\omega + \Omega + t_0), \\ D_3 &= a - 1, \\ D_4 &= 0, \quad (\text{as } \tau = 0), \\ D_5 &= a\sin\omega\sin i, \\ D_6 &= a\cos\omega\sin i. \end{aligned} \tag{3.187}$$

Denote the angle between $\boldsymbol{X}$ axis and m_3 at its encounter with m_2 by λ_c, then one has

$$\lambda_c = -t_0, \tag{3.188}$$

and the relation between D_i and the orbital elements reads:

$$\begin{aligned} D_1 &= -ae\cos(\omega + \Omega - \lambda_c), \\ D_2 &= -ae\sin(\omega + \Omega - \lambda_c), \\ D_3 &= a - 1, \\ D_4 &= 0, \\ D_5 &= a\sin\omega\sin i, \\ D_6 &= a\cos\omega\sin i. \end{aligned} \tag{3.189}$$

Finally, the change of orbital elements of m_3 after an encounter with m_2 can be deduced from Eqs. (3.171), (3.179) and (3.189).

3.9.4   *Size of chaotic region*

Suppose m_3 suffers the perturbation from m_2 only in a short duration around the encounter with m_2, just as a "kick", and apart from the encounter the perturbation is so negligible that there is only a jump of orbital elements $(a, e, \omega, \Omega, \lambda_c, i)$ of m_3 due to the encounter and the new elements $a + \Delta a, e + \Delta e, \omega + \Delta\omega, \Omega + \Delta\Omega, \lambda_c + \Delta\lambda_c, i + \Delta i$ will be kept unchanged until the next encounter. Let λ'_c, a' be the new elements, then according to previous discussions

$$\Delta a = \Delta D_3 = -\mu h^{-3} e_p \sin t_0 \, \mathrm{sgn}(h) \left[\frac{64}{27} k_0 \left(\frac{2}{3} \right) + \frac{32}{27} k_1 \left(\frac{2}{3} \right) \right]$$
$$+ 3.34 \mu^2 h^{-5} \mathrm{sgn}(h), \tag{3.190}$$

$$\lambda'_c = \lambda_c + 2\pi g(q'), \tag{3.191}$$

where

$$q = \frac{a - a_p}{a_p}, \qquad q' = \frac{a' - a_p}{a_p}, \tag{3.192}$$

with a_p being the semi-major axis of m_2, and the function $g(x)$ is defined as

$$g(x) = \frac{1}{\left| (x + 1)^{-3/2} - 1 \right|}. \tag{3.193}$$

If in one encounter, the perturbation of m_2 on m_3 is sufficiently large, so that the change of λ_c exceeds $180°$, the perturbation effect in the consecutive encounters would be unpredictable, i.e. the system "forgot" the λ_c for proceeding encounter, indicating the occurrence of chaotic motion. Based on this idea, the size of the chaotic region of motion can be estimated analytically.

As q' is small, one has from Eq. (3.191)

$$\lambda'_c = \lambda_c + \frac{4\pi}{3} \frac{1}{q} - \frac{4\pi}{3} \frac{\Delta q}{q^2}, \tag{3.194}$$

where $\Delta q = q - q'$. Then the criterion for chaotic motion is

$$\frac{4\pi}{3} \frac{\Delta q}{q^2} \geq \pi. \tag{3.195}$$

Substituting Eq. (3.190) into this inequality, one has

$$\left| \frac{3.34\mu^2}{|q|^7} - \frac{3.02\mu e_p}{|q|^5} \sin t_0 \right| \geq \frac{3}{4}. \tag{3.196}$$

The maximum of possible chaotic region is attained at $\sin t_0 = -1$, thus the criterion becomes

$$\frac{3.34\mu^2}{|q|^7} + \frac{3.02\mu e_p}{|q|^5} \geq \frac{3}{4}, \tag{3.197}$$

when $e_p = 0$, i.e. in a circular restricted three-body problem, the corresponding criterion is

$$|q| \leq 1.237 \mu^{2/7}. \tag{3.198}$$

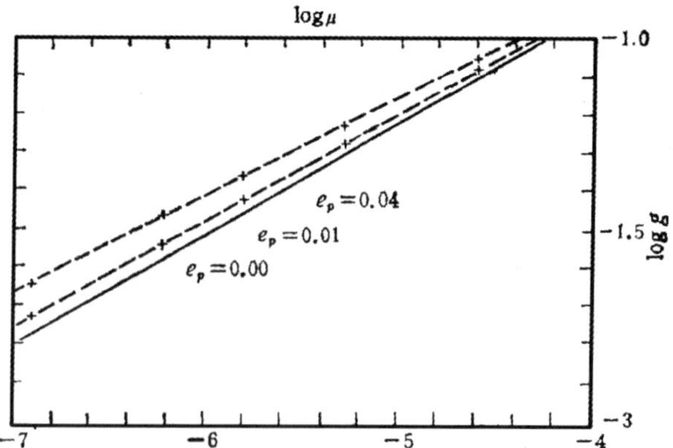

Fig. 3.36 q versus μ. Crosses denote the points satisfying Eq. (3.198), and solid lines satisfy function $q = c(e_p)\mu^{\nu(e_p)}$. Three eccentricity values have been employed as indicated in the plot. From Liu & Sun (1991).

For $e_p \neq 0$, take equality in Eq. (3.197),

$$\frac{3.34\mu^2}{|q|^7} + \frac{3.02\mu e_p}{|q|^5} = \frac{3}{4}. \tag{3.199}$$

In a $\log q$–$\log \mu$ diagram and for given e_p, this equation is approximately a straight line, as shown in Fig. 3.36. Hence, Eq. (3.197) can be approximated to a further simplified form

$$|q| \leq c(e_p)\mu^{\nu(e_p)}, \tag{3.200}$$

where $c(e_p)$ and $\nu(e_p)$ are functions of m_2's eccentricity e_p. Several example values of c and ν for different e_p are given below.

$$\begin{aligned}
\nu(0.00) &= 0.286, \quad \nu(0.01) = 0.278, \quad \nu(0.04) = 0.254, \\
c(0.00) &= 1.237, \quad c(0.01) = 1.216, \quad c(0.04) = 1.125.
\end{aligned} \tag{3.201}$$

In fact, $\nu(e_p)$ decreases monotonically with increasing e_p, thus the chaotic region expands when e_p increases.

For the spatial case, the influence of small inclination on the semi-major axis is a small quantity of second order, so its influence on the size of chaotic region is of the second order too. The eccentricity of the perturbing body still plays the major role in influencing the orbital evolution, leading to a wider chaotic region, while the small inclination brings only a negligibly small effect.

Chapter 4

Orbit diffusion

The diffusion of orbits in the phase space of a dynamical system is a very important issue, tightly related to the stability and the long-term evolution of the system. It has many direct applications in different subjects of physics, astronomy, chemistry and even economics. This chapter devotes to the orbit diffusion phenomenon, mainly in different mapping systems. The existence of invariant tori in a three-dimensional mapping, the diffusion in high-dimensional system and the stickiness effects of the hyperbolic structures will be discussed in detail. Since the orbit diffusion is a general phenomenon, nearly all these discussions are made in some mapping models, without intending of any immediate application to a specific problem.

4.1 Diffusion in comet motion

The orbital dynamics of comets have been investigated in previous sections (Sec. 3.5, Sec. 3.7, Sec. 3.9). The subjects such as the chaotic motion of comets, the transfer of comets from near-parabolic orbits to short-period orbits, etc. are all related to the "diffusion" of orbits in the phase space. In fact, orbital diffusion in the phase space is the basis of many topics in Hamiltonian systems.

For a nearly integrable Hamiltonian system of two-degree freedom, the KAM theorem guarantees the stability of most orbits under sufficiently small perturbation. Generally the phase space of a Hamiltonian system is a mixture of regular (invariant tori or islands) and chaotic regions. For an orbit near an invariant torus, it usually stays near the torus for a long time before it diffuses away. Such a slow orbital diffusion is a general phenomenon and has been widely studied. Karney (Karney, 1983) introduced the term "sticky" to describe such effects around tori, and the phenomenon was then called the "stickiness effect". In this section, the stickiness effect in the diffusion motion of comets will be discussed using mapping method.

The mapping models in Eqs. (3.75) and (3.76) are used to study the diffusion motion of comets on direct and retrograde orbits. For convenience, the mapping is copied again here, but in a uniform form:

$$
\begin{cases}
K_{n+1} = K_n + \mu \sum_{n=1}^{7} b_i \sin i g_n, \\[2ex]
g_{n+1} = g_n + \dfrac{2\pi}{(-K_{n+1})^{3/2}} \quad (\text{mod } 2\pi),
\end{cases}
\tag{4.1}
$$

where, $K = -1/a$ is the energy of comet motion, g is the phase angle of Jupiter at the time when the comet is passing the perihelion, and the coefficients b_i for comets on direct orbits (denoted by b_i' for retrograde orbits) are listed in Table 3.1. Equation (4.1) is an area-preserving mapping and it describes the evolution of comets in direct as well as in retrograde orbits. Just as in Sec. 3.5, the parameters $q = 1.5, \mu = 0.01$ are chosen to investigate the diffusion characters in both direct and retrograde orbits.

4.1.1   *Comets in direct orbits*

For comets in direct orbits, Fig. 4.1 displays a part of the phase space of the mapping. The invariant curves construct a central island, above which there is the chaotic sea imbedded with many island-chains, with energy ranging from $K = -0.94$ to $K = -0.91$.

To see how the orbits in the chaotic sea escape to infinity $(K > 0)$, one orbit starting from the initial point $(\pi, -0.9542)$ in the (g, K) plane (as Fig. 4.1) is calculated using the mapping. The evolution of K of this orbit is shown in Fig. 4.2. Apparently, the orbit oscillates for a long time in the interval of energy $(-1.06, -0.94)$, where a central island and a chaotic zone near it bounded by the island-chains can be seen. And in the interval $(-0.94, -0.91)$ there is a chaotic sea imbedded and bounded by island-chains (see Fig. 4.1). Since the phase angles g of these island-chains occupy the full interval of $(0, 2\pi]$, an orbit with $K < -0.94$ diffusing to $K > -0.91$ must pass through the island-chains, and it spends much of its time wandering along these island-chains one after another. Near the central island, no large island-chain exists, thus the orbit can move from the region above the island to below it, causing large variations of K. Only after it

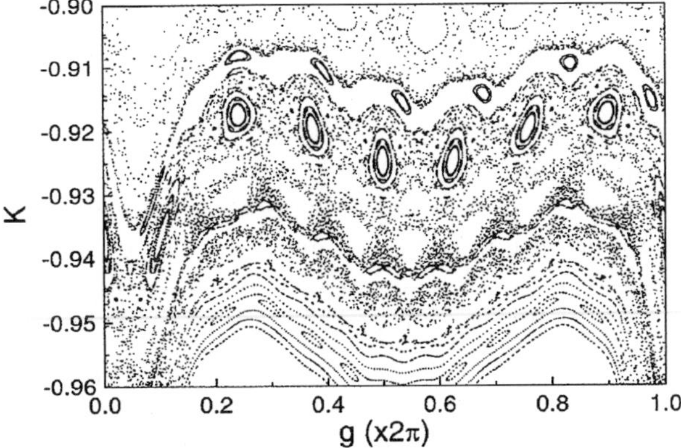

Fig. 4.1   Phase portrait of mapping Eq. (4.1) for comets in direct orbits near the central island. From Sun *et al.* (2002).

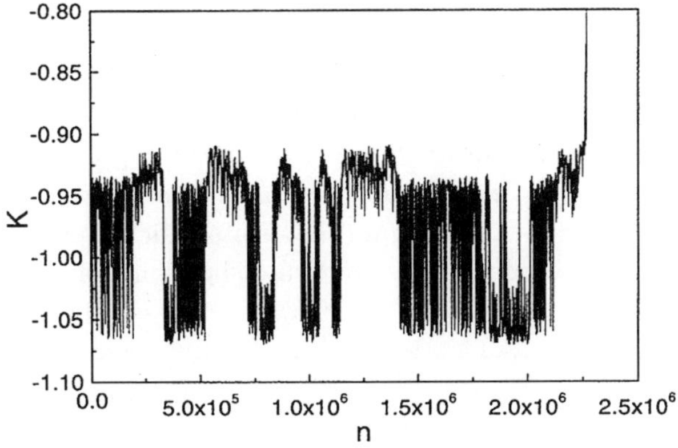

Fig. 4.2   Energy (K) evolution of a typical orbit starting from $(g, K) = (\pi, -0.9542)$ in Fig. 4.1. The abscissa is the iteration number, serving as "time" here. From Sun *et al.* (2002).

succeeds in passing the final barrier (with $K_{\max} \approx -0.91$) can it escape to infinity. Other orbits in the chaotic sea with $K < -0.91$ show similar evolutions. In this way these island-chains now serve as the partial barriers to the escape of orbits, which may greatly enhance the stability of the comet motion with energy lower than $K \approx -0.91$.

The above results show that the orbits with energy larger than $K_c \approx -0.91$ corresponding to the final barrier, will quickly escape to infinity, while those with energy lower than K_c will stay for a long time before it escapes. To see more clearly the different diffusion characters between the two kinds of chaotic regions below and above the final energy barrier, numerous initial points uniformly distributed on a line $K = $ constant are iterated up to 10^5 iterations using the mapping. The number of orbits having not escaped at iteration number n is called the "remain number", which after a modulation is just the probability of a comet remaining in that region at iteration number n (equivalently, the evolution time). Since it reflects the local diffusion characters of phase space, the calculation of remain number is widely used in many literatures.

First, 1,000 orbits in the chaotic region with initial points uniformly on the lines $K = -0.90$ and $K = -0.60$ are selected, and their evolutions will be followed. Some initial points on the lines are located in some small islands (perhaps invisible in Fig. 4.1) thus they will never escape. In the counting of remain numbers, these points are discarded. The calculations show that the remain number N decreases exponentially with the iteration number. Such an exponential law is similar to the previously discussed diffusion rule for asteroids (Eq. (2.64) in Sec. 2.3).

However, if the initial orbits are chosen in the region with energy $K < -0.91$, the situation will be quite different. In fact, this region is a mixture of chaotic sea and invariant curves. Since orbits in this region are more difficult to escape, 10,000 initial points are chosen uniformly on each of the lines $K = -0.920$ and $K = -0.935$. Again the orbits which do not escape before 10^5 iterations are discarded. The results are shown in Fig. 4.3. Clearly, the

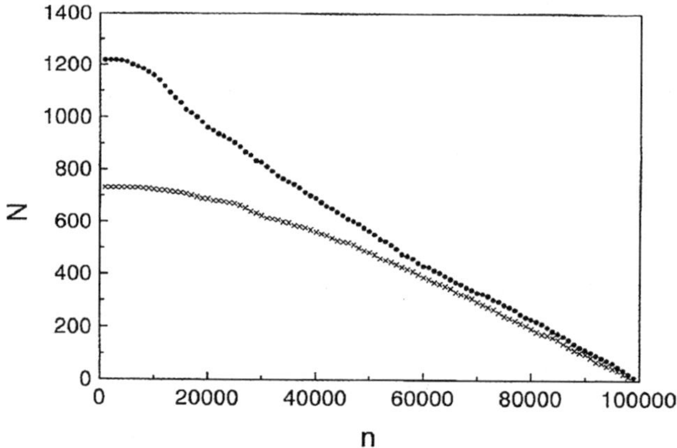

Fig. 4.3   Variations of remain number N with the iteration number n. The orbits are located originally on the lines $K = -0.920$ (circles) and $K = -0.935$ (crosses), respectively. From Sun *et al.* (2002).

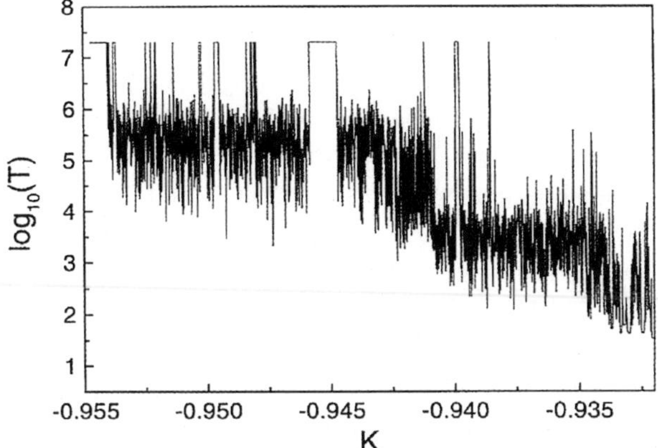

Fig. 4.4　Variations of escape time with initial energy of comets originated on the line $g = \pi$ and $K \in (-0.955, -0.931)$ in Fig. 4.1. From Sun *et al.* (2002).

remain number decays linearly with the iteration number in this case. It is believed that this slower diffusion rule is due to the existence of partial barriers (island-chains).

To study the diffusion character of orbits near the invariant tori, 2,300 orbits initialized on the line $g = \pi, K \in (-0.955, -0.931)$ are selected. The outmost invariant torus of the central island is generated with the initial point $(g, K) = (\pi, -0.954727)$. If an orbit evolves to a position where its distance to the torus is greater than the sum of its original distance plus 0.01, it is considered as escaped from the torus, and the iteration number at this moment, regarded as the escape time, is denoted by T. The maximum iteration number in this calculation is 2×10^7. The variation of escape time T with the initial energy (or, equivalently the distance from the torus) is shown in Fig. 4.4.

As well known, the plateau structure with the escape time 2×10^7 in Fig. 4.4 correspond to some islands in the phase space. Except these plateaus, the T versus K curve fluctuates strongly, due to the complicated structures in the vicinity of the torus. However, one can still see roughly two different regimes with different scales of escape time:

$$\begin{aligned} T &\sim 10^2 - 10^4 \quad (-0.941 < K < -0.932), \\ T &\sim 10^5 - 10^6 \quad (-0.954 < K < -0.941). \end{aligned} \tag{4.2}$$

The critical value $K = -0.941$ corresponds just to the location of the island-chain (see Fig. 4.4). So, once again it displays the importance of island-chains in the orbital diffusion.

According to the character of the stable and unstable manifolds in the vicinity of hyperbolic points, around the hyperbolic points the stickiness effect should appear. The exact location of hyperbolic fixed points of mapping Eq. (4.1) can be found by setting $K_{n+1} = K_n$ and $g_{n+1} = g_n$, and one of the solutions is $g_0 = \pi$, $K_0 = (-0.25)^{\frac{1}{3}}$.

Similar to the above calculations, 2,000 initial points on a line passing through the hyperbolic point and with $g = \pi$, $K \in (K_0, K_0 + DK)$ are studied (where DK is a given number). The escape time T of an orbit is defined as the iteration number at which the orbit leaves the box: $g/(2\pi) \in [0.05, 0.95]$, $K \in [-0.655, -0.605]$ that encloses the studied hyperbolic structure. Calculations show that near the hyperbolic structure ($DK < 0.01$) all the orbits diffuse in about the same order of escape time $T \sim 10^{2.5}$, though large fluctuations are present. This is a clear hint of the existence of stickiness effect in the neighborhood of the hyperbolic structure.

4.1.2   *Comets in retrograde orbits*

For comets in retrograde orbits, both orbits in the vicinity of invariant torus and in the vicinity of hyperbolic structure are studied. For the former case, Fig. 4.5 displays the phase diagram near a central island. The curves in the lower left part of the figure are the edges of the island. To show the diffusion characters near the invariant tori, 1,000 orbits initially distributed

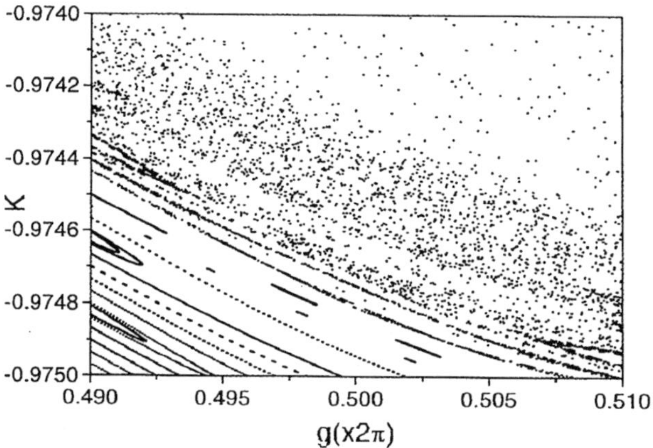

Fig. 4.5   Phase portrait of the mapping for comets in retrograde orbits near the central island. From Sun *et al.* (2002).

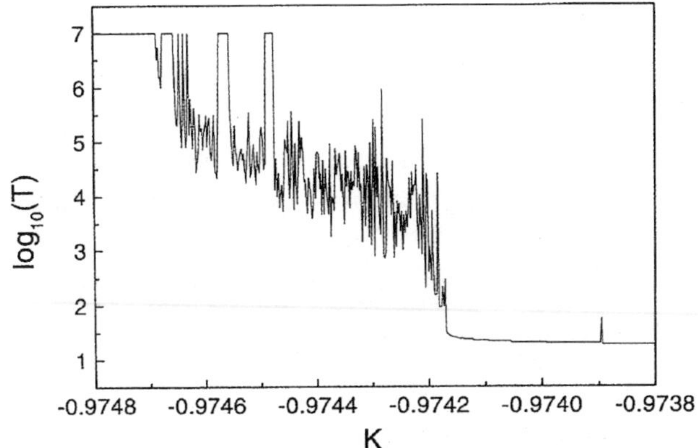

Fig. 4.6 Variations of escape time with initial energy of comets originated on the line $g = \pi$ and $K \in (-0.9748, -0.9738)$ in Fig. 4.5. From Sun *et al.* (2002).

uniformly on a line $g = \pi$ with $K \in (-0.9748, -0.9738)$ are followed and the number of escape orbits from the tori are counted. The escape time T is defined as the iteration number at which an orbit gains an energy increment of 0.01. The variation of the escape time T with the initial energy K is shown in Fig. 4.6.

Apparently, for $K > -0.9742$, T remains almost constant, while for $K < -0.9742$ T increases exponentially with the approaching to the torus. As it is believed from the general theory of orbital diffusion in Hamiltonian system that the diffusion speed of orbits in the vicinity of an invariant torus depends on the distance d from the torus. Explicitly, the diffusion is superexponentially slow as $1/\exp[\exp(1/d)]$ for very small d (Morbidelli & Giorgilli, 1995). But it is difficult to distinguish between the superexponential and exponential regimes by numerical computations, thus the exponential law of orbits diffusion near the tori can be regarded as the evidence of stickiness effects of tori. The boundary between the "sticky layer" and the "non-sticky layer" is about $K = -0.9742$. Perhaps some cantori exist there so that the two sides of the boundary belong to different diffusion regimes. This kind of stickiness effect caused by cantori was also discussed in Contopoulos *et al.* (1997).

For the orbital diffusion in the vicinity of hyperbolic structures, one of the hyperbolic fixed points of the mapping in the retrograde case has the same location as that in the direct motion, i.e. $K_0 = (-0.25)^{1/3}, g_0 = \pi$. This hyperbolic fixed point and the phase diagram around it are shown in Fig. 4.7, in which other two hyperbolic fixed points can be easily recognized.

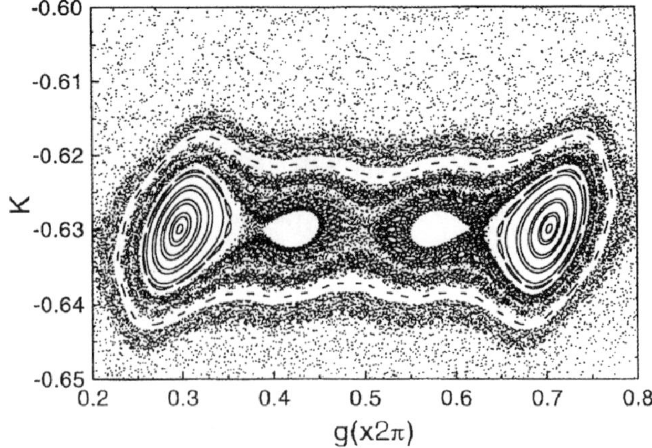

Fig. 4.7   Phase portrait of the mapping for comets in retrograde orbits near the hyperbolic fixed point at $K_0 = -0.25^{1/3}$, $g_0 = 0.3858752 \times 2\pi$. From Sun *et al.* (2002).

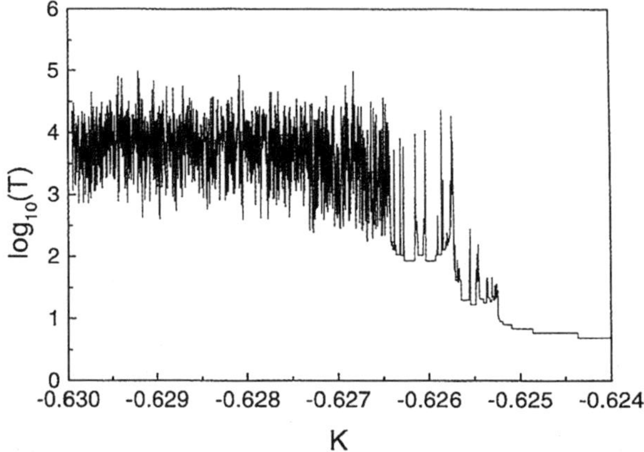

Fig. 4.8   Variations of escape time with initial energy of comets originated on the line $g = g_0 = 0.3858752 \times 2\pi$ and $K \in (K_0, K_0 + 0.006)$ in Fig. 4.7. From Sun *et al.* (2002).

The orbital diffusion characters in the neighborhood of the hyperbolic fixed point at $K_0 = (-0.25)^{1/3}$, $g_0 = 0.3858752 \times 2\pi$ (the left one in Fig. 4.7) will be investigated below. Just like before, 2,000 orbits initialized on the line $g = g_0$, $K \in (K_0, K_0 + 0.006)$ are calculated. This time, the escape time T of an orbit is defined as the iteration number at which the orbit leaves the box $g/2\pi \in [0.225, 0.5]$, $K \in [-0.64, -0.62]$, and the results of the escape time versus initial energy are plotted in Fig. 4.8. In this figure,

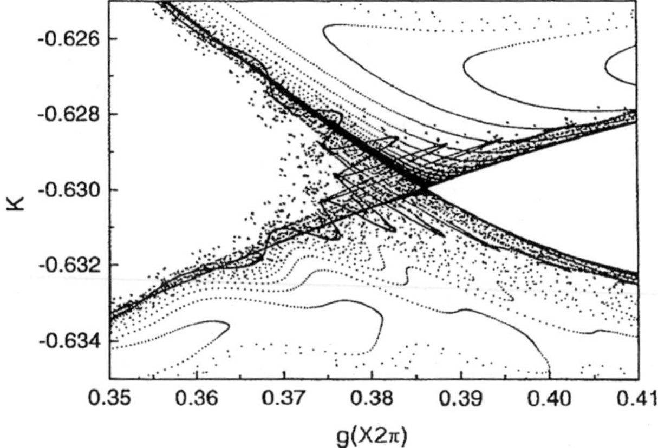

Fig. 4.9　Diagram of hyperbolic structure and orbit diffusion near it. The crosses denote the orbit with initial value $(-0.6295, 0.3858752 \times 2\pi)$. From Sun *et al.* (2002).

four different velocities of diffusion can be identified, which correspond to different zones in the vicinity of hyperbolic structure (see Fig. 4.8).

In the zone closest to the hyperbolic fixed point, the velocity of diffusion is the slowest. This corresponds to the layer in the closest neighborhood of the hyperbolic structure of the phase space. Outside of the layer the diffusion possesses the exponential law. Comparing Fig. 4.6 with Fig. 4.8, one can find similar law for orbital diffusion near the invariant tori and near the hyperbolic structures, implying again the stickiness effect of hyperbolic structure. In Fig. 4.9, the diagram of hyperbolic structure and the orbit diffusion near it are plotted.

One implication of the above results is that the stability of comet motion is energy dependent, especially for the Jupiter-family comets. As well-known, comets are usually categorized into two classes including long period comets with period greater than 200 years and short period comets with period less than 200 years, while the short period comets are further classified into Halley-family (with period greater than 20 years) and Jupiter-family (with period less than 20 years). In the unit adopted here, the relation between energy K and the comet period P is $K = -(11.86\,\mathrm{yr}/P)^{2/3}$, thus the boundary between Jupiter family and Halley family is about $K = -0.70$. In this study, the mass ratio is taken as $\mu = 0.01$, this is not the real case for the Sun–Jupiter–comet system where $\mu \approx 0.001$. Nevertheless, the results represent qualitatively the diffusion characters of comet orbits. The above results also indicate that the energy chain (i.e. the island-chains in phase space) may greatly enhance the stability of Jupiter-family comets (see Figs. 4.1 and 4.2).

Moreover, some comets with initial energies locating near the invariant tori or near the hyperbolic structure in the phase plane (g, K) will wander in the Solar System for long time before finally escaping.

About the phenomenon of orbit diffusion, concluding remarks can be made from the above discussions include:

1. The island-chains surrounding the invariant torus can serve as a partial barrier for the orbit diffusion to infinity.
2. In the vicinity of hyperbolic structures, similar stickiness effects exist as near the invariant tori.
3. Concerning the stability of comet motion, it is found to be energy dependent. Due to the stickiness effect, some comets with special initial energies will wander in the Solar System for long time before escaping.

In fact, not only the cantori may cause the stickiness effect to orbits around the invariant tori (e.g. Contopoulos *et al.* (1997)), but also the island-chains enclosing a central island may also results in the similar effect. Because the cantori consist of hyperbolic invariant sets, and there are hyperbolic structures too between the islands in island-chains, it is reasonable to suggest that both of these two kinds of stickiness effect are caused by the orbits' spending a long time around the hyperbolic structures, thus they can also be included into the stickiness effect of the hyperbolic structures. The relation between the stickiness effect of invariant tori and of the hyperbolic structures will be discussed in detail in the following sections.

4.2 KS entropy of area-preserving mapping

The orbital diffusion and chaotic motion discussed in previous sections are related to disordered or stochastic motions. It is well-known that the Kolmogorov-Sinai (KS) entropy can be regarded as a measure of disordered motion (Ott, 2002). Here in this section, a mapping based on the very well-known "standard mapping" (Chirikov, 1979) is used as the model to discuss the entropy of dynamical systems. The generalized standard mapping is as follows

$$\phi_{(n,a)}: \quad \begin{cases} x_{i+1} = x_i + a f_n(x_i + y_i), \\ y_{i+1} = x_i + y_i, \end{cases} \quad (\text{mod } 2\pi), \qquad (4.3)$$

where $f_n(x)$ is a periodic function of period 2π.

Using the Lyapunov Characteristic Numbers (LCN's) to compute the KS entropy h of a mapping through Pesin's integral, the variations of h against a and for a sequence of functions of the form

$$f_n(x) = \sin x + \sin 2x + \cdots + \sin nx$$

will be studied numerically by a Monte Carlo method.

4.2.1 *Basics about LCN's*

Let's recall the definition of LCN's first. Suppose M is an n-dimensional compact differentiable manifold, μ a normalized measure on it and ϕ^t a measure-preserving flow, i.e. a one-parameter measure-preserving diffeomorphism $M \to M$, satisfying the law of combination

$$\phi^{t+s} = \phi^t \circ \phi^s.$$

Let $P \in M$, $T_p(M)$ denote the tangent space of M at the point P and $D\phi_P^t$ the tangent mapping of ϕ^t from $T_P(M)$ onto $T_{\phi^t(P)}(M)$.

Given a nonzero vector $\boldsymbol{w} \in T_P(M)$, one may define the quantity

$$\gamma_{\boldsymbol{w}}^t(P) = \ln \frac{\|D\phi_P^t(\boldsymbol{w})\|}{\|\boldsymbol{w}\|}, \tag{4.4}$$

where $\|\cdot\|$ denotes the Euclidean norm in the tangent space. If the limit

$$X(P, \boldsymbol{w}) = \lim_{t\to\infty} \frac{1}{t}\gamma_{\boldsymbol{w}}^t(P) \tag{4.5}$$

exists, it is called the LCN of the flow ϕ^t relative to P and $\boldsymbol{w}$. The limit is known to exist for almost all initial points P and all vectors $\boldsymbol{w}$.

LCN's have the following properties.

(1) When $\boldsymbol{w}$ varies in $T_P(M)$, $X(P, \boldsymbol{w})$ takes n values at most. According to their magnitudes, they are set the order as $X_1(P), \ldots, X_n(P)$, and among them it is possible to have the same magnitude.

(2) Let $\boldsymbol{w}^{(1)}, \ldots, \boldsymbol{w}^{(k)}$ be k linear independent vectors in $T_P(M)$, $V^{(k)}(\boldsymbol{w}^{(1)}, \ldots, \boldsymbol{w}^{(k)})$ be the k-dimensional volume of the parallelotope consisted of $\boldsymbol{w}^{(1)}, \ldots, \boldsymbol{w}^{(k)}$. If the following limit exists

$$X^{(k)}(P, \boldsymbol{w}^{(1)}, \ldots, \boldsymbol{w}^{(k)}) = \lim_{t\to\infty} \frac{1}{t}V^{(k)}\left(D\phi_P^t(\boldsymbol{w}^{(1)}), \ldots, D\phi_P^t(\boldsymbol{w}^{(k)})\right), \tag{4.6}$$

it is called the LCN's of k order, correspondingly the $X(P, \boldsymbol{w})$ defined by Eq. (4.5) is called the LCN of one order. This limit exists for almost all $P \in M$.

(3) Between LCN's of k order and one order, there is a relation

$$X^{(k)}(P, \boldsymbol{w}^{(1)}, \ldots, \boldsymbol{w}^{(k)}) = \sum_{i=1}^{k} X_i(P), \quad (k = 1, 2, \ldots, n), \qquad (4.7)$$

and it will hold for almost all $\boldsymbol{w}^{(1)}, \ldots, \boldsymbol{w}^{(k)} \in T_P(M)$.

(4) For conservative systems, one has for almost all $P \in M$

$$\sum_{i=1}^{n} X_i(P) = 0. \qquad (4.8)$$

4.2.2 *Numerical estimation*

Pesin's formula has given the relation between the LCN's and the KS entropy of a system (Pesin, 1976). Let $\rho(P)$ denote the sum of all positive LCN's, the formula states that the total entropy is given by

$$h = \int_M \rho(P)\mathrm{d}\mu \qquad \left(\mathrm{d}\mu = \frac{\mathrm{d}x\mathrm{d}y}{4\pi^2}\right). \qquad (4.9)$$

Consequently the local quantity $\rho(P)$ defines a density of KS entropy which is related to the exponential stretch of small volume of the phase space in the directions corresponding to the positive LCN's.

In practice, h may be computed using a Monte Carlo method, i.e. randomly N_p initial points on the torus $M = \{(x, y) \pmod{2\pi}\}$ are selected and h is estimated by

$$h \approx \frac{1}{N_P} \sum_{j=1}^{N_P} X(P_j, \boldsymbol{w}). \qquad (4.10)$$

The mapping Eq. (4.3) depends on two parameters: a indicating the magnitude of the nonlinear perturbation and n the number of Fourier components. Below, the KS entropy of the mapping will be computed using the Monte Carlo method and then the variations of h with respect to a and n will be investigated.

Figure 4.10 displays the typical diagrams of the mappings $\phi_{(n,a)}$ for $a = -0.3, -0.7, -1.5$ and $n = 1, 2, 3, 4$. These figures exhibit all the characteristics and the well-known features of dynamical systems with two degrees of freedom, such as invariant curves, islands and also wild chaotic sea. Note that the size of these chaotic seas in this case increases with a and n. This trend is in complete agreement with the results in Fig. 4.11, which displays the variation of the estimated KS entropy with different values of n. The estimation is made through formula in Eq. (4.10) with $N_P = 100$

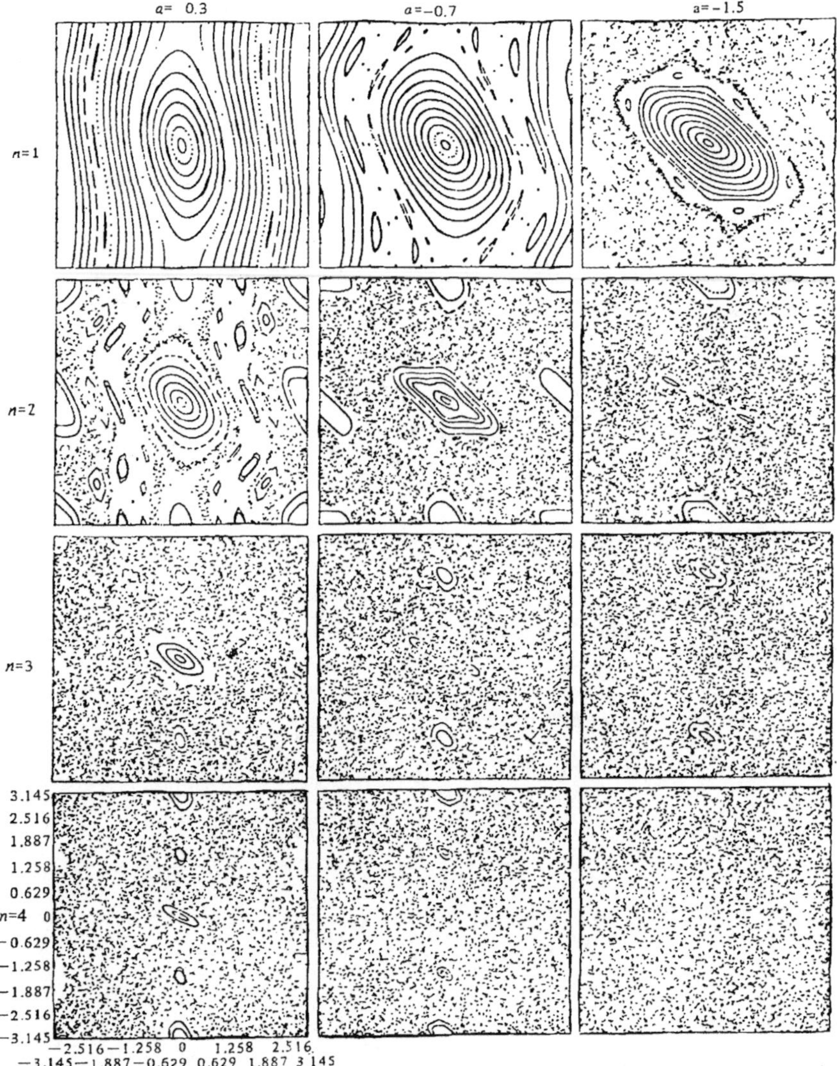

Fig. 4.10 Diagram of the mapping $\phi_{(n,a)}$ for $a = (-0.3, -0.7, -1.5)$ and $n = 1, 2, 3, 4$. From Sun & Froeschlé (1982).

where $X(P_j, \boldsymbol{w})$ is estimated through Eq. (4.5), in which t is the number of iterations for each orbit, i.e. 10,000.

Actually, the computed values of h (shown by crosses in Fig. 4.11) increase with n and $|a|$.

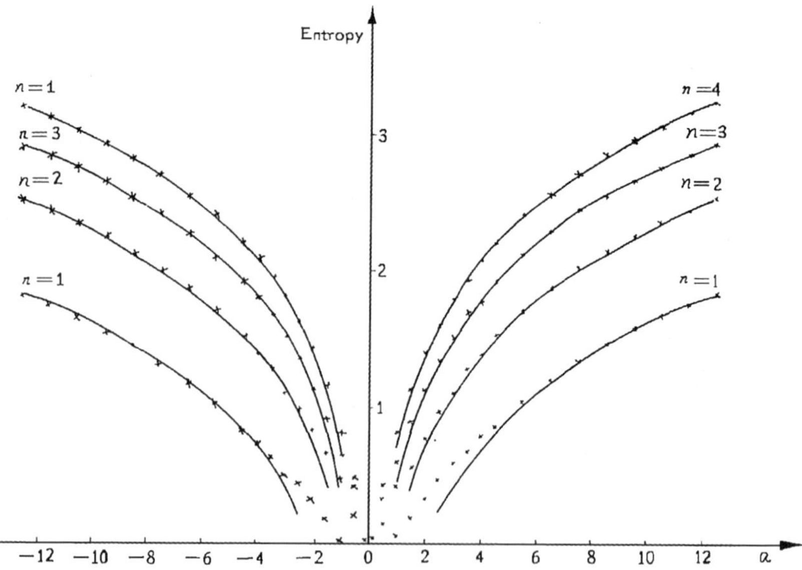

Fig. 4.11   Variation of the entropy h for $n = 1, 2, 3, 4$ plotted against a. The crosses show the numerical estimations and the curves give the analytical results. From Sun & Froeschlé (1982).

4.2.3   *Analytical estimation*

The analytical expression of h may be derived as below. Let

$$\boldsymbol{w} = \begin{pmatrix} u_1 \\ v_1 \end{pmatrix} \in T_P(M),$$

and

$$D(\phi_{(n,a)})^i_P(\boldsymbol{w}) = \begin{pmatrix} u_i \\ v_i \end{pmatrix}, \tag{4.11}$$

then one has

$$u_i = \left(1 + a\sum_{j=1}^{n} j\cos jy_i\right) u_{i-1} + \left(a\sum_{j=1}^{n} j\cos jy_i\right) v_{i-1}, \tag{4.12}$$

$$v_i = u_{i-1} + v_{i-1}.$$

For large values of $|a|$,

$$u_i \approx a\left(\sum_{j=1}^{n} j\cos jy_i\right)(u_{i-1} + v_{i-1}) = a\left(\sum_{j=1}^{n} j\cos jy_i\right) v_i. \tag{4.13}$$

Therefore,

$$|u_i| \geq |v_i|. \tag{4.14}$$

By a second approximation, neglecting v_{i-1} with respect to u_{i-1}, one obtains

$$u_i \approx \left(1 + a \sum_{j=1}^{n} j \cos j y_i\right) u_{i-1}. \tag{4.15}$$

Then

$$u_i \approx \left[\prod_{k=2}^{i} \left(1 + a \sum_{j=1}^{n} j \cos j y_k\right)\right] u_1, \tag{4.16}$$

and

$$X(P, \boldsymbol{w}) \approx \lim_{i \to \infty} \frac{1}{i} \ln \frac{\|D(\phi_{(n,a)})_P^i(\boldsymbol{w})\|}{\|\boldsymbol{w}\|},$$

will be approximated by

$$X(P, \boldsymbol{w}) \approx \lim_{i \to \infty} \frac{1}{i} \ln \left|\prod_{k=2}^{i+1} \left(1 + a \sum_{j=1}^{n} j \cos j y_k\right)\right|.$$

Figure 4.11 and the distribution of the $\frac{1}{10000}\gamma_{\boldsymbol{w}}^{10000}(P_i)$ $(i = 1, 100)$ shown in Fig. 4.12 indicate clearly that for large $|a|$ invariant curves zones vanish and that the motions are chaotic or the mapping is "ergodic". Hence, one can use approximately the ergodic theorem

$$\frac{1}{2\pi} \int_0^{2\pi} \ln \left|1 + a \sum_{j=1}^{n} j \cos j y\right| dy,$$

instead of

$$\lim_{i \to \infty} \frac{1}{i} \sum_{k=2}^{i+1} \ln \left|1 + a \sum_{j=1}^{n} j \cos j y_k\right|.$$

Then one has

$$1 + a \cos y = a\left(\frac{1}{a} + \cos y\right),$$

$$1 + a \sum_{j=1}^{2} j \cos j y = 4a(\cos y - k_{12})(\cos y - k_{22}),$$

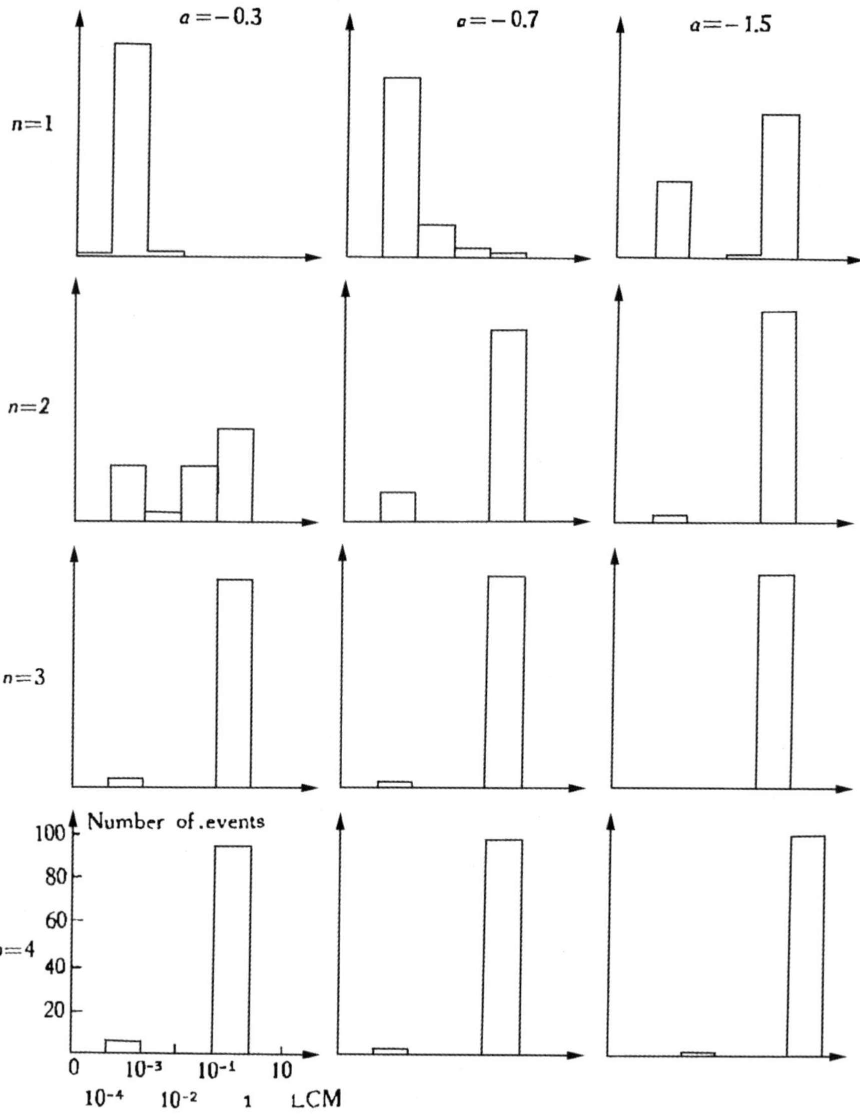

Fig. 4.12   Distributions of the numerical values of LCN's for the same mappings as shown in Fig. 4.10. From Sun & Froeschlé (1982).

$$1 + a \sum_{j=1}^{3} j \cos jy = 12a(\cos y - k_{13})(\cos y - k_{23})(\cos y - k_{33}),$$

$$1 + a \sum_{j=1}^{4} j \cos jy = 32a(\cos y - k_{14})(\cos y - k_{24})(\cos y - k_{34})(\cos y - k_{44}),$$

and when $|a| > 1$ all the k_{ij} are real and $|k_{ij}| < 1$, therefore the integration can be performed exactly. Finally, one obtains for $n \leq 4$:

$$h \approx \ln \frac{n|a|}{2}. \tag{4.17}$$

The curves in Fig. 4.11 give the estimated values of h through such analytical computations. The agreement between the curves and the numerical computed values are quite good for large $|a|$.

For any large natural number n, as

$$\sum_{j=1}^{n} j \cos jy = \frac{\cos \frac{y}{2} - \cos \left(n + \frac{1}{2}\right) y}{2 \sin \frac{y}{2}},$$

one has

$$\left| 1 + a \sum_{j=1}^{n} j \cos jy \right| = \left| 1 + \frac{a}{2} \frac{-\frac{1}{2} + \frac{1}{2} \cos ny + n \sin \frac{y}{2} \sin \left(n + \frac{1}{2}\right) y}{\sin^2 \frac{y}{2}} \right|$$

$$\approx \left| \frac{an \sin \left(n + \frac{1}{2}\right) y}{2 \sin \frac{y}{2}} \right|,$$

then finally

$$h \approx \ln \frac{n|a|}{2},$$

i.e. for any large n the above approximation formula for h is still correct.

The above results show that for a peculiar family of area-preserving mapping as Eq. (4.3), the KS entropy is well approximated by $h \approx \ln \frac{n|a|}{2}$, the agreement between the numerical and analytical computations is quite good.

4.3 Invariant tori in volume-preserving mapping

The phase space of a Hamiltonian system of n degrees of freedom has $2n$ dimensions, which can be reduced to the measure-preserving mappings with $2n - 2$ dimensions by using energy integral and Poincaré section. Hence the measure-preserving mappings with even dimension have been widely studied. On the other hand, the measure-preserving mappings with odd dimension have never been studied, either analytically or numerically. For the odd dimensional mapping and its linearized mapping in the vicinity of a fixed point, at least one eigenvalue is real and the product of all eigenvalues is equal to one, therefore in general there are some eigenvalues whose moduli are larger than one. Thus fixed points are generally unstable in the linear approximation and one may conjecture that the mapping is chaotic in the domain of definition D. The first nontrivial odd dimensional mapping is dimension 3. This section presents the first exploration ever made on this topic.

Two three-dimensional mappings T_1 and T_2 have been used as models. The first mapping reads

$$T_1 : \quad \begin{cases} x_{n+1} = x_n + z_n, \\ y_{n+1} = y_n + x_{n+1}, \\ z_{n+1} = z_n + A\sin y_{n+1}, \end{cases} \quad (\text{mod } 2\pi) \qquad (4.18)$$

where A is a parameter. Such a mapping is constructed following some basic ideas: First, the odd-dimensional mapping should be as simple as possible but it should contain a parameter so that a family of mappings can be studied; second, the mapping is set in such a way that the escape cannot occur and thus the numerical investigation on it will not be limited. The mapping T_1 is restrained on the torus $M_3 = \{(x, y, z), (\text{mod } 2\pi)\}$, and it is easy to verify that T_1 is measure-preserving.

The origin $(0, 0, 0)$ is a fixed point. For different values of parameter A, the eigenvalues λ_i $(i = 1, 2, 3)$ of the linear part of T_1 near the origin are different. Three values of A, $A = -7.0, A = -1.0$ and $A = 0.1$ are taken, which correspond to three different situations for the eigenvalues, i.e.

(a) $A = -7.0$, $\lambda_{1,2,3}$ are all real, and $|\lambda_{1,2}| > 1, |\lambda_3| < 1$;
(b) $A = -1.0$, λ_1 is real, $\lambda_{2,3}$ are complex and $|\lambda_1| < 1, |\lambda_2| = |\lambda_3| > 1$;
(c) $A = 0.1$, λ_1 is real, $\lambda_{2,3}$ are complex and $|\lambda_1| > 1, |\lambda_2| = |\lambda_3| < 1$.

As will be seen later, for the mapping T_1 the invariant manifolds do not exist in any of the three cases (a), (b) and (c).

The second mapping T_2 is

$$T_2 : \quad \begin{cases} x_{n+1} = x_n + y_n + B\sin z_n, \\ y_{n+1} = y_n + A\sin x_{n+1}, \\ z_{n+1} = z_n + B\sin y_{n+1}, \end{cases} \quad (\text{mod } 2\pi) \qquad (4.19)$$

with A, B being two parameters. Mapping T_2 is an extension of the two-dimensional standard mapping

$$\begin{cases} x_{n+1} = x_n + y_n, \\ y_{n+1} = y_n + A \sin x_{n+1}, \end{cases} \pmod{2\pi}$$

(where A is a parameter) to a three-dimensional mapping, in which $A = -1.5$ and B is a small parameter. Therefore, mapping T_2 is a three-dimensional mapping containing such a small perturbing term that when this term vanishes the mapping reduces to one that possesses not only chaotic region but also ordered regions.

4.3.1 *Existence of invariant manifolds in T_1 and T_2*

Using LCN's method and slice cutting method, the following results are obtained.

For mapping T_1, the whole domain D of definition is chaotic for all three cases (a), (b) and (c), and the chaotic regions are connected.

For mapping T_2, first a transversal exploration along the y-axis is performed to explore the properties of mapping T_2 with $B = 0.03$. The values of LCN's $\gamma_i^N(P_0)$ (N: number of iterations, $i = 1, 2, 3$) are computed, where the initial points are $P_0 = (0.1, y, 0.0)$ with y taking 60 values uniformly spaced between $-\pi$ and $+\pi$. In almost all cases, the values of LCN's are approximately the same, as plotted in Fig. 4.13. However, one set of values $|\gamma_i^{100000}(P_0)|$ where $P_0 = (0.1, -2.5, 0.0)$ is found to be particularly smaller than others.

According to the usual relation between the LCN's and the ordered or chaotic behavior of the orbit, this point $P_0 = (0.1, -2.5, 0.0)$ with smaller LCN's may be on an invariant manifold. To confirm it, the variation of $\gamma_i^N(P_0)$ ($i = 1, 2, 3$) as the function of the iteration number N up to $N = 10^6$ is calculated. As shown clearly in Fig. 4.14, $\lim_{N\to\infty} |\gamma_i^N(P_0)| = 0$ ($i = 1, 2, 3$). On the other hand, the slice cuttings of mapping T_2 with the same initial point P_0 are plotted in Fig. 4.15. The slice cuttings are defined by

$$|z_n - Q| < 0.01, \quad Q = \frac{(k-5)\pi}{4},$$

$$-\pi \le x_n \le +\pi, \quad -\pi \le y_n \le +\pi, \tag{4.20}$$

$$(k = 1, 2, \ldots, 9, \quad n = 1, 2, \ldots, 10^5),$$

and as k increases from 1 to 9, the "slices" are seen from left to right and top to bottom in Fig. 4.15.

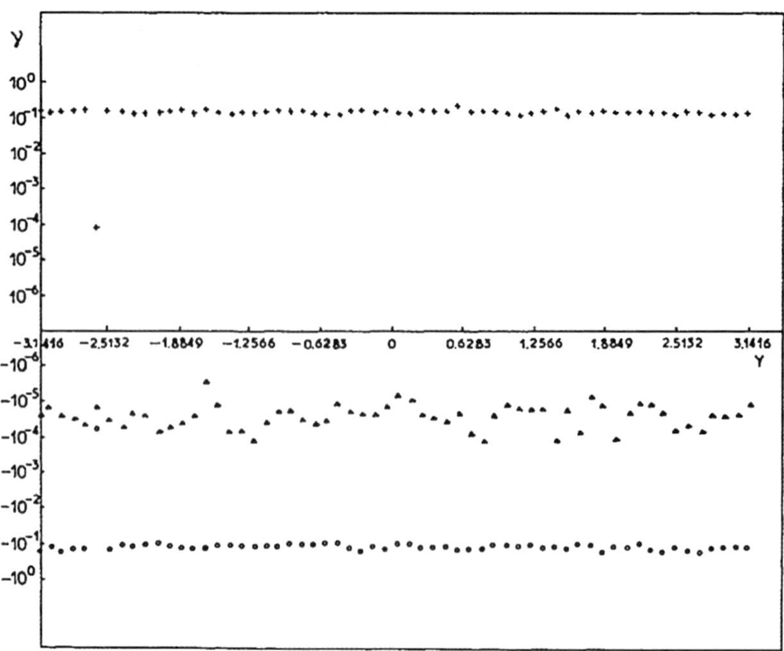

Fig. 4.13 Variation of the $\gamma_i^{100000}(P_0)$ ($i = 1, 2, 3$) of mapping T_2 ($A = -1.5, B = 0.03$) for the initial points $P_0 = (0.1, y, 0.0)$ with y uniformly distributed in $[-\pi, +\pi]$. The crosses, open triangles and open circles represent $\gamma_1^{100000}(P_0), \gamma_2^{100000}(P_0)$ and $\gamma_3^{100000}(P_0)$, respectively. From Sun (1983a).

In the slices shown in Fig. 4.15, the points lie on the invariant tubes (which will be called "island-tubes" later) with period 3. To investigate the mappings on a finer scale, the leftmost island-tube is enlarged in Fig. 4.16.

Figure 4.17 displays orbits near the island-tube studied above. These slice cuttings exhibit features analogous to those of the standard mapping, such as invariant curves, islands, and chaotic region, which will be seen more clearly in Fig. 4.18 and Fig. 4.19. As illustrated in those figures, the size of chaotic region increases with the parameter B, i.e. with the increasing perturbing term. This feature is analogous to the standard mapping.

Finally, Fig. 4.20 and Fig. 4.21 show slice cuttings of mapping T_2 with $B = 0.03$ for the same initial point $P_0 = (0.1, 0.1, 0.0)$, but with different iteration numbers $N = 10^4$ and $N = 10^5$ respectively. From these two figures, a very slow diffusion away from the z-axis can be seen.

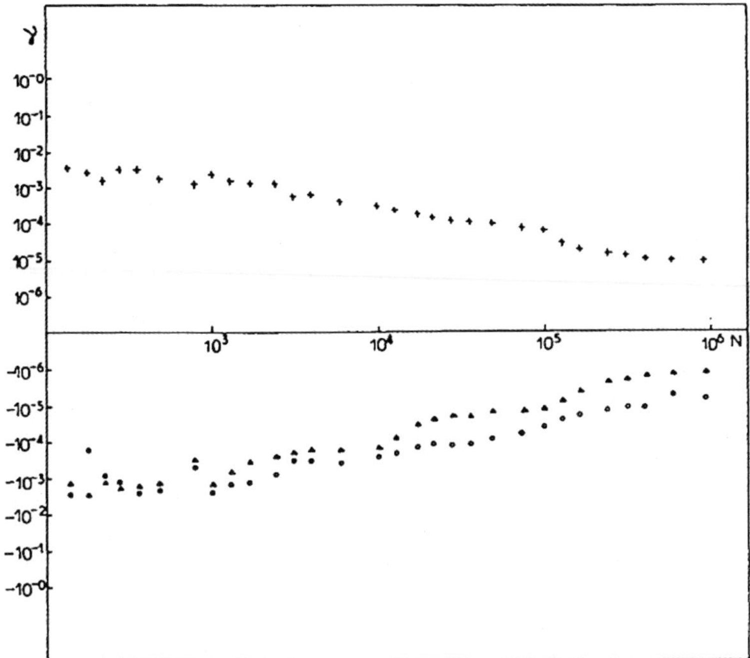

Fig. 4.14   Variation of the $\gamma_i^N(P_0)$ $(i = 1, 2, 3)$ of mapping T_2 $(A = -1.5, B = 0.03)$ for $P_0 = (0.1, -2.5, 0.0)$ as functions of the iteration number N. The crosses, open triangles and open circles represent $\gamma_1^N(P_0), \gamma_2^N(P_0)$ and $\gamma_3^N(P_0)$, respectively. From Sun (1983a).

In fact, the mapping T_2 is a perturbed extension of the area-preserving standard mapping to the volume-preserving one, with B being the perturbation parameter. When $B = 0$, the mapping T_2 is reduced to a standard mapping multiplied by a mapping along z-axis. It is known that in the standard mapping near the origin there is a large ordered region (the region of invariant curves) for small A, and far away from the origin there exist some small islands and the chaotic sea. So a question arises: Under perturbation, why the tubes near the z-axis disappear but the tubes near the small island-tubes persist?

Focus on a region denoted by G of the island-tubes of mapping T_2, where G is defined as $G : -2.55 \le x \le -2.3, \ -1.6 \le y \le -1.35, \ -\pi \le z \le +\pi$. When the perturbation is small, one has $|y| \gg B$. And according to the principle of mean value, the small perturbing term $B \sin z$ in the formulae for x_{n+1} and y_{n+1} can be replaced by its mean value, which however, is zero because in the domain G the values z_n $(i = 1, 2, \ldots)$ fill uniformly the whole

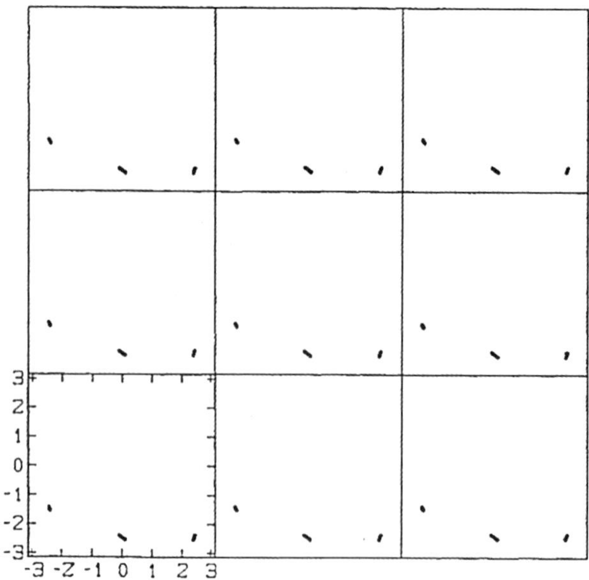

Fig. 4.15   Sections of mapping T_2 ($A = -1.5, B = 0.03$), defined as in Eq. (4.20). From Sun (1983a).

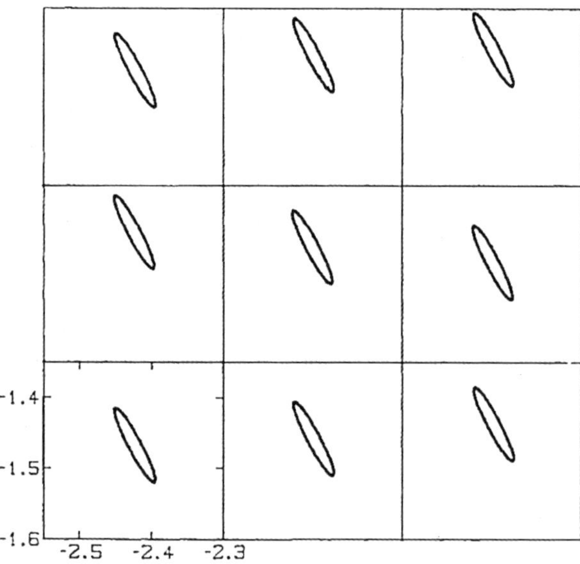

Fig. 4.16   Enlargement of the leftmost island-tube in Fig. 4.15. From Sun (1983a).

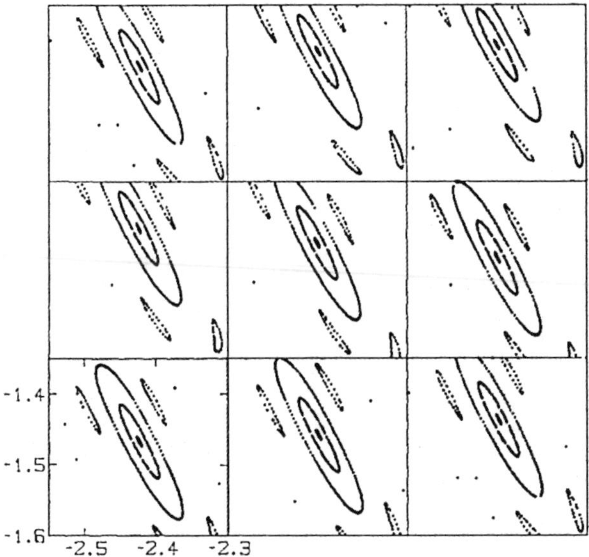

Fig. 4.17   Sections of mapping T_2 $(A = -1.5, B = 0.03)$. From Sun (1983a).

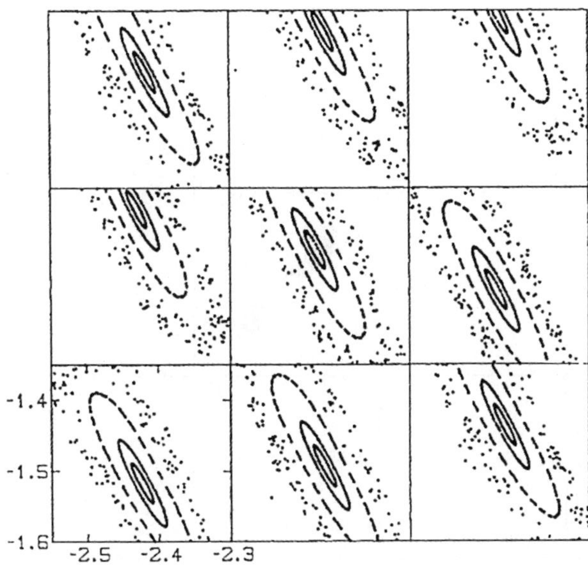

Fig. 4.18   Sections of mapping T_2 $(A = -1.5, B = 0.08)$. From Sun (1983a).

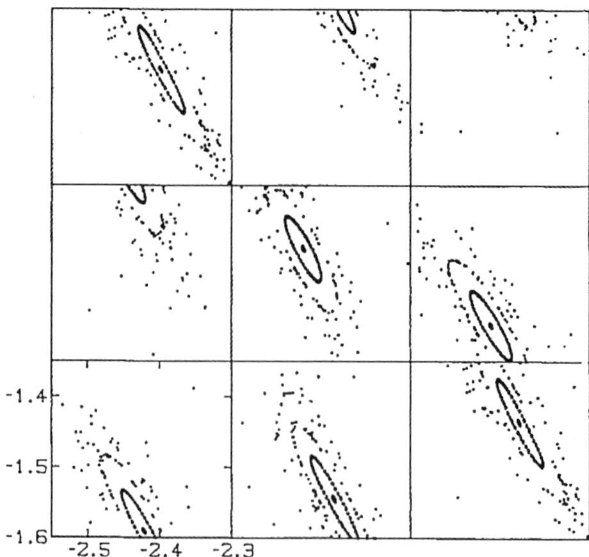

Fig. 4.19 Sections of mapping T_2 $(A = -1.5, B = 0.15)$. From Sun (1983a).

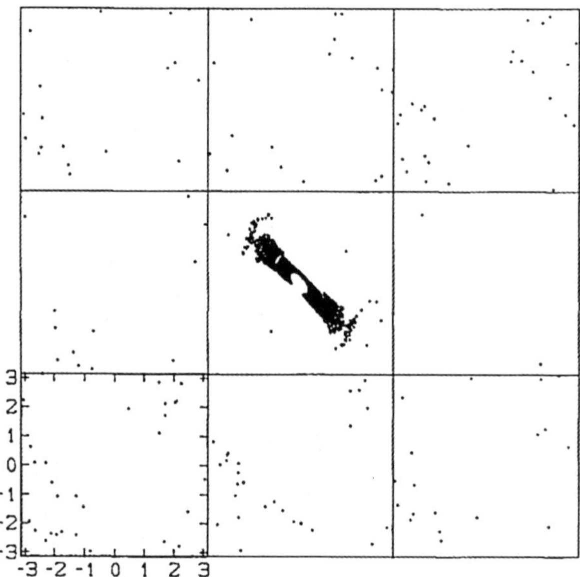

Fig. 4.20 Sections of mapping T_2 $(A = -1.5, B = 0.03)$ for initial point $P_0 = (0.1, 0.1, 0.0)$ with the iteration number $N = 10^4$. From Sun (1983a).

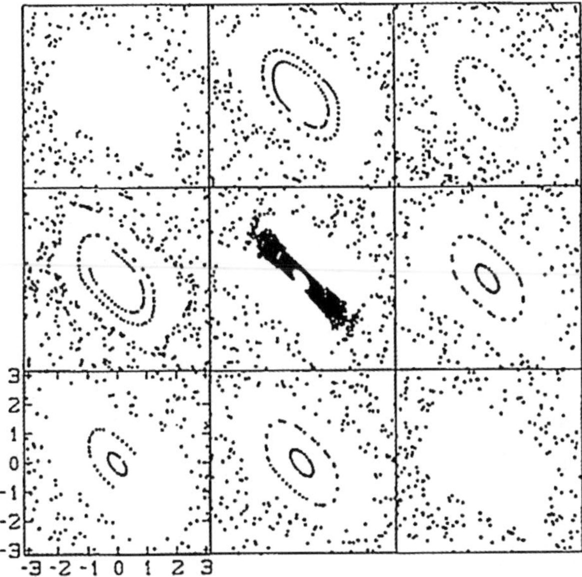

Fig. 4.21 Sections of mapping T_2 $(A = -1.5, B = 0.03)$ for initial point $P_0 = (0.1, 0.1, 0.0)$ with the iteration number $N = 10^5$. From Sun (1983a).

z-axis. In other words, in the domain G, the mapping $\bar{T}(\equiv T_2 \circ T_2 \circ T_2 = T_2^3)$ can be approximated by $\bar{T}' = \bar{T}_2 \circ \bar{T}_2 \circ \bar{T}_2$, where

$$\bar{T}_2 : \begin{cases} x_{n+1} = x_n + y_n, \\ y_{n+1} = y_n + A \sin x_{n+1}, \quad (\text{mod } 2\pi). \\ z_{n+1} = z_n + B \sin y_{n+1}, \end{cases} \qquad (4.21)$$

And the mapping $\bar{T}_2$ can be written as $\bar{T}_2 = \bar{T}_2^{(1)} \circ \bar{T}_2^{(2)}$ with

$$\bar{T}_2^{(1)} : \quad z_{n+1} = z_n + B \sin y_{n+1}, \quad (\text{mod } 2\pi),$$

$$\bar{T}_2^{(2)} : \begin{cases} x_{n+1} = x_n + y_n, \\ y_{n+1} = y_n + A \sin x_{n+1}, \end{cases} \qquad (\text{mod } 2\pi).$$

So it can explain approximately the existence of island-tubes in Fig. 4.16.

4.3.2 *Theoretical analysis and criterion of tube existence*

The above discussions imply the existence of tubes of the mapping T_2 is related to the variation of variable z in mapping T_2. In order to justify the above conjecture, another parameter D is added in the mapping, by which the variation of the variable z can be controlled. The new mapping T now reads

$$T: \begin{cases} x_{n+1} = x_n + y_n + B \sin z_n, \\ y_{n+1} = y_n + A \sin x_{n+1}, \qquad (\text{mod } 2\pi), \\ z_{n+1} = z_n + C \sin y_{n+1} + D, \end{cases} \qquad (4.22)$$

where $A = -1.5$, and B, C, D are all positive small parameters. As pointed out above, for the measure-preserving mappings of three-dimensions, as for the linearized mapping in the vicinity of a fixed point, there are, in general, some eigenvalues whose moduli are larger than one, i.e. there is at least a dilating characteristic direction. Naturally, the existence of invariant surfaces in the neighborhood of the fixed point cannot be expected, but the existence of invariant tubes along the direction parallel to z-axis is possible, as shown in Fig. 4.15 and Fig. 4.16. According to the KAM theorem on the existence of tori, if the fixed point of an area-preserving mapping is of a general elliptic type, in general, the invariant curves exist in the neighborhood of the fixed point. It may be conjectured that generally the invariant tubes probably exist together with the invariant curves for the three-dimensional mappings. And they play an analogous role to that of the invariant curves and the fixed points in the two-dimensional mappings. Here, the existence of the invariant curves for the mapping T will be analyzed first.

Let

$$x = F(z), \quad y = G(z) \qquad (4.23)$$

be the invariant curve of the mapping T, one has

$$\begin{aligned} F(z_{n+1}) &= F(z_n) + G(z_n) + B \sin z_n, \\ G(z_{n+1}) &= G(z_n) + A \sin F(z_{n+1}), \\ z_{n+1} &= z_n + C \sin G(z_{n+1}) + D. \end{aligned} \qquad (4.24)$$

Because C, D are small, one can take approximately

$$F(z) = 0, \quad G(z) = -B \sin z. \qquad (4.25)$$

Therefore, along this invariant curve, one has

$$\Delta z = z_{n+1} - z_n = D + C \sin(-B \sin z_n) \approx D - BC \sin z_n. \qquad (4.26)$$

When $D > BC$, $\Delta z > 0$, i.e. z varies monotonously between $-\pi$ and $+\pi$ (see Fig. 4.22) and the invariant curve exists. when $D < BC$, z tends to a fixed value $\bar{z}$ (see Fig. 4.23). In this case, the invariant curve degenerates to a point.

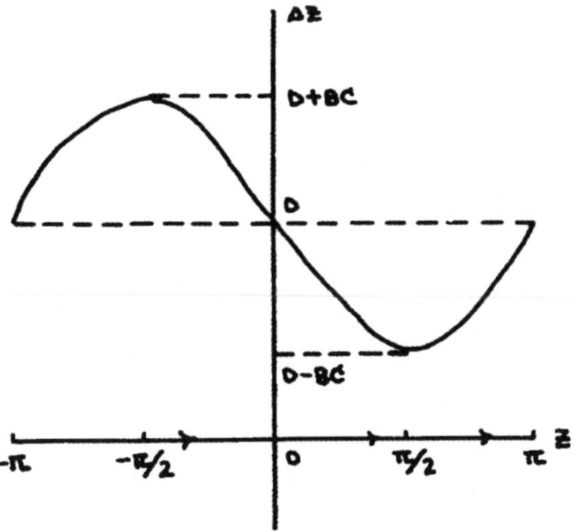

Fig. 4.22   Variation of z when $D > BC$. From Sun (1984).

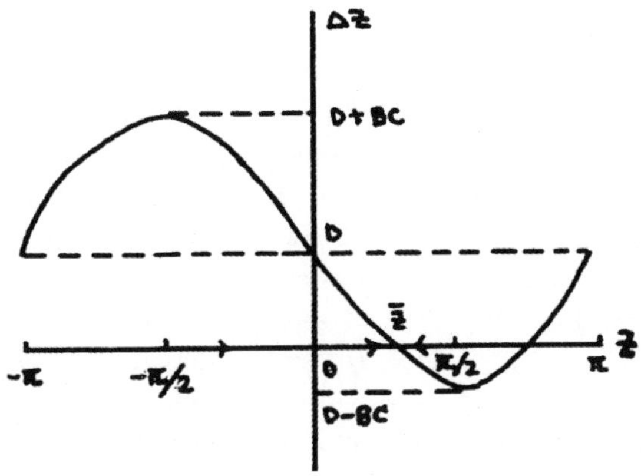

Fig. 4.23   Variation of z when $D < BC$. From Sun (1984).

Now one may look for the invariant tubes of the mapping T in the neighborhood of the invariant curve. Transform the variables to suitable ones as

$$S: \quad \begin{cases} X = x, \\ Y = y + B\sin z, \\ Z = z. \end{cases} \qquad (4.27)$$

It is evident that S conserves volumes. The mapping T now becomes

$$\bar{T}: \begin{cases} X_{n+1} = X_n + Y_n, \\ Y_{n+1} = Y_n + A\sin(X_n + Y_n) + B\left(\sin Z_{n+1} - \sin Z_n\right), \\ Z_{n+1} = Z_n + C\sin(Y_n - B\sin Z_n \\ \qquad\qquad + A\sin(X_n + Y_n)) + D, \end{cases} \quad (\text{mod } 2\pi).$$

$$(4.28)$$

In the neighborhood of Z-axis, $|X|, |Y|$ are small, thus

$$\Delta Z = Z_{n+1} - Z_n \approx D - BC\sin Z_n. \tag{4.29}$$

When $D \gg BC$ and $2\pi/D$ is an irrational number, Z varies rapidly between $-\pi$ and $+\pi$ and the values Z_n ($n = 1, 2, \ldots$) fill uniformly the whole Z-axis. According to the principle of the mean value, in the mapping $\bar{T}$ the small perturbing terms that are approximately periodic functions of Z can be replaced by their mean values that are all zero, i.e. the mapping $\bar{T}$ can be replaced by

$$\bar{T}': \begin{cases} X_{n+1} = X_n + Y_n, \\ Y_{n+1} = Y_n + A\sin(X_n + Y_n), \\ Z_{n+1} = Z_n + D, \end{cases} \quad (\text{mod } 2\pi). \tag{4.30}$$

According to the properties of the standard mapping, the invariant tubes parallel to the Z-axis exist in the neighborhood of the Z-axis for the mapping $\bar{T}'$, in other words, there exist invariant tubes in the neighborhood of the invariant curve for the mapping $\bar{T}$ with $D \gg BC$. From the above discussions, the mapping T will possess the same qualitative behaviors as the mapping $\bar{T}$. Figure 4.24 displays the sections of mapping T for this case.

For $D > BC$ but $D \approx BC$, Z still varies between $-\pi$ and $+\pi$, but now $\Delta Z = Z_{n+1} - Z_n \approx D - BC\sin Z_n$ depends strongly on Z_n and $\Delta Z|_{Z_n<0} > \Delta Z|_{Z_n>0}$ $(-\pi \le Z_n \le +\pi)$. Thus when $Z_n = \pm\pi/2$, ΔZ reaches its minimum and maximum respectively. Since the mapping $\bar{T}$ conserves volumes, the sections parallel to the (X, Y)-plane of invariant tubes will possess the maximum and minimum area respectively. Hence the invariant tubes will possess a bottle-like structure. Figure 4.25 displays the sections of mapping T with $A = -1.5, B = C = 0.03, D = 0.001$, and the initial point is $P_0 = (0.00, -0.01, 0.00)$. This is just the case where $D > BC$ but $D \approx BC$. Clearly in Fig. 4.25 the invariant tubes have a bottle-like structure, just as expected. And the center of the tubes is in fact an invariant curve (see Fig. 4.26).

These results are quite analogous to the ones of the area-preserving mappings in a plane, in which the roles played by the fixed point and the

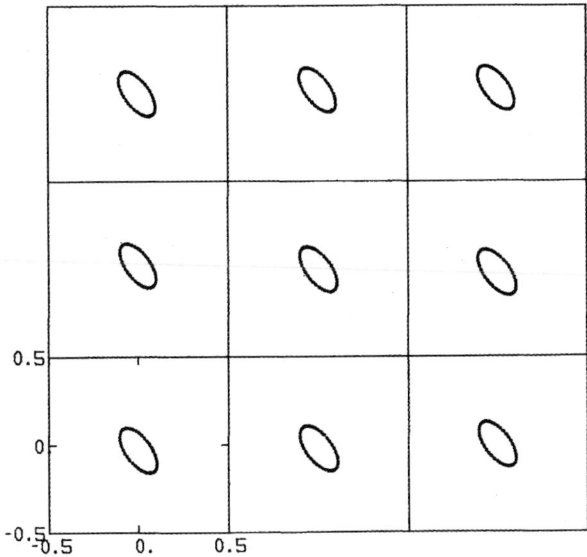

Fig. 4.24   Sections of mapping T for $A = -1.5, B = C = 0.03, D = 0.02$. The initial point is $P_0 = (0.0, -0.1, 0.0)$. From Sun (1984).

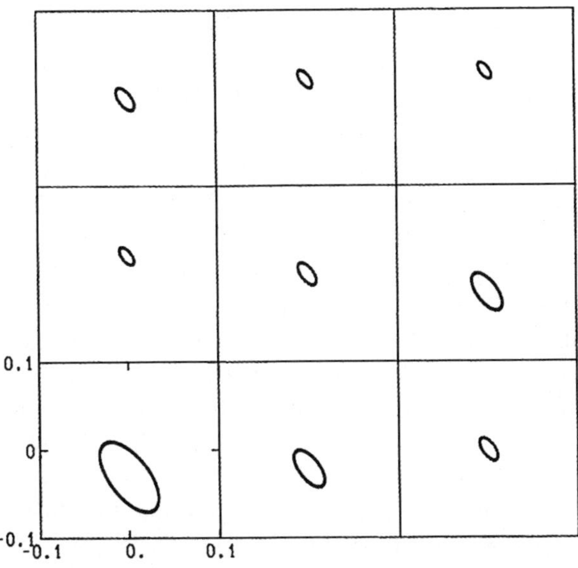

Fig. 4.25   Sections of mapping T for $A = -1.5, B = C = 0.03, D = 0.001$. The initial point is $P_0 = (0.00, -0.01, 0.00)$. From Sun (1984).

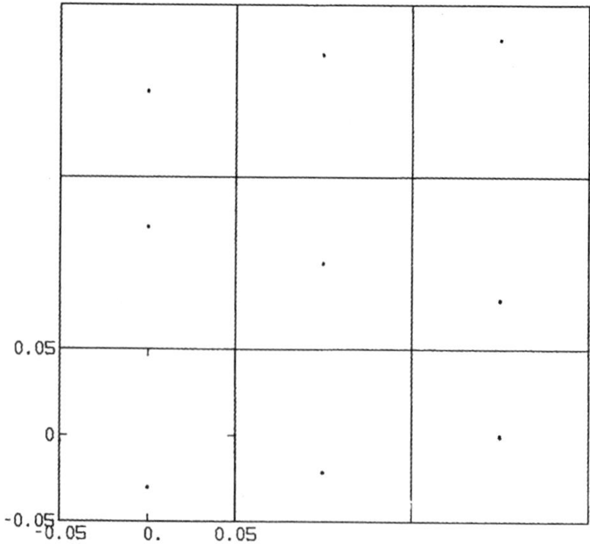

Fig. 4.26 Sections of mapping T for $A = -1.5, B = C = 0.03, D = 0.001$. The initial point is $P_0 = (0.00, -0.03, \pi/2)$. From Sun (1984).

invariant curves are now instead played by the invariant curve and invariant tubes respectively.

When $D < BC$, the invariant curve degenerates to a fixed point and in the neighborhood of Z-axis Z_n tends to fixed value $\bar{Z}$. As the mapping T conserves volumes, the points in this neighborhood will leave from there along the plane $Z = \bar{Z}$. Hence, a slow diffusion away from the neighborhood of z-axis appears (see Figs. 4.27 and 4.28). In Fig. 4.27 and Fig. 4.28, $D = 0.0008$ while $BC = 0.0009$, i.e. D is a little smaller than BC, and clearly the diffusion appears, implying that the criterion for the existence of invariant tubes of mapping T is quite precise.

Numerical experiments also reveal some more results as follows. When B or C is not small but their product BC is still small and $D > BC$ is fulfilled, the invariant tubes still exist. However, when BC is no longer small, although $D > BC$, the invariant tubes do not exist and chaotic motions appear.

The last case will be discussed here is the one in which $D > BC$ but $2\pi/D$ is a rational number. In this case, although Z varies between $-\pi$ and $+\pi$, Z_n values do not fill uniformly the whole Z-axis. The invariant curves and invariant tubes in preceding cases might degenerate to fixed points

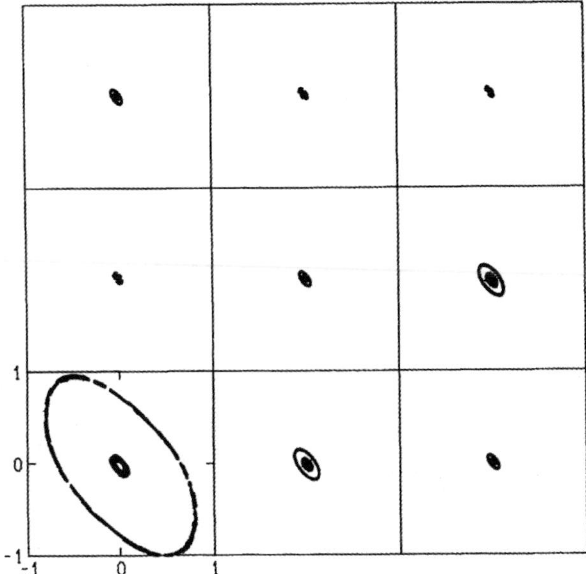

Fig. 4.27   Sections of mapping T for $A = -1.5, B = C = 0.03, D = 0.0008$. The initial point is $P_0 = (0.0, 0.0, 1.0951442)$. From Sun (1984).

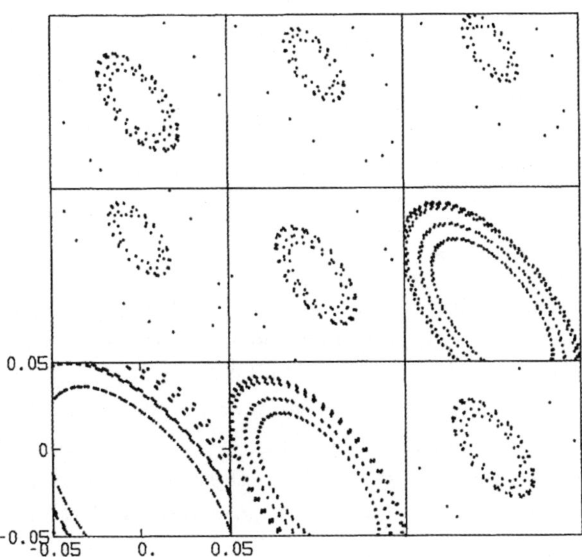

Fig. 4.28   Enlargement near the center part of Fig. 4.27. From Sun (1984).

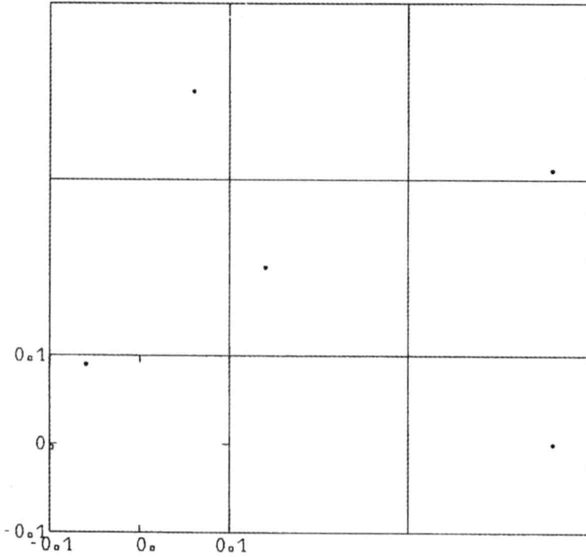

Fig. 4.29   Sections of mapping T for $A = -1.5, B = C = 0.03, D = \pi/2$. The initial point is $P_0 = (-0.06, 0.00, 0.00)$. From Sun (1984).

and isolated invariant curves. As expected, the fixed points and isolated invariant curves are found in the numerical experiments. Figures 4.29 and 4.30 show these structures.

In both of these two figures (Figs. 4.29 and 4.30) with two different initial points, $D = \pi/2$ thus $2\pi/D = 4$ is a rational number. The fixed points and invariant curves are seen in some cycles.

4.3.3   *Implications*

All the above numerical results, i.e. the existence of invariant tubes in the three-dimensional mapping, have been proven theoretically (Cheng & Sun, 1990a; Cheng & Sun, 1990b). And from these results, the following important consequences can be derived.

The first is the failure of the quasi-ergodic hypothesis. The ergodic (quasi-ergodic) hypothesis states that the generic Hamiltonian flow is ergodic (has a dense orbit) on the generic (compact, connected) energy surface.

Another important consequence is the failure of a conjecture raised by Pesin: On any manifold M (of dimension ≥ 3) there exists a nonempty

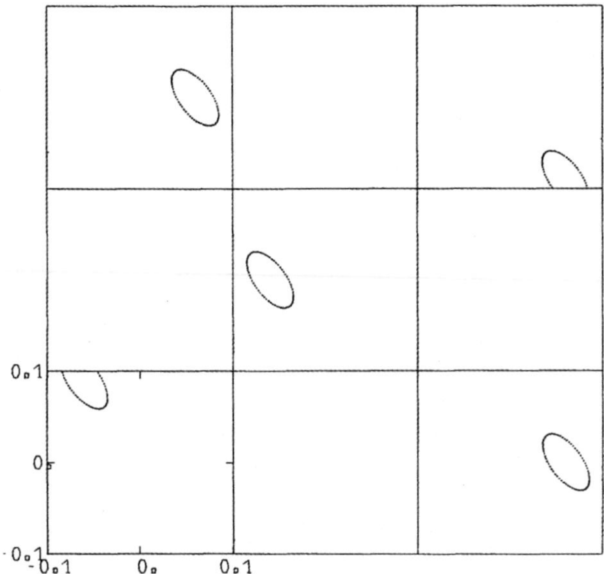

Fig. 4.30   Sections of mapping T for $A = -1.5, B = C = 0.03, D = \pi/2$. The initial point is $P_0 = (-0.04, 0.00, 0.00)$. From Sun (1984).

open set of volume-preserving diffeomorphisms whose Lyapunov exponents are all zero on a set of positive volume. In dimension 2, this follows from Moser's twist theorem (Yoccoz, 1994).

The above results have also been confirmed by physical experiments.

4.4   Perturbed extension of area-preserving mapping

The volume-preserving mapping studied in the previous section (Sec. 4.3) is in fact the perturbed extension of a two-dimensional mapping to a three-dimensional one. In the two-dimensional area-preserving mapping, at the origin there is an elliptic fixed point that is stable. In this section, a different mapping model will be discussed, in which the area-preserving mapping has an unstable fixed point at the origin.

The mapping reads:

$$T: \begin{cases} x_{n+1} = s\left(x_n \cos\varphi_n - y_n \sin\varphi_n\right) + A\cos z_n, \\ y_{n+1} = s^{-1}\left(x_n \sin\varphi_n + y_n \cos\varphi_n\right) + B\sin z_n, \\ z_{n+1} = z_n + C\cos\left(x_{n+1} + y_{n+1}\right) + D, \quad (\mathrm{mod}\, 2\pi), \end{cases} \tag{4.31}$$

where $\varphi_n = (x_n^2 + y_n^2)^k$ with s and k being parameters, and A, B, C, D are the perturbation parameters. Obviously, the Jacobian of mapping T

$$\left| \frac{\partial (x_{n+1}, y_{n+1}, z_{n+1})}{\partial (x_n, y_n, z_n)} \right| \equiv 1,$$

hence the mapping T is measure-preserving. When $A = B = C = D = 0$, the mapping is reduced to a hyperbolic twist mapping $\bar{T}$

$$\bar{T} : \quad \begin{cases} x_{n+1} = s(x_n \cos \varphi_n - y_n \sin \varphi_n), \\ y_{n+1} = s^{-1}(x_n \sin \varphi_n + y_n \cos \varphi_n), \end{cases} \tag{4.32}$$

where the parameters s ($s \neq 1$) implies an instability, i.e. hyperbolicity at the fixed point $(0,0)$ for $k > 0$, because the eigenvalues at this fixed point are s and s^{-1}. For $k < 0$, the mapping $\bar{T}$ is not defined at the origin and the eigenvalues at the origin do not exist. This is a new class of three-dimensional extension of the two-dimensional measure-preserving mapping, in which the unperturbed two-dimensional one possesses a hyperbolic fixed point. It is different from those mappings studied in previous section, where the fixed point at the origin of the unperturbed two-dimensional measure-preserving mapping is of elliptic type.

4.4.1 *Two-dimensional mapping $\bar{T}$*

For $k = 1.5$, the origin $(0,0)$ is a hyperbolic fixed point and it is not analytic but belongs to C^1 class at the origin. Obviously it is analytic in an annulus surrounding the origin and is symmetric with respect to the origin. The one-parameter (ϵ) family of mapping $\bar{T}$ with $s = 1 + \epsilon, k = 1.5$ can be considered as a family of perturbed twist mappings. For small ϵ, according to the KAM theorem, the invariant curves of the mapping $\bar{T}$ in the annulus will exist and there the stable and unstable manifolds of the origin will also exist. For $k = -1.5$, the above discussion will be still valid but in this case the mapping $\bar{T}$ is not defined at the origin.

In order to study the behavior of the mapping $\bar{T}$ in the annulus and the variation of ordered regions with the parameters, two values of $k = 0.5$ and $k = -1.5$ are taken, and for each k value, the parameter s is set to be $1.02, 1.05, 1.08$ and 1.1 respectively. By inspecting the values of the LCN's, whether the points are on the invariant manifolds can be determined. For each set of parameters k, s, points along the x-axis and with a step of 0.01 are investigated, the corresponding LCN's are calculated up to N iterations. For $k = 1.5$, points uniformly distributed along the line $y = x, z = 0$ from $(x, y) = (0, 0)$ to $(0.1, 0.1)$ with a step 0.01 have also been investigated.

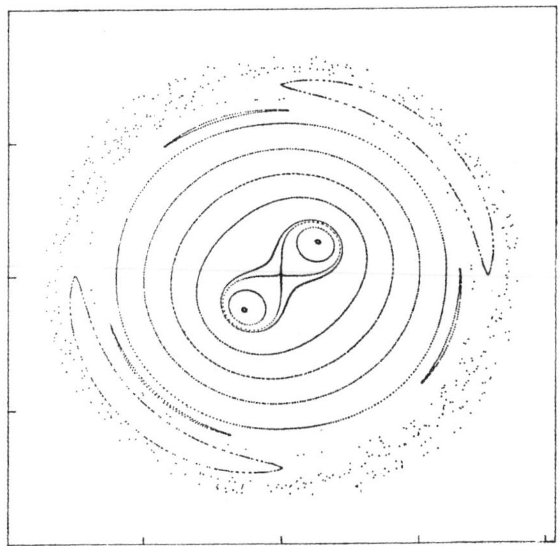

Fig. 4.31   The mapping $\bar{T}$ for $k = 1.5$, $s = 1.05$. From Sun & Yan (1988).

The results show that for $k = 1.5$ and $s = 1.02, 1.05, 1.08, 1.10$ there exist no chaotic regions in the neighborhood of the origin, but the dimensions of corresponding ordered regions decrease when s increases, with the ranges on x-axis being $(-2.1, 2.1), (-1.3, 1.3), (-1.1, 1.1)$ and $(-1.1, 1.1)$ respectively.

Figure 4.31 is the diagram of the mapping $\bar{T}$ with $k = 1.5, s = 1.05$ for several initial points. The invariant curves surrounding the origin, 2-periodic islands, and in addition to the origin two other fixed points near it can been seen clearly. Around these two fixed points, there are two families of invariant curves. Also in this figure, there is a figure eight-like invariant curve passing near the origin. For $k = 1.5, s = 1.1$ with initial point $P_0 = (0.1, 0.0)$, the same curve exists (see Fig. 4.32). To confirm that it is an invariant curve, the variation of the corresponding positive LCN $\gamma^N(P_0)$ with the iteration number N up to $N = 10^6$ is calculated and plotted in Fig. 4.33. Apparently $\gamma^N(P_0)$ is a decreasing function, tending to zero. The above results indicate that the point $P_0 = (0.1, 0.0)$ is indeed on an invariant curve. In fact, the above figure eight-like curve is the invariant curve closing two homoclinic orbits of the origin.

For $k = -1.5$, the same explorations along the x-axis as for $k = 1.5$ have been carried out. For $k = -1.5$ and $s = 1.02, 1.05, 1.08, 1.10$ respectively, no invariant curve has been found for $\bar{T}$ in the vicinity of the origin, i.e. the

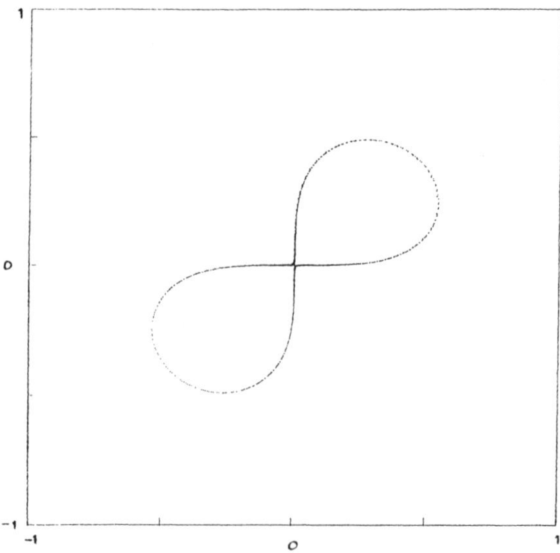

Fig. 4.32   Invariant curve of mapping $\bar{T}$ for $k = 1.5, s = 1.1$ with initial point $P_0 = (0.1, 0.0)$. From Sun & Yan (1988).

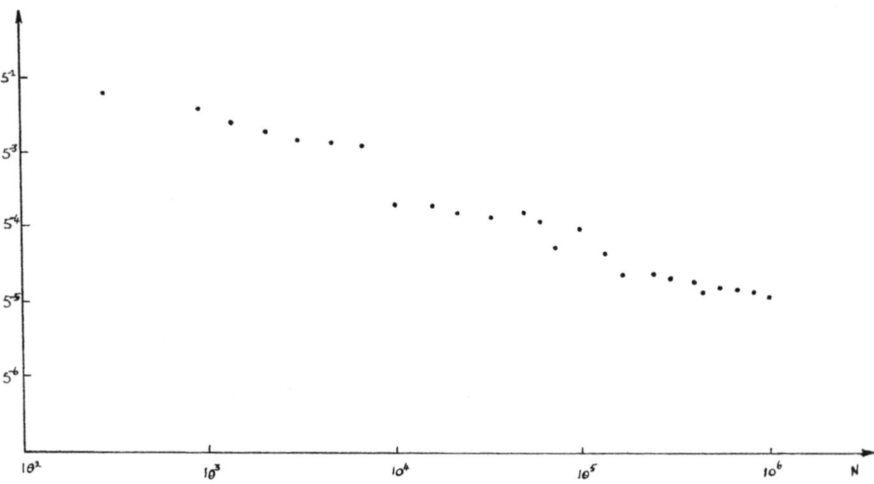

Fig. 4.33   Variation of the positive LCN's $\gamma^N(P_0)$ of the mapping $\bar{T}$ for $k = 1.5, s = 1.1$ as function of the iteration number N for initial point $P_0 = (0.1, 0.0)$. From Sun & Yan (1988).

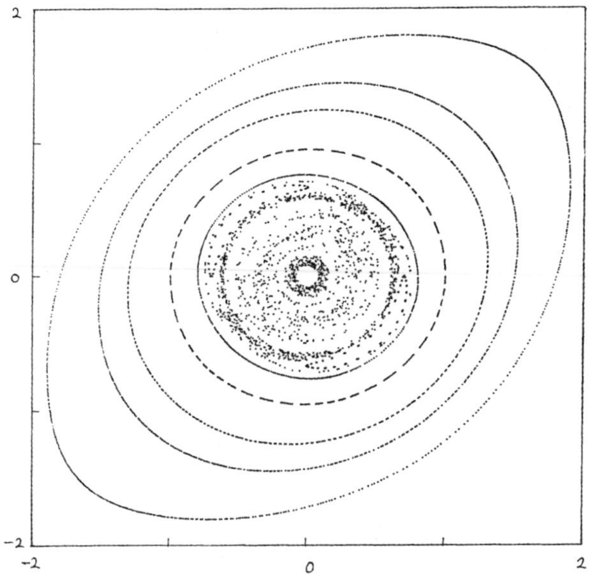

Fig. 4.34   The mapping $\bar{T}$ for $k = -1.5, s = 1.05$. From Sun & Yan (1988).

whole region in the vicinity of the origin is chaotic. This is perhaps because for $k = -1.5$ the angle $\varphi_n = (x_n^2 + y_n^2)^k$ near the origin rotates very quickly, implying the sensitive dependence of orbit on the initial condition and thus chaotic motion. The ordered region is found to be an annulus surrounding the origin, and there are 2-periodic islands near the inner boundary of the ordered region. The area of the ordered regions decreases with increasing s and the corresponding ranges on the x-axis are: $0.6 \leq |x| \leq 2.4$, $0.7 \leq |x| \leq 1.8$, $0.9 \leq |x| \leq 1.6$ and $1.0 \leq |x| \leq 1.5$ for $s = 1.02, 1.05, 1.08$ and $s = 1.10$ respectively. As these ranges tell, when s increases the corresponding annulus shrinks from both inner and outer boundaries. And the farther the invariant curve is from the origin the more oblate it becomes (see Fig. 4.34). The behavior of the mapping $\bar{T}$ with $k = 1.5$ and $k = -1.5$ is very different.

4.4.2   *Three-dimensional mapping T*

For the perturbed extension of mapping $\bar{T}$, i.e. the mapping T, the parameters k, s will be kept fixed below as $s = 1.05$, $k = 1.5$ and $k = -1.5$. Then the behaviors of mapping T with different perturbation parameters

A, B, C, D will be investigated. As in Sec. 4.3, two methods, i.e. the LCN's method and the slice-cutting method will be used.

The slices are defined by

$$|z_n - Q| < 0.01, \quad Q = \frac{(j-5)\pi}{4},$$

$$-\infty \le x_n \le +\infty, \quad -\infty \le y_n \le +\infty,$$

$$(j = 1, 2, \ldots, 9, \quad n = 1, 2, \ldots, N),$$

where N is the number of iterations. And as j increases from 1 to 9, the slices are seen from left to right and from top to bottom in figures.

For $k = 1.5, s = 1.05$ and $A = B = C = 0.03$, $D = -0.04$, one has

$$\Delta z = z_{n+1} - z_n = C \cos(x_{n+1} + y_{n+1}) + D < 0,$$

i.e. z varies between $-\pi$ and $+\pi$. According to the results obtained in the previous section (Sec. 4.3), some two-dimensional invariant manifolds (invariant tubes) may be generated in the ordered regions of the mapping $\bar{T}$. The slices and LCN's both show that in the vicinity of a curved axis near the z-axis there exist no two-dimensional manifolds and the region is chaotic, but there is a family of two-dimensional invariant manifolds distant from the z-axis (see Fig. 4.35). That is, in the ordered region of mapping $\bar{T}$ the part near the origin is destroyed by the perturbation introduced in mapping T more easily than the part farther from the origin. This is different from the case in which the unperturbed two-dimensional mappings possess an elliptic fixed point. This may be because that near the fixed point the unperturbed two-dimensional mapping is very close to the linear mapping with hyperbolic fixed point which is unstable at the fixed point, and under the perturbation the stable and unstable manifolds will intersect each other, leading to the chaotic motion.

Moreover, from Fig. 4.35 one can see that there are anti-symmetries between the slices of $j = 1, 2, 3, 4$ and $j = 5, 6, 7, 8$ (slice 1 and slice 9 are the same). This is because when z is replaced by $z + \pi$, x by $-x$ and y by $-y$ the mapping T will not be changed. It can also be seen that the outermost diagram of each slice corresponding to the orbit with initial point $P_0 = (1.2, 0, 0)$ seems to be on an invariant tube. In fact it is in a chaotic region, but the image points diffuse very slowly around a tube, as four hyperbolic structures can be seen there and the spectrum of the LCN's $\gamma_i^N(P_0)$ $(i = 1, 2, 3; N = 10^5)$ is $(0.1570472 \times 10^{-1}, -0.3597040 \times 10^{-4}, -0.1566875 \times 10^{-1})$ whose maximum does not tend to zero, although quite small.

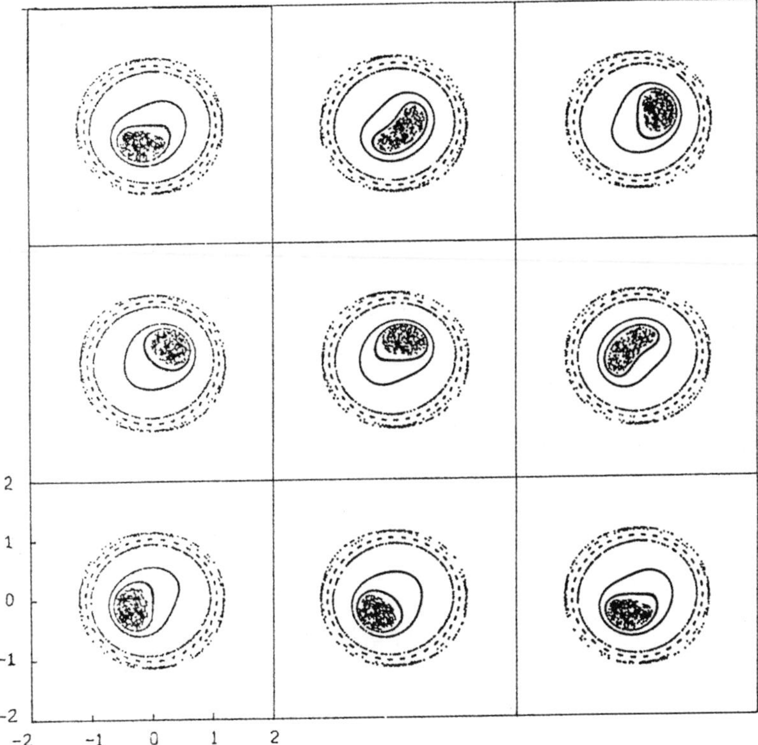

Fig. 4.35   Sections of mapping T for $k = 1.5, s = 1.05, A = B = C = 0.03, D = -0.04$. From Sun & Yan (1988).

For $k = 1.5$ and $A = B = C = 0.03, D = 0$, one has

$$\Delta z = z_{n+1} - z_n = C \cos(x_{n+1} + y_{n+1}).$$

It will not always vary monotonously. The mode of the variation of Δz depends on the values of $|x_{n+1} + y_{n+1}|$. Evidently, when $|x_{n+1} + y_{n+1}|$ is small, the variation of Δz will be monotonous, thus the two-dimensional invariant manifolds may be expected. For $\sqrt{x^2 + y^2} = 1.1$, one has $|x + y| < \pi/2$, consequently $\cos(x + y)$ is always positive. This explains why for $A = B = C = 0.03, D = 0$ there exists an outermost invariant tube passing the point $(1.1, 0, 0)$, as its slice parallel to the (x, y)-plane is nearly a circle with radius $d = 1.1$ (see Fig. 4.36).

Figure 4.36 displays the slices of several orbits of the mapping T with $k = 1.5, A = B = C = 0.03, D = 0$. It clearly exhibits the existence of invariant tubes between two cylindroids, of a chaotic region, and of four

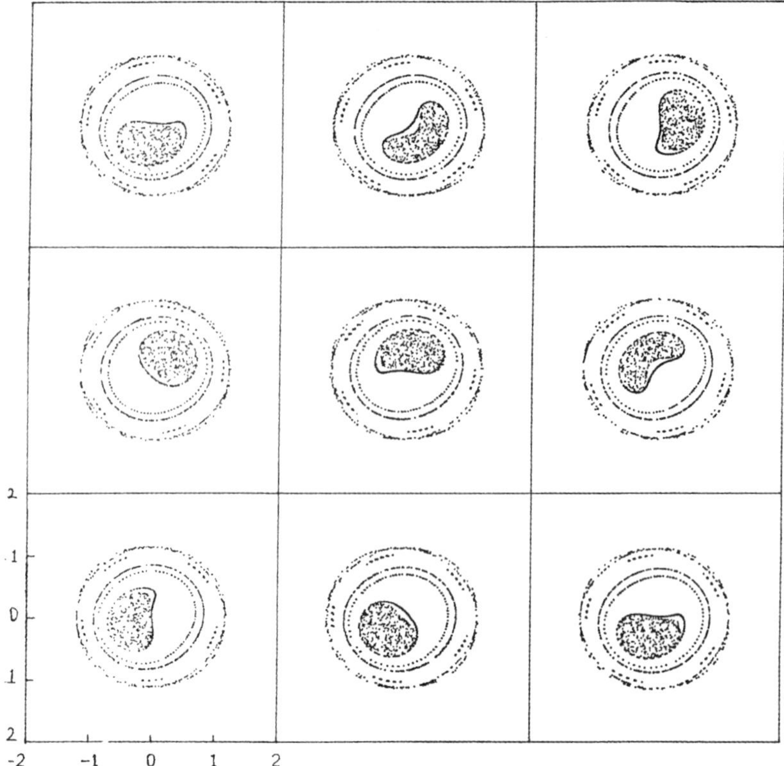

Fig. 4.36 Sections of mapping T for $k = 1.5, s = 1.05, A = B = C = 0.03, D = 0.0$. From Sun & Yan (1988).

hyperbolic structures. It can also be seen that the chaotic region is larger than that in Fig. 4.35.

For $A = B = C = 0.003$, $D = -0.004$ and $D = 0$, the LCN's of orbits initialized along the line $x = y$, $z = 0$ are calculated. The results show that for both cases the points $(0, 0, 0)$ and $(0.1, 0, 0)$ are both in the chaotic region. The corresponding LCN's $\gamma_i^N(P_0)$ $(i = 1, 2, 3)$ oscillate and do not tend to zero, and for each case of $D = -0.004$ and $D = 0$, the behaviors and the values of two sets of $\gamma_i^N(P_0)$ $(i = 1, 2, 3)$ for $P_0 = (0, 0, 0)$ and $P_0 = (0.1, 0, 0)$ are almost the same. It shows that the invariant curve passing the origin of the mapping $\bar{T}$ with $k = 1.5, s = 1.05$ is destroyed by the perturbed extension. It becomes a chaotic zone that is also the chaotic zone generated on the outer invariant curves surrounding two fixed points other than the origin (see Fig. 4.31), because they have the same values of

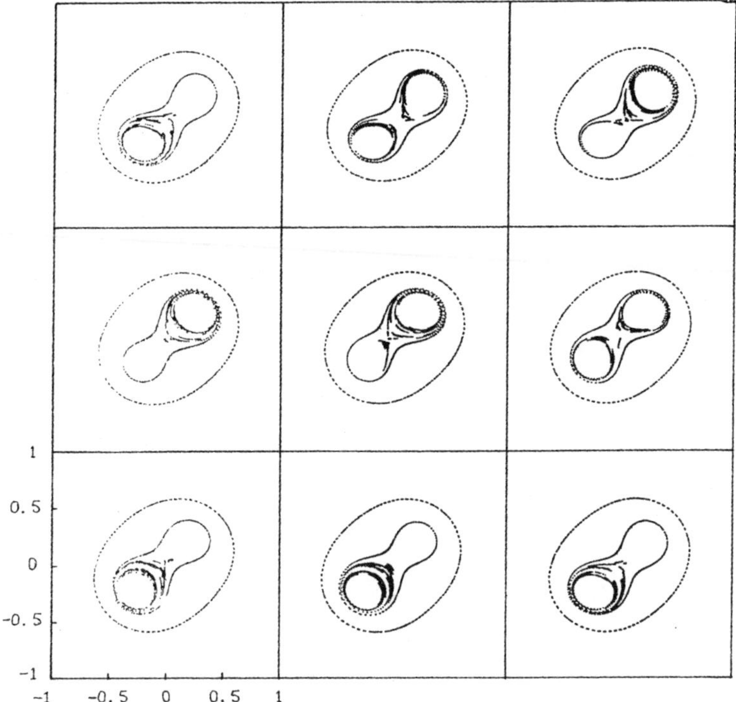

Fig. 4.37   Sections of mapping T for $k = 1.5, s = 1.05, A = B = C = 0.003, D = -0.004$. From Sun & Yan (1988).

LCN's. The image points will diffuse slowly around a tube with a shape like a dumbbell (see Fig. 4.37 and Fig. 4.38).

The outermost boundaries of the ordered regions for the two cases $D = -0.004$ and $D = 0$ all pass the point $(1.1, 0)$ in the plane $z = 0$. This implies that the areas of the ordered regions for the two cases are the same. For $A = B = C = 0.003, D = -0.004$, the 2-periodic invariant tubes with a crescent shape (see Fig. 4.39) can be seen, which is generated by perturbed extension on the 2-periodic invariant curves of the mapping $\bar{T}$.

The case $k = -1.5$ is very different from $k = 1.5$. For this case, the same exploration of LCN's along the x-axis as for $k = 1.5$ have been carried out. It is found that for $A = B = C = 0.03, D = -0.04$ and $D = 0$, the ranges of ordered regions on the positive x-axis in the plane $z = 0$ are $1.0 \le x \le 1.6$ and $1.0 \le x \le 1.7$ respectively, and in the vicinity of the z-axis the regions are all chaotic for both of the above cases (see Fig. 4.40 and Fig. 4.41). When the perturbation parameters A, B, C, D are decreased

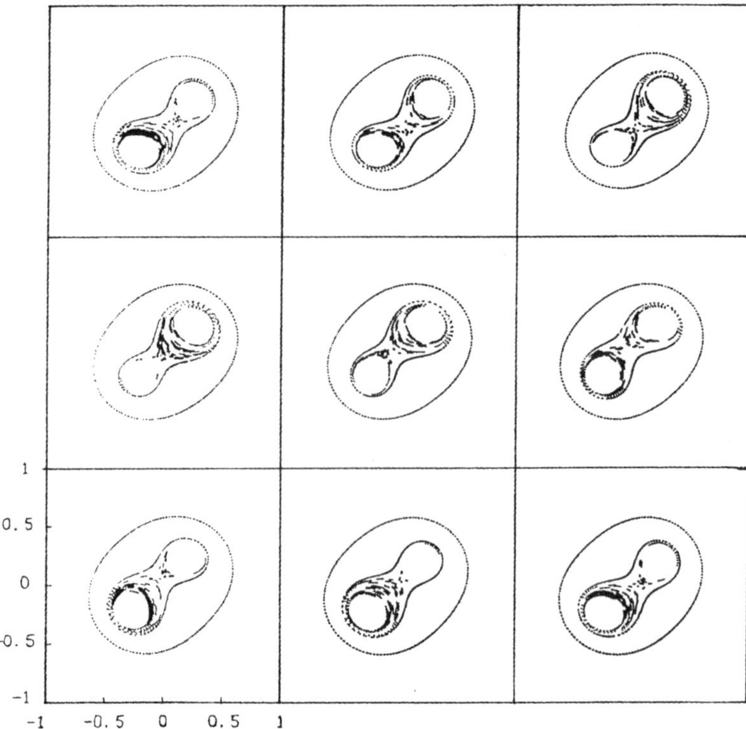

Fig. 4.38 Sections of mapping T for $k = 1.5, s = 1.05, A = B = C = 0.003, D = 0.0$. From Sun & Yan (1988).

to $A = B = C = 0.003$, $D = -0.004$ and $D = 0$, the corresponding ranges of ordered regions on the positive x-axis in the plane of $z = 0$ become $0.7 \leq x \leq 1.8$ and $0.7 \leq x \leq 1.8$, respectively. These results are similar to those for $k = 1.5$, i.e. the magnitude of the ordered region increases with decreasing perturbation parameters and for quite small perturbation parameters the size of ordered region is insensitive to the parameter D.

Something a little strange is also found: For $A = B = C = 0.03, D = 0$, the area of ordered region is larger than that for $A = B = C = 0.03, D = -0.04$. The outmost invariant tube is quite remote from the z-axis (see Figs. 4.40 and 4.41). In the case $D = 0$, $|x_{n+1} + y_{n+1}|$ does not remain small, hence, it is not necessary that the variation of Δz will be monotonous. This implies that for the existence of invariant tubes in the perturbed extension of area-preserving mappings, the monotonous variation of Δz is only a sufficient condition, but not a necessary condition.

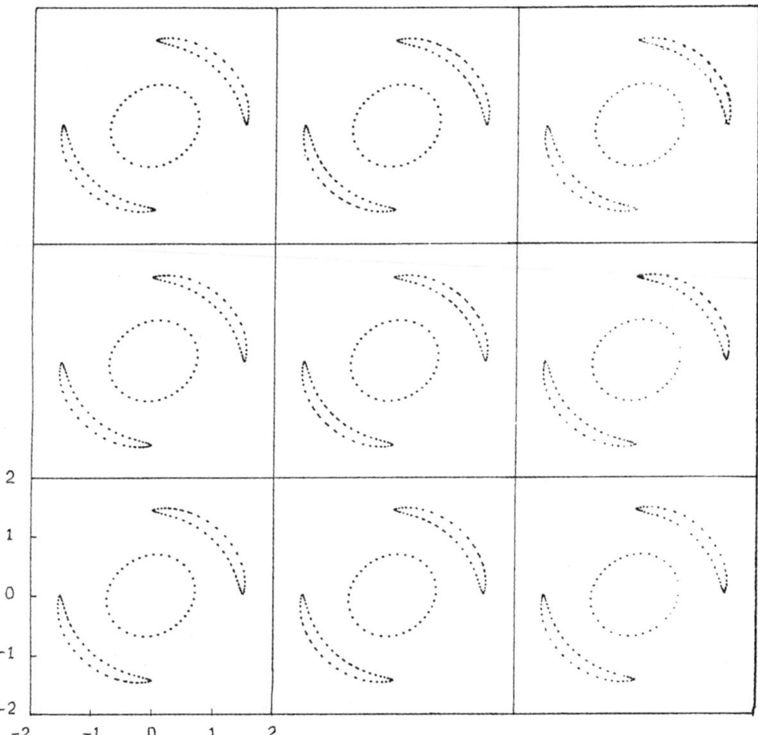

Fig. 4.39   Sections of mapping T for $k = 1.5, s = 1.05, A = B = C = 0.003, D = -0.004$. From Sun & Yan (1988).

The investigations presented in this section reveal:

(a) The two-dimensional invariant manifolds of three-dimensional perturbed extension can only be generated in the subset of the invariant manifolds in the two-dimensional measure-preserving mappings.

(b) The ordered region near the hyperbolic fixed point is destroyed by the perturbed extension more easily than the one distant from the fixed point. This is apparently different from the case with an elliptic fixed point.

Another class of perturbed extension of measure-preserving mappings that have a parabolic fixed point have also been investigated (Zhang & Sun, 1989), and the results are very similar to the perturbed extension of hyperbolic twist mappings discussed in this section where the mapping is characterized by the existence of a parabolic fixed point.

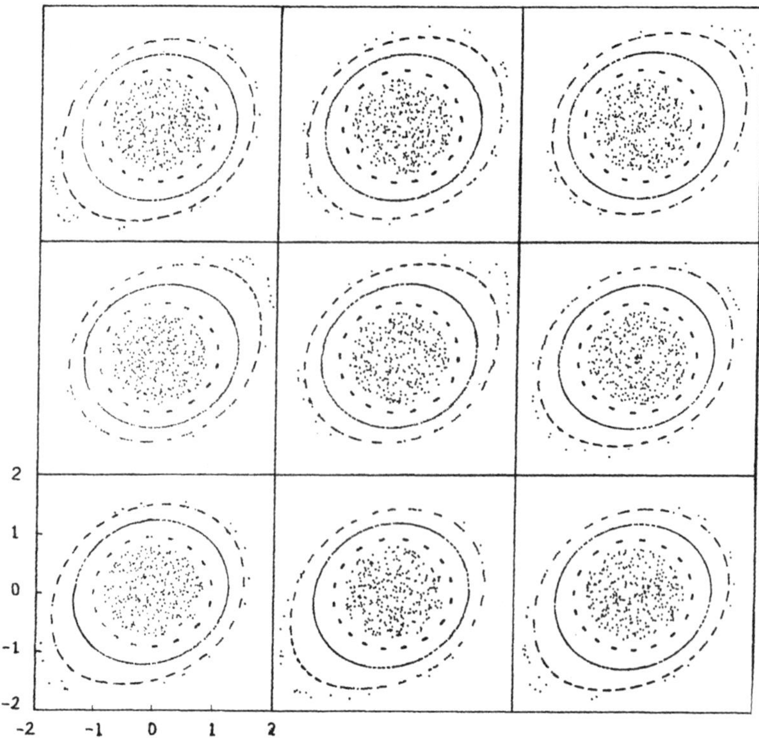

Fig. 4.40 Sections of mapping T for $k = -1.5, s = 1.05, A = B = C = 0.03, D = -0.04$. From Sun & Yan (1988).

4.5 KS entropy of volume-preserving mapping

In previous section (Sec. 4.4) the three-dimensional extension of the two-dimensional mapping have been discussed. The ordered (tube) region and chaotic region in the phase space are found in the three-dimensional mappings. As pointed out in Sec. 4.2, KS entropy is a measure of chaotic motion. And also in Sec. 4.2 the KS entropy of the generalized standard mappings was calculated and an analytical expression of estimating the KS entropy was obtained and found to be consistent quite well with numerical results. In this section, the stochasticity (or KS entropy) of a three-dimensional extension of the two-dimensional mapping will be studied in detail.

The mapping model adopted in this section is the three-dimensional mapping T introduced in Sec. 4.3 as Eq. (4.22). For convenience, the

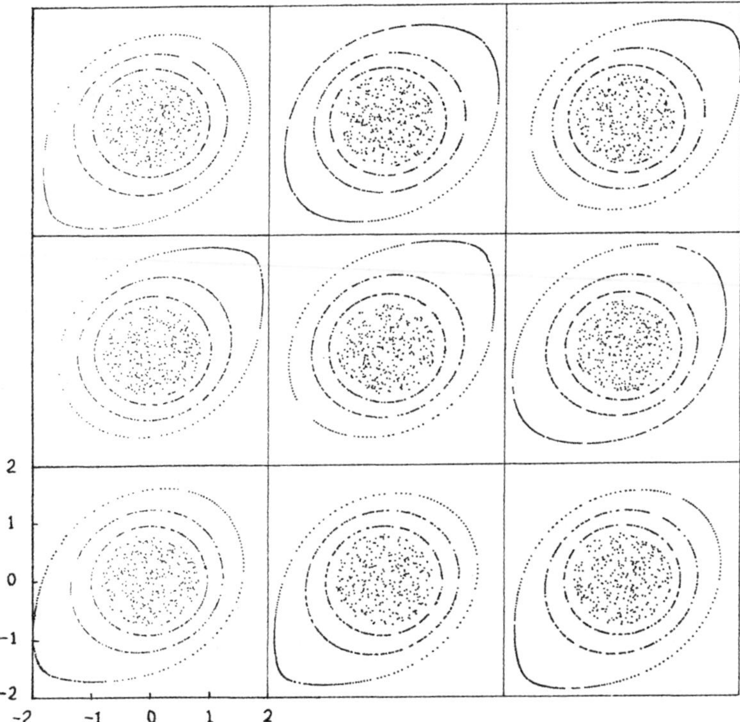

Fig. 4.41   Sections of mapping T for $k = -1.5, s = 1.05, A = B = C = 0.03, D = 0.0$.
From Sun & Yan (1988).

mapping is rewritten below, and it was denoted by ϕ in this section:

$$\phi: \begin{cases} x_{n+1} = x_n + y_n + B\sin z_n, \\ y_{n+1} = y_n + A\sin x_{n+1}, \\ z_{n+1} = z_n + C\sin y_{n+1} + D, \end{cases} \quad (\text{mod } 2\pi), \qquad (4.33)$$

with B, C, D being three positive parameters, $A = -1.5$. In an n-dimensional dynamical system, there are n LCN's, hence, one needs to calculate $n, n-1, \ldots, 1$ order LCN's. For the three-dimensional mapping, it needs to calculate the spectrum of three LCN's. This makes the calculation more complicated than in the two-dimensional system where it is enough to calculate the maximum of LCN's.

Based on the definition and properties of LCN's (see Sec. 4.2), the LCN's of the three-dimensional mapping ϕ will be investigated. For linear independent vectors $\boldsymbol{w}^{(1)}, \ldots, \boldsymbol{w}^{(k)}$ in the n-dimensional tangent space $T_P(M)$, the volume consisted of them is

$$V^{(k)}\left(\boldsymbol{w}^{(1)},\ldots,\boldsymbol{w}^{(k)}\right) = \sqrt{G\left(\boldsymbol{w}^{(1)},\ldots,\boldsymbol{w}^{(k)}\right)}, \; k = 1,2,\ldots,n, \qquad (4.34)$$

where $G\left(\boldsymbol{w}^{(1)},\ldots,\boldsymbol{w}^{(k)}\right)$ is the Gram determinant

$$G\left(\boldsymbol{w}^{(1)},\ldots,\boldsymbol{w}^{(k)}\right) = \begin{vmatrix} (\boldsymbol{w}^{(1)},\boldsymbol{w}^{(1)}) & (\boldsymbol{w}^{(1)},\boldsymbol{w}^{(2)}) & \cdots & (\boldsymbol{w}^{(1)},\boldsymbol{w}^{(k)}) \\ (\boldsymbol{w}^{(2)},\boldsymbol{w}^{(1)}) & (\boldsymbol{w}^{(2)},\boldsymbol{w}^{(2)}) & \cdots & (\boldsymbol{w}^{(2)},\boldsymbol{w}^{(k)}) \\ \vdots & \vdots & \vdots & \vdots \\ (\boldsymbol{w}^{(k)},\boldsymbol{w}^{(1)}) & (\boldsymbol{w}^{(k)},\boldsymbol{w}^{(2)}) & \cdots & (\boldsymbol{w}^{(k)},\boldsymbol{w}^{(k)}) \end{vmatrix},$$

$$(4.35)$$

with $(\,,\,)$ denoting the inner product in $T_P(M)$. If $\bar{\boldsymbol{w}}^{(1)},\ldots,\bar{\boldsymbol{w}}^{(k)}$ is a set of orthogonal vectors induced by $\boldsymbol{w}^{(1)},\ldots,\boldsymbol{w}^{(k)}$, then

$$G\left(\boldsymbol{w}^{(1)},\ldots,\boldsymbol{w}^{(k)}\right) = \left(\bar{\boldsymbol{w}}^{(1)},\bar{\boldsymbol{w}}^{(1)}\right)\left(\bar{\boldsymbol{w}}^{(2)},\bar{\boldsymbol{w}}^{(2)}\right)\cdots\left(\bar{\boldsymbol{w}}^{(k)},\bar{\boldsymbol{w}}^{(k)}\right),$$
$$k = 1,2,\ldots,n. \qquad (4.36)$$

For three-dimensional mapping, one needs to calculate the spectrum of LCN's, i.e. one order LCN's $X_1(P), X_2(P), X_3(P)$ (see Sec. 4.2 for definition). According to the relation between LCN's of k order and one order, i.e. Eq. (4.7), one needs to calculate the following expressions

$$X_1(P) = \lim_{t\to\infty} \frac{1}{t} \ln \|D\phi_P^t(\boldsymbol{w}^{(1)})\|,$$

$$X_1(P) + X_2(P) = \lim_{t\to\infty} \frac{1}{t} \ln V^{(2)}\left(D\phi_P^t\left(\boldsymbol{w}^{(1)}\right), \; D\phi_P^t\left(\boldsymbol{w}^{(2)}\right)\right),$$

$$X_1(P) + X_2(P) + X_3(P) =$$

$$(4.37)$$

$$\lim_{t\to\infty} \frac{1}{t} \ln V^{(3)}\left(D\phi_P^t\left(\boldsymbol{w}^{(1)}\right), \; D\phi_P^t\left(\boldsymbol{w}^{(2)}\right), \; D\phi_P^t\left(\boldsymbol{w}^{(3)}\right)\right),$$

where $\boldsymbol{w}^{(1)}, \boldsymbol{w}^{(2)}, \boldsymbol{w}^{(3)}$ are linear independent vectors in $T_P(M)$. Let

$$\boldsymbol{w} = \begin{pmatrix} u_1 \\ v_1 \\ w_1 \end{pmatrix}, \qquad D\phi_P^i(\boldsymbol{w}) = \begin{pmatrix} u_{i+1} \\ v_{i+1} \\ w_{i+1} \end{pmatrix},$$

where i denotes the iteration number for each orbit. Recall the mapping ϕ and its tangent mapping, one has

$$u_{i+1} = u_i + v_i + (B\cos z_i)w_i,$$
$$v_{i+1} = (A\cos x_{i+1})u_i + (1 + A\cos x_{i+1})v_i + (AB\cos x_{i+1}\cos z_i)w_i,$$

$$w_{i+1} = (AC \cos y_{i+1} \cos x_{i+1})u_i + C \cos y_{i+1}(1 + A \cos x_{i+1})v_i$$
$$+ (1 + ABC \cos x_{i+1} \cos y_{i+1} \cos z_i)w_i. \tag{4.38}$$

When B, C are large enough, one has

$$w_{i+1} \approx (C \cos y_{i+1})v_{i+1},$$

so $|w_{i+1}| \gg |v_{i+1}|$, and thus one has approximately

$$u_{i+1} = u_i + (B \cos z_i)w_i,$$
$$v_{i+1} = (A \cos x_{i+1})u_i + (AB \cos x_{i+1} \cos z_i)w_i,$$
$$w_{i+1} = (AC \cos y_{i+1} \cos x_{i+1})u_i + (ABC \cos x_{i+1} \cos y_{i+1} \cos z_i)w_i$$
$$= (C \cos y_{i+1})v_{i+1} = (AC \cos x_{i+1} \cos y_{i+1})u_{i+1}.$$

In the same way there is also $|w_{i+1}| \gg |u_{i+1}|$, therefore, one has approximately

$$\left\| D\phi_P^i\left(\boldsymbol{w}\right) \right\| = |w_{i+1}|, \tag{4.39}$$

and

$$w_{i+1} = (ABC \cos x_{i+1} \cos y_{i+1} \cos z_i)\, w_i$$
$$= \left[\prod_{k=1}^{i} (ABC \cos x_{k+1} \cos y_{k+1} \cos z_k) \right] w_1.$$

Hence,

$$X_1(P) = \lim_{i \to \infty} \frac{1}{i} \ln \left\| D\phi_P^i\left(\boldsymbol{w}^{(1)}\right) \right\|$$
$$\approx \lim_{i \to \infty} \frac{1}{i} \ln \prod_{k=1}^{i} |ABC \cos x_{k+1} \cos y_{k+1} \cos z_k| \cdot \left| w_1^{(1)} \right|.$$

From the results in Sec. 4.3, when B, C are large, the mapping ϕ is chaotic or "ergodic". According to the ergodic theorem, the limit

$$\lim_{i \to \infty} \frac{1}{i} \sum_{k=1}^{i} \ln |ABC \cos x_{k+1} \cos y_{k+1} \cos z_k|,$$

can be approximately replaced by

$$\frac{1}{8\pi^3} \int_0^{2\pi} \int_0^{2\pi} \int_0^{2\pi} \ln |ABC \cos x \cos y \cos z|\, \mathrm{d}x\mathrm{d}y\mathrm{d}z.$$

Hence,

$$X_1(P) \approx \ln \frac{|ABC|}{8}.$$

The $X_2(P)$ can be calculated using Eq. (4.37)

$$X_1(P) + X_2(P) = \lim_{i \to \infty} \frac{1}{i} \ln V^{(2)}\left(D\phi_P^i\left(\boldsymbol{w}^{(1)}\right), D\phi_P^i\left(\boldsymbol{w}^{(2)}\right)\right).$$

From Eqs. (4.34) and (4.35), one has

$$V^{(2)}\left(D\phi_P^i\left(\boldsymbol{w}^{(1)}\right), D\phi_P^i\left(\boldsymbol{w}^{(2)}\right)\right)$$
$$= \left\|D\phi_P^i\left(\boldsymbol{w}^{(1)}\right)\right\|$$
$$\cdot \left\|D\phi_P^i\left(\boldsymbol{w}^{(2)}\right) - \frac{\left(D\phi_P^i\left(\boldsymbol{w}^{(1)}\right), D\phi_P^i\left(\boldsymbol{w}^{(2)}\right)\right)}{\left\|D\phi_P^i\left(\boldsymbol{w}^{(1)}\right)\right\|^2} D\phi_P^i\left(\boldsymbol{w}^{(1)}\right)\right\|,$$

then

$$X_2(P) = \lim_{i \to \infty} \frac{1}{i}$$
$$\times \ln \left\|D\phi_P^i\left(\boldsymbol{w}^{(2)}\right) - \frac{\left(D\phi_P^i\left(\boldsymbol{w}^{(1)}\right), D\phi_P^i\left(\boldsymbol{w}^{(2)}\right)\right)}{\left\|D\phi_P^i\left(\boldsymbol{w}^{(1)}\right)\right\|^2} D\phi_P^i\left(\boldsymbol{w}^{(1)}\right)\right\|.$$

Approximately,

$$D\phi_P^i\left(\boldsymbol{w}^{(1)}\right) = \begin{pmatrix} u_{i+1}^{(1)} \\ v_{i+1}^{(1)} \\ w_{i+1}^{(1)} \end{pmatrix} \approx \begin{pmatrix} 0 \\ 0 \\ w_{i+1}^{(1)} \end{pmatrix},$$

$$D\phi_P^i\left(\boldsymbol{w}^{(2)}\right) = \begin{pmatrix} u_{i+1}^{(2)} \\ v_{i+1}^{(2)} \\ w_{i+1}^{(2)} \end{pmatrix} \approx \begin{pmatrix} 0 \\ 0 \\ w_{i+1}^{(2)} \end{pmatrix}.$$

Therefore, when $i \to \infty$,

$$\left\|D\phi_P^i\left(\boldsymbol{w}^{(2)}\right) - \frac{\left(D\phi_P^i\left(\boldsymbol{w}^{(1)}\right), D\phi_P^i\left(\boldsymbol{w}^{(2)}\right)\right)}{\left\|D\phi_P^i\left(\boldsymbol{w}^{(1)}\right)\right\|^2} D\phi_P^i\left(\boldsymbol{w}^{(1)}\right)\right\| \to 0,$$

and thus $X_2(P) < 0$. Furthermore, because $X_1(P) \geq X_2(P) \geq X_3(P)$, one has $X_3(P) < 0$.

The Pesin's formula of KS entropy ($h = \int_M \rho d\mu$, with ρ being the sum of all positive LCN's) has been given in Eq. (4.9) in Sec. 4.2, but in this section for 3-dimensional mapping, $d\mu$ is

$$d\mu = \frac{dxdydz}{8\pi^3}.$$

As $\rho(P) = X_1(P) \approx \ln \frac{|ABC|}{8}$, one has

$$h = \int_M X_1(P)\mathrm{d}\mu \approx \ln \frac{|ABC|}{8}. \qquad (4.40)$$

Theoretically, the LCN's can be calculated using Eq. (4.37), but in practice it does not work in numerical calculations. In the chaotic region, the norms of tangent vectors generally increase exponentially and the angle between two tangent vectors becomes very small quickly, resulting in overflow and termination of the calculation by computer. To overcome this difficulty, one can use the Gram–Schmidt orthonormalization.

Choose p orthonormal tangent vectors $\boldsymbol{w}_0^{(1)}, \boldsymbol{w}_0^{(2)}, \ldots, \boldsymbol{w}_0^{(p)}$ as the initial vectors and do the orthonormalization after every s iterations, that is, calculate the following formula recurrently

$$\alpha_k^{(1)} = \left\| D\phi_{P_{(k-1)s}}^s \left(\boldsymbol{w}_{k-1}^{(1)} \right) \right\|,$$

$$\boldsymbol{w}_k^{(1)} = \frac{D\phi_{P_{(k-1)s}}^s \left(\boldsymbol{w}_{k-1}^{(1)} \right)}{\alpha_k^{(1)}},$$

and

$$\alpha_k^{(j)} = \left\| D\phi_{P_{(k-1)s}}^s \left(\boldsymbol{w}_{k-1}^{(j)} \right) - \sum_{i=1}^{j-1} \left(\boldsymbol{w}_k^{(i)}, D\phi_{P_{(k-1)s}}^s \left(\boldsymbol{w}_{k-1}^{(j)} \right) \right) \boldsymbol{w}_k^{(i)} \right\|,$$

$$\boldsymbol{w}_k^{(j)} = \frac{1}{\alpha_k^{(j)}} \left[D\phi_{P_{(k-1)s}}^s \left(\boldsymbol{w}_{k-1}^{(j)} \right) - \sum_{i=1}^{j-1} \left(\boldsymbol{w}_k^{(i)}, D\phi_{P_{(k-1)s}}^s \left(\boldsymbol{w}_{k-1}^{(j)} \right) \right) \boldsymbol{w}_k^{(i)} \right].$$

$$(j = 2, 3, \ldots, p; \quad 2 \leq p \leq n)$$

According to Eqs. (4.34), (4.36) and (4.37) and taking into account the linearity and the law of combination, one gets

$$X_j(P) = \lim_{k \to \infty} \frac{1}{ks} \sum_{i=1}^{k} \ln \alpha_i^{(j)}. \quad (j = 1, 2, \ldots, p; \quad 1 \leq p \leq n).$$

For each mapping (corresponding to a set of parameters), the KS entropy h is calculated by Monte Carlo method in a similar way introduced in Sec. 4.2, i.e. take 10 initial points randomly on the torus $M = \{(x, y, z) \pmod{2\pi}, z = 0\}$, iterate each orbit up to 10,000 times, and estimate $h = \int_M \rho(P)\mathrm{d}\mu$ using the following formula

$$h \approx \frac{1}{10} \sum_{j=1}^{10} \rho(P_j).$$

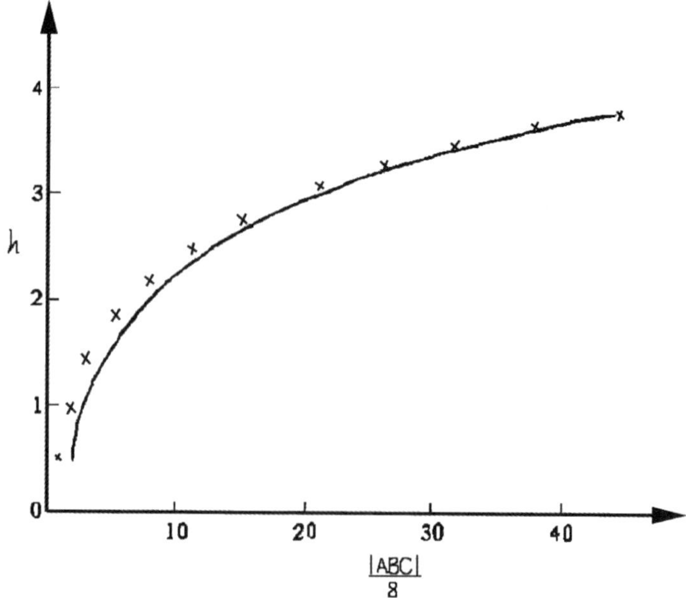

Fig. 4.42 Variation of h with B, C. Crosses are numerical results and the curve is from the theoretical expression. From Sun (1983b).

Totally, the LCN's of 13 sets of parameters and KS entropy are calculated. Some results are shown in Fig. 4.42.

These results show

(1) In each of the 13 specific mappings of mapping ϕ, the LCN's calculated from the randomly selected 10 orbits are nearly the same non-zero values, implying that the chaotic regions are connected.

(2) Among three LCN's $X_1(P), X_2(P), X_3(P)$, only $X_1(P) > 0$ for large B, C, while the other two are negative. For large B, C, the approximate expression of h is $\ln \frac{|ABC|}{8}$, and it agrees with the numerical results quite well (see Fig. 4.42). From the results in this section and in Sec. 4.2, one finds that the entropies of the two-dimensional generalized standard mapping and the three-dimensional extension of the standard mapping have the similar expression, as the logarithm function of parameters.

4.6 Attractor in three-dimension mapping

In previous sections Secs. 4.3 and 4.4, the perturbed extension of the two-dimensional area-preserving mappings that have different kinds of fixed points, have been discussed. For invariant manifolds and chaotic regions in the two-dimensional area-preserving mappings and their perturbed extensions of three-dimensional mappings, some relations between these two cases have been found.

In dissipative dynamical systems there exist "attractors". In this section, the attractors in the perturbed extension of the two-dimensional dissipative mappings will be discussed. The questions as follows will be answered. How the attractors will change? Whether and under what conditions the attractors of the original two-dimensional systems will lead to attractors in the extended three-dimensional systems? What relations exist for their dimensions?

4.6.1 *Mapping model*

The mapping is based on the well-known Hénon mapping as below

$$\begin{cases} x_{i+1} = y_i + 1 - Ax_i^2, \\ y_{i+1} = Bx_i, \end{cases} \tag{4.41}$$

where A, B are positive parameters. By adding perturbing terms, this mapping is extended to three-dimension:

$$T: \quad \begin{cases} x_{i+1} = y_i + 1 - Ax_i^2 - C\cos z_i, \\ y_{i+1} = Bx_i + D\sin z_i, \\ z_{i+1} = z_i + E\sin y_{i+1} + F. \quad (\text{mod } 2\pi) \end{cases} \tag{4.42}$$

The Jacobian of this mapping is a constant

$$\left| \frac{\partial (x_{i+1}, y_{i+1}, z_{i+1})}{\partial (x_i, y_i, z_i)} \right| = -B, \tag{4.43}$$

and consequently, this mapping is a dissipative system.

It is well-known that for different values of parameters A, B in the Hénon mapping Eq. (4.41) different attractors may appear. For example, when $A = 0.2, B = 0.3$ the mapping has a stable fixed point; when $A = 0.65, B = 0.3$ and $A = 2.6577, B = 0.3$, it has respectively a period-two and a period-three attractor, but for $A = 1.4, B = 0.3$, there exist the unstable fixed point and the strange attractor (see for example Hénon (1976), Feit (1978)).

To discuss the behavior of attractors in three-dimensional mapping as the perturbed extension of the Hénon mapping, three sets of values of parameters A, B as mentioned above are taken, and other parameters C, D, E, F (especially F is a parameter controlling the variation of Z_n)

are set to be small. In addition, when C, D, E, F are sufficiently small, an approximate expression of the invariant curves of the mapping T, namely

$$x = \frac{B-1}{2A} \pm \frac{\sqrt{(1-B)^2 + 4A}}{2A} \pm \frac{C\cos z + D\sin z}{\sqrt{(1-B)^2 + 4A}}, \qquad (4.44)$$

$$y = Bx + D\sin z,$$

can be obtained. In fact, the invariant curves pass closely by the fixed points of the mapping defined by Eq. (4.44),

$$x = \frac{1}{2A}\left[(B-1) \pm \sqrt{(1-B)^2 + 4A}\right],$$

$$y = Bx.$$

These points are real for $A > -(1-B)^2/4$. In this case, one of them is always linearly unstable. The other one is also unstable when $A > 3(1-B)^2/4$, but it is stable for $A < 3(1-B)^2/4$ (Feit, 1978). Using these conclusions, one can choose the parameters A, B to decide the stability of the fixed points.

4.6.2 *LCN's and fractal dimension*

LCN's can be used to indicate the ordered region and the chaotic region for the conservative dynamical system. For dissipative systems the different attractors can be characterized by the spectrum of LCN's. For instance, in three-dimensional system an attractor with all negative LCN's is a fixed point, i.e. with an LCN's spectrum $\gamma_i^N(P)$ ($i = 1, 2, 3$ and N is the iteration number) of $(-,-,-)$. A limit cycle attractor has an LCN's spectrum of $(0, -, -)$, a two-torus attractor has an LCN's spectrum of $(0, 0, -)$ and a strange attractor has an LCN's spectrum of $(+, 0, -)$ (Froehlin *et al.*, 1981). Moreover, the LCN's spectrum can roughly give a measure of the dimension of an attractor. Evidently, the fixed point is zero-dimensional, the limit cycle is one-dimensional and the two-torus has two dimensions. It was also pointed out (Hénon, 1976) that a strange attractor in the two-dimensional dynamical system appears to be the product of a one-dimensional manifold by a cantor set. To determine the dimension of strange attractor, the notion of fractal dimension was proposed.

Imagine a box that contains a small part of an attractor. If this box is subdivided into smaller boxes, some of them will contain the pieces of the attractor. For example, if the small part of the attractor is a simple plane, a three-dimensional box containing it is divided by 10 in each dimension, then roughly 100 of these small boxes will contain the pieces of the plane.

The number of piece-containing boxes will scale as L^d, where L is the factor by which each dimension of the box is divided. This construction defines the fractal dimension

$$d = \lim_{l \to 0} \frac{\log N(l)}{\log (1/l)}, \tag{4.45}$$

where $N(l)$ is the number of boxes, whose sides have length l, necessary to cover the attractor. For a plane, d is 2 by this construction, the same as the topological dimension. By means of Eq. (4.45) one can determine any local dimension of an attractor.

Numerical results support the following relation between an attractor's fractal dimension and its spectrum of LCN's (Kaplan & Yorke, 1979):

$$d = j + \frac{\sum_{i=1}^{j} \lambda_i}{-\lambda_{j+1}}, \tag{4.46}$$

where the corresponding LCN's are ordered by $\lambda_1 > \lambda_2 > \cdots > \lambda_k$ and j is the largest integer so that $\lambda_1 + \lambda_2 + \cdots + \lambda_j > 0$. Obviously, formula in Eq. (4.46) can be used to determine the fractal dimension of an attractor by considering it as an entire. By means of Eq. (4.46), the fractal dimension of the strange attractor in the mapping Eq. (4.41) with $A = 1.4, B = 0.7$ can be obtained, $d = 1.26$.

4.6.3 *Numerical results*

In order to study numerically the behavior of attractor in the extended mapping T in Eq. (4.42), two usual methods i.e. LCN's method and slice-cutting method are used, as in Secs. 4.3 and 4.4. The spectrum of LCN's can be used to characterize different attractors and to estimate the dimension of attractor, and the slice-cutting method can display directly the diagram of the mapping. Hence these two methods can verify the results of each other.

In this part, the "slices" are defined as

$$|z_n - Q| < 0.01, \quad Q = \frac{(k-5)\pi}{4},$$
$$-\infty < z_n < +\infty, \quad -\infty < y_n < +\infty,$$
$$(k = 1, 2, \ldots, 9; \quad n = 1, 2, \ldots, N)$$

where N is the iteration number. As k increases from 1 to 9 these "slices" are seen from left to right and from top to bottom in all the figures presented below.

The first to be discussed is the behavior of strange attractor of Hénon mapping in the extended mapping T. The parameters are taken values as follows: $A = 1.4, B = 0.3, C = D = 0.03, E = 0.01$ and $F = 0.03$. In this case, $\Delta z = z_{n+1} - z_n = E \sin y_{n+1} + F > 0$. According to results in Secs. 4.3 and 4.4, there probably exists a "strange attractor torus", and this will be a torus with fractal dimension and the sections perpendicular to it are strange attractors in the two-dimensional plane. The results show that after $N = 183$ iterations the points have escaped from the trapping region and have diverged to infinity. If keeping A, B, E, F and decreasing C, D to $C = D = 0.003$, the "strange attractor torus" appears (see Fig. 4.43), and the corresponding LCN's are $\gamma_1^{100000}(P) = 0.4147586, \gamma_2^{100000}(P) = -0.3349091 \times 10^{-4}, \gamma_3^{100000}(P) = -0.1618698 \times 10^1$ where point $P = (0.0, 0.0, 0.0)$, thus the the LCN's spectrum is $(+, 0, -)$.

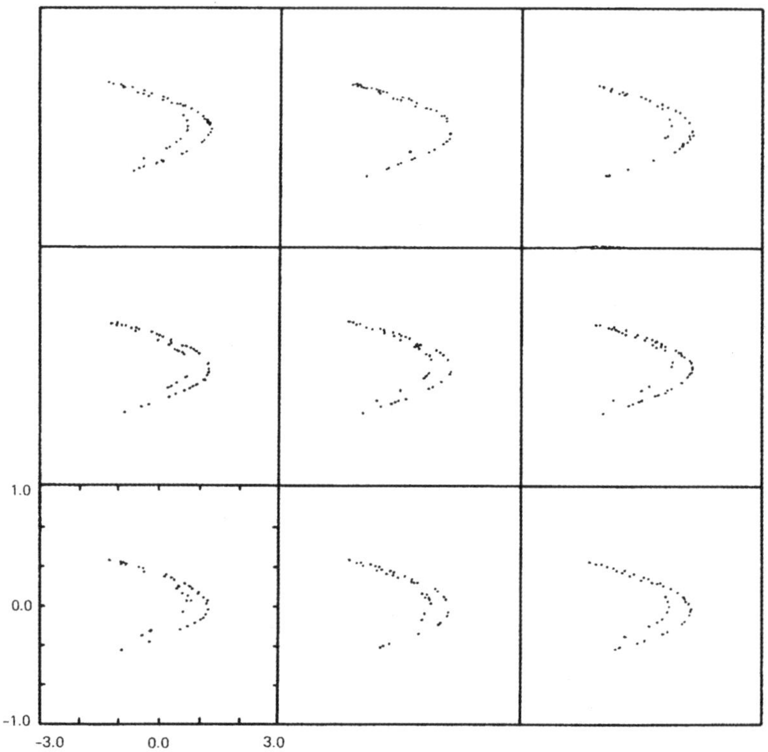

Fig. 4.43   Sections of mapping T for $A = 1.4, B = 0.3, C = D = 0.003, E = 0.01, F = 0.03$ with initial point $P = (0.0, 0.0, 0.0)$. From Sun (1985).

Both the slice-cuttings and the LCN's spectrum lead to the conclusion that it is a strange attractor torus in the three-dimensional space. Formula in Eq. (4.46) gives the fractal dimension $d = 2.26$, which is just bigger by 1 than the dimension of the strange attractor in the Hénon mapping in Eq. (4.41), confirming once again that the strange attractor obtained above is a special kind of strange attractor, i.e. it is the product of a one-dimensional manifold by a 1.26-dimensional strange attractor. As a module 2π is performed on z in the mapping T, the strange attractor becomes a strange attractor torus here. Because z_n forms a continuum, according to the principle of mean value, the existence of strange attractor torus can be approximately explained, just in a similar way as in Secs. 4.3 and 4.4.

Moreover, the invariant curves Eq. (4.44) has been numerically verified for the same A, B, C, D values as above. They pass two points $P_1 = (0.6315714, 0.1894714, 0.0)$ and $P_2 = (-1.1315714, -0.3394714, 0.0)$. The slice-cuttings of mapping T show that the successive points of these two points are not on the corresponding two invariant curves but are trapped into the strange attractor torus rapidly. Since the corresponding LCN's possess the same spectrum as that of the above strange attractor torus, they should belong to the same attractor. Perhaps, this happens because of the instability of the fixed points of the mapping Eq. (4.41) with $A = 1.4, B = 0.3$. To verify this, another set of parameters are chosen as $A = 0.2, B = 0.3, C = D = 0.03, E = 0.01, F = 0.03$, so that one of the fixed points of mapping Eq. (4.41) is linearly stable. Now it is possible to verify numerically the invariant curve passes the above stable fixed point. Figure 4.44 shows the variations of the corresponding $\gamma_i^N(P)$ $(i = 1, 2, 3)$ as function of the iteration number N up to $N = 10^6$, where $P = (1.224709, 0.3367412, 0.0)$. Apparently in this figure, the limits of $\gamma_i^N(P)$ $(i = 1, 2, 3)$ when $N \to \infty$ have an LCN's spectrum as $(0, -, -)$, indicating it is a limit cycle attractor, i.e. the invariant curve of mapping T. The slice-cuttings in Fig. 4.45 verify this too.

A period-two attractor of mapping Eq. (4.41) is known to exist when $3(1 - B)^2/4 < A < (1 - B)^2 + (1 + B)^2/4$ (Feit, 1978). To discuss the behavior of this period-two attractor in the extended mapping T, parameters $A = 0.65, B = 0.3$ satisfying the above criterion and $C = D = 0.03, E = 0.01, F = 0.03$ are chosen. Again, the LCN's γ_i^N $(i = 1, 2, 3)$ up to $N = 10^6$ are calculated under such situation and with initial point $P = (0.0, -1.0, 0.0)$, and they are plotted in Fig. 4.46. Clearly, the corresponding LCN's have also a spectrum of $(0, -, -)$, that is, it is also a limit cycle attractor. The slice-cuttings shown in Fig. 4.47 indicate that it is a

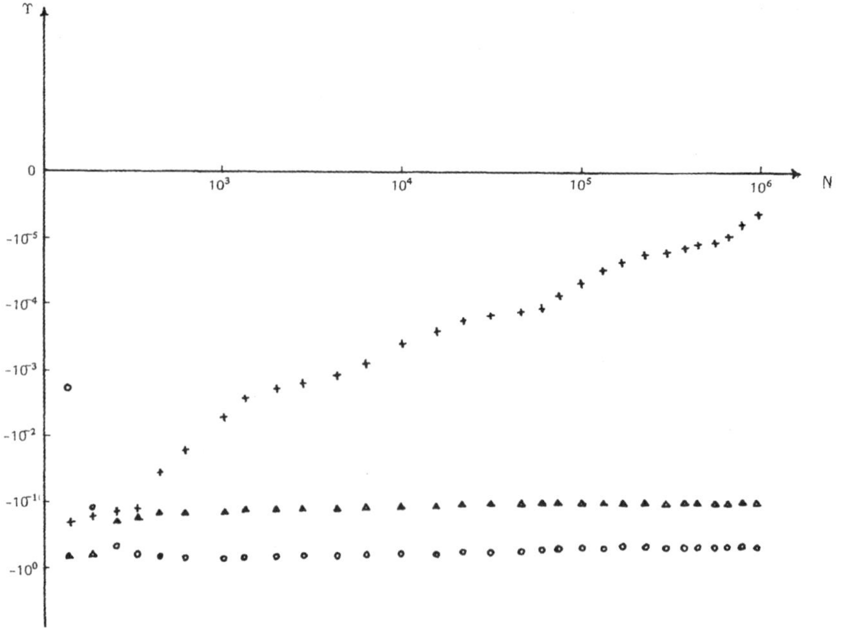

Fig. 4.44 Variations of $\gamma_i^N(P)$ $(i = 1, 2, 3)$ of mapping T (with $A = 0.2, B = 0.3, C = D = 0.03, E = 0.01, F = 0.03$) as function of the iteration number N. The initial point $P = (1.1224709, 0.3367412, 0.0)$. The crosses, open triangles and open circles stand for γ_1^N, γ_2^N and γ_3^N respectively. From Sun (1985).

symmetrically separating bifurcation of the limit cycle attractor, with its period doubled.

For $A = 2.6577, B = 0.3$, the Hénon mapping has a period-three attractor (Feit, 1978) as $(0.0104\cdots, -0.2039\cdots) \to (0.7957\cdots, 0.0031\cdots) \to (-0.6798\cdots, 0.2387\cdots)$. With parameters $C = D = 0.03, E = 0.01, F = 0.03$ and from the initial point $P = (0.0104, -0.2039, 0.0)$, even after only $N = 12$ iterations the orbit has diverged to infinity. When C, D decrease to $C = D = 0.003$, the escape appears a little later at $N = 123$ iterations.

As shown above, the fixed point and the period-two attractor of the Hénon mapping Eq. (4.41) can generate the limit cycle attractors of the extended mapping T with small C, D, but for the period-three attractor, the orbit diverges to infinity even when C, D are much smaller than in the former case. Possibly, for the Hénon mapping with $B = 0.3$, there exists a large interval of A, in which the Hénon mapping has a stable fixed point and a period-two attractor, but for the period-three attractor the corresponding interval of A is a very small periodic window in the chaotic belt (Feit, 1978).

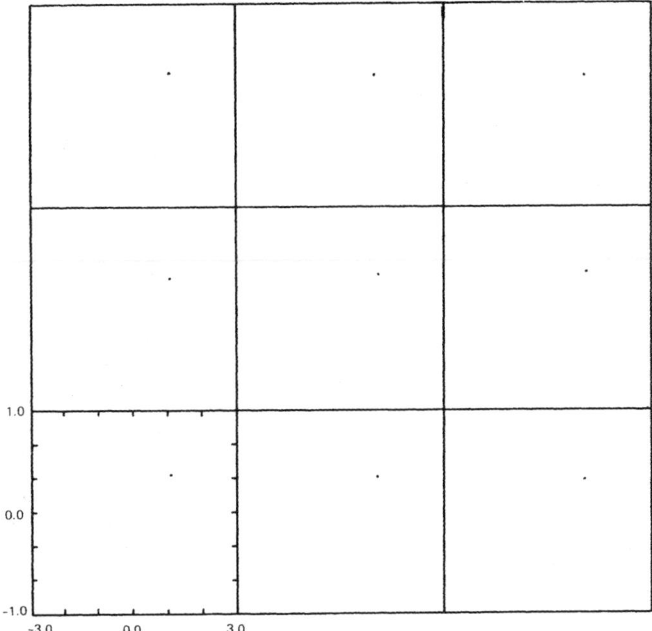

Fig. 4.45 Sections of mapping T for $A = 0.2, B = 0.3, C = D = 0.03, E = 0.01, F = 0.03$ with initial point $P = (1.1224709, 0.3367412, 0.0)$. From Sun (1985).

In summary, when perturbation parameters are small, the stable fixed point and the period-two attractor of the Hénon mapping will generate the limit cycle attractor and the symmetrically separating bifurcation of the limit cycle attractor. This is analogous to the case in the conservative system. For the strange attractor, the escape appears rapidly, but when the perturbation parameters decrease further, there exists a "strange attractor torus" generated by the strange attractor of the Hénon mapping. It seems that the strange attractor in the dissipative system is destroyed by the perturbed extension more easily than the invariant manifold in the conservative system and the trivial attractors except the period-three attractor in the Hénon mapping. This is probably because the latter is in the small periodic window, a little bit similar to the case in the conservative system where the "islands" or rational torus are destroyed more easily than the irrational torus by the perturbed extension.

Figures 4.44 and 4.46 display the variations of $\gamma_i^N(P)$ $(i = 1, 2, 3)$ as functions of the iteration number N. They correspond to the limit cycle attractor and its symmetrically separating bifurcation, a limit cycle

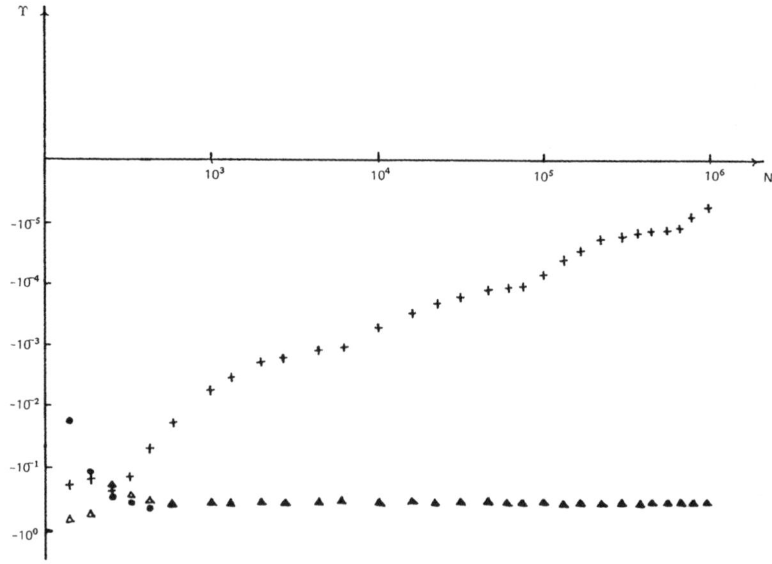

Fig. 4.46 Variations of $\gamma_i^N(P)$ ($i = 1, 2, 3$) of mapping T (with $A = 0.65, B = 0.3, C = D = 0.03, E = 0.01, F = 0.03$) as function of the iteration number N. The initial point $P = (0.0, -1.0, 0.0)$. The crosses, open triangles and open circles stand for γ_1^N, γ_2^N and γ_3^N respectively. From Sun (1985).

attractor with double period. It is found that the behaviors and the values of these two sets of $\gamma_i^N(P)$ ($i = 1, 2, 3$) are almost the same. Whether this feature is an occasional coincidence or a necessary outcome becomes a question. If the latter is true, it would show that the limit cycle attractor and its symmetrically separating bifurcation have the same spectrum of LCN's. Inversely, one can make use of the behavior and the LCN's spectrum to look for the connection among the attractor. Of course, the conjecture presented above is only based on the numerical results. To confirm it, one needs a rigorous theoretical proof.

4.7 Stickiness effect and hyperbolic structure (I)

The phase space of a nearly integrable Hamiltonian system typically consists of regular and chaotic regions. The study of orbital diffusion in the phase space is the basis of many topics in Hamiltonian dynamics. A chaotic orbit initialized close to a KAM torus may wander for a long time before it finally leaves the vicinity of the torus. This phenomenon, called "stickiness

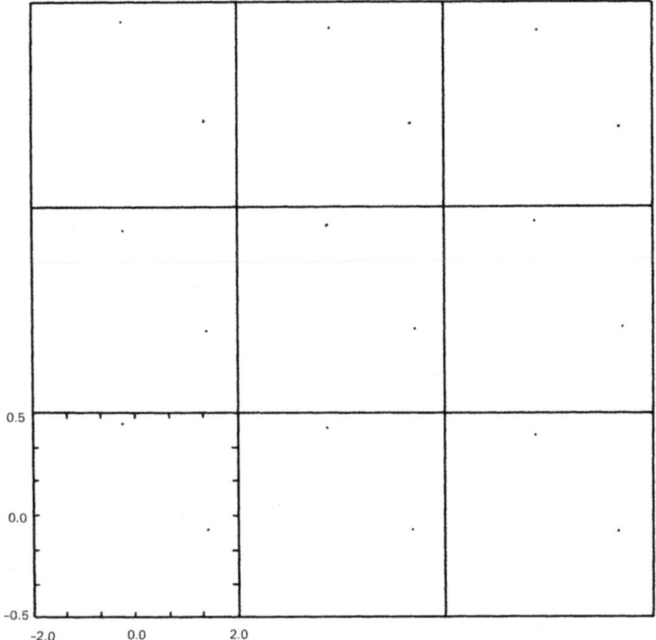

Fig. 4.47   Sections of mapping T for $A = 0.65, B = 0.3, C = D = 0.03, E = 0.01, F = 0.03$ with initial point $P = (0.0, -1.0, 0.0)$. From Sun (1985).

effect", has been introduced in Sec. 4.1, where the stickiness effect is used to describe the motion of comets. And also in Sec. 4.1, it is shown that the stickiness effects may arise from different invariant sets such as invariant tori, island-chains and Cantori. This is a "generalized stickiness effect". Moreover, it is also known that hyperbolic invariant sets possess the stickiness effect too (Froeschlé & Lega, 1998; Contopoulos *et al.*, 1999). In a two-dimensional area-preserving map, periodic island-chains accumulate in the neighborhood of torus. The tangent directions of the stable and unstable manifold at the hyperbolic periodic orbit will be more and more parallel to the torus. Thus, the orbits beginning very close to the torus will diffuse along the hyperbolic structures between islands in the island-chain for a long time before going outward or entering the adjoining island-chains. Considering the above geometrical picture, it is suggested that the hyperbolic invariant sets is essential to the stickiness effect (Zhou, Zhou & Sun, 2002a). In this section, this conclusion is clarified and confirmed numerically, adopting both two-dimensional and three-dimensional mappings as models.

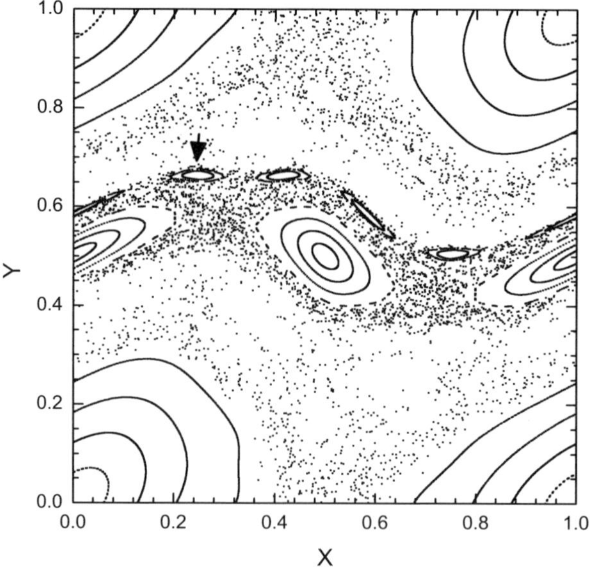

Fig. 4.48   Phase diagram of map T with $k = 1.0$. From Sun *et al.* (2005).

4.7.1   *Two-dimensional model*

The mapping

The well-known standard map, which has been mentioned in Secs. 4.2 and 4.3, is adopted again in this section as the model. For convenience, the standard mapping is copied again below.

$$
T: \quad \begin{cases} x_{n+1} = x_n + y_{n+1}, \\[2mm] y_{n+1} = y_n - \dfrac{k}{2\pi}\sin(2\pi x_n), \end{cases} \quad (\mathrm{mod}\ 1) \qquad (4.47)
$$

where k is the perturbation parameter. The phase space of this mapping, shown in Fig. 4.48 for $k = 1$, is well-known.

More or less arbitrarily, an island from the period-5 island-chain (indicated by the arrow in Fig. 4.48) and with its center at $(x_0, y_0) = (0.2476544, 0.6638289)$ is chosen. This torus island selected above merits a careful consideration because it is closely surrounded by a period-23 island chain, and the Birkhoff theorem guarantees the existence of 23 hyperbolic points between the 23 islands, therefore, this region in a near vicinity of this torus is diversified with invariant torus, island chain, hyperbolic fixed points, higher order (secondary) islands and chaotic sea. It will be called the "chief torus"

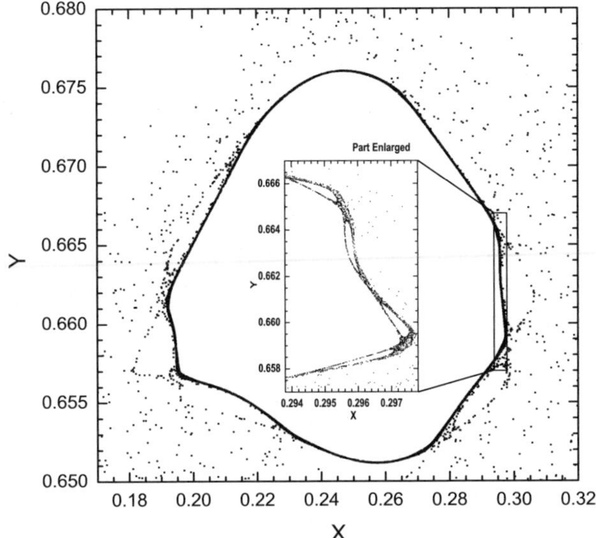

Fig. 4.49   Diagram of the period-23 secondary islands around the chief torus. From Sun *et al.* (2005).

hereafter. The stickiness effect of this chief torus as well as of the periodic islands and the hyperbolic invariant sets around it will be analyzed. Particularly the role of the hyperbolic invariant sets in the generalized stickiness effect will be discussed in detail.

Diffusion speeds

In the vicinity of the chief torus, there is an island-chain with 23 periodic islands and a hyperbolic invariant set consisting of 23 periodic hyperbolic fixed points as predicted by the Poincaré-Birkhoff fixed point theorem. This structure is shown in Fig. 4.49.

Generally the stickiness effect can be qualitatively revealed and quantitatively described by the slow diffusion of orbits. In order to study the stickiness effects of the chief torus and of the hyperbolic invariant sets, the diffusion speeds in the vicinity of the torus and in the vicinity of the hyperbolic invariant sets are compared. To do this, a "neighboring zone" of the chief torus is defined first. All orbits that will be discussed are initially located in this neighboring zone and if an orbit diffuses away from this zone to a definite distance, it is regarded as "escaped". Such a neighboring zone is defined as follows.

First the length L of the "boundary" curve l of the chief torus is calculated, while the "boundary" can be approximatively defined as the possible outermost invariant curve l of the chief torus. The area A surrounded by the boundary l is also computed. Then the neighboring zone around the chief torus is defined by expanding the curve l outwards a width $D = \frac{A}{10L}$, i.e. the area of the neighboring zone is taken as $\frac{A}{10}$, where $A = 1.8341447 \times 10^{-3}$, $L = 0.2221069$. As soon as the distance of an orbit to the chief torus is larger than $2D$, it is regarded as escaped.

In the vicinity of the chief torus, 20,000 initial points spread uniformly in the neighboring zone, P_0^{ij}: $(r_0^i + \frac{j}{20}D, \theta_0^i = \frac{2\pi}{1000}i)$, $i = 1, 2, ..., 1000, j = 1, 2, ..., 20$, where (r_0^i, θ_0^i) are the polar coordinates of the points P_0^i on l. In the same way another 20,000 initial points near one of the 23 periodic hyperbolic fixed points $(x_h, y_h) = (0.2972704, 0.6590667)$ are taken. These initial points are all in a segment of the above mentioned neighboring zone with angle coordinates from $\theta_1 = 6.1851$ to $\theta_2 = 6.1903$. Interpretatively, this fixed point and its two neighboring fixed points respectively have angle coordinates of $\theta = 6.1875$, 6.1478 and 6.2337.

After the above definitions, Fig. 4.50 displays the variations of the surviving orbit number with time (iteration number) for both cases. In this

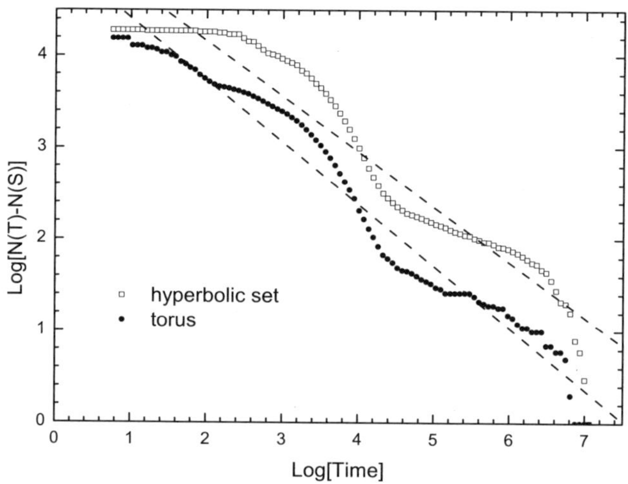

Fig. 4.50 Diffusion of orbits started close to the chief torus and close to the hyperbolic fixed point. The total number of points that have not escaped at time T is denoted by $N(T)$ and the number of points that will never escape is denoted by $N(S)$. Solid circles and open squares stand for the cases close to the chief torus and close to the hyperbolic fixed points, respectively. The dashed lines are the linear fittings. From Sun *et al.* (2005).

figure two curves are very similar to each other, implying that the stickiness effect of the chief torus and of the hyperbolic sets have the same origin. For the case of orbits starting near the hyperbolic fixed point, the diffusion is slower than those starting from the vicinity of the chief torus.

As shown in Sec. 4.1 and also mentioned in Sec. 2.3, the surviving number N of orbits starting from a mixed region in the phase space of a Hamiltonian system will decrease in a power law with respect to time T: $N \sim T^{-z}$, where z is a positive exponent. In Fig. 4.50, each curve may be divided into several segments, which can be linearly fitted with different slopes. Particularly, when $\log[\text{Time}] \in [3.6, 4.4]$ both of the two curves can be well fitted by lines with $z = 1.5$, consistent with the exponent value suggested by e.g. Ding *et al.* (1990). However, the most interesting feature in this figure is the similar profiles of two varying curves. The average exponents over the whole time range from 0 to 10^7 are $z_1 = 0.6756$ and $z_2 = 0.6028$ respectively for the two cases. The average times of escape $\bar{T} = \frac{1}{N_0} \int_0^{N_0} T \mathrm{d}N$ are also numerically calculated and they are $\bar{T}_1 = 4815$ and $\bar{T}_2 = 18677$ (iteration times) for the cases of torus and hyperbolic set respectively.

Both the exponent z and the average time of escape $\bar{T}$ imply that, the orbits with initial points in the vicinity of the chief torus, will diffuse faster than those started near the hyperbolic invariant set. These results also imply the hyperbolic invariant sets will play a major role in slowing down the diffusion of orbits near the torus, or in the stickiness effect of tori.

Stickiness effect of hyperbolic invariant sets

By following an orbit that is "stuck" in the vicinity of the chief torus, one can see directly where the orbit spends the longest time duration before its final escape. Here, an orbit with an initial point $(x, y) = (0.2972650, 0.6591000)$, which is very close to one of the period-23 hyperbolic fixed points $(x_h, y_h) = (0.2972704, 0.6590667)$, is selected. The diffusion process of this orbit is shown in Fig. 4.51. Because the orbit diffuses to a secondary island-chain (or a secondary hyperbolic invariant set) when the iterative number $n \sim 3 \times 10^5$, the orbit is traced up to $n = 2 \times 10^5$ iterations, before that the orbit diffuses around the hyperbolic invariant set. Since the orbit is very close to the boundary l of the chief torus, the distribution of points on the orbit can be calculated along l. Every point near l can be projected to l with respect to the center of the chief torus, so that gets a reference coordinate from the projection on l. In such a way, one can see where the orbit is "stuck" during its diffusion.

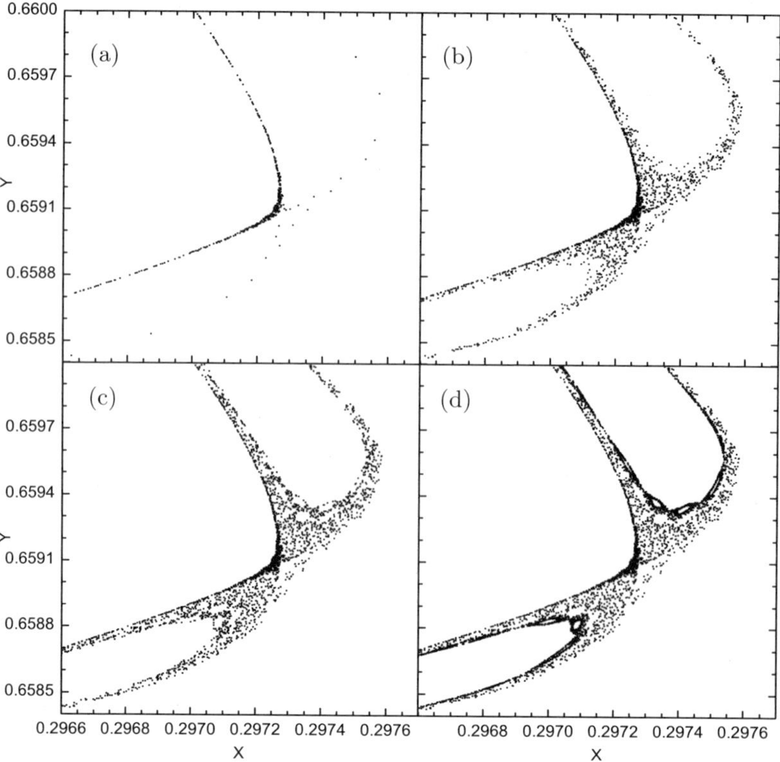

Fig. 4.51   Diffusion of an orbit around a hyperbolic invariant fixed point. (a), (b), (c) and (d) are the snapshots of the orbit up to 5×10^4, 2×10^5, 3×10^5 and 6×10^5 iterations, respectively. From Sun *et al.* (2005).

The distribution of orbit points with respect to the length of l is plotted in Fig. 4.52, and the positions of the 23 hyperbolic fixed points are marked too. The outstanding feature of Fig. 4.52 is that every peak evidently corresponds just to one of the positions of the period-23 hyperbolic fixed points (hyperbolic invariant set). The distribution of points on other orbits starting from points not very close to the hyperbolic fixed point, but lying between two neighboring hyperbolic fixed points in the same hyperbolic invariant set, are found to have very similar distribution to the one in Fig. 4.52.

It should be stressed that an orbit spreads in fact in a two-dimensional zone, so that the density of orbit points in the sticky zone should be counted in definite areas rather than along a curve. However, from a practical view,

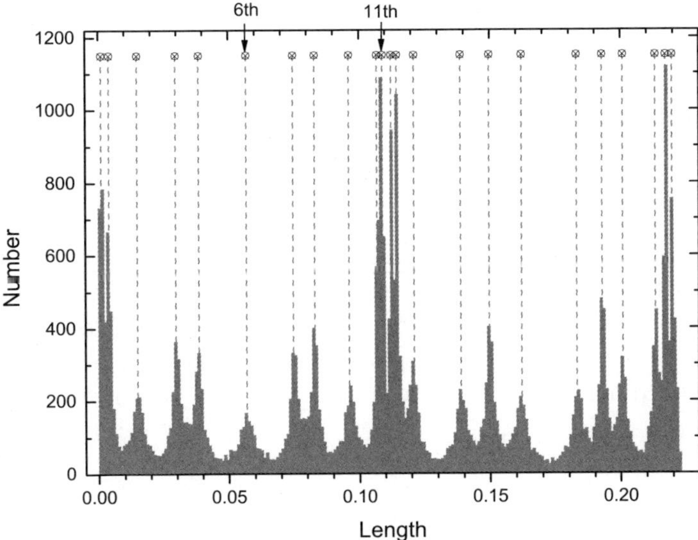

Fig. 4.52   Distribution of points of an orbit. The crosses inside circles correspond to the positions of the 23 hyperbolic fixed points. From Sun *et al.* (2005).

there is no straightforward method to define such areas, in which higher order islands are embedded as holes and around them there are secondary hyperbolic invariant sets related with the stickiness effects of the higher order islands. On the other hand, the sticky zone occupied by an orbit shown in Fig. 4.51 as an example shows a little thinner width around the hyperbolic points than around the rest of the zone, and in fact, from Fig. 4.51 one can see that the orbit points near the hyperbolic fixed point are denser than in the zones far from it. Therefore, even if the sticky zone is considered roughly as a uniformly wide strip around the chief torus, the distribution along a curve in Fig. 4.52 would reflect approximately the distribution of orbit points in the sticky area near the chief torus.

It is well-known that there are many different unstable periodic orbits in a chaotic region. In Fig. 4.52 the match between the positions of the outstanding peaks and the hyperbolic fixed points reveals that this hyperbolic invariant set is much more important than others in this region. It also implies that such orbits spend more time around this hyperbolic invariant set than elsewhere, therefore the hyperbolic invariant set plays an important role in the stickiness effects. Moreover, an orbit on the "outermost" invariant curve also has a similar accumulation of points (the distribution is very similar to Fig. 4.52). Since this invariant curve is very close to the hyperbolic set located among the island-chain, this accumulation can be explained by the continuous dependence of the density of orbits on the initial conditions.

Note the peaks in Fig. 4.52 have different heights. Meanwhile, the locations of hyperbolic fixed points along l are not uniformly distributed, that is, the distances between every adjacent fixed points are not equal. For instance, the sixth (counting from left to right, indicated by an arrow in Fig. 4.52) hyperbolic points is relatively farther away from its neighboring fixed points, as a result, orbit points spread around the sixth point in a wider region and the corresponding peak is lower. As a contrast, the eleventh fixed point is closer to its neighbors so that the corresponding peak is higher.

Size of islands

The orbits locating in the islands in the phase space will stay in the islands for ever, the orbits in the vicinity of islands show significant stickiness effects, while the cantori and hyperbolic structures may accumulate around the islands. Moreover, the orbits cannot cross the islands, thus the size of islands in the phase space of the two-dimensional mapping is also an important issue when considering the orbital diffusion in the phase space. The variations of sizes of Fibonacci islands with the distance to the chief torus, and the variations of islands sizes with perturbation parameter, have been studied (Efthymiopoulos *et al.*, 1997; Froeschlé & Lega, 1998). Here the sizes of different islands including the chief torus in the phase space are measured, and then the variations of islands' area with respect to the periods of island-chains and to the system perturbation parameter k are investigated. With these results one may understand further the orbital diffusion in phase space and the role of hyperbolic invariant sets in the stickiness effect.

The area of an island is approximated by the summation of areas of 2,000 triangles inside the "outermost" invariant curve surrounding the island. Figure 4.53 displays the areas of islands of different periods (the chief torus is a period-5 island) in the main island sequences (island-chain sequence surrounding the central island) for $k = 0.90, 0.95, 1.0, 1.00, 1.05$ and 1.10 respectively. In this figure, each square denotes the total area of an island-chain of a definite period. Apparently, the decay of island area with respect to the period obeys roughly a power law, and the absolute values of the

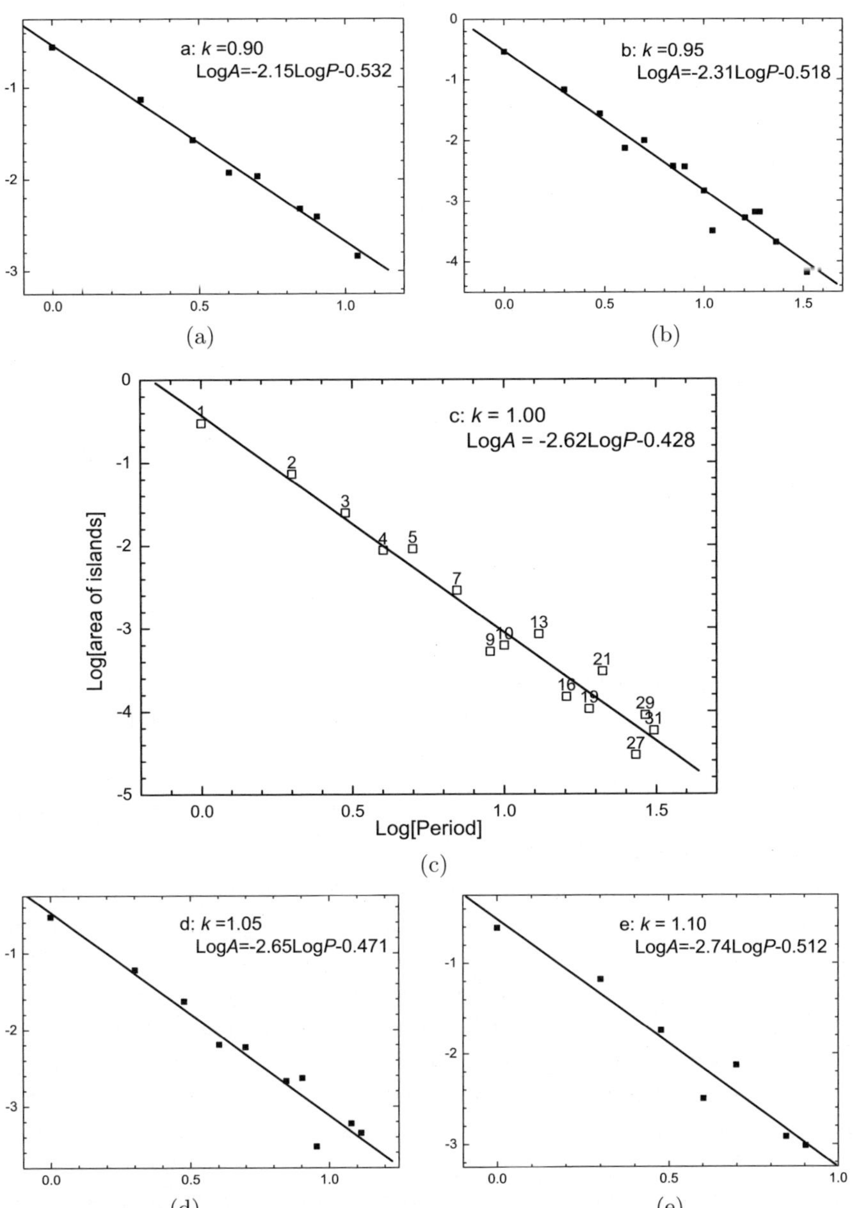

Fig. 4.53 Variations of area of islands in the main island sequence with respect to the period of island. The linear fit of the data is indicated by both a line and a linear function. (a), (b), (c), (d) and (e) are the situations of $k = 0.90, 0.95, 1.00, 1.05$ and 1.10, respectively. The numbers above the squares in (c) denote the periods of corresponding islands. From Sun *et al.* (2005).

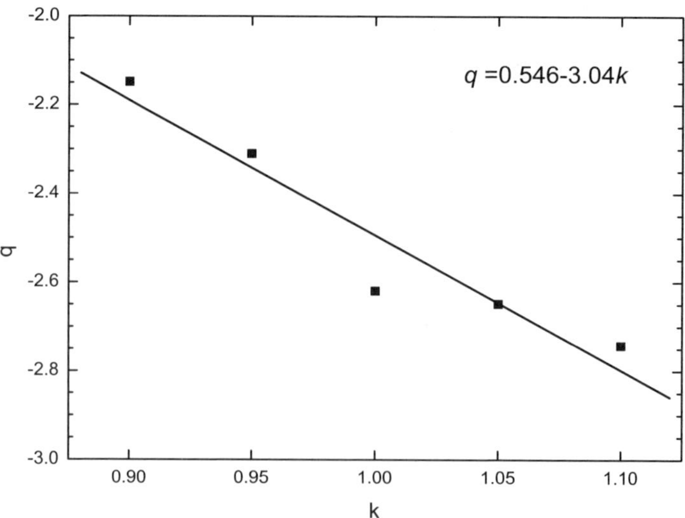

Fig. 4.54 Variation of slopes q of fitting lines in Fig. 4.53 with the perturbation parameter k. From Sun *et al.* (2005).

slopes of fitting lines increase monotonically with k. This means that the decay rate of the area for island sequences increases with the perturbation parameter k (Fig. 4.54).

As well-known, the self-similarity property indicates that there are secondary islands around an island in the main sequence, and then higher order islands around this secondary islands, and so on. To study the area of such "islands around islands", one may select the most outstanding secondary island-chain around an island, and successively repeat such selecting to higher and higher order. Taking $k = 1.0$, and selecting several "hierarchical (island) sequence" starting respectively from islands with periods of 4, 16 and 29 in the main sequence, the areas of islands in these hierarchical sequences are computed. The numerical results displayed in Fig. 4.55 show that the decay of the area of islands in the hierarchical sequence possesses a power law too, but compared with the case for main sequence, the absolute values of the slopes q of fitting lines are smaller. This implies the decay of areas in the hierarchical sequences is slower than that in the main sequence.

With the rule of the change of islands' area with respect to the perturbation parameter and islands' periods, the upper bound of the total area of islands in phase space can be estimated. As indicated above, the variation of island area with period possesses roughly a power law, i.e.

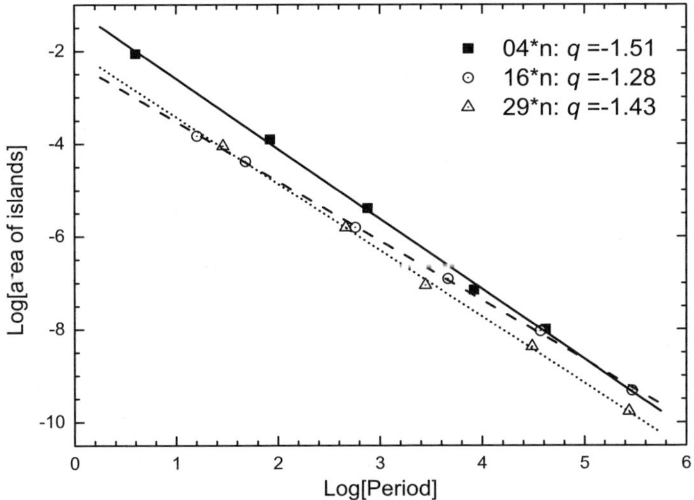

Fig. 4.55 Variations of area of island in the hierarchical island sequences with the period of islands, and the corresponding fitting lines. The filled squares, circles and triangles represent the situations in the sequence starting from the period-4, 16, and 29 islands in the main sequence. q is the slope of the fitting line. From Sun *et al.* (2005).

$\log A_P(k) = q(k) \log P + C(k)$, where $A_P(k)$ is the total area of period-P islands for a given k, $q(k)$ the slope and $A_1(k) = 10^{C(k)}$ the area of the central island (period-1 island). According to the result in Fig. 4.54, one has

$$q(k) = 0.546 - 3.04k,$$
$$A_P(k) = P^{0.546-3.04k} A_1(k), \quad P \in \{N\} \tag{4.48}$$

while $\{N\}$ is the set of period of islands in the main sequence for $k \in [0.90, 1.10]$. Therefore, the total area of islands in the main island sequence is

$$A(k) = \sum_{P \in \{N\}} A_P(k) \le \sum_{P \in [1,\infty)} A_P(k) = A_1(k) \sum_{P \in [1,\infty)} P^{q(k)}. \tag{4.49}$$

The series $\sum_{n=1}^{\infty} \frac{1}{n^s}$ is convergent if $s > 1$, and the sum is called Riemann function $\zeta(s)$. After elementary calculus, one gets

$$\zeta(s) < \frac{1 - s - 2^{-s}(1+s)}{1-s} = 1 - \frac{2^{-s}(1+s)}{1-s}. \tag{4.50}$$

Finally, one obtains

$$A(k) < A_1(k) \left[1 - \frac{2^{q(k)}(1 - q(k))}{1 + q(k)}\right] = U(k), \quad k \in [0.90, 1.10]. \tag{4.51}$$

Generally the area of central island decreases as k increases in a long term for large variation of k, but the varying is not smooth. It can even increase temporarily (Efthymiopoulos *et al.*, 1997), as happening to occur in the present case. However, the average for the five values of k can be taken, $\bar{C}(k) = -0.502$.

Substituting $q(k) = 0.546 - 3.04k$ and $A_1(k) = 10^{-0.502}$, one may finally get an upper bound $U(k)$ of the total area of islands in the main island sequence. Because the total area of higher order islands in the hierarchical island sequences is much smaller than that in the main island sequence, $U(k)$ is roughly an upper bound of the total area of islands in phase space. Calculating of Eq. (4.51) shows $U(k) = 0.500, 0.469, 0.445, 0.426, 0.410$ when $k = 0.90, 0.95, 1.00, 1.05, 1.10$, respectively.

The whole phase space in this model consists of islands and other invariant sets. If the total area of islands is smaller than 1.0, the residue of them must have a positive measure. This may also imply the importance of hyperbolic sets in the orbital diffusion. Although the above calculation is only a coarse estimate, it still can give some valuable hints.

4.7.2 *Three-dimensional model*

It has been shown in Secs. 4.3 and 4.4 that in a three-dimensional mapping there may exist invariant tori. It is interesting to check whether the stickiness effect exists in such a three-dimensional phase space, and whether the hyperbolic invariant sets play a similar role in causing stickiness effect as in the two-dimensional mapping. In this part, a three-dimensional mapping model is adopted to analyze the stickiness effect.

The model

The three-dimensional mapping that was first developed by Sun and Yan (Sun & Yan, 1988) and has been shown in Sec. 4.4 as Eq. (4.31) is adopted again here. The volume-preserving mapping reads:

$$M: \begin{cases} x_{n+1} = s(x_n \cos \varphi_n - y_n \sin \varphi_n) + C \cos z_n, \\ y_{n+1} = s^{-1}(x_n \sin \varphi_n + y_n \cos \varphi_n) + C \sin z_n, \\ z_{n+1} = z_n + C \cos(x_{n+1} + y_{n+1}), \quad (\mathrm{mod}\, 2\pi) \end{cases} \tag{4.52}$$

where $\varphi_n = (x_n^2 + y_n^2)^k$ and s, k, C are parameters. Comparing with Eq. (4.31), the mapping Eq. (4.52) is simplified a little by letting $A = B = C$ (denoting it as C) and $D = 0$. This mapping can be regarded as a perturbed extension of a hyperbolic twist mapping:

$$M_0 : \begin{cases} x_{n+1} = s(x_n \cos\varphi_n - y_n \sin\varphi_n) \\ y_{n+1} = s^{-1}(x_n \sin\varphi_n + y_n \cos\varphi_n). \end{cases} \tag{4.53}$$

In this part, the parameters s, k are set as $s = 1.05, k = 1.5$. Thus, there is now only one parameter C to control the perturbation. Except for specific indication, the parameter $C = 0.04$ in this section.

There exist invariant tori and hyperbolic sets in the two-dimensional phase plane of the mapping M_0, and under certain conditions these tori and hyperbolic sets will be stretched to the three-dimensional phase space of the extended three-dimensional mapping M. Such phenomenon has been shown in Sec. 4.4. For example, one can see in Fig. 4.36 the structure of the phase space of mapping M: A chaotic region located in the center is surrounded by invariant surfaces and the far outer region is a chaotic sea again. Note that the "invariant curve" in the slices represent in fact a two-dimensional "invariant tube" in the three-dimensional phase space, and an "island" is a section of an "island tube".

As in the case of two-dimensional model, the "outermost" surface (farthest from the center) of an island tube can be located among the family of invariant surfaces surrounding the central chaotic region. This outermost surface is called "Main Island Tube" (MIT). Figure 4.56(a) shows a slice

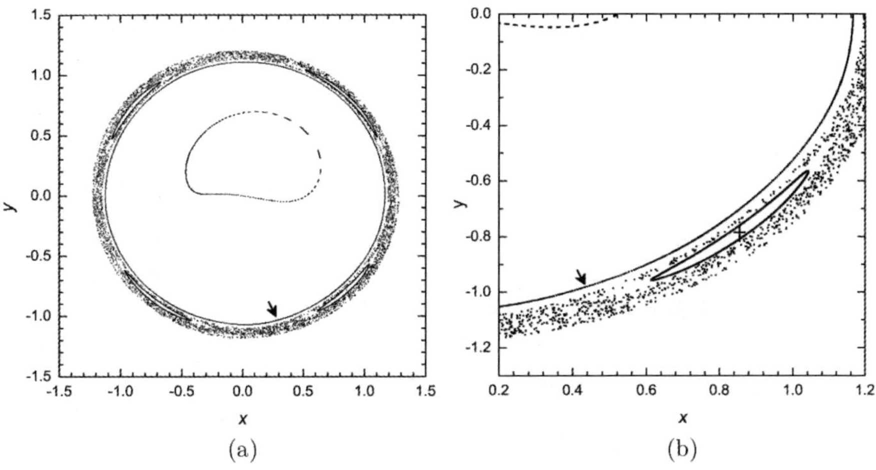

Fig. 4.56 The MIT and the SIT in a slice ($|z| < 10^{-3}$). **a**. The outermost surface of the MIT and the SIT buried in the chaotic sea. **b**. An enlargement of **a** to show the details. The cross in the right panel represents the position of the center of the SIT. The arrows indicate the MIT. From Sun & Zhou (2009).

of the MIT and its close vicinity, where four other invariant tubes besides the MIT can be seen clearly. Since in the phase plane of M_0 these surfaces will reduce to a period-4 secondary island chain around the primary island, they are called the "secondary island tubes" (SIT) here. Figure 4.56(b) is an enlargement of one of the island tubes. According to the Poincaré-Birkhoff fixed points theorem, between every two secondary islands in the mapping M_0, there should be a hyperbolic fixed point. As M is a perturbed extension of M_0, due to the structural stability of hyperbolic invariant sets, there must exist one-dimensional hyperbolic invariant sets of mapping M between any two adjacent secondary island tubes shown in Fig. 4.56. Unfortunately, the hyperbolic invariant sets cannot be illustrated directly.

Although the positions of the one-dimensional hyperbolic invariant sets are not easy to be located precisely, they must lie between the neighboring island tubes whose positions can be accurately spotted. Especially, the coordinates of the centers of the SITs can be calculated. These centers appear as continuous "central curves" in the three-dimensional phase space (see Fig. 4.57). A point in the three-dimensional phase space can be located either by its orthogonal coordinates (x, y, z) or by the cylindrical coordinates (r, θ, z), where r and θ are the radius and azimuth respectively. The (r, θ, z) of the central curves of the island tubes are also shown in Fig. 4.57. Note that r changes only a little, indicating that the central curves of the island tubes lie on a nearly cylindrical surface. Thus one can examine on the (θ, z)-plane the relative position of any point near the MIT with respect to the locations of the SITs.

Diffusion speeds

Knowing the invariant tubes and the hyperbolic invariant sets in the phase space, their stickiness effects can be compared as done in the two-dimensional model. Below the diffusion rules of orbits starting from the vicinity of the invariant tube (the MIT) and from the vicinity of the hyperbolic invariant sets will be computed. For convenience, hereafter the former case will be called case A and the latter case B.

The MIT can be regarded approximately as a cylindrical surface, and its radius is in a relatively small range satisfying $1.060 < r_{\mathrm{MIT}} < 1.175$. So the "vicinity of the MIT" for case A is defined as a boundary layer with a thickness of $d = 0.1$, which is a little smaller than one tenth of r_{MIT}. Then from this region, 20,480 points are randomly selected as the initial points. For case B, with the knowledge of the locations of the SIT (Fig. 4.57),

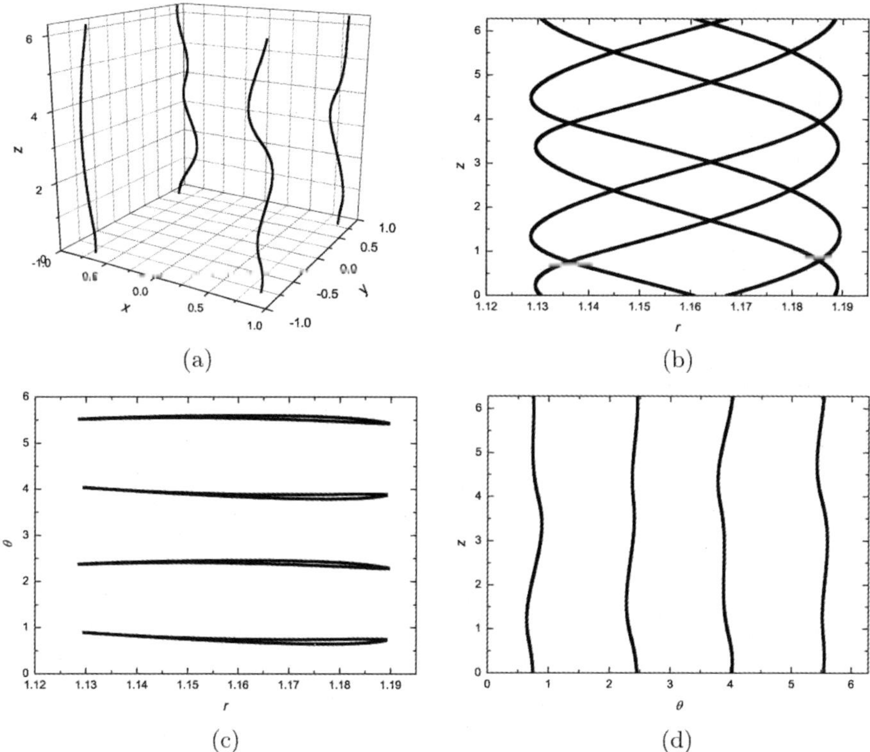

Fig. 4.57   The central curves of the SITs. **a**. The curves in an orthogonal coordinate system. **b**, **c**, and **d** show the cylindrical coordinates of curves in **a**. From Sun & Zhou (2009).

20,480 initial points that are in the vicinity of the hyperbolic invariant sets can also be chosen, in a relatively more complicated way (Sun & Zhou, 2009).

The evolutions of these two sets of initial points for both case A and case B are followed, up to 10^8 iterations. An orbit is regarded as escaped and discarded from further iterations if $|x| + |y| > 20$ during its evolution, and as stable if it does not satisfy this criterion until 10^8 iterations. The criterion for escapes is chosen in a sense arbitrarily, but a point satisfying $|x| + |y| > 20$ is quite far from the MIT, and in fact after an orbit leaves the thin sticky layer around the MIT, the "hyperbolic parameter" $s = 1.05$ of the mapping M will drive it outwards very quickly to large values of $|x| + |y|$. The stickiness time T, when the escape of an orbit happens, of each escaped orbit is recorded thus the total number of orbits $N(T)$ at time T

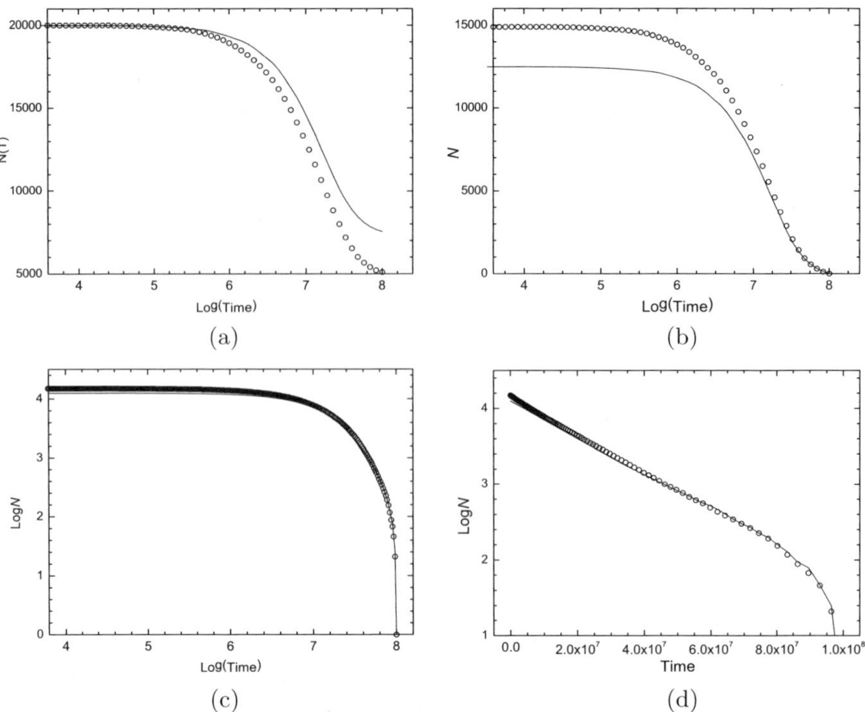

Fig. 4.58 The variations of the number of diffusion orbits with respect to time. The solid lines and open circles represent case A and case B (see text), respectively. **a**. The total number of not-escaped orbits versus time: $N(T) \sim \log T$; **b**, **c** and **d** are the surviving numbers of orbits versus time in different scales. **b**: $N \sim \log T$, **c**: $\log N \sim \log T$ and **d**: $\log N \sim T$. In **c** and **d** the two cases nearly superpose each other. Adapted from Sun & Zhou (2009).

can be counted. According to the definitions, the number of stable orbits is $N(S) = N(T = 10^8)$, and as usual the surviving number $N = N(T) - N(S)$ is used to describe the diffusion rule.

Figure 4.58 summarizes the variations of these numbers of orbits with respect to the time T. As these plots show, both the diffusion of orbits starting from the vicinity of the invariant tube (case A) and the diffusion of orbits from the vicinity of the hyperbolic invariant sets (case B) are very slow. The decrease of surviving number can only be distinguished for T larger than 10^6 iterations. From Fig. 4.58c and Fig. 4.58d, it seems that the diffusion accelerates a little after 9×10^7 iterations. This is probably resulting from a partial barrier surrounding the region, for example, a network of the hyperbolic invariant sets, which may arise from a three-dimensional extension of a cantorus in the two-dimensional mapping.

Although, the power law diffusion is not so distinct in these two cases, the most important feature in Fig. 4.58 is that the diffusion in both case A and case B obeys the same rule, implying again that the stickiness effects of the invariant surface and of the invariant hyperbolic sets in three-dimensional phase space arise from the same origin. In fact, the orbits in case A go towards the hyperbolic invariant sets before escaping, and after such an approach the diffusion routes for these two cases are the same. This will be shown more clearly below.

Stickiness effects of hyperbolic invariant sets

The MIT is surrounded by 4 SITs as shown in Fig. 4.57, and between the SITs there are hyperbolic invariant sets. Like in the two-dimensional model, one can investigate the diffusion of an orbit, check where it is "stuck" during its diffusion route and finally clarify the mechanism that causes stickiness effects in the three-dimensional phase space.

An orbit starting close to the MIT from $(x_0, y_0, z_0) = (0.7818, -0.8084, 0)$ is traced. This orbit will escape $(|x|+|y| > 20)$ after $\sim 3.89 \times 10^7$ iterations. The first 10^5 points of the orbit on the (θ, z)-plane is shown in Fig. 4.59. The positions of the central curves of the SITs are plotted

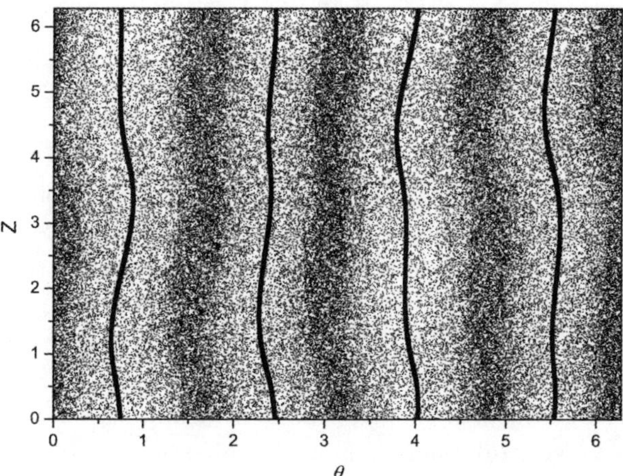

Fig. 4.59   Orbit points of an orbit starting from $(x_0, y_0, z_0) = (0.7818, -0.8084, 0)$. The thick curves are the central curves of the SITs as in Fig. 4.57d. From Sun & Zhou (2009).

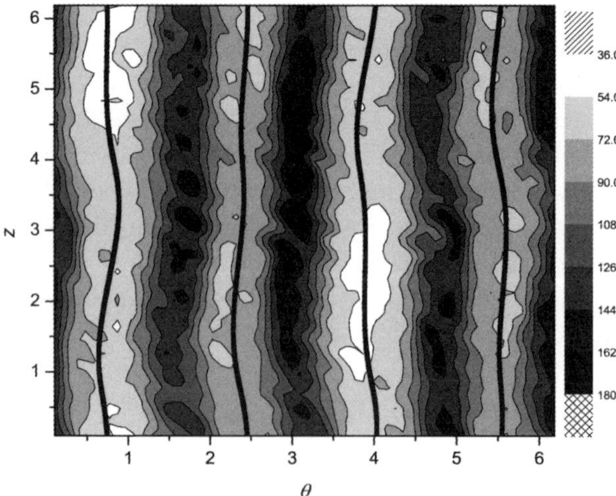

Fig. 4.60   The distribution of orbital points for the orbit shown in Fig. 4.59. The color depth scale indicates the density of points in the (θ, z)-plane. From Sun & Zhou (2009).

simultaneously (thick curves). It is clear that the denser regions of orbit points appear between the thick curves. Checking the positions of the orbit points in different time intervals (e.g. 10^5 points with iterations between $T = 1.0 \times 10^6$ and $T = 1.1 \times 10^6$) or the averaged positions (one point out of every 100 points along this orbit), it is found that all of them have the same appearance as in Fig. 4.59. This means that the orbit fills very densely the narrow chaotic domain determined by the multiplicity-4 hyperbolic invariant sets, and it escapes very soon after it leaves this sticky region.

By counting the number of the orbit points on the (θ, z)-plane, one gets in Fig. 4.60 a contour indicating their spatial distribution. This contour shows more clearly the concentration of points in four density local maxima along the azimuth θ, a fact which indicates that the manifolds of the multiplicity-4 hyperbolic set cause the most dominant stickiness effects from all the hyperbolic sets embedded in the narrow chaotic volume of stickiness shown in Fig. 4.56.

The stickiness effects of the hyperbolic invariant sets shown in Figs. 4.59 and 4.60 are a general phenomenon for all sticky orbits in this three-dimensional mapping. Many different orbits with different initial points in the vicinity of the MIT have been examined and they all behave in the same way. In fact, for an arbitrary selection of initial conditions in the vicinity of the MIT, one always gets a similar diffusion route and a similar spatial distribution of the orbital points as above. Thus one may convincingly conclude that the stickiness effects in the three-dimensional mapping are caused mainly by the hyperbolic invariant sets, just as in the two-dimensional case.

Volume of the island tubes

Since the stickiness effects are most significant phenomenon around the invariant manifolds in the phase space, the size of the island tubes in the three-dimensional phase space is an important factor for orbital diffusion as in two-dimensional case and it is closely related to the stickiness effects. The volume of the island tubes, which will act as the barriers for the orbital diffusion, is calculated too, in a similar way as in the two-dimensional case. The volume of an island tube here is the volume surrounded by the "outermost" invariant surface around an island.

To this end, the island tube is cut into 1,000 thin slices, which are perpendicular to the z-axis. Each of the slices has a definite thickness Δz. The volume of the island tube is then the sum of these slices, which are approximated by cylinders. The volume of one cylinder equals the product of its bottom area by its thickness. Because of the complexity of the profile of the invariant tube in phase space, the contour of the slice bottom can be convex or concave. One needs to take into account the convexity in this particular type of calculation. If the bottom slice contour is convex, the bottom may be divided into numerous triangles sharing common vertices and the area is equal to the sum of these triangles. A concave bottom must be divided into several convex subregions before computing its total area.

The volumes of the MIT and the SIT are calculated for different values of the perturbation parameter C. The values are listed in Table 4.1, and also illustrated in Fig. 4.61. The volume of the MIT exhibits an abrupt decrease at $C \approx 0.078$ when the outer boundary invariant surface breaks

Table 4.1   The volumes of the MIT (V_m) and the SIT (V_s). The control parameter C in mapping M is adopted from 0.01 to 1.00, but those values with $C > 0.07$ when the SIT submerges in the chaotic sea, are neglected.

C	0.01	0.02	0.03	0.04	0.05	0.06	0.07
V_m	33.77	30.68	25.44	24.47	22.48	21.46	19.70
V_s	1.314	0.8604	0.7171	0.6071	0.3826	0.3170	
$V_\mathrm{m}/V_\mathrm{s}$	25.69	35.65	35.48	40.93	58.76	67.69	

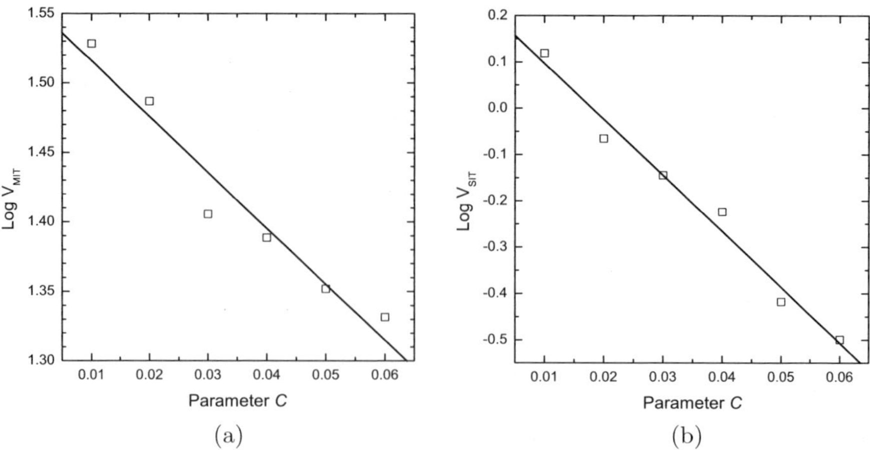

Fig. 4.61   The volumes of the MIT (a) and SIT (b) versus the perturbation parameter C. The fitted lines described as Eqs. (4.54) and (4.55) are also plotted to guide the eyes. From Zhou & Sun (2009).

and the outer chaotic region communicates with the inner region in which the chaos was developed before (for smaller C, and those values $C \geq 0.07$ have been abandoned).

In a two-dimensional mapping, the area of a "period-P" island A_P depends on its period and the perturbation parameter k as described by Eq. (4.48). As an extension of the two-dimensional mapping, one may expect the volume of the island tube in this three-dimensional model varies in a similar way with respect to the perturbation parameter C. In Fig. 4.61, the linear fit gives

$$\log V_{\mathrm{m}} = 1.55 - 3.81C, \tag{4.54}$$

$$\log V_{\mathrm{s}} = 0.217 - 12.1C, \tag{4.55}$$

for the MIT and the SIT, respectively. The V_{m} and V_{s} in above equations and hereafter denote the volumes of the MITs and SITs. Subtract Eq. (4.55) from Eq. (4.54),

$$\log \frac{V_{\mathrm{m}}}{V_{\mathrm{s}}} = 1.33 + 8.29C. \tag{4.56}$$

This is consistent with the linear fit given in Fig. 4.62

$$\log \frac{V_{\mathrm{m}}}{V_{\mathrm{s}}} = 1.34 + 8.04C. \tag{4.57}$$

For the variation of an island tube's volume V with respect to its period P, one may suppose that a general expression in a similar way as

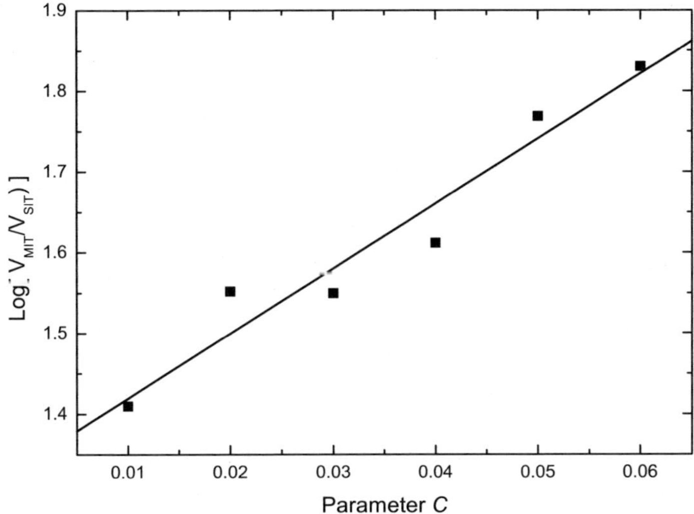

Fig. 4.62 The variation of the ratio between the volumes of MIT and SIT with respect to the perturbation parameter C. The line represents the linear fit given in Eq. (4.57). From Zhou & Sun (2009).

Eq. (4.48) for the two-dimensional case can be applied:

$$V(P,C) = V_1(C)P^{q_0+\alpha C}, \tag{4.58}$$

where V_1 is the volume of period-1 island tube, C is the perturbation parameter, q_0 and α are free parameters. Then the ratio between the volumes of two island tubes with periods P and P' respectively reads

$$R_{P,P'} = \frac{V(P,C)}{V(P',C)} = \left(\frac{P}{P'}\right)^{q_0+\alpha C}. \tag{4.59}$$

Since the MIT is period-1 ($P = 1$) and the SIT is period-4 ($P' = 4$), one has

$$R_{4,1} = 4^{q_0+\alpha C}. \tag{4.60}$$

Comparing this with Eq. (4.57), which gives the ratio between the volumes of MIT and SIT, one has

$$\begin{aligned} q_0 &= -1.34/\log 4 = -2.25, \\ \alpha &= -8.04/\log 4 = -13.4. \end{aligned} \tag{4.61}$$

It would be reasonable to assume that the size variation of the invariant tori in the phase space with respect to the perturbation parameter C and period P possesses the same rule in both the two-dimensional (island's area) and three-dimensional (island tube's volume) cases. Recalling the variation of island's area A with respect to the parameter k and period P in Eq. (4.48), $A_P(k) = A_1(k)P^{0.546-3.04k}$, and comparing it with Eq. (4.58) (q_0 and α given in Eq. (4.61)), it seems that the volume of the island tubes decreases much faster with increasing period and perturbation in three-dimensional model than the area with the period and the value of the perturbation in the two-dimensional case.

From the above numerical experiments in two-dimensional and three-dimensional mapping models, one may draw conclusions as follows.

(1) An orbit started close to invariant tori will spend most of time near the hyperbolic invariant sets surrounding the tori, before escaping from the vicinity. This is reasonable because of the following fact: According to the character of the hyperbolic invariant set, an orbit would spend very long time in approaching to (leaving from) a hyperbolic fixed point along the stable (unstable) manifold.

(2) The orbits started close to tori and those started close to hyperbolic invariant sets obey the same diffusion rules, suggesting a common mechanism that causing the stickiness effects for both situations. The hyperbolic invariant sets play an essential role in causing the stickiness effects, both in the two-dimensional and in the three-dimensional phase space. The above conclusions about the essential role of hyperbolic invariant sets in the stickiness effects are only in a qualitative way, in next section, some quantitative results will be presented.

(3) The space occupied by invariant islands in two-dimensional phase space (or island tubes in three-dimensional case) cannot be penetrated by chaotic orbits started outside it. Due to this fact, the size of the islands or island tubes influences the stickiness effects and the diffusion velocity. The size can be expressed as an exponential function of the nonlinearity parameter and it decreases quickly with the periods of the islands or the island tubes for given nonlinearity. From these results it seems that when an orbit diffuses outwards from the central tori more and more, the probability of its encounter with the islands or island tubes should be smaller and smaller, and this could be a reason for the gradually faster diffusion of an orbits farther from the central invariant torus. And the speed of diffusion increases with perturbation parameter.

4.8 Stickiness effect and hyperbolic structure (II)

In Sec. 4.7, it is shown qualitatively that the hyperbolic structures in the phase space play an essential role in causing the stickiness effect. In this section, a quantitative relation between the stickiness effect and the geometric property of hyperbolic structures will be presented.

In a two-dimensional phase space, the hyperbolic periodic orbits are included in cantori and island chains. When an orbit crosses these structures on its diffusion route, it approaches from one direction to the hyperbolic periodic point along the stable manifold and then it may leave from the other direction along the unstable manifold. Therefore, the geometric features of the phase space influence strongly the characteristic of diffusing trajectories. Using a two-dimensional area-preserving twist mapping as the model, the angle between the stable and unstable manifolds of the hyperbolic periodic orbit will be calculated and be related to the orbital diffusion speed.

4.8.1 *Mapping model*

The mapping model used in this section is a modification of the standard mapping. Contrary to the standard mapping whose "twist" depends linearly on the action-variable, the mapping model here has a nonlinear twist. Such a nonlinearity in twist is common in physics.

It is well-known that the standard mapping

$$\begin{cases} \tilde{x}' = \tilde{x} - \tilde{y}', \\ \tilde{y}' = \tilde{y} + \frac{k}{2\pi} \sin(2\pi\tilde{x}), \end{cases} \tag{4.62}$$

can be generated from a generating function

$$\tilde{S} = \frac{1}{2}(\tilde{x} - \tilde{x}')^2 + \frac{k}{4\pi^2} \cos(2\pi\tilde{x}), \tag{4.63}$$

through equations

$$\begin{cases} \tilde{y} = \tilde{S}_{\tilde{x}} = \frac{\partial \tilde{S}}{\partial \tilde{x}} = \tilde{x} - \tilde{x}' - \frac{k}{2\pi} \sin(2\pi\tilde{x}), \\ \tilde{y}' = -\tilde{S}_{\tilde{x}'} = -\frac{\partial \tilde{S}}{\partial \tilde{x}'} = \tilde{x} - \tilde{x}'. \end{cases} \tag{4.64}$$

Modifying the generating function Eq. (4.63) by adding an extra high-order term $\frac{1}{4}(x - x')^4$,

$$S = \frac{1}{2}(x - x')^2 + \frac{1}{4}(x - x')^4 + \frac{k}{4\pi^2} \cos(2\pi x), \tag{4.65}$$

one gets the corresponding area-preserving mapping through

$$\begin{cases} y = S_x = (x - x') + (x - x')^3 - \frac{k}{2\pi}\sin(2\pi x), \\ y' = -S_{x'} = (x - x') + (x - x')^3. \end{cases} \qquad (4.66)$$

Explicitly, the mapping reads

$$\begin{cases} x' = x - \sqrt[3]{\frac{y'}{2} + \sqrt{\frac{y'^2}{4} + \frac{1}{27}}} - \sqrt[3]{\frac{y'}{2} - \sqrt{\frac{y'^2}{4} + \frac{1}{27}}} \quad \mathrm{mod}(1), \\ y' = y + \frac{k}{2\pi}\sin(2\pi x). \end{cases} \qquad (4.67)$$

This mapping is defined on the cylinder $0 \leq x \leq 1$, $-\infty < y < +\infty$, and k is the only one perturbation parameter. As in the standard mapping, the y in this mapping is the action variable, and the x is the angle variable that can perform a modulo over 1. As usual, orbital diffusion refers to the drifting of the action variable y in the phase space. Note here the new iteration variable is denoted by a prime, not like in other sections in this book by subscripts of iteration numbers (for convenience, the subscript space is reserved for the sequence of periodic fixed points as will be seen below). This mapping behaves chaotically around $|y| < M$ (where $M > 0$), and the KAM curves exist at large $|y|$ for any given k (Cheng & Sun, 1996). When k is small, the phase space of the mapping is mainly full of (horizontal) invariant tori. As k increases, the KAM tori break, first at low $|y|$ value, and chaos sets in. The area occupied mainly by chaotic orbits extends outward from the region around $y = 0$. This feature is illustrated in Fig. 4.63 where the phase space is plotted for $k = 20.75$. The negative y half of the phase space is symmetric to the positive half, so that only the positive one is shown in Fig. 4.63.

The mapping is an area-preserving monotone twist mapping. Thus along a vertical line in the phase space the rotation number increases monotonically. The rotation numbers along two vertical lines are computed and shown in Fig. 4.64. Clearly, the scattering distribution in the lower left corner of Fig. 4.64 is a reflection of the chaotic motion at low y value in the phase space. The "plateaus" in both cases arise from the fact that the vertical lines cross stable islands, on which the rotation numbers are constants.

Any chaotic orbit in the connected chaotic sea of the phase space will wander ergodically as time tends to infinity. The stickiness effect happens when an orbit is in the close vicinity around an embedded island, the region where a KAM torus is newly broken, or the region occupied by hyperbolic structures, as is shown in previous sections (Secs. 4.1, 4.7).

More or less arbitrarily, three initial points $P_1 = (0.509, 21.667)$, $P_2 = (0.509, 21.672)$ and $P_3 = (0.509, 21.677)$ are selected, and their orbits are

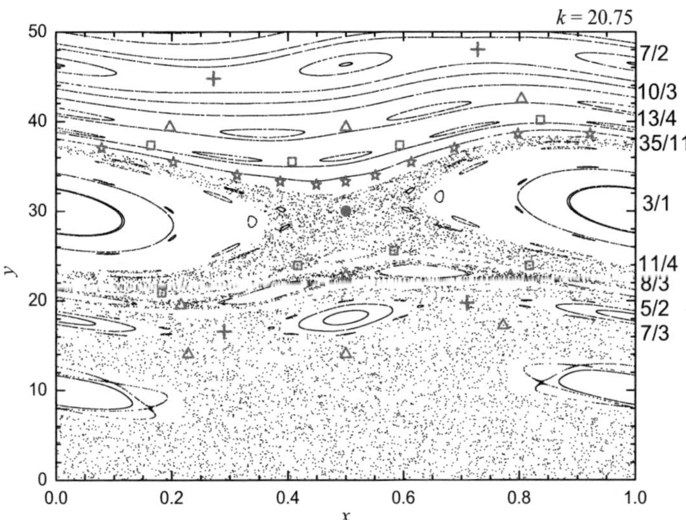

Fig. 4.63   The phase space of the mapping with parameter $k = 20.75$. Some hyperbolic periodic orbits (fixed points) are located and plotted. For a clean appearance, the period-2 orbits are indicated by crosses, and the period-3, 4 orbits are indicated by triangles and squares, respectively. The 3/1 and the 35/11 periodic orbits are indicated by solid dots and stars. The rotation number of the orbits are labelled on the right side. From Zhou *et al.* (2014).

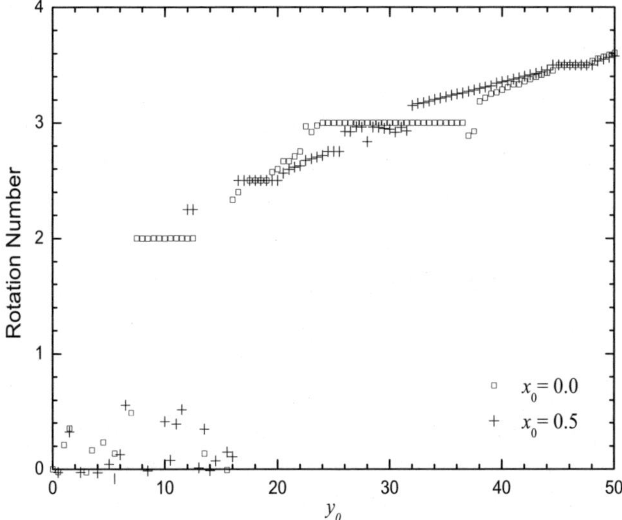

Fig. 4.64   The rotation number of the mapping ($k = 20.75$). Open squares show the change of rotation number along the vertical line $x = 0.0$, while the crosses are for the line $x = 0.5$. From Zhou *et al.* (2014).

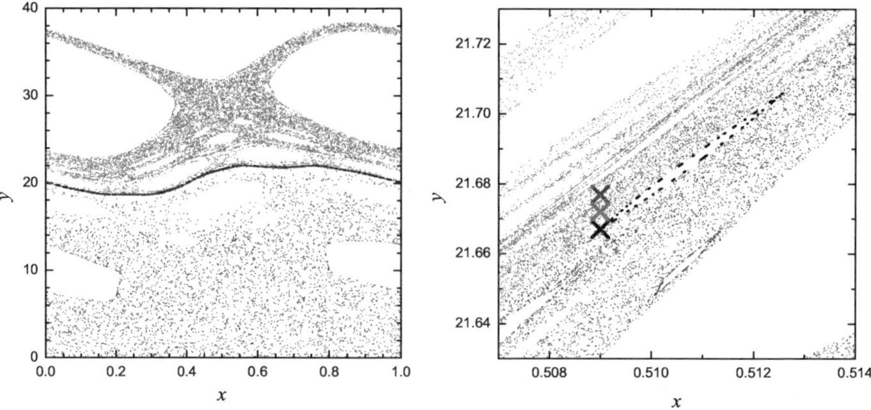

Fig. 4.65 The most sticky region. Three initial points (crosses) and the corresponding trajectories (dots) are shown. In the right panel, the region containing the initial points is magnified. From Zhou *et al.* (2014).

shown in Fig. 4.65, where the orbit of P_3 appears like a thick "curve" and the orbits of P_1 and P_2 spread at both sides of this "curve" respectively. And in fact this thick "curve" is an island chain of high periods, as shown in the magnified part (right panel). It is found that a point initially above the region containing this high-periodic island chain will spend very long time ($\sim 10^9$ iterations) to travel down through this sticky region, implying the existence of a compact partial barrier in the vicinity of this island chain. The rotation number around this obstacle is around 2.6 (see Fig. 4.64). And in fact, this difficult-to-cross barrier is the remnant of the last and most robust KAM torus that has the "golden rate" rotation number, as will be verified later. Since the most significant stickiness effect appears around this region, the following investigation will focus on this region.

4.8.2 *Angle between stable and unstable manifolds*

Generally the diffusion of an orbit is through the hyperbolic structures between the islands in an island-chain, which is in fact the relic of the KAM tori if the rotation number in this position does not satisfy the Diophantine condition (Siegel & Moser, 1971). Conversely, if the rotation number is sufficiently irrational but KAM theorem fails, there would be a cantorus left after the breaking of the KAM torus. And an orbit's diffusion is through the gaps in the cantorus. Moreover, the cantorus itself is composed of hyperbolic points and in the gaps exist the hyperbolic structures. Evidently, the geometric property of the hyperbolic structures will affect the diffusion speed. The angle between the stable and unstable manifolds of the hyperbolic structure is among the most important geometrical properties of the hyperbolic structures.

The locations of hyperbolic periodic points

Usually a recursive algorithm is employed to compute the locations of periodic orbits. But it does not work for the hyperbolic ones because these orbits are unstable. However, the hyperbolic periodic orbits in the mapping model Eq. (4.67) are of minimal action, so the gradient algorithm could be used to find these orbits. The existence of these minimal periodic orbits is guaranteed by Aubry-Mather theory (Bangert, 1988).

Suppose $x_0, x_1, \ldots, x_{p-1}, x_p$ are the angle variables in the configuration space of a minimal periodic orbit of rotation number q/p, with q and p being coprime integers. Obviously, $x_p = x_0 + q$. The action functional f corresponding to this periodic orbit is:

$$f(x_0, x_1, \ldots, x_{p-1}) = S(x_0, x_1) + S(x_1, x_2) + \cdots + S(x_{p-1}, x_p), \quad (4.68)$$

where S is the generating function Eq. (4.65). Then the minimal periodic configuration $(x_0, x_1, \ldots, x_{p-1}, x_p)$ is the minimum of f. Therefore, a periodic orbit of rotation number q/p can be found by locating the minimum of the above mentioned functional.

The gradient algorithm can be applied to find the minimum. Consider the following differential equation:

$$\dot{\Xi} = -\nabla f(\Xi), \tag{4.69}$$

where $\Xi = (\xi_0, \xi_1, \ldots, \xi_{p-1})^T$ is a column vector and the dot over it indicates time derivative. A trajectory defined by this differential equation converges to a minimum of the functional f as $t \to \infty$. Thus, a minimal periodic orbit can be obtained through integrating this differential equation. Specifically in the model adopted here, taking into account the property of the mapping in Eq. (4.66), one has

$$f_{x_k} = S_{x_k}(x_{k-1}, x_k) + S_{x_k}(x_k, x_{k+1}) = -y_k + y_k = 0, \tag{4.70}$$

where the subscript attached to S denotes the partial derivative over the corresponding variable. The corresponding differential equations have a simple form. The nth $(n = 0, 1, \ldots, p-1)$ line of these equations reads:

$$\begin{aligned}
\dot{\xi}_n = {} & -(\xi_n - \xi_{n+1}) - (\xi_n - \xi_{n+1})^3 + \frac{k}{2\pi}\sin(2\pi\xi_n) \\
& + (\xi_{n-1} - \xi_n) + (\xi_{n-1} - \xi_n)^3,
\end{aligned} \tag{4.71}$$

The solution to these differential equations can be computed numerically with any proper numerical integrators, e.g. the 7th order Runge-Kutta-Fehlberg algorithm (RKF7(8)). Integrate the differential equations until the solution reaches the convergent value within a given error (namely 10^{-14} here), then the solution attains the required accuracy.

With the x values of the minimal periodic orbit, the corresponding y values can be computed using the generating function, via Eq. (4.65). And finally the positions of the minimal periodic orbit, i.e. the hyperbolic periodic orbits in the phase space, are obtained. Some of these calculated points on the phase space have been over-plotted in Fig. 4.63.

Characteristic angle of the hyperbolic structure

One of the most important characteristics of the hyperbolic structure is the angle making by directions of the stable and unstable manifolds, which will be called "characteristic angle" for short hereafter. The directions of stable and unstable manifolds are determined by two special tangent vector fields along the periodic orbit, while the tangent vector fields can be obtained by a limit processes. Along an orbit, the projection of the tangent vector field onto the horizontal direction gives a Jacobi field, which can be easily calculated by using the generating function of the mapping.

The definition of Jocobi fields is as follows. Assume a hyperbolic periodic orbit $\{x_n\}_{n=-\infty}^{\infty}$ of rotation number q/p has been found, and it satisfies

$$x_0, x_1, \ldots, x_{p-1} \quad \text{and} \quad x_{n+p} = x_n + q \quad (n = 0, \pm 1, \pm 2, \ldots).$$

Then, $\{\xi_n\}_{n=-\infty}^{\infty}$ is said to be a "Jacobi field" along the orbit $\{x_n\}_{n=-\infty}^{\infty}$ if

$$\begin{aligned} S_{12}(x_{n-1}, x_n)\xi_{n-1} + [S_{22}(x_{n-1}, x_n) + S_{11}(x_n, x_{n+1})]\,\xi_n \\ + S_{12}(x_n, x_{n+1})\xi_{n+1} = 0, \end{aligned} \tag{4.72}$$

where S is the generating function Eq. (4.65) and the subscripts '1' and '2' indicate the partial derivatives with respect to the first and second variables of the function. Actually, the above equations can be written in a matrix form as:

$$\begin{pmatrix} \ddots & \ddots & \ddots & \ddots & \ddots & \\ \cdots & 0 & B_{n-1} & A_n & B_n & 0 & \cdots \\ & \ddots & \ddots & \ddots & \ddots & \ddots \end{pmatrix} \begin{pmatrix} \vdots \\ \xi_{n-1} \\ \xi_n \\ \xi_{n+1} \\ \vdots \end{pmatrix} = 0, \tag{4.73}$$

where $A_n = S_{22}(x_{n-1}, x_n) + S_{11}(x_n, x_{n+1})$, $B_n = S_{12}(x_n, x_{n+1})$ ($n = 0, \pm1, \pm2, \ldots$). Any finite piece of this triangular matrix is positive definite if the corresponding configuration $\{x_n\}$ is minimal. A Jacobi field along a minimal configuration induces an orbit of the corresponding tangent map (Cheng, 2004).

To compute the Jacobi fields along a minimal periodic orbit $\{x_n\}$, which correspond to the stable or unstable directions, the real Jacobi fields can be approximated by solving the following linear equations:

$$
\begin{pmatrix}
A_1 & B_1 & 0 & \cdots & & 0 \\
B_1 & A_2 & B_2 & & \ddots & \vdots \\
\vdots & \ddots & \ddots & \ddots & & \vdots \\
\vdots & & \ddots & B_{n-2} & A_{n-1} & B_{n-1} \\
0 & & \cdots & 0 & B_{n-1} & A_n
\end{pmatrix}
\begin{pmatrix}
\xi_1 \\ \xi_2 \\ \vdots \\ \xi_{n-1} \\ \xi_n
\end{pmatrix}
=
\begin{pmatrix}
-B_0 \\ 0 \\ \vdots \\ 0 \\ 0
\end{pmatrix}.
\tag{4.74}
$$

As $n \to \infty$, the real Jacobi field is obtained. Note that ξ_1 in above equations is monotonically increasing as n goes to infinity. A recursive algorithm may be applied to compute $\xi_1(n)$.

In fact, the limit value of ξ_1 is the projection on x-axis of the stable direction at point (x_1, y_1). With $\xi_0 = 1$ and ξ_1, the projection on y-axis of the stable direction at point (x_0, y_0) can be computed from

$$
\eta_0 = S_{11}(x_0, x_1)\xi_0 + S_{12}(x_0, x_1)\xi_1.
\tag{4.75}
$$

And the stable direction at (x_0, y_0) is defined by $\boldsymbol{r}_s = (1, \eta_0)^T$. Note that this Jacobi field could be used to calculate the Lyapunov exponent λ of the orbit, where

$$
\lambda = -\lim_{n \to \infty} \frac{1}{n} \ln \frac{\xi_n}{\xi_0}.
$$

In a similar way, reversing the direction of calculation, one obtains the projections of the unstable direction at point (x_0, y_0), $\boldsymbol{r}_u = (1, \eta'_0)^T$. And finally, the angle between the stable and unstable directions, i.e. the characteristic angle of the hyperbolic structure, can be calculated easily

$$
\alpha = \arccos\left(\frac{\boldsymbol{r}_s \cdot \boldsymbol{r}_u}{|\boldsymbol{r}_s||\boldsymbol{r}_u|}\right).
\tag{4.76}
$$

As examples of the above algorithm, the positions of a series of hyperbolic periodic orbits and their characteristic angles are calculated. Generally, the more irrational a rotation number is, the more robust the corresponding KAM torus is, and the stronger the stickiness effect is around the relics of the KAM torus after its breaking. People often write the rotation number in the form of continued fraction, on account of the merit of its direct relation to the "irrationality". A continued fraction

$$a_0 + \cfrac{1}{a_1 + \cfrac{1}{a_2 + \cfrac{1}{a_3 + \cdots}}},$$

can be denoted by $[a_0; a_1, a_2, a_3, \ldots]$, where a_0 is the integer part of the number and a_i are the positive integers. Any rational number can be expressed as a truncated continued fraction $[a_0; a_1, a_2, a_3, \ldots, a_n]$. And the integer n will be called the "order" of the continued fraction hereafter. Note an nth order continued fraction ended with $a_n = 2$ is the same as an $(n+1)$th order one ended with $a_n = 1, a_{n+1} = 1$.

Several hyperbolic periodic orbits are calculated and the results are listed in Table 4.2. An obvious tendency derived from Table 4.2 is that, the larger period number (the integer p of the rotation number q/p) leads to the smaller characteristic angle. As the rotation number is getting more and more irrational the angle becomes smaller and smaller. It is well-known that the last KAM tori are those with the "noble" rotation numbers, i.e. the continued fractions that have $a_k = 1$ for all k above a certain number. And the torus with the "golden rate" rotation number $(\sqrt{5}-1)/2 = [0; 1, 1, 1, \ldots, 1, \ldots]$ is surely the most robust one. In Table 4.2, the angle becomes smaller when the rotation number approaches this noble value. And also, the "difficult-to-cross partial barrier" are just in the region corresponding to this noble rotation number.

Table 4.2   Angles between the stable and unstable manifolds of hyperbolic periodic orbits. The first three columns are the rotation numbers given in different forms, the forth column is the order of the continued fraction, and the last column is the angle (in degrees). From Zhou *et al.* (2014).

Rotation Number	Fraction	Continued Fraction	Order	Angle
2.5	5/2	[2;2]	1	169.573
2.6666$\cdots$	8/3	[2;1,2]	2	116.235
2.6	13/5	[2;1,1,2]	3	9.034
2.625	21/8	[2;1,1,1,2]	4	2.094
2.6153$\cdots$	34/13	[2;1,1,1,1,2]	5	1.042
2.6190$\cdots$	55/21	[2;1,1,1,1,1,2]	6	1.657
2.6176$\cdots$	89/34	[2;1,1,1,1,1,1,2]	7	0.742
2.6186$\cdots$	144/55	[2;1,1,1,1,1,1,1,2]	8	0.583

4.8.3 *Stickiness effect and characteristic angle of hyperbolic structure*

When the trajectory of an orbit is plotted on the phase space, the trajectory points accumulate in some specific areas, indicating the existence of some structures causing stickiness effect, as shown in Sec. 4.7.

Typical diffusion route

A chaotic orbit in its diffusion route may diffuse toward any direction at a moment. Since the upper part of the phase space is not open to chaotic orbits but barred by KAM tori (Fig. 4.63), an orbit will diffuse downward overall. To monitor the diffusion, one may check the minimal y value that an orbit attains on its trajectory in the phase space. During the diffusion, each time when the y value reaches a new minimum $y_{\min}$, record the time and the minimum value $(t, y_{\min})$, so that the "back and forth" motion in the diffusion is abandoned, and the "one-way" diffusion is emphasized.

Arbitrarily, two initial points $(x_0, y_0) = (0.49, 29.0)$ and $(0.50, 30.00001)$ are selected and their diffusions are followed. The temporal evolutions of their $y_{\min}$ are shown in Fig. 4.66. Note only part of the orbits, when the minimums are smaller than $y = 20.5$, is displayed. Each point $(t, y_{\min})$ in Fig. 4.66 tells that at time t the orbit attains a new minimal value $y_{\min}$.

On the $(t, y_{\min})$ curve in Fig. 4.66, an orbit makes some steep "steps" and flat "plateaus". A steep step indicates that the orbit diffuses very quickly at this stage, probably crossing a well-developed chaotic region or jumping over a wide island-chain, while a plateau indicates that the orbit spends a long time achieving a small advance in y direction, i.e., it is facing an obstacle on its diffusion route. And the wider the plateau is, the more effective the obstacle is (the stronger stickiness effect it has).

Owing to the chaotic character, two orbits follow different diffusion routes and may suffer different stickiness effects, which is why in Fig. 4.66 two curves differ from each other. But in the same phase space, the sticky regions that they have to pass through are the same. Thus the plateaus appear at the same $y_{\min}$ on two curves. Note the logarithmic scale of the abscissa affects the apparent widths of the plateaus.

When an orbit reaches a new minimum $y_{\min}$ at moment t, a temporary (local) rotation number ω of the orbit around this point is calculated and these records $(\omega, y_{\min})$ are plotted in Fig. 4.66 too, as the curves crossing the panel from lower left to upper right. Again, the jumping of ω implies that the orbit is temporarily in the vicinity of a big island, while the (nearly)

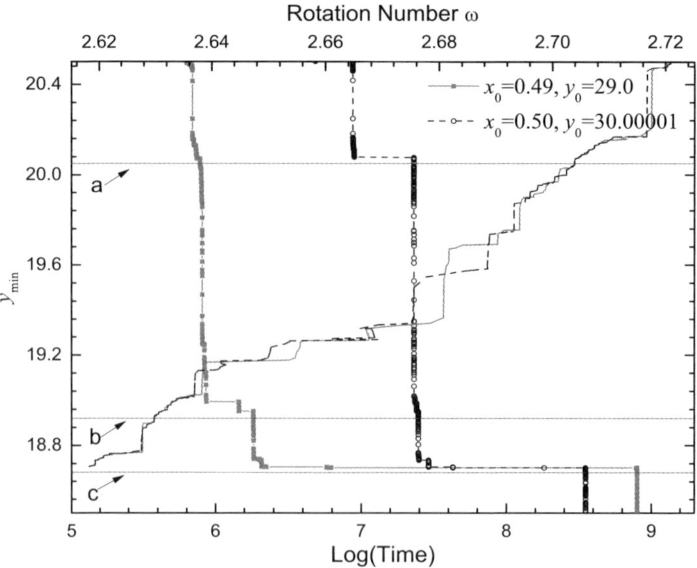

Fig. 4.66   The variation of minimal y of an orbit with respect to time (curves from upper left to lower right). The solid squares are for the orbit initialized at $(0.49, 29.0)$ and the open circles for $(0.50, 30.00001)$. The rotation numbers (indicated on the top x-axis) of orbits around the corresponding local lowest position are plotted (two curves crossing the panel from the lower left to upper right). Three regions (see text for explanations) that will be analyzed below are indicated by horizontal lines. From Zhou *et al.* (2014).

continuous variation of ω implies that the orbit is crossing a series of finely ground island-chains, or a cantorus. The consistence between the continuous variation of ω and the wide plateau of y_{min} is evident in Fig. 4.66.

A diffusion orbit in the phase space meets considerable resistance when crossing some regions, where the (t, y_{min}) curves make plateaus and the (ω, y_{min}) curves vary continuously. Three such "sticky regions" are chosen to investigate in detail. Three horizontal lines in Fig. 4.66 show the lower bounds of these regions.

As an example, the sticky region labelled Region **c** is defined as follows. A lower bound $y_c^l = 18.68$, as shown in Fig. 4.66, is set. For an orbit starting above this lower bound, if its y value gets smaller than y_c^l for the first time during its diffusion, it is regarded as having crossed Region **c**. In addition, an upper bound y_c^u is selected as well. When the y value of an orbit gets smaller than y_c^u for the first time, the orbit is defined as having entered Region **c**. The difference between the upper and lower bound $\delta y_c = y_c^u - y_c^l$ is the width of this sticky region.

Characteristic angle of hyperbolic structure and diffusion speed

The hyperbolic periodic points can be located and the characteristic angles of the hyperbolic structures can be calculated, using the algorithm introduced above. Moreover, the diffusion speeds in these regions can also be measured numerically. Therefore, a quantitative relation between the characteristic angle and the diffusion speed may be obtained.

By scanning the Regions **a**, **b** and **c**, the most sticky hyperbolic structures are searched. This is possible because in a specific region the significant stickiness effect always appears when the orbits cross the island chains with high period numbers (multiplicities), while the period has a direct relation with the rotation number. So, by looking for those hyperbolic periodic orbits with a relatively irrational rotation numbers (equals to relatively longer continued fractions), those most sticky hyperbolic structures can be determined. Some characteristic angles of such hyperbolic structures in Regions **a**, **b** and **c** are calculated and summarized in Table 4.3.

It has been shown in Table 4.2 that the characteristic angles of hyperbolic structures with higher period numbers are smaller. This trend can be seen again in each region in Table 4.3. But when the characteristic angles are considered with nearly the same period number in different regions, for example, the characteristic angles corresponding to the rotation number 192/71 in Region **a**, 192/73 in **b** and 199/76 in **c**, the angle in Region **c** (0.506°) is significantly smaller than the other two in Regions **a** (0.772°) and **b** (0.777°), while the latter two are nearly equal. This result is expectable since Region **c** makes the widest plateau in Fig. 4.66 and the diffusion time through this region is in the order of $10^8 \sim 10^9$, much longer than that in Regions **a** and **b**.

The characteristic angle may decrease as the period of orbits increases, but it is impossible in practice to compute the characteristic angles for all the hyperbolic structures. Fortunately, the information about the smallest characteristic angle in a region can still be obtained. In Region **c**, the rotation numbers of the hyperbolic periodic orbits are close to the "golden rate", corresponding to the most robust KAM torus. Meanwhile, it is natural to assume that the "smallest" characteristic angle should be found in the hyperbolic structures of the new-born cantorus just after the breaking of KAM torus. Since the golden rate can be approached by continued fractions, the characteristic angles of hyperbolic structures with rotation numbers of truncated continued fractions may approach the characteristic

Table 4.3 Characteristic angles of hyperbolic structures in three sticky regions. The hyperbolic periodic orbits are denoted by their rotation numbers, given in normal fraction numbers and continued fractions. The angles are in degrees. From Zhou *et al.* (2014).

Rotation Number		
Fraction	Continued Fraction	Angle (α)
Region **a**		
165/61	[2;1,2,2,1,1,3]	0.718653
119/44	[2;1,2,2,1,1,2]	0.845155
192/71	[2;1,2,2,1,1,1,2]	0.772372
73/27	[2;1,2,2,1,2]	1.094884
Region **b**		
171/65	[2;1,1,1,2,2,3]	0.717274
121/46	[2;1,1,1,2,2,2]	0.863938
192/73	[2;1,1,1,2,2,1,2]	0.777229
71/27	[2;1,1,1,2,3]	1.179671
Region **c**		
199/76	[2;1,1,1,1,1,1,1,3]	0.506420
343/131	[2;1,1,1,1,1,1,1,2,2]	0.472460
144/55	[2;1,1,1,1,1,1,1,2]	0.582802
377/144	[2;1,1,1,1,1,1,1,1,1,2]	0.477654
233/89	[2;1,1,1,1,1,1,1,1,2]	0.534822
89/34	[2;1,1,1,1,1,1,2]	0.741975

angle in the cantorus. Following this idea, the calculations listed in Table 4.2 can be continued to higher order of the continued fraction. The variation of the characteristic angle with respect to the order of the continued fraction is illustrated in Fig. 4.67. Note, as the order increases the corresponding periodic orbits tend to gather around the golden rate cantori, and no later than the 7th order (89/34) all the hyperbolic structures under consideration accumulate in Region **c**.

As shown in Fig. 4.67, the calculations are stopped at the 16th order where the rotation number is 10946/4181. The calculations have to be terminated because of two reasons as follows. First, the calculations so far have shown that the characteristic angle has a lower bound when the truncated continued fractions approach the golden rate. Even a higher order does not decrease this lower bound. Second, calculations for the high order continued fractions (therefore, high period number) are very expensive in

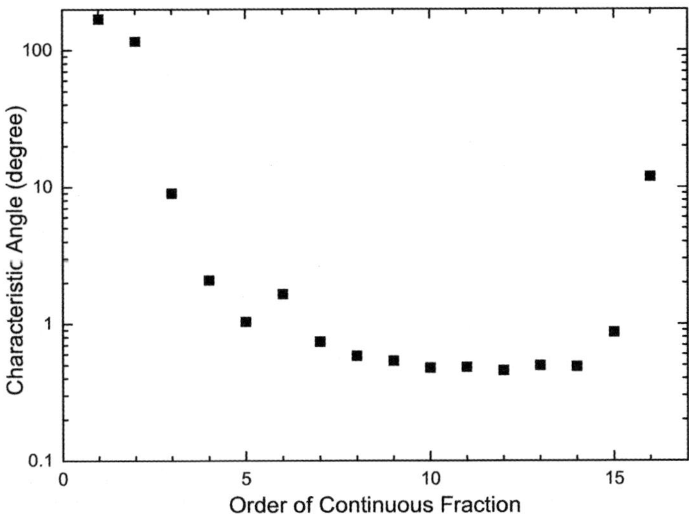

Fig. 4.67 The lower bound of characteristic angles in Region **c**. The abscissa is the order of the truncated continued fraction, which is the rotation number of the hyperbolic periodic orbit. The ordinate is the corresponding characteristic angle, in degrees. The last point of order 16 represents the hyperbolic periodic orbit with rotation number 10946/4181. From Zhou *et al.* (2014).

computer time, and it is difficult to control the error in this case due to the large number of linear equations as Eq. (4.74). So the smallest angle in Fig. 4.67 at the 12th order, $\alpha_c = 0.457°$, is adopted as the lower bound of the characteristic angle in Region **c**.

Although there is no golden rate rotation number for Regions **a** and **b**, a similar process as for Region **c** can be followed to calculate the lower bound of characteristic angles. Expand the continued fractions listed in Table 4.3 by inserting number '1' into them right before the last number, a series of truncated continued fractions of different orders is obtained. Then adopt these continued fractions as the rotation numbers, the characteristic angles of the corresponding hyperbolic structures can be calculated, and finally the lower bound of the characteristic angles $\alpha_a = 0.665°$ in Region **a** and $\alpha_b = 0.667°$ in Region **b** are obtained.

Since $\alpha_a \approx \alpha_b > \alpha_c$, one may derive directly from these values that the diffusion in Region **c** is slower than that in Regions **a** and **b**. In other words, the stickiness effect in Region **c** is more intensive than that in Regions **a** and **b**, while Regions **a** and **b** have nearly the same stickiness effects. This can be verified by some further calculations of diffusion speed below.

The diffusion speed, generally is measured by the decay rate of the survival numbers of a bunch of orbits in a defined area. Set randomly $40,000$ initial points in a rectangle from $(0.45, 26.5)$ to $(0.55, 31.5)$ around the hyperbolic fixed point of rotation number $3/1$ that is embedded in a developed chaotic area and above the Regions **a**, **b** and **c** in the phase space (see Fig. 4.63). All the orbits are iterated for 10^9 times using the mapping. The evolution of these 40,000 orbits are then followed, and especially the variation of the action variable y of each orbit is monitored.

The the lower boundaries of sticky Regions **a**, **b** and **c** are chosen to be $y_a^l = 20.05$, $y_b^l = 18.92$, $y_c^l = 18.68$ respectively, as indicated by the horizontal lines in Fig. 4.66. After some tests, the width of the sticky region is set as $\delta y = 0.1$ for all three regions. Thus the upper boundaries of the sticky regions are then defined as $y_{a,b,c}^u = y_{a,b,c}^l + 0.1$.

An orbit may wander for quite a while before it enters a specific area. To exclude the evolution history before an orbit entering the vicinity of Regions **a**, **b** or **c**, the moment t_a^u (or t_b^u, t_c^u) when for the first time the y value of an orbit is smaller than the upper boundary y_a^u (or y_b^u, y_c^u) is defined as the moment of entering Region **a** (or Region **b**, **c**). On the other hand, when an orbit satisfies for the first time $y < y_a^l$ (or y_b^l, y_c^l) at moment t_a^l (or t_b^l, t_c^l), it is regarded as having escaped from the sticky region. The time duration $\Delta t_{a,b,c} = t_{a,b,c}^l - t_{a,b,c}^u$ then is defined as the surviving time of an orbit in the corresponding sticky region. Finally the variation of surviving number of orbits (normalized to unit) with respect to time in the three regions is summarized in Fig. 4.68.

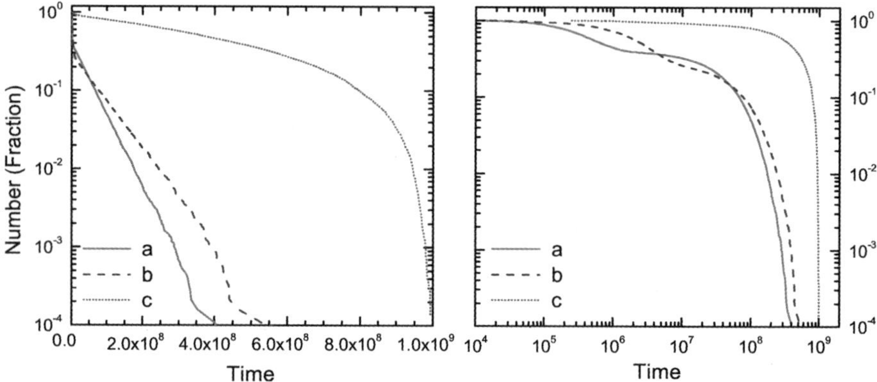

Fig. 4.68   The diffusion speeds of orbits in Regions **a**, **b** and **c**. The time is given in linear scale in the left panel, and logarithmic scale in the right panel. From Zhou *et al.* (2014).

Apparently, the diffusion is quick in Regions **a** and **b**, but slow in Region **c**, reflecting that the stickiness effect in Region **c** is much stronger than in other regions. On the other hand, although Regions **a** and **b** occupy different areas in the phase space, they have nearly the same diffusion speed. The results are consistent with the conclusion derived from the characteristic angle values.

Characteristic angle and perturbation parameter

For a given perturbation parameter k, the stickiness effects of different sticky regions are related to the characteristic angles of hyperbolic structures embedded in the regions. Smaller characteristic angles lead to stronger stickiness effects and slower diffusions. Take Region **c** as an example, where the diffusing orbits suffer the most significant stickiness effects, the relation between the stickiness effect and the characteristic angle when the parameter k changes is investigated below.

Five hyperbolic periodic orbits, staying close to each other in a narrow area in Region **c** and with rotation numbers 377/144, 199/76, 343/131, 144/55 and 233/89 are selected. Their characteristic angles at different k values are calculated and the results are illustrated in Fig. 4.69 and some of them are listed in Table 4.4 too.

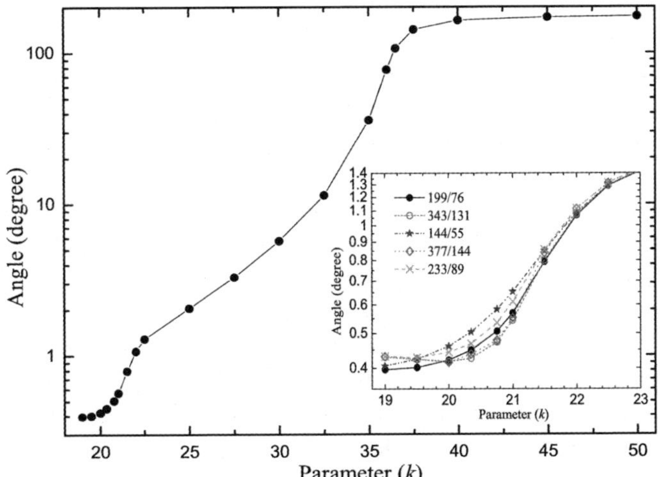

Fig. 4.69 The variation of characteristic angle with respect to the parameter k. The solid circles show the angle of the hyperbolic structure with rotation number 199/76. The embedded panel shows the characteristic angles of several hyperbolic structures with different rotation numbers. From Zhou *et al.* (2014).

Table 4.4 The characteristic angles at different parameter k. The angles are in degrees. Only five hyperbolic structures in Region **c** are listed here. From Zhou *et al.* (2014).

	Angle (α)						
Rot. Num.	$k = 19.00$	20.35	21.50	22.50	27.50	32.50	36.00
199/76	0.39616	0.44887	0.79279	1.29196	3.28884	11.4248	77.0651
343/131	0.42951	0.42608	0.79933	1.29455	3.28884	11.4248	77.0651
144/55	0.40566	0.50425	0.85153	1.29567	3.28884	11.4248	77.0651
377/144	0.43103	0.43743	0.84026	1.31475	3.28884	11.4248	77.0651
233/89	0.43126	0.46775	0.85682	1.31481	3.28884	11.4248	77.0651

When k is small, the global KAM tori exist in Region **c** and these hyperbolic structures are bounded by these tori. In the vicinity of an invariant torus in a two-dimensional phase space, the resonances (island-chains and hyperbolic periodic orbits) accumulate geometrically toward the torus with their increasing orders (period numbers). Owing to continuity, the directions of the stable and unstable manifolds of the hyperbolic structures are "forced" to be nearly parallel to the torus. Thus the characteristic angles are quite small in the close neighborhood of a torus. The angles in the embedded panel in Fig. 4.69 seem to approach a limit slowly as k decreases. Consequently, orbits starting from the close vicinity of an invariant torus will diffuse along the resonance line (high-order island-chain) for a long time before its leaving for adjoining resonances. As k increases, the chaos is developed and the characteristic angle increases. Particularly, after the global KAM tori disappear in this region (when $k > 20.75$) the characteristic angle increases dramatically. As a result, the diffusion speed is expected to be enhanced dramatically.

Another interesting phenomenon is the difference between the characteristic angles of different rotation numbers. When k is small, the characteristic angles are small and the relative difference between different rotation numbers is considerable. But the relative difference becomes smaller as the angles increase and almost invisible when $\alpha \sim 2°(k = 25)$. This is clearly illustrated in the embedded picture in Fig. 4.69. The reason is that the difference of hyperbolicity between different periodic orbits is apparent when k is small, but it becomes ignorable when k is large. In the latter case, chaos has developed, making the whole region uniformly hyperbolic. Thanks to this observation, to show the angle variation in a wide k range, it is enough to follow just one of the hyperbolic structures, e.g. the one shown in Fig. 4.69 with rotation number 199/76.

The stickiness effect (or the diffusion speed) changes when the characteristic angle changes with k. For several different k values, 40,000 points

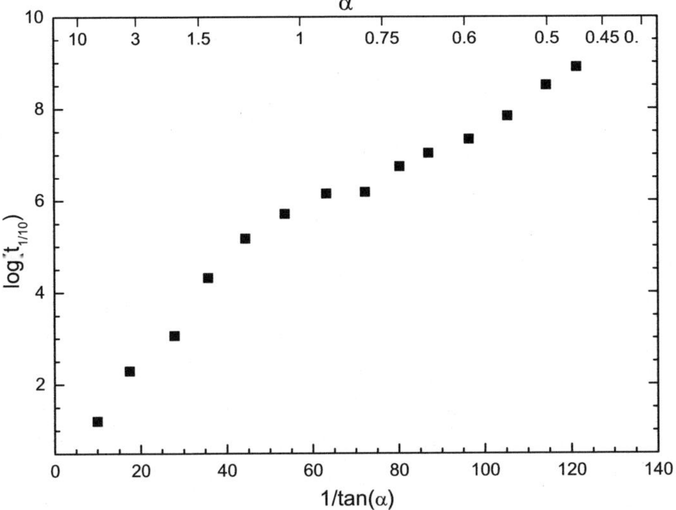

Fig. 4.70   The variation of escaping time with respect to the characteristic angle (α). The abscissa is $\cot\alpha = 1/\tan\alpha$ and the ordinate is the escaping time ($t_{1/10}$, see text for definition) in logarithm. Several α values are indicated along the top axis too. From Zhou *et al.* (2014).

initialized in the rectangle area for each k are followed up to 10^9 times iterations. Their surviving times in Region **c** are recorded.

The diffusion speeds up as k increases. The time when 90 percent of orbits have escaped, denoted by $t_{1/10}$, is used to measure the diffusion speed (or inversely, the intensity of stickiness effect). A quick diffusion has a small $t_{1/10}$ while a large $t_{1/10}$ indicates a slow diffusion (and strong stickiness effect). For k varying from 20.75 to 30.00, the $t_{1/10}$ and the characteristic angles are computed, and the results are illustrated in Fig. 4.70. To visualize the variation of diffusion speed at small characteristic angle, $1/\tan\alpha$ instead of α itself is set to be the abscissa in Fig. 4.70, while several α values are indicated along the top axis. The ordinate is the diffusion speed $t_{1/10}$ in logarithm scale. For a given parameter k, the characteristic angle in Fig. 4.70 is the lower bound of the characteristic angles in Region **c**, calculated in the same way as above done for $k = 2.075$.

As indicated by Fig. 4.70, the stickiness effect is significant only when the characteristic angle is small. When the characteristic angle $\alpha \sim 5°$ ($1/\tan\alpha = 11.5$), most of orbits have escaped just after tens of iterations, implying a weak stickiness effect. On the other end of smaller α in Fig. 4.70, the escaping time will be longer than 10^9 when α is smaller than $0.45°(1/\tan 0.45° = 127.3)$. An orbit with such a long diffusion time can be practically regarded as stable (never escape). In fact, the global KAM curves exist in Region **c** when $k < 20.5$. And the characteristic angle is smaller than $0.45°$ when $k < 20.5$ (Fig. 4.69). Last but maybe not least, it is worth noting that the points in Fig. 4.70 may be fitted by two linear functions with a crossover around $\alpha \sim 1°$.

4.9 Diffusion in four-dimension mapping

The diffusion behavior of orbits in two-dimensional mapping has been studied previously in Sec. 4.1, Sec. 4.7. The stickiness effects of invariant tori, island-chains and cantori are found to influence greatly the diffusion velocities of orbits, and the hyperbolic structures play a main role in the stickiness effects. According to the KAM theorem for Hamiltonian systems with two degrees of freedom, the two-dimensional invariant tori will prevent the escape of orbits on a three-dimensional energy surface. When the degree of freedom n exceeds two, the n-dimensional invariant tori cannot divide the $(2n-1)$-dimensional energy surface into two disconnected parts, so the escape will appear. Because of the existence of n-dimensional invariant tori, the escape route across a net of invariant tori will be complicated and the velocity of escape will be very slow. This is called Arnold diffusion. Such kind of diffusion can hardly be detected by numerical methods.

In the phase space of a dynamical system, there are invariant manifolds, chaotic sea, fixed points, etc. For higher dimensional system, the investigation on the diffusion characters in different regions will be presented here. The four-dimensional volume-preserving mapping is the simplest system that corresponds to a higher-dimensional Hamiltonian system. The diffusion characters in different regions of the four-dimensional mapping will be discussed below.

4.9.1 *Mapping model*

The following volume-preserving mapping of four-dimension is adopted in this section:

$$T_{hp} : \begin{cases} x_{n+1} = s(x_n \cos \varphi_n - y_n \sin \varphi_n), \\ y_{n+1} = s^{-1}(x_n \sin \varphi_n + y_n \cos \varphi_n) + c\sin(x_{n+1} + z_{n+1}), \\ z_{n+1} = z_n - bt_n^3, \\ t_{n+1} = z_n + t_n - bt_n^3 + c\sin(x_{n+1} + z_{n+1}). \end{cases} \quad (4.77)$$

where φ_n is defined as

$$\varphi_n = (x_n^2 + y_n^2)^k. \tag{4.78}$$

This mapping T_{hp} is the coupling of the following two area-preserving mappings

$$T_h : \begin{cases} x_{n+1} = s(x_n \cos\varphi_n - y_n \sin\varphi_n), \\ y_{n+1} = s^{-1}(x_n \sin\varphi_n + y_n \cos\varphi_n), \end{cases} \tag{4.79}$$

and

$$T_p : \begin{cases} z_{n+1} = z_n - bt_n^3, \\ t_{n+1} = z_n + t_n - bt_n^3. \end{cases} \tag{4.80}$$

Here b, s, k are parameters ($s \neq 1$) and c the perturbation parameter (coupling parameter). When $k > 0$ the $(x, y) = (0, 0)$ is a hyperbolic fixed point of T_h, and when $k < 0$ the mapping T_h is not defined at $(0, 0)$ and $(z, t) = (0, 0)$ is a parabolic fixed point of T_p. In this section, the parameters are taken as $s = 1.05, k = -1.5, b = 1.5, c = 0.03$, in which case the mapping T_{hp} is not defined at the origin. Obviously the mapping T_{hp} is symmetric with respect to the origin, and it is easy to verify that the mapping T_{hp} is volume-preserving but not symplectic.

The structure of phase space, i.e. the spread of ordered and chaotic regions and the hyperbolic structures, of the mapping T_{hp} is explored by plotting the projections of the phase space, by computing the LCN's and by exploring the fixed points and their stabilities. Figure 4.71 are the projections of some invariant tori (roughly approximated by quasi-periodic orbits) and chaotic sea onto the planes (x, y) and (z, t) respectively. These figures display the global structure of phase space. It is known that the dimension of the invariant tori is two and these tori cannot divide the four-dimensional phase space into two disconnected parts. As a result, the chaotic orbits diffuse around the invariant tori, as easily seen in Fig. 4.71.

4.9.2 *Diffusion characters in different regions*

In order to clarify the diffusion of orbits in the phase space globally, the escape time of points along a selected line in the phase space is investigated. The selected line passes through the origin and the point of which the projections on the (x, y) and (z, t) planes are respectively the farthest

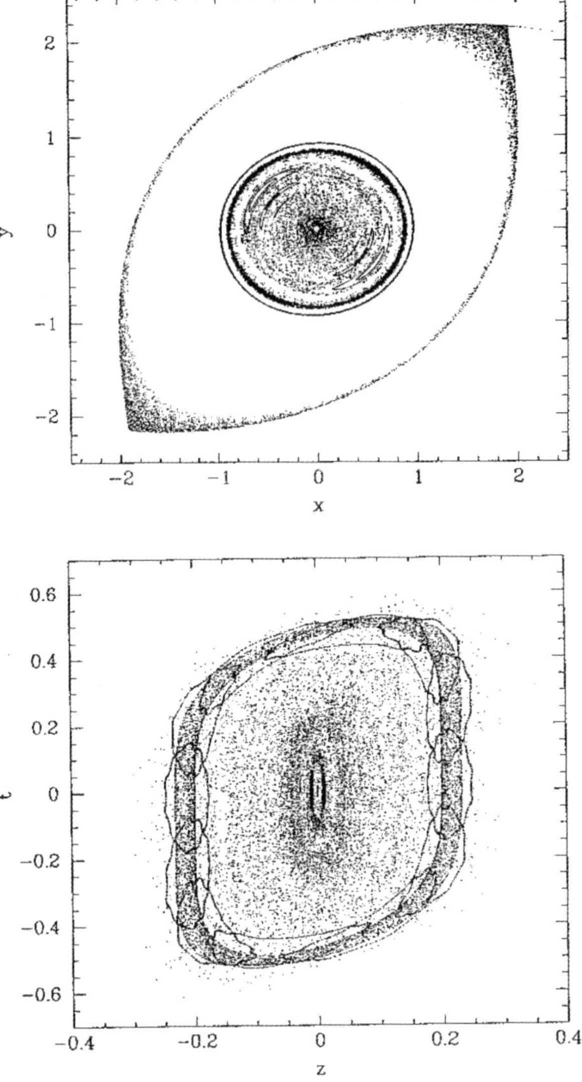

Fig. 4.71   The projections on the two coordinate planes of the quasi-periodic orbits starting from the following initial points: (0.376, 0.582, −0.0002, 0.008), (0.232, 0.658, 0.00019, 0.0038), (0.68, 0.68, 0.2125, 0.4250), (0.356, −0.246, −0.012, 0.003), (0.392, −0.268, 0, 0.0058), (0.2, −0.5, −0.0127, 0.0016), (0.40, −0.37, −0.0127, 0.0016), (0.16, −0.65, 0, 0.009), (0.61, 0.61, 0.190625, 0.38125) and their symmetric points with respect to the origin for the mapping T_{hp}, and the chaotic orbits starting from the following initial points: (0.261101993556, 0.341296119830, −0.007526015390, 0.001646639192), (1.918131071904, 2.156373097694, −0.028487926171, −0.000415023486), (0.68, 0.53, 0.190625, 0.381250). The parameters are $s = 1.05, k = −1.5, b = 1.5$ and $c = 0.03$. From Sun & Fu (1999).

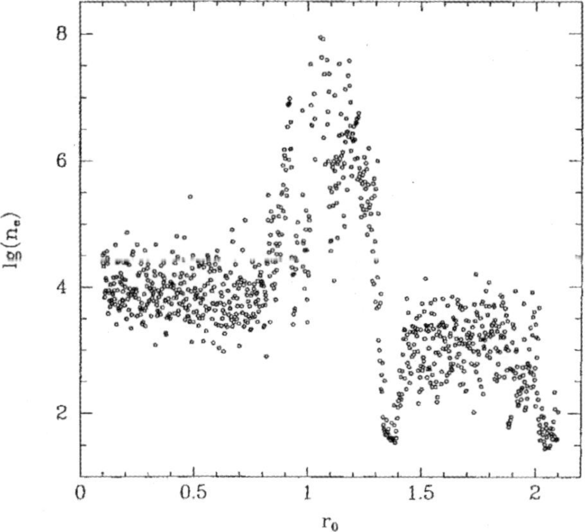

Fig. 4.72   Distribution of escape time of orbits with respect to the distance r_0 along the line Eq. (4.81). From Sun & Fu (1999).

boundary point of the ordered region for the mappings T_h and T_p. The equation of this line is

$$5x = 5y = 16z = 8t. \tag{4.81}$$

The 1,001 initial points on this line are so selected that their distances to the origin are as follows

$$r_0 = 0.1 + 0.002i, \quad i = 0, 1, \ldots, 1000. \tag{4.82}$$

An orbit is defined as escape if $r = \sqrt{x^2 + y^2 + z^2} > 2.5$. The escape time of each orbit is calculated, and the results (not including the orbits on the invariant tori) are plotted in Fig. 4.72. In this figure, a slow escape region near $r = 1.05$ can be found, where most of the explored initial points on invariant tori locate. This slow escape implies that the escape velocity of orbits near the invariant tori is much slower than that in the chaotic region. This is due to the stickiness effect of invariant tori as can be seen below. In Fig. 4.72 one can also see that even in the "dense" region of invariant tori, there exist chaotic zones characterized by faster escape. In the following, the diffusion characters in different regions will be investigated.

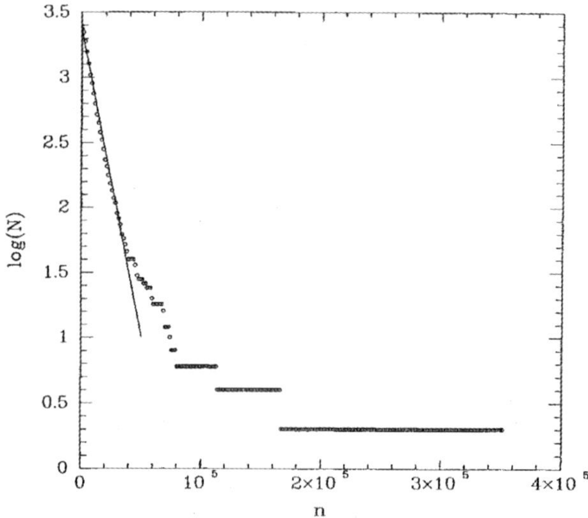

Fig. 4.73 Diagrams of the number of orbits having not escaped before the iteration number n for the case of in chaotic sea. From Sun & Fu (1999).

Chaotic sea

In the chaotic region, 2352 initial points (x_0, y_0, z_0, t) near the origin are selected as

$$(x_0, y_0, z_0, t_0) = (i, j, l, m)q,$$
$$q = 0.1, \quad i, j, l, m = 0, \pm 1, \pm 1.5, \pm 2 \ (\text{excluding } i = j = 0). \tag{4.83}$$

As before, define the escape of orbits as $r = \sqrt{x^2 + y^2 + z^2} > 2.5$. The number N, of orbits that have not escaped yet before the iteration number n, is counted and the variation of N with respect to n is plotted in Fig. 4.73 as open circles (and N in a logarithm scale). Because there are positive LCN's for the chaotic orbits, the numerical results are fitted by an exponential law as

$$N = N_0 \times 10^{\alpha n}. \tag{4.84}$$

However, this law cannot be applied for large n, as can be easily seen in Fig. 4.73. Actually, there should exist some island chains that exert the stickiness effect on the orbits near them as pointed out in Sec. 4.1. Therefore, only those points with $n < 40,000$ (including about 99% of the explored orbits) are fitted in Fig. 4.73 and the fitting gives $N_0 = 2,532, \alpha = 4.735163 \times 10^{-5}$ for the fitted line. Apparently, this analytical result is in good agreement with the numerical values for $n < 40,000$.

From Fig. 4.73, one can also see that there exist three "lines" parallel to the abscissa, on which the orbits diffuse slowly. On the lowest line, the orbits even do not diffuse. Nevertheless, the number of orbits on these three lines is less than 10. Probably these orbits are very near to the islands, some of them may be on the islands. Accordingly, it is reasonable to suspect that the diffusion of orbits in a "complete" chaotic sea, i.e. the region where the area occupied by islands can be neglected, should possess the exponential law. As it is difficult to find out small tori (islands) in the four-dimensional phase space, the above point is just a conjecture.

Vicinity of invariant torus

There exists a slow escape region near the point on the line Eq. (4.81), of which the distance to the origin is $r = 1.05$ (see Fig. 4.72).

By computing the LCN's and drawing the projective figures of the orbit starting from this point, one finds that the point is on an invariant torus. Below, the diffusion character of orbits in the vicinity of this torus will be investigated by taking 2,401 initial points (x_0, y_0, z_0, t_0) as follows

$$(x_0, y_0, z_0, t_0) = (x_c, y_c, z_c, t_c) + (i, j, l, m)q,$$
$$q = 0.005, \quad i, j, l, m = 0, \pm 1, \pm 2, \pm 3. \tag{4.85}$$

First, a region in which the orbit will stay for quite a long time ($n \geq 4096$) will be chosen. This region is chosen as small as possible so that it contains as less points far from the torus as possible. After some testing calculations, the escape of orbits is defined as they leave the following zone.

$$G = \{(x, y, z, t) : \ x \in (-0.98, 0.98), y \in (-0.94, 0.94),$$
$$z \in (-0.25, 0.26), t \in (-0.55, 0.55), \sqrt{x^2 + y^2} \in (0.86, 0.99)\}. \tag{4.86}$$

(Note: $\sqrt{x^2 + y^2}$ is an invariant of mapping T_{hp} with $s = 1, c = 0$.)

Figure 4.74 shows the diagram of $\log N$ versus n with 605 orbits escaped before $n = 10^8$ iterations, where N is the number of orbits in the region G at the iteration number n. Note that only a small fraction of the orbits escapes before $n = 10^7$ iterations, implying that the diffusion in the vicinity of invariant tori is very slow. This is due to the stickiness effect of invariant tori. But, here in this case the diffusion is different from the usual one that has an algebraic decay. Considering the method used to study the usual stickiness effect in related papers (Meiss & Ott, 1985; Ding *et al.*, 1990), one may find that the algebraic decay law in fact is the "averaged" law of diffusion over the chaotic region with invariant islands and cantori

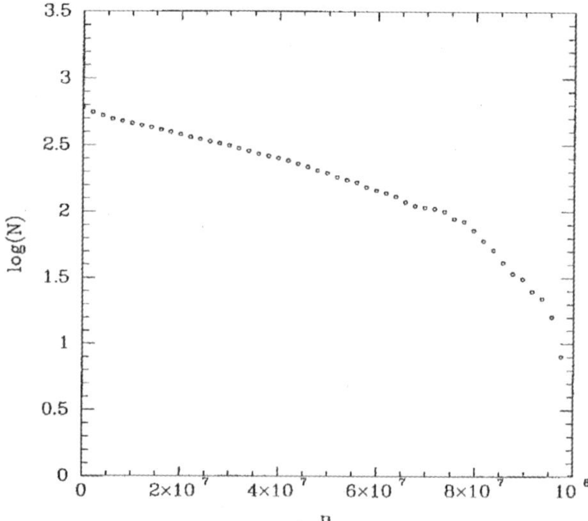

Fig. 4.74   Diagrams of the number of orbits having not escaped before the iteration number n for the case of in the vicinity of invariant torus. From Sun & Fu (1999).

embedded in. The orbits diffuse sometimes "freely" through chaotic sea, and sometimes by crossing cantori and passing around islands, so the usual "stickiness effect of invariant tori" is actually modified by the diffusion process in chaotic sea. If a region of higher dimensional phase space is filled "densely" by invariant tori, the stickiness effect would appear to be that in this case presented here.

Vicinity of hyperbolic-elliptic fixed point

By exploring the character of fixed points, the fixed point

$$(x_h, y_h, z_h, t_h) = (1.918131071904, 2.15637309767694,$$
$$- 0.028487926171, -0.000415023486)$$

is a hyperbolic-elliptic fixed point, i.e. it is hyperbolic in the subspace of (x, y) and is elliptic in the subspace of (z, t). Choose 2,401 initial points (x_0, y_0, z_0, t_0) in its vicinity as follows

$$
\begin{aligned}
(x_0, y_0, z_0, t_0) &= (x_h, y_h, z_h, t_h) + (i, j, l, m)q, \\
q &= 10^{-5}, \ i, j, l, m = 0, \pm 1, \pm 2, \pm 3,
\end{aligned}
\tag{4.87}
$$

and the diffusion character in the vicinity of the above fixed point are studied through these points. The orbit with initial point sufficiently close

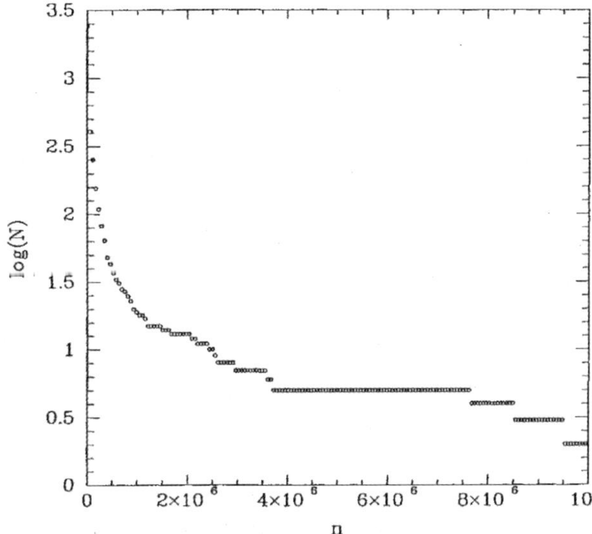

Fig. 4.75 Diagrams of the number of orbits having not escaped before the iteration number n for the case of in the vicinity of hyperbolic-elliptic fixed point. From Sun & Fu (1999).

to the hyperbolic-elliptic fixed point should wander near the hyperbolic structure on the plane (x, y) and near the origin on the plane (z, t) for quite a long time. After some testing calculations, it is found that as soon as the distance d between the fixed point and the returning point of orbit to the vicinity of the fixed point exceeds $\bar{d} = 0.670213829709$ or the distance d' of the orbit to the origin exceeds 10, the orbit will diffuse to very distant point from the origin $(d' > 10^5)$ with no return. Thus $d > \bar{d}$ or $d' > 10$ is regarded as escape.

Figure 4.75 shows the variation of N, number of orbits having not escaped yet, with respect to the number of iteration n. This result is similar to that in chaotic sea, except that there are some orbits with much slower diffusion, of which the initial points are very close to the hyperbolic-elliptic fixed point. This implies that the hyperbolic-elliptic fixed point also has stickiness effect on some nearby orbits. Another interesting feature is the appearance of fast escape orbits starting near the hyperbolic-elliptic fixed point, causing fast decreasing of N for small n (about 49% of the explored orbits escaped before $n = 145$). It was found that there exists the "hole" of escape embedded in stickiness region near an invariant curve of two-dimensional area-preserving mapping. The above mentioned feature shows

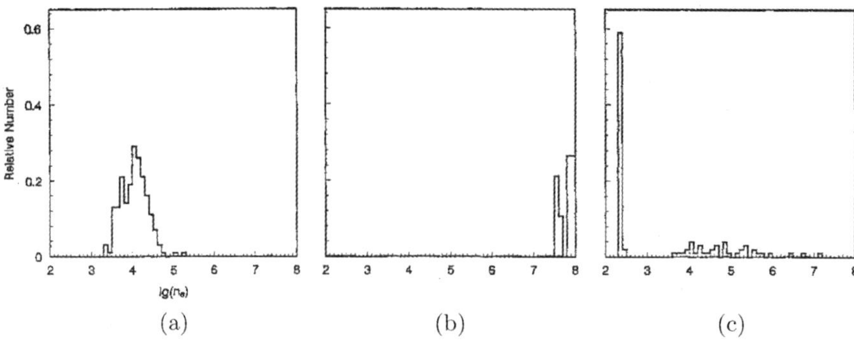

Fig. 4.76   Distribution of the relative number of orbits escaping at the iteration number n for (a) chaotic sea, (b) vicinity of invariant torus and (c) vicinity of hyperbolic-elliptic fixed point. From Sun & Fu (1999).

that the "holes" of escape also exist in the vicinity of the hyperbolic-elliptic fixed point of four-dimensional volume-preserving mapping.

In the above studies, the diffusion characters in three different regions have been studied separately. Below the diffusion velocities for different cases, with the same definition of escape, will be compared. For this purpose, randomly 250 initial points in each region as above are chosen and their orbital evolutions are followed. An orbit is regarded as escape if the distance of an orbit point to the origin exceeds 100. In the chaotic region and the vicinity of hyperbolic-elliptic fixed point, all the 250 orbits escape, but in the vicinity of invariant torus there are only 45 orbits escaping before $n = 10^8$ iterations. All the results are illustrated in Fig. 4.76. Apparently, the diffusion velocities in different regions are different. The diffusion in chaotic sea is the fastest, while the slowest one is in the vicinity of invariant torus. In the vicinity of hyperbolic-elliptic fixed point the diffusion velocity is between th above two cases, because of the existence of narrow weak stickiness zone nearby the perturbed hyperbolic-elliptic structure.

In a higher dimensional volume-preserving mapping, due to the existence of invariant tori, chaotic sea and hyperbolic structure, the diffusion of orbits will be much more complicated than the two-dimensional cases. The work presented above in this section, i.e. the diffusion characters in different regions, is the base of realizing the global diffusion in phase space. In three different regions: The chaotic sea, the vicinity of hyperbolic-elliptic fixed point and the vicinity of invariant torus, the diffusion velocity in the vicinity of invariant torus is the slowest. The stickiness effect would take the main responsibility for very slow diffusion in higher dimensional phase space.

Bibliography

Bangert V.: 1988, Mather sets for twist maps and geodesics on tori, *Dyn. Rep.*, 1, 1–56.

Beaugé C., Ferraz-Mello S. & Michtchenko T. A.: 2003, Extrasolar planets in mean motion resonance: apses alignment and asymmetric stationary solutions, *Astroph. J.*, 593, 1124–1133.

Beaugé C. & Michtchenko T.: 2003, Modelling the high-eccentricity planetary three-body problem. Application to the GJ876 planetary system, *Mon. Not. R. Astron. Soc.*, 341, 760–770.

Borderies N. & Longaretti P. Y.: 1987, Description and behavior of streamlines in planetary rings, *Icarus*, 103, 301–318.

Brož M., Vokrouhlický, Roig F., Nesvorný D., Bottke W. F. & Morbidelli A.: 2005, The population of asteroids in the 2:1 mean motion resonance with Jupiter revised, in Dynamics of Populations of Planetary Systems, *Proceedings IAU Colloquium 197*, eds. Knežević Z. & Milani A., 179–186.

Chen X.-Y., Sun Y.-S. & Luo Ding-Jun: 1978, Some problems of the regions of motion of general three-body problem (in Chinese), *Acta Astronomica Sinica*, 19, 119–130.

Chen X.-Y.: 1979, The topology of manifold M_8 of the general three-body problem, *Chinese Astronomy*, 3, 1–23.

Cheng J.: 2004, Variational approach to homoclinic orbits in twist maps and an application to billiard systems. *Z. Angew. Math. Phys.*, 55, 400–419.

Cheng Q.-C. & Sun Y.-S.: 1990a, Existence of invariant tori in three-dimensional measure-preserving mapping, *Celest. Mech. Dyn. Astron.*, 47, 275–292.

Cheng Q.-C. & Sun Y.-S.: 1990b, Existence of periodically invariant curves in three-dimensional measure-preserving mappings, *Celest. Mech. Dyn. Astron.*, 47, 293–303.

Cheng J. & Sun Y.-S.: 1996, Stable and unstable properties of the standard-liked map, *Random Comput. Dyn.*, 4, 73–85.

Chirikov B.: 1979, A universal instability of many-dimensional oscillator systems, *Physics Reports*, 52, 263–379.

Contopoulos G., Harsoula M., Voglis N. & Dvorak R.: 1999, Destruction of islands of stability, *J. Phys. A: Math. Gen.*, 32, 5213–5232.

Contopoulos G., Voglis N., Efthymiopoulos C., Froeschlé Cl., Gonczi R., Lega E., Dvorak R. & Lohinger E.: 1997, Transition spectra of dynamical systems, *Celest. Mech. Dyn. Astron.*, 67, 293–317.

Ding M., Bountis T. & Ott E.: 1990, Algebraic escape in higher dimensional Hamiltonian systems, *Phys. Lett. A*, 151, 395–400.

Duncan M., Quinn T. & Tremaine S.: 1989, The long-term evolution of orbits in the Solar system: A mapping approach. *Icarus*, 82, 402–418.

Dvorak R. & Sun Y.-S.: 1997, The phase space structure of the extended Sitnikov problem, *Celest. Mech. Dyn. Astron.*, 67, 87–106.

Efthymiopoulos C., Contopoulos G., Voglis N. & Dvorak R.: 1997, Stickiniess and cantori, *J. Phys. A: Math. Gen.*, 30, 8167–8186.

Feit D. S.: 1978, Characteristic exponents and strange attractors, *Commun. Math. Phys.*, 61, 249–260.

Fernandez J. A. & Gallardo T.: 1994, The transfer of comets from from parabolic orbits to short-period orbits: numerical studies, *Astron. Astrophy.*, 281, 911–922.

Fernández J.A. & Ip W.-H.: 1984, Some dynamical aspects of the accretion of Uranus and Neptune: The exchange of orbital angular momentum with planetesimals, *Icarus*, 58, 109–120.

Froehlin H., Crutchfiled J. P., Farmer D., Packard N. H. & Shaw R.: 1981, On determining the dimension of chaotic flows, *Physica D: Nonlinear Phenomena*, 3, 605–617.

Froeschlé Cl. & Lega E.: 1998, Modelling mappings: an aim and a tool for the study of dynamical systems, In: Benest D. and Froeschlé Cl. (ed.), *Analysis and Modelling of Discrete Dynamical Systems*, Gordon and Breach Science Publishers, Netherlands, pp. 3–54.

Fu Y.-N. & Sun Y.-S.: 1992, Central configurations of the n-body problem with a general force function (I) — Fundamental equations and theoretical analysis (in Chinese), *Acta Astronomica Sinica*, 33, 186–193.

Fu Y.-N. & Sun Y.-S.: 1993, The homographic solutions in the N-body problem with general attraction (in Chinese), *Acta Astronomica Sinica*, 34, 202–208.

Gladman B.: 1993, Dyanamics of systems of two close planets, *Icarus*, 106, 247–263.

Gladman B., Kavelaars J. J., Holman M. J. & Loredo T.: 2001, The stucture of the Kuiper Belt: Size distribution and radial extent, *Astron. J.*, 122, 1051–1066.

Hadjidemetriou J. D.: 1991, Mapping models for Hamiltonian systems with application to resonant asteroid motion, In: Roy A. F. (ed.), Predictability, Stability and Chaos in N-Body Dynamical Systems, Kluwer Acad Publication Netherlands, pp.157–175.

Hagihara Y.: 1970, Celestial Mechanics (I). MIT Press, Cambridge.

Hanslmeier A. & Dvorak R.: 1984, Numerical integration with Lie series, *Astron. Astrophy.*, 132, 203–207.

Hénon M.: 1976, A two-dimensional mapping with a strange attractor. *Commun. Math. Phys.*, 50, 69–77.

Hénon M. & Heiles C.: 1964, The applicability of the third integral of motion: some numerical experiments, *Astron. J.*, 69, 73–79.

Hénon M. & Petit J. M.: 1986, Series expansions for encounter-type solutions of Hill's problem. *Celest. Mech.*, 38, 67–100.

Holman M. J. & Wisdom J.: 1993, Dynamical stability in the outer Solar System and the delivery of short period comets, *Astron. J.*, 105, 1987–1999.

Jewitt D. C. & Luu J. X.: 1993, Discovery of the candidate Kuiper belt object 1992QB$_1$, *Nature*, 362, 730–732.

Kaplan J. L. & Yorke J. A.: 1979, Chaotic behavior of multidimensional difference equations, Springer, Lecture Notes in Mathematics, 730, 204–222.

Karney C. F.: 1983, Long-time correlations in the stochastic regime, *Physica D*, 8, 360–380.

Kuiper G. P.: 1951, On the origin of the Solar System, *Proc. Natl. Acad. Sci.*, 37, 1–14.

Lai Y. C., Ding M. Z., Grebogi C. & Blumel R.: 1992, Algebraic decay and fluctuations of the decay exponent in Hamiltonian systems, *Phys. Rev. A*, 46, 4661–4669.

Laughlin G., Chambers J. & Fischer D.: 2002, A dynamical analysis of the 47 Ursae Majoris planetary system, *Astrophy. J.*, 579, 455–467.

Lee M.H. & Peale S.J.: 2002, Dynamics and origin of the 2:1 orbital resonances of the GJ876 planets, *Astroph. J.*, 567, 596–609.

Li J., Fu Y.-N. & Sun Y.-S.: 2010, The Hill stability of low mass binaries in hierarchical triple systems, *Celest. Mech. Dyn. Astron.*, 107, 21–34.

Li J., Zhou L.-Y. & Sun Y.-S.: 2007, The origin of the high-inclination Neptune Trojan 2005 TN53, *Astron. Astrophys.*, 464, 775–778.

Lichtenberg A. J. & Lieberman M. A.: 1983, Regular and Stochastic Motion, Springer-Verlag, New York, 383.

Liu Jie & Sun Y.-S., 1990, On the Sitnikov problem, *Celest. Mech. Dyn. Astron.*, 49, 285–302.

Liu J. & Sun Y.-S.: 1991, The chaotic region of an encounter-type orbit. *Acta Astron. Sinica*, 32, 277–287.

Liu J. & Sun Y.-S.: 1994, Chaotic motion of comets in near-parabolic orbit: mapping approaches, *Celest. Mech. Dyn. Astron.*, 60, 3–28.

Luk'Yanov L.G.: 2005, Energy conservation in the restricted elliptical three-body problem, *Astronomy Reports*, 49, 1018–1027.

Malhotra R.: 1993, The origin of Pluto's peculiar orbit, *Nature*, 365, 819–821.

Malhotra R.: 1995, The origin of Pluto's orbit: implications for the solar system beyond Neptune, *Astron. J.*, 110, 420–429.

Malhotra R.: 2002, A dynamical mechanism for establishing apsidal resonance, *Astrophy. J.*, 575, L33–L36.

Marchal C. & Sun Y.-S., 1985, Evolution of the moment of inertia in the many-body problem, *Scientia Sinica (A)*, Vol. XXVIII, 638–647.

Meiss J. D. & Ott E.: 1985, Markov-tree model of intrinsic transport in Hamiltonian systems, *Phys. Rev. Lett.*, 55, 2741–2744.

Michtchenko T. & Ferraz-Mello S.: 1995, Comparative study of the asteroidal motion in the 3:2 and 2:1 resonances with Jupiter I. Planar model, *Astron. Astrophys.*, 303, 945–963.

Mikkola S. & Palmer P.: 2000, Simple derivation of Symplectic Integrators with First Order Correctors, *Celest. Mech. Dyn. Astron.*, 77, 305.

Morbidelli A. & Giorgilli A.: 1995, Superexponential stability of KAM tori, *J. Stat. Phys.*, 78, 1607–1617.

Morbidelli A. & Levison H. F.: 2004, Scenarios for the Origin of the Orbits of the Trans-Neptunian Objects 2000 CR105 and 2003 VB12 (Sedna), *Astron. J.*, 128, 2564–2576.

Murray C. D. & Dermott S. F.: 1999, Solar system dynamics, Cambridge University Press, Cambridge, UK.

Ott E.: 2002, Chaos in Dynamical Systems (2nd Ed.), Cambridge University Press, Cambridge, UK.

Pesin Y. B.: 1976, Ergodic properties of smooth dynamic systems with invariant measure and nonzero Lyapunov indicators, *Dokl. Akad. Nauk. SSSE*, 226, 774–777 (English translation: *Sov. Math. Dokl.* 17, 196, 1976).

Pittichova J. *et al.*: 2003, 2001 QR322, *Minor Planet Electronic Circ.*, 2003-A55.

Routh E. J.: 1875, On Laplace's three particles, with a supplement on the stability of steady motion, *Proc. Lond. Math. Soc.*, 6, 86–97.

Saari D. G.: 1980, On the role and properties of n body central configurations, *Celets. Mech.*, 21, 9–20.

Sheppard S. & Trujillo C.: 2006, *Science*, 313, 511.

Shu F. H.: 1984, Waves in planetary rings, eds. Greenberg R. & Brahic R. A., Arizona Unversity Press, Tucson, 531–561.

Siegel C. L. & Moser J. K.: 1971, Lectures on Celestial Mechanics, Springer-Verlag, Berlin.

Sun Y.-S.: 1983a, On the measure-preserving mappings with odd dimension, *Celest. Mech.*, 30, 7–19.

Sun Y.-S.: 1983b, Stochasticity of the measure-preserving mappings with three-dimensions (in Chinese), *Acta Astronomica Sinica*, 24, 128–136.

Sun Y.-S.: 1984, Invariant manifolds in the measure-preserving mappings with three dimensions, *Celest. Mech.*, 33, 111–125.

Sun Y.-S.: 1985, Attractors in a dissipative dynamical system with three dimensions, *Celest. Mech.*, 37, 171–181.

Sun Y.-S. & Froeschlé C.: 1982, On the Kolmogrov entropy of two-dimensional area-preserving mappings. *Scientia Sinica (Series A)*, 25, 750–757.

Sun Y.-S. & Fu Y.-N.: 1999, Diffusion character in four-dimensional volume-preserving maps, *Celest. Mech. Dyn. Astron.*, 73, 249–258.

Sun Y.-S. & Marchal C.: 1985, Behaviours of the motion of isolated body in the three-body problem (in Chinese), *Acta Astronomica Sinica*, 26, 1–12.

Sun Y.-S. & Luo D.-J.: 1980a, Further analysis about the regions of motion of general three-body problem (in Chinese), *Acta Astronomica Sinica*, 21, 96–103.

Sun Y.-S. & Luo D.-J.: 1980b, The ranges of change of the configurations and positions of orbits in general three-body problem (in Chinese), *J. Nanjing Univ.*, 4, 74–82.

Sun Y.-S. & Yan Z.-M.: 1988, A perturbed extension of hyperbolic twist mappings, *Celest. Mech.*, 42, 369–383.

Sun Y.-S. & Zhou J.-L.: 1996, Global applicability of the symplectic integrator method of Hamiltonian systems, *Celest. Mech. Dyn. Astron.*, 64, 185–195.

Sun Y.-S. & Zhou J.-L.: 2008, Introduction to Modern Celestial Mechanics (In Chinese), Higher Education Press, China.

Sun Y.-S & Zhou L.-Y.: 2009, Stickiness in three-dimensional preserving mappings, *Celest. Mech. Dyn. Astron.*, 103, 119–131.

Sun Y.-S., Zhou L.-Y. & Zhou J.-L.: 2005, The role of hyperbolic invariant sets in stickiness effects, *Celest. Mech. Dyn. Astron.*, 92, 257–272.

Sun Y.-S., Zhou J.-L., Zheng J.-Q. & Valtonen M.: 2002, Diffusion in comet motion, *Contemporary Mathematics*, 292, 229–238.

Szebehely V. & Zare K.: 1977, Stability of classical triples and of their hierarchy, *Astron. Astrophys.*, 58, 145–152.

Thommes E. W., Duncan M. J. & Levison H. F.: 1999, The formation of Uranus and Neptune in the Jupiter–Saturn region of the Solar System, *Nature*, 402, 635–638.

Tsiganis K., Gomes R., Morbidelli A. & Levison H. F.: 2005, Origin of the orbital architecture of the giant planets of the Solar System, *Nature*, 435, 459–461.

Wisdom J.: 1983, The origin of the Kirkwood gaps — A mapping for asteroidal motion near the 3/1 commensurability, *Astron. J.*, 87, 577–593.

Wisdom J.: 1983, Chaotic behavior and the origin of the 3/1 Kirkwood gap, *Icarus*, 56, 51–74.

Yan Z.-M. & Sun Y.-S.: 1988, The central configurations of general 4-body problem. Scientia Sinica (Series A) XXXI, 724–733.

Yoccoz J. C.: 1994, Recent developments in dynamics, Proceedings of the international congress of mathematicians, Ed. by Chatterji S. D., Zurich, Switzerland, Birkhauser Verlag, Basel, Switzerland.

Zare K.: 1977, Bifurcation points in the planar problem of three bodies, *Celest. Mech.*, 16, 35–38.

Zhang T.-L. & Sun Y.-S.: 1989, Behaviours of a class of perturbed measure-preserving mappings. *Journal of Nanjing University (Natural Sciences edition)* (in Chinese), 25, 187–198.

Zhou L.-Y., Dvorak R. & Sun Y.-S.: 2009, The dynamics of Neptune Trojan — I. The inclined orbits, *Mon. Not. R. Astron. Soc.*, 398, 1217–1227.

Zhou L.-Y., Dvorak R. & Sun Y.-S.: 2011, The dynamics of Neptune Trojan — II. Eccentric orbits and observed objects, *Mon. Not. R. Astron. Soc.*, 410, 1849–1860

Zhou J.-L., Fu Y.-N. & Sun Y.-S.: 1994, Mapping models for near conservative systems with applications, *Celest. Mech. Dyn. Astron.*, 60, 471–487.

Zhou L.-Y., Lehto H., Sun Y.-S. & Zheng J.-Q.: 2004, Apsidal corotation in mean motion resonance: the 55 Cancri system as an example, *Mon. Not. R. Astron. Soc.*, 350, 1495–1502.

Zhou L.-Y., Li J., Cheng J. & Sun Y.-S.: 2014, Hyperbolic structure and stickiness effect: A case of 2D area-preserving twist mapping, *Sci. China — Phys. Mech. Astron.*, 57, 1737–1750.

Zhou J.-L. & Sun Y.-S.: 1995, Hamiltonian description of streamline formalism with application in planetary rings, *Celest. Mech. Dyn. Astron.*, 62, 99–109.

Zhou J.-L. & Sun Y.-S.: 2001, Lévy flight in comet motion and related chaotic systems, *Phys. Lett. A*, 287, 217–222.

Zhou J.-L. & Sun Y.-S.: 2003, Occurrence and stability of apsidal resonance in multiple planetary systems, *Astrophys. J.*, 598, 1290–1300.

Zhou J.-L. & Sun Y.-S.: 2004, Criteria for the occurence of apsidal and nodal secular resonance in planetary systems., in Stars as Suns: Activity, Evolution, and Planets. *Proceedings of IAU Symposium 219*, eds. Dupree A. K. & Benz A. O., CD-820–824.

Zhou J.-L., Sun Y.-S. & Hu B.: 2000, On the Shepherding of Uranian epsilon ring, *Science in China (Series A)*, 43, 666–672.

Zhou J.-L., Sun Y.-S., Zheng J.-Q. & Valtonen M. J.: 2000, The transfer of comets from near-parabolic to short period orbits: Mapping approach. *Astron. & Astrophys.*, 364, 887–893.

Zhou L.-Y., Sun Y.-S. & Zhou J.-L.: 2000, Structure of the phase space near Lagrange's triangular equilibrium points, *Chin. Astron. Astrophys.*, 24, 119–126.

Zhou L.-Y., Sun Y.-S. & Zhou J.-L.: 2001, Diffusion characters of the orbits in the asteroid motion, *Appl. Math. Mech.*, 22, 808–819.

Zhou L.-Y., Sun Y.-S., Zhou J.-L., Zheng J.-Q. & Valtonen M.: 2002, Stochastic effects in the planet migration and orbital distribution of the Kuiper belt, *Mon. Not. R. Astron. Soc.*, 336, 520–526.

Zhou J.-L., Zhou L.-Y. & Sun Y.-S.: 2002a, Hyperbolic structure and Stickiness effect, *Chin. Phys. Lett.*, 19, 1254–1256.

Zhou L.-Y., Zhou J.-L. & Sun Y.-S.: 2002b, Stabilities of asteroids orbits in resonances, *Science in China (Series A)*, 45, 122–128.

Index